Birkhäuser

Operator Theory: Advances and Applications

Volume 310

Founded in 1979 by Israel Gohberg

Alexander Meskhi • Humberto Rafeiro •
Stefan Samko

Integral Operators in Non-Standard Function Spaces

A Sequel: Inequalities, Sharp Estimates, Bounded Variation, and Approximation

Birkhäuser

Alexander Meskhi
A. Razmadze Mathematical Institute
I. Javakhishvili Tbilisi State Univ
Tbilisi, Georgia

Kutaisi International University
Kutaisi, Georgia

Stefan Samko
Department of Mathematics
University of Algarve
Faro, Portugal

Humberto Rafeiro
Department of Mathematical Sciences
United Arab Emirates University
Al Ain, United Arab Emirates

ISSN 0255-0156 ISSN 2296-4878 (electronic)
Operator Theory: Advances and Applications
ISBN 978-3-032-03690-2 ISBN 978-3-032-03691-9 (eBook)
https://doi.org/10.1007/978-3-032-03691-9

Mathematics Subject Classification: 26A45, 47H30, 47H50, 26B05, 34A40, 42B25, 42B35, 47B34, 26B30, 42B20, 46E30, 47H60

This book is published under the imprint Birkhäuser, www.birkhauser-science.com by the registered company Springer Nature Switzerland AG
The registered company address is: Gewerbestrasse 11, 6330 Cham, Switzerland

Preface

This volume, as a sequel to Volumes I–IV, is devoted to the authors' most recent advances in harmonic analysis and their applications. The primary topics of this work include: Rellich inequalities in the variable exponent and multilinear settings, trace inequalities for linear and multilinear fractional integrals, sharp weighted estimates for norms of operators of harmonic analysis such as singular, fractional, one-sided, and multiple integral operators, boundedness/compactness characterizations of linear and multilinear fractional integrals defined with respect to non-doubling measures, Sobolev-type inequalities criteria for (generalized) fractional integrals associated with non-doubling measures, sharp Olsen-type inequalities; studies on Morrey-type and Herz spaces (including embeddings and coincidences of spaces); approximation in subspaces of Morrey spaces, and the preservation of certain vanishing properties of generalized Morrey spaces by some classical operators; introduction of variable exponent bounded variation spaces in the Riesz sense; boundedness of Nemytskii operator on these spaces, linear functionals on variable exponent Bocher–Lebesgue spaces, and characterization of weighted Sobolev spaces via weighted Riesz bounded variation spaces.

In Chap. 1, we present Rellich-type inequalities in classical and variable exponent Lebesgue spaces. These problems are studied in both linear and multilinear settings. Finally, we present the corresponding results for the higher-dimensional case in variable exponent Lebesgue spaces.

In Chap. 2, we discuss the problem of finding an appropriate Lorentz space $L^{p,s}$ such that the well-known D. Adams-type condition on a weight function v is both necessary and sufficient for the validity of the trace inequality $\|I_\alpha f\|_{L^p_v} \leqslant C\|f\|_{L^{p,s}}$, where I_α is the Riesz potential. We show that the desired space is $L^{p,1}$. This investigation is motivated by the fact that the inequality fails for $p = s$ under the D. Adams-type condition on v (the D. Adams condition is necessary and sufficient for the validity of the trace inequality $\|I_\alpha f\|_{L^q_v} \leqslant C\|f\|_{L^p}$ if and only if $1 < p < q < \infty$). Furthermore, in this chapter, we give a complete characterization of the trace inequality for multilinear Riemann-Liouville operators.

Chapter 3 is devoted to the characterization of measures μ that guarantee the boundedness/ compactness of fractional-type integral operators defined with respect to μ in (grand) Lebesgue spaces. Similar problem is studied in Lorentz and Lorentz-Morrey spaces. A necessary and sufficient condition is derived for a non-doubling measure μ to guarantee the boundedness of a generalized fractional integral operator defined with respect to μ. Furthermore, we establish estimates for the measure of non-compactness (i.e., the essential norm) of multilinear integral operators in rearrangement-invariant Banach function spaces. In particular, we derive lower estimates for the essential norms of multilinear variants of the Hilbert and Riesz transforms on rearrangement-invariant spaces. We also provide two-weight estimates for the essential norm of multilinear integral operators acting in classical Lebesgue spaces.

Sharp weighted bounds (Buckley-type estimates) for various integral operators of harmonic analysis are presented in Chap. 4. In particular, we deal with one-sided and multiple integral operators of fractional-type. More precisely, in this chapter, sharp weighted bounds in weighted norm inequalities for various integral operators are discussed in terms of weighted characteristics. The same problem is studied for positive kernel operators defined on homogeneous type spaces, fractional integral operators, one-sided integral operators, Hilbert transform on the classes of odd (resp. even) functions, strong maximal functions, multiple singular and fractional integrals.

In Chap. 5, we deal with Olsen-type norm estimates in the multilinear setting. In particular, we derive sharp Olsen-type and other weighted inequalities for multilinear fractional integrals of various types in Morrey spaces. As a consequence of these estimates, we obtain necessary and sufficient conditions governing the trace inequality for multilinear fractional integrals in Morrey spaces.

Chapter 6 is a continuation of Kokilashvili et al. [183, Section 13.4]. We begin by establishing the boundedness of the Marcinkiewicz integral on continual Herz spaces with variable exponent, where all defining parameters are variable. Utilizing the atomic decomposition of Herz-type Hardy spaces with variable smoothness and integrability, we then derive boundedness results for central Calderón–Zygmund operators on these spaces. The chapter proceeds to extend these results to grand variable Herz spaces, demonstrating the boundedness of the Marcinkiewicz integral in this more general context. We further introduce a new class of grand variable Herz-type spaces, constructed via discrete grand spaces, and prove the boundedness of sublinear operators on them. Finally, we study multilinear Calderón–Zygmund operators acting on products of grand variable Herz spaces, thereby generalizing earlier boundedness results for products of variable exponent Lebesgue and Herz spaces.

Chapter 7 addresses the approximation in subspaces of Morrey spaces by introducing a new subspace that allows the approximation by infinitely differentiable compactly supported functions. It provides an explicit description of the closure of these functions in Morrey spaces and generalizes embeddings of Morrey spaces into weighted Lebesgue spaces. The chapter also examines the approximation of functions in generalized Morrey spaces using similar methods. The introduced subspaces are shown to remain invariant under classical harmonic analysis operators, such as the Hardy–Littlewood maximal operator, singular integral operators, and Hardy operators. Furthermore, the vanishing properties defining these subspaces are still preserved under the action of Riesz potential and fractional maximal operators.

The final chapter, Chap. 8, begins with an introduction to variable exponent bounded variation spaces in the Riesz sense. The chapter also investigates linear functionals on variable exponent Bochner–Lebesgue spaces, discusses weighted Riesz bounded variation spaces and their relation to the Nemytskii operator, and concludes with a characterization of weighted Sobolev spaces via weighted Riesz bounded variation spaces. This final result is also established in the unweighted setting for variable exponent Riesz bounded variation spaces.

Notations for the same objects may differ across chapters—for example, in the case of function spaces, integral operators, weighted characteristics, and similar concepts. For the reader's convenience, some definitions and statements are repeated, and we have aimed to make the chapters as self-contained and independent of each other as possible.

The book is aimed at a wide audience, ranging from researchers in operator theory and harmonic analysis to experts in applied mathematics and postgraduate students. In particular, we hope that this book will serve as a source of inspiration for researchers in abstract harmonic analysis, function spaces, PDEs, and BVPs. This book can be read without frequent reference to other sources; however, a basic knowledge of function theory and functional analysis is assumed. Therefore, the book can also be used for teaching purposes.

Acknowledgments A. Meskhi was partially supported by the Shota Rustaveli National Science Foundation Grant FR-25-1693.

H. Rafeiro was supported by the UPAR grants G00004572 and G00005021, United Arab Emirates University.

Tbilisi, Georgia — Alexander Meskhi
Al Ain, United Arab Emirates — Humberto Rafeiro
Faro, Portugal — Stefan Samko

Contents

Chapter 1
Rellich Inequality

The famous Rellich inequality states that if $u \in C_0^\infty(\mathbb{R}^n \setminus \{0\})$ and $n \neq 2$, then

$$\frac{n^2(n-4)^2}{16} \int_{\mathbb{R}^n} \frac{|u(x)|^2}{|x|^4} dx \leqslant \int_{\mathbb{R}^n} |\Delta u(x)|^2 dx, \tag{1.1}$$

and the constant $n^2(n-4)^2/16$ is sharp. When $n = 2$, the inequality still holds but only for a restricted class of functions: see Balinsky et al. [23]. In these results, the underlying set is $\mathbb{R}^n \setminus \{0\}$ and $|x|$ may be thought of as the distance of a point $x \in \mathbb{R}^n$ from the boundary $\{0\}$ of $\mathbb{R}^n \setminus \{0\}$. With this in mind, it is natural to consider functions defined on a more general open subset Ω of $\mathbb{R}^n$ and to seek an inequality involving the distance function δ defined by $\delta(x) = \operatorname{dist}(x, \partial\Omega)$. This approach led to the inequality:

$$\frac{9}{16} \int_{\Omega} \frac{|u(x)|^2}{\delta(x)^4} dx \leqslant \int_{\Omega} |\Delta u(x)|^2 dx, \quad u \in C_0^\infty(\Omega), \tag{1.2}$$

valid when Ω is convex and even under rather more general conditions: see Balinsky et al. [23, Corollary 6.2.7]. Extensions of these inequalities into the framework of L^p functions were obtained in Davies and Hinz [69] and Edmunds and Evans [86] (see Sect. 1.5, for exact formulation of the statements).

This chapter deals mainly with the Rellich inequality in the framework of variable exponent Lebesgue spaces, but includes also a multilinear version in the classical Lebesgue spaces.

Section 1.1 is devoted to the study of the Rellich inequality in the one-dimensional case in variable exponent Lebesgue spaces.

In Sect. 1.2 we establish bounds for the norm of the two-weighted Hardy operator $H_{v,w}$ in variable exponent Lebesgue spaces. The results derived there are applied to

A. Meskhi et al., *Integral Operators in Non-Standard Function Spaces*,
Operator Theory: Advances and Applications 310,
https://doi.org/10.1007/978-3-032-03691-9_1

obtain estimates of the measure of noncompactness for $H_{v,w}$, and to prove two-weight variant of the Rellich inequality in $L^{p(\cdot)}$ space.

A multilinear variant of the Rellich inequality on the real line is studied in Sect. 1.3. In particular, we establish the weighted inequality:

$$\left(\int_a^b w(\delta(x)) \left| \prod_{k=1}^m u_k(x) \right|^q dx \right)^{1/q} \leqslant C \prod_{k=1}^m \|u_k''\|_{L^{p_k}(a,b)},$$

$$u_k \in C_0^2(a,b), k = 1, \ldots, m,$$

with a positive function w on $(0, b-a))$, where $-\infty \leqslant a < b \leqslant +\infty$, m is a positive integer, and $\delta(x) = \min\{x-a, b-x\}$ is the distance function on (a,b).

In Sect. 1.4 we prove weighted Rellich–Sobolev and Hardy–Sobolev inequalities in variable exponent Lebesgue spaces $L^{p(\cdot)}(\mathbb{G})$ defined on stratified homogeneous groups $\mathbb{G}$. To prove the main statements of this section, we rely on weighted estimates for the Riesz potential operators in $L^{p(\cdot)}(\mathbb{G})$, where $\mathbb{G}$ is a general homogeneous group. The results are new even for the Abelian (Euclidean) case $\mathbb{G} = \left(\mathbb{R}^d, +\right)$ and the Heisenberg groups $\mathbb{G} = \mathbb{H}^n$. The main statements are obtained for variable exponents satisfying the condition that the Hardy–Littlewood maximal operator is bounded in appropriate variable exponent Lebesgue spaces. We also give some quantitative estimates for the norms of integral operators involved in derived estimates.

1.1 One-Dimensional Rellich Inequality in Variable Exponent Lebesgue Spaces ($VELSs$)

Our aim in this section is to establish the one-dimensional Rellich inequality in $VELS$ defined on an interval; see (1.7) for the corresponding definition. The main result states that the inequality

$$\left\| u(\cdot)\delta(\cdot)^{-2} \right\|_{L^{p(\cdot)}(a,b)} \leqslant C \left\| u''(\cdot) \right\|_{L^{p(\cdot)}(a,b)} \tag{1.3}$$

holds for all $u \in C_0^2(a,b)$, where $-\infty < a < b \leqslant \infty$ and $\delta(t) = \min\{t-a, b-t\}$; see Theorem 1.14 for exact formulation.

1.1.1 Preliminaries

Before formulating the main statements, we introduce some useful notation and results.

For a positive continuous exponent $p(\cdot)$ defined on a set $E \subset \mathbb{R}$, we set $p_-(E) := \inf_E p(\cdot)$ and $p_+(E) := \sup_E p(\cdot)$. In the case when $E = (c, d)$, we use the notation:

$$p_-(E) := p^-_{c,d}; \quad p_+(E) := p^+_{c,d}.$$

If $d = \infty$, then we denote

$$p^-_{c,d} := p^-_c; \quad p^+_{c,d} := p^+_c.$$

We will write p_- and p_+ for $p_-(E)$ and $p_+(E)$, respectively, if E is an interval on which $VELS$ is defined.

For a set $E = (a, b)$, we denote by $\delta(\cdot)$ the following function:

$$\delta(x) = \text{dist}(x, \partial E),$$

when E is clear from the context.

Let $P(E)$ be the class of continuous exponents $p(\cdot)$ defined on E, $E \subset \mathbb{R}$, such that

$$1 < p_-(E) \leqslant p_+(E) < \infty.$$

If E is an interval with endpoints a and b, then we will assume that

$$P_{a,b} := P(E).$$

For $E = [a, \infty)$, we denote:

$$P_a := P(E).$$

Further, let E be an interval in $\mathbb{R}$. We denote by $\mathcal{P}(E)$ the class of all positive continuous exponents on E satisfying the condition: there is a positive constant A such that for all $x, y \in \overline{E}$: $0 < |x - y| \leqslant \frac{1}{2}$,

$$|p(x) - p(y)| \leqslant \frac{A}{\ln \frac{1}{|x-y|}}. \tag{1.4}$$

It is known that the condition $p \in P(E) \cap \mathcal{P}(E)$ guarantees the boundedness of various important operators of harmonic analysis in $VELS$s (see Diening [71], Diening et al. [74], Cruz-Uribe and Fiorenza [58], Kokilashvili et al. [182, 183]).

Definition 1.1 Let p be a positive continuous function on $E \subset \mathbb{R}$:

(i) Let $E = [a, b)$ with $-\infty < a < b \leqslant \infty$. We say that $p \in \mathcal{P}(a, A_a)$, where A_a is a positive constant, if for all y, $y \in [a, b)$, $a \leqslant y < a + 1/2$, the inequality

$$|p(y) - p(a)| \leqslant \frac{A_a}{-\ln(y-a)}$$

holds.

(ii) Suppose that $E = (a, b]$ with $-\infty < a < b < \infty$. We say that $p \in \mathcal{P}(b, A_b)$, where A_b is a positive constant, if for all z with $z \in (a, b]$ and $b - 1/2 < z \leqslant b$, the inequality

$$|p(z) - p(b)| \leqslant \frac{A_b}{-\ln(b-z)}$$

holds.

Remark 1.2 It is immediate to see that:

(i) if $p(\cdot) \in \mathcal{P}(a, A_a)$ (resp. $p(\cdot) \in \mathcal{P}(b, A_b)$) and l is a positive constant, then $p(\cdot)/l \in \mathcal{P}(a, A_a/l)$ (resp. $p(\cdot)/l \in \mathcal{P}(b, A_b/l)$).

(ii) the condition $p(\cdot) \in \mathcal{P}(a, A_a)$ (resp. $p(\cdot) \in \mathcal{P}(b, A_b)$) implies that $1/p(\cdot) \in \mathcal{P}\Big(a, A_a/p_-^2\Big)$ (resp. $p'(\cdot) \in \mathcal{P}\Big(b, A_b/p_-^2\Big)$);

(iii) the condition $p(\cdot) \in \mathcal{P}(a, A_a)$ (resp. $p(\cdot) \in \mathcal{P}(b, A_b)$) implies that $p'(\cdot) \in \mathcal{P}\Big(a, A_a/(p_- - 1)^2\Big)$ (resp. $p'(\cdot) \in \mathcal{P}\Big(b, A_b/(p_- - 1)^2\Big)$.

The next statement, similar to those in Diening [71], is true (see also Kokilashvili et al. [183, p. 834] for an analogous statement).

Lemma 1.3 *Let* $-\infty < a < b \leqslant \infty$ *and let* $s(\cdot)$ *be a positive exponent on* $[a, b)$ *such that* $s \in \mathcal{P}(a, A_a)$. *Then, for all* $x \in (a, b)$, *the following inequality holds:*

$$(x-a)^{s^-_{a,x} - s^+_{a,x}} \leqslant c_{s,A_a},$$

where

$$c_{s,A_a} = \max\left\{e^{2A_a}, 2^{s^+_{a,b} - s^-_{a,b}}\right\}. \tag{1.5}$$

Proof If $x - a \geqslant 1/2$, then

$$(x-a)^{s^-_{a,x} - s^+_{a,x}} \leqslant 2^{s^-_{a,b} - s^+_{a,b}},$$

while for $x - a < 1/2$, we find that

$$
\begin{aligned}
(x-a)^{s^-_{a,x}-s^+_{a,x}} &= \left(\frac{1}{x-a}\right)^{|s(y)-s(z)|} \\
&\leqslant \left(\frac{1}{x-a}\right)^{|s(y)-s(a)|+|s(a)-s(z)|} \\
&\leqslant \left(\frac{1}{x-a}\right)^{\frac{A_a}{-\ln(y-a)}+-\frac{A_a}{-\ln(z-a)}} \\
&\leqslant \left(\frac{1}{x-a}\right)^{\frac{2A_a}{-\ln(x-a)}}.
\end{aligned}
$$

Hence,

$$(x-a)^{s^-_{a,x}-s^+_{a,x}} \leqslant e^{2A_a},$$

and the result follows. □

The next statement follows just in the same manner; therefore, we omit the proof.

Lemma 1.4 *Let* $-\infty \leqslant a < b < \infty$ *and let* $s(\cdot)$ *be an exponent on* $(a, b]$ *such that* $s \in \mathcal{P}(b, A_b)$. *Then, for all* $x \in (a, b)$, *the following inequality holds:*

$$(b-x)^{s^-_{x,b}-s^+_{x,b}} \leqslant c_{s,A_b},$$

where

$$c_{s,A_b} = \max\left\{e^{2A_b}, 2^{s^+_{a,b}-s^-_{x,b}}\right\}. \tag{1.6}$$

Let $p(\cdot) \in \mathcal{P}(E)$. The variable exponent Lebesgue spaces ($VELS$), denoted by $L^{p(\cdot)}(E)$, is the linear space of all (Lebesgue) measurable functions f on E for which

$$S_{p(\cdot)}(f) := \int_E |f(x)|^{p(x)} dx < \infty. \tag{1.7}$$

The norm in $L^{p(\cdot)}(E)$ is defined as follows:

$$\|f\|_{L^{p(\cdot)}(E)} = \inf\left\{\lambda > 0 : S_{p(\cdot)}(f/\lambda) \leqslant 1\right\}.$$

If $p(\cdot) \equiv p_c \equiv$ *const*, then $L^{p(\cdot)}(E)$ coincides with the classical Lebesgue space $L^{p_c}(E)$.

It is known (see, e.g., Kováčik and Rákosník [198]) that $L^{p(\cdot)}(E)$ is a Banach space. For other properties of $L^{p(\cdot)}$ spaces, we refer also to Sharapudinov [311] and Samko [292].

Throughout this chapter, we denote by $p'(\cdot)$ the conjugate of $p(\cdot)$, i.e., the function defined on E such that $\frac{1}{p(x)} + \frac{1}{p'(x)} = 1$. In variable exponent Lebesgue space Hölder's inequality holds also in the following form (see, e.g., Samko [292], Kováčik and Rákosník [198]):

$$\left| \int_E f(x)g(x)dx \right| \leqslant \left(\frac{1}{p_-} - \frac{1}{p_+} + 1 \right) \|f\|_{L^{p(\cdot)}(E)} \|g\|_{L^{q(\cdot)}(E)}. \tag{1.8}$$

The following inequality is also true (see, e.g., Kováčik and Rákosník [198]):

$$\|f\|_{L^{r(\cdot)}(E)} \leqslant (1 + |E|) \|f\|_{L^{s(\cdot)}(E)}, \quad 1 \leqslant r(x) \leqslant s(x), \tag{1.9}$$

where $|E|$ denotes the Lebesgue measure of a set E.

The following one-dimensional weighted Hardy inequality is known (see, e.g., Muckenhoupt [253]). We need estimates of the constant (see, e.g., Kufner and Persson [202], pp. 3-4):

Proposition 1.5 *Let r be a constant such that $1 < r < \infty$ and let $-\infty < a < b \leqslant \infty$. Suppose that v and w are weights on (a, b). Then the inequality*

$$\left(\int_a^b v(x) \left(\int_a^x f(t)dt \right)^r dx \right)^{1/r} \leqslant C \left(\int_a^b (f(x))^r w(x)dx \right)^{1/r}, \quad f \geqslant 0, \tag{1.10}$$

holds if and only if

$$A_r = \sup_{a<t<b} \left(\int_t^b v(x)dx \right)^{1/r} \left(\int_a^t w^{1-r'}(x)dx \right)^{1/r'} < \infty.$$

Moreover, for the best constant C in (1.10)*, we have that*

$$A_r \leqslant C \leqslant (r')^{1/r'} r^{1/r} A_r. \tag{1.11}$$

Denoting by H_a the following weighted Hardy operator:

$$H_a f(x) = \frac{1}{x-a} \int_a^x f(t)dt, \quad x \in (a, b), \ -\infty < a < b \leqslant \infty, \tag{1.12}$$

the following Hardy-type estimate holds:

Proposition 1.6 *Let r be a constant such that $1 < r < \infty$ and let $-\infty < a < b \leqslant \infty$. Then*

$$\|H_a f\|_{L^r(a,b)} \leqslant r' \|f\|_{L^r(a,b)}, \quad f \in L^p(a,b).$$

This inequality follows from Proposition 1.5 by calculating A_r in (1.11).

We will also need the next Hardy-type inequality, where the operator H_b' is defined by

$$H_b' f(x) = \frac{1}{b-x} \int_x^b f(t)dt.$$

Proposition 1.7 *Let r be a constant such that $1 < r < \infty$ and let $-\infty < a < b \leqslant \infty$. Then*

$$\|H_b' f\|_{L^r(a,b)} \leqslant r' \|f\|_{L^r(a,b)}, \quad f \in L^p(a,b).$$

Proof Observe that the boundedness of H_b' in $L^r(a,b)$ is equivalent to the inequality:

$$\left(\int_a^b \left(\int_a^x g(t)dt \right)^{r'} dx \right)^{1/r'} \leqslant \|H_b' f\|_{L^r(a,b)} \left(\int_a^b (g(x))^{r'} dx \right)^{1/r'}, \quad g \geqslant 0.$$

By using Proposition 1.5 and the estimate (1.11), we find that the proposition holds. □

For the Hardy inequalities in $VELS$s, we refer, e.g., to Kokilashvili et al. [182, Chapter 3] and references therein (see also Mamedov and Mammadova [224] and Mamedov and Zeren [222] for relevant results).

1.1.2 Rellich Inequality

To prove the main statements of this section, we need some auxiliary statements.

Lemma 1.8 *Let $-\infty < a < b \leqslant \infty$ and let $p(\cdot)$ be a continuous exponent on $[a,b)$ such that $p \in P_{a,b} \cap \mathcal{P}(a, A_a)$. Then for all non-negative f with $\|f\|_{L^{p(\cdot)}(a,b)} \leqslant 1$,*

$$\left(H_a f(x)\right)^{p(x)/p_-} \leqslant C(p, A_a)\left(H_a(f^{p/p_-}) + 1\right), \quad x \in (a,b), \tag{1.13}$$

where

$$C(p, A_a) = 3^{p_+/p_-} + 2^{p_+/p_- -1}\Big[\max\{e^{2A_a/p_-}; 2^{(p_+-p_-)/p_-}\}\Big]^{p_+/p_-}. \quad (1.14)$$

Proof We use the arguments in Diening [71, Lemma 3.3]. Let $q := p/p_-$ and let $\|f\|_{L^{p(\cdot)}} \leqslant 1$. Then $S_p(f) \leqslant 1$. Suppose first that $x \in (a, b)$ be such that $x - a > 1/2$. Then for all $x \in (a, b)$:

$$\begin{aligned}
\big(H_a f\big)^{q(x)}(x) &\leqslant \frac{1}{(x-a)^{q(x)}}\Bigg[\int\limits_{(a,x)\cap\{f\geqslant 1\}} f(x)dx + \int\limits_{(a,x)\cap\{f<1\}} f(x)dx\Bigg]^{q(x)} \\
&\leqslant \frac{1}{(x-a)^{q(x)}}\Bigg(\int\limits_a^x (f(x))^{q(y)}dy + x - a\Bigg)^{q(x)} \\
&\leqslant \frac{1}{(x-a)^{q(x)}}\Bigg(1 + x - a\Bigg)^{q(x)} \\
&\leqslant \Bigg[\frac{1}{x-a} + 1\Bigg]^{q_+} \leqslant 3^{p_+/p_-}.
\end{aligned}$$

Let now $0 \leqslant x - a \leqslant 1/2$. By Jensen's inequality, Remark 1.2, Lemma 1.3, and other simple arguments, we find that

$$\begin{aligned}
&\big(H_a f\big)^{q(x)}(x) \leqslant \Bigg(\frac{1}{x-a}\int\limits_a^x (f(y))^{q^-_{a,x}}dy\Bigg)^{q(x)/q^-_{a,x}} \\
&\leqslant \Bigg[\frac{1}{x-a}\int\limits_{(a,x)\cap\{f>1\}} (f(y))^{q^-_{a,x}}dy + \frac{1}{x-a}\int\limits_{(a,x)\cap\{f\leqslant 1\}} \leqslant (f(y))^{q^-_{a,x}}dy\Bigg]^{q(x)/q^-_{a,x}} \\
&\leqslant \Bigg(\frac{1}{x-a}\int\limits_a^x (f(y))^{q(y)}dy + 1\Bigg)^{q(x)/q^+_{a,x}} \\
&\leqslant (x-a)^{-\frac{q(x)}{q^-_{a,x}}}\Bigg(\int\limits_a^x (f(y))^{q(y)}dy + x - a\Bigg)^{q(x)/q^-_{a,x}} \\
&= (x-a)^{-\frac{q(x)}{q^-_{a,x}}}2^{\frac{q(x)}{q^-_{a,x}}}\Bigg(\frac{1}{2}\int\limits_a^x (f(y))^{q(y)}dy + x - a\Bigg)^{q(x)/q^-_{a,x}} \\
&\leqslant (x-a)^{-\frac{q(x)}{q^-_{a,x}}}2^{\frac{q(x)}{q^-_{a,x}}}\Bigg(\frac{1}{2}\int\limits_a^x (f(y))^{q(y)}dy + x - a\Bigg)
\end{aligned}$$

$$\leqslant (x-a)^{-\frac{q(x)}{q^-_{a,x}}+1} 2^{\frac{q(x)}{q^-_{a,x}}-1} \left(\frac{1}{x-a} \int_a^x (f(y))^{q(y)} dy + 1 \right)$$

$$\leqslant (x-a)^{\frac{q^-_{a,x}-q^+_{a,x}}{q^-_{a,x}}} 2^{\frac{q(x)}{q^-_{a,x}}-1} \left(\frac{1}{x-a} \int_a^x (f(y))^{q(y)} dy + 1 \right)$$

$$\leqslant \left[\max\{e^{2A_a/p_-}; 2^{(p_+-p_-)/p_-}\} \right]^{1/q_-} 2^{\frac{p_+}{p_-}-1} \left(\frac{1}{x-a} \int_a^x (f(y))^{q(y)} dy + 1 \right).$$

Summarizing these estimates we get (1.13) with constant (1.14). □

The next statement can be proved just in the same manner as the previous one; therefore, the proof is omitted.

Lemma 1.9 *Let* $-\infty < a < b < \infty$ *and let* $p(\cdot)$ *be a continuous exponent on* $(a, b]$ *such that* $p \in P_{a,b} \cap \mathcal{P}(b, A_b)$. *Then, for all non-negative* f *with* $f \in L^{p(\cdot)}(a, b)$: $\|f\|_{L^{p(\cdot)}(a,b)} \leqslant 1$,

$$\left(H_b' f(x)\right)^{p(x)/p_-} \leqslant C(p, A_b)\left(H_b'(f^{p/p_-}) + 1\right), \quad x \in (a, b), \tag{1.15}$$

where

$$C(p, A_b) = 3^{p_+/p_-} + 2^{p_+/p_- -1}\left[\max\{e^{2A_b/p_-}; 2^{(p_+-p_-)/p_-}\} \right]^{p_+/p_-}. \tag{1.16}$$

The next statement deals with a similar estimate but only for the case when $b = \infty$. The proof is similar to that of Lemma 3.4 in Diening [71].

Lemma 1.10 *Let* $-\infty < a < \infty$ *and let* $p(\cdot)$ *be a continuous exponent on* $[a, \infty)$ *such that* $p \in P_a \cap \mathcal{P}(a, A_a)$. *Let* p *be constant* $p(\cdot) \equiv p_c$ *outside some large interval* $[a, a+R]$. *Then, for all non-negative* f *satisfying the conditions* $f \in L^{p(\cdot)}(a, \infty)$: $\|f\|_{L^{p(\cdot)}(a,\infty)} \leqslant 1$,

$$\left(H_a f(x)\right)^{p(x)/p_-} \leqslant \widetilde{C}(p, R, a)\left(H_a(f^{p/p_-})\right)(x) + h(x) \quad x \in (a, \infty), \tag{1.17}$$

where

$$\widetilde{C}(p, a, R) = \max\left\{C(p, A_a), 2^{p_+/p_- -1}\right\}, \tag{1.18}$$

$$h(x) = C(p, A_a)\chi_{(a,a+2R)}(x) + \frac{2^{p_c/p_- -1}(1+R)^{p_c/p_-}\chi_{(a+2R,\infty)}}{x-a}, \tag{1.19}$$

$C(p, A_a)$ *is defined by* (1.14).

Proof Let $\|f\|_{L^{p(\cdot)}} \leqslant 1$. Then $S_p(f) \leqslant 1$. Let us split f as follows: $f = f_0 + f_1$, where $f_0 = \chi_{(a,a+R]}f$, $f_1 = \chi_{(a+R,\infty)}$. We denote:

$$q = p/p_-, \quad q_c := p_c/p_-.$$

By Lemma 1.8, for all $x \in (a, a+R]$, we have:

$$(H_a f(x))^{q(x)} \leqslant C(p, A_a)\Big[H_a(f^q) + 1\Big],$$

where $C(p, A_a)$ is defined by (1.14).

We have:

$$\begin{aligned}
\left(\frac{1}{x-a}\int_a^x f_0(y)dy\right)^{q(x)} &\leqslant \left[\frac{1}{x-a}\int_a^{a+R} (f(y))^{p(y)}dy + \frac{R}{x-a}\right]^{q(x)} \\
&\leqslant \left(\frac{1}{x-a} + \frac{R}{x-a}\right)^{q(x)} \\
&= \frac{(1+R)^{q_c}}{(x-a)^{q(x)}} \\
&\leqslant \frac{(1+R)^{q_c}}{x-a},
\end{aligned}$$

where $q_c \equiv p_c/p_-$ and $x > a + 2R$.

Thus

$$\Big(H_a f_0(x)\Big)^{q(x)} \leqslant \frac{(1+R)^{q_c}}{x-a}, \qquad x \in (a+2R, \infty).$$

Further, by Jensen's inequality, for the same x, we find that

$$\begin{aligned}
\Big(H_a f_1(x)\Big)^{q(x)} = \Big(H_a f_1(x)\Big)^{q_c} &\leqslant (H_a f_1^{q_c})(x) \\
&= \frac{1}{x-a}\int_{a+R}^{x} (f_1(y))^{q_c} \\
&\leqslant (H_a f^{q(\cdot)})(x).
\end{aligned}$$

Summarizing these estimates we get:

$$
\begin{aligned}
&\Big(H_a f(x)\Big)^{q(x)} \\
&\leqslant \chi_{(a,a+2R)}(x)\Big(H_a f(x)\Big)^{q(x)} + \chi_{(a+2R,\infty)}(x)\Big[H_a f_0(x) + H_a f_1(x)\Big]^{q(x)} \\
&\leqslant \chi_{(a,a+2R)}(x)\Big(H_a f(x)\Big)^{q(x)} + 2^{q_c-1}\chi_{(a+2R,\infty)}(x)\Big[(H_a f_0(x))^{q_c} + (H_a f_1(x))^{q_c}\Big] \\
&\leqslant C(p, A_a)\chi_{(a,a+2R)}(x)\Big(H_a f^{q(\cdot)}(x)\Big)(x) + C(p, A_a)\chi_{(a,a+2R)}(x) \\
&\qquad + 2^{q_c-1}\chi_{(a+2R,\infty)}(x)\frac{(1+R)^{q_c}}{x-a} + 2^{q_c-1}\chi_{(a+2R,\infty)}(x)\big(H_a f^{q(\cdot)}\big)(x) \\
&\leqslant \max\{C(p, A_a), 2^{q_c-1}\}\big(H_a f^{q(\cdot)}\big)(x) + C(p, A_a)\chi_{(a,a+2R)}(x) \\
&\quad + \chi_{(a+2R,\infty)}(x)\frac{2^{q_c-1}(1+R)^{q_c}}{x-a} \leqslant \max\{C(p, A_a), 2^{q_+-1}\}\big(H_a f^{q(\cdot)}\big)(x) + h(x),
\end{aligned}
$$

where h is defined by (1.19). □

The following statements give bounds for the norm of the Hardy operators in $VELS$.

Proposition 1.11 *Let $-\infty < a < b < \infty$ and let $p(\cdot)$ be an exponent on $[a, b)$ such that $p \in P_{a,b} \cap \mathcal{P}(a, A_a)$. Then, the following inequality holds for the norm of H_a in $L^{p(\cdot)}(a, b)$:*

$$
\|H_a\|_{L^{p(\cdot)}(a,b)} \leqslant \overline{c}(p, a, b), \tag{1.20}
$$

where

$$
\overline{c}(p, a, b) := C(p, A_a)^{p_-} 2^{(p_- -1)/p_-}\Big(\|H_a\|^{p_-}_{L^{p_-}(a,b)} + b - a\Big)^{1/p_-}, \tag{1.21}
$$

and the constant $C(p, A_a)$ is defined by (1.14).

Proof Let $\|f\|_{L^{p(\cdot)}(a,b)} \leqslant 1$. Then $S_p(f) \leqslant 1$. Denoting (as before) $q := p/p_-$ and applying Lemma 1.8, we have that

$$
\begin{aligned}
S_p(H_a f) &= \|(H_a f)^{q(\cdot)}\|^{p_-}_{L^{p_-}(a,b)} \\
&\leqslant (C(p, A_a))^{p_-}\|H_a(f^{q(\cdot)}) + 1\|^{p_-}_{L^{p_-}(a,b)} \\
&\leqslant (C(p, A_a))^{p_-} 2^{p_- -1}\Big(\|H_a(f^{q(\cdot)})\|^{p_-}_{L^{p_-}(a,b)} + \|1\|^{p_-}_{L^{p_-}}\Big) \\
&\leqslant (C(p, A_a))^{p_-} 2^{p_- -1}\Big(\|H_a\|^{p_-}_{L^{p_-}(a,b)}\|f^{q(\cdot)}\|^{p_-}_{L^{p_-}(a,b)} + b - a\Big) \\
&\leqslant (C(p, A_a))^{p_-} 2^{p_- -1}\Big(\|H_a\|^{p_-}_{L^{p_-}(a,b)} + b - a\Big).
\end{aligned}
$$

Now the result follows easily (see also Kokilashvili et al. [183, p. 836] for a similar argument). □

Proposition 1.12 *Let* $-\infty < a < b < \infty$ *and let* $p(\cdot)$ *be an exponent on* $(a, b]$ *such that* $p \in P_{a,b} \cap \mathcal{P}(b, A_b)$. *Then, the following inequality holds for the norm of* H_b *in* $L^{p(\cdot)}(a, b)$:

$$\|H_b'\|_{L^{p(\cdot)}(a,b)} \leqslant \overline{c}'(p, a, b), \tag{1.22}$$

where

$$\overline{c}'(p, a, b) = \Big[C(p, A_b)^{p_-} 2^{p_- - 1}\Big(\|H_b'\|^{p_-}_{L^{p_-}(a,b)} + b - a\Big)\Big]^{1/p_-} \tag{1.23}$$

and the constant $C(p, A_b)$ *is defined by* (1.16).

The following statement deals with the bound of H_a defined in variable exponent Lebesgue space but defined on on (a, ∞).

Proposition 1.13 *Let* $-\infty < a < \infty$ *and let* $p(\cdot)$ *be an exponent on* (a, ∞) *such that* $p \in P_a \cap \mathcal{P}(a, A_a)$. *Suppose that* $p \equiv p_c \equiv$ *const on* $(a + R, \infty)$ *for some large* R. *Then, the following estimate holds:*

$$\|H_a\|_{L^{p(\cdot)}(a,b)} \leqslant \overline{C}(p, a, R), \tag{1.24}$$

where

$$\overline{C}(p, a, R) := 2^{(p_- - 1)/p_-}\Big[\big(\widetilde{C}(p, a, R)\big)^{p_-} \|H_a\|_{L^{p_-}} + 2^{p_-} R \cdot C(p, A_a)^{p_-} + 2^{p_+ - 1}(1 + R)^{p_+} \frac{(2R)^{1-p_-}}{p_- - 1}\Big]^{1/p_-}, \tag{1.25}$$

and the constant $\widetilde{C}(p, a, R)$ *is defined by* (1.18).

Proof Assuming again that $f \geqslant 0$ and $\|f\|_{L^{p(\cdot)}} \leqslant 1$. Denoting $q := p/p_-$ and applying Lemma 1.10, we have:

$$\begin{aligned} S_p(H_a) &= \|(H_a f)^{q(\cdot)}\|^{p_-}_{L^{p_-}} \leqslant \|\widetilde{C}(p, a, R)\Big(H_a(f^{p/p_-})\Big)(x) + h(x)\|^{p_-}_{L^{p_-}} \\ &\leqslant 2^{p_- - 1}\Big(\|\widetilde{C}(p, a, R) H_a f^{q(\cdot)}\|^{p_-}_{L^{p_-}} + \|h\|^{p_-}_{L^{p_-}}\Big) \\ &\leqslant 2^{p_- - 1}\Big(\big(\widetilde{C}(p, a, R)\big)^{p_-} \|H_a\|^{p_-}_{L^{p_-}} \|f^{q(\cdot)}\|^{p_-}_{L^{p_-}} + \|h\|^{p_-}_{L^{p_-}}\Big) \\ &\leqslant 2^{p_- - 1}\Big(\big(\widetilde{C}(p, a, R)\big)^{p_-} \|H_a\|^{p_-}_{L^{p_-}} \end{aligned}$$

$$+ 2^{p_- -1}(2R)C(p, A_a)^{p_-} + 2^{p_- -1}\Big[2^{q_+ -1}(1+R)^{q_+}\Big]^{p_-} \int\limits_{a+2R}^{\infty} \frac{dx}{(x-a)^{p_-}}\Big)$$

$$= 2^{p_- -1}\Big[\big(\widetilde{C}(p, a, R)\big)^{p_-} \|H_a\|_{L^{p_-}} + 2^{p_-} R \cdot C(p, A_a)^{p_-}$$

$$+ 2^{p_- -1}\Big[2^{q_+ -1}(1+R)^{q_+}\Big]^{p_-} \frac{(2R)^{1-p_-}}{p_- - 1}\Big].$$

Thus

$$S_p(H_a f) \leqslant \Big(\overline{C}(p, a, R)\Big)^{p_-},$$

where $\overline{C}(p, a, R)$ and $\widetilde{C}(p, a, R)$ are defined by (1.25) and (1.18), respectively. This leads to the desired result. □

Now we are in the position to formulate and prove the main statements of this section.

Theorem 1.14 *Let* $-\infty < a < b < \infty$ *and let* $p(\cdot)$ *be a exponent on* $[a, b]$ *such that* $p \in P_{a,b} \cap \mathcal{P}(a, A_a) \cap \mathcal{P}(b, A_b)$. *Then, the following inequality holds for all* $u \in C_0^2(a, b)$:

$$\|u(\cdot)\delta^{-2}(\cdot)\|_{L^{p(\cdot)}(a,b)} \leqslant (\overline{c}(p, a, b) + \overline{c}'(p, a, b))\|u''(\cdot)\|_{L^{p(\cdot)}(a,b)}, \quad u \in C_0^2(a, b),$$

where the constants $\overline{c}(p, a, b)$ *and* $\overline{c}'(p, a, b)$ *are defined by* (1.21) *and* (1.23) *respectively, and*

$$\delta(x) = \min\{x - a, b - x\}.$$

Proof The result is a consequence of Propositions 1.11, 1.12 and the following observation:

$$\|u(\cdot)\delta^{-2}(\cdot)\|_{L^{p(\cdot)}(a,\infty)} \leqslant \|u(\cdot)(x-a)^{-2}(\cdot)\|_{L^{p(\cdot)}(a,c)} + \|u(\cdot)(b-x)^{-2}(\cdot)\|_{L^{p(\cdot)}(c,b)}$$

$$\leqslant \Big\|(x-a)^{-2}\Big(\int\limits_a^x (x-t)u''(t)dt\Big)\Big\|_{L^{p(\cdot)}(a,c)}$$

$$+ \Big\|(b-x)^{-2}\Big(\int\limits_x^b (t-x)u''(t)dt\Big)\Big\|_{L^{p(\cdot)}(c,b)}$$

$$\leqslant \Big\|(x-a)^{-1}\Big(\int\limits_a^x u''(t)dt\Big)\Big\|_{L^{p(\cdot)}(a,b)}$$

$$+\left\|(b-x)^{-1}\left(\int_x^b (t-x)(x-t)u''(t)dt\right)\right\|_{L^{p(\cdot)}(c,b)}$$

$$\leqslant (\|H_a\|_{L^{p(\cdot)}(a,b)} + \|H_b'\|_{L^{p(\cdot)}(a,b)})\|u''\|_{L^{p(\cdot)}(a,b)}$$

$$\leqslant (\overline{c}(p,a,b) + \overline{c}'(p,a,b))\|u''\|_{L^{p(\cdot)}(a,b)},$$

which ends the proof. □

From Proposition 1.13 we have the following.

Theorem 1.15 *Let* $-\infty < a < \infty$ *and let* $p(\cdot)$ *be a exponent on* $[a,\infty)$ *such that* $p \in P_{a,b} \cap \mathcal{P}(a, A_a)$. *Suppose that* $p(x) \equiv p_c \equiv$ *const outside some large interval* $(a, a+R)$. *Then the following inequality holds for all* $u \in C_0^2(a,\infty)$*:*

$$\|u(\cdot)\delta^{-2}(\cdot)\|_{L^{p(\cdot)}(a,\infty)} \leqslant \overline{C}(p,a,R)\|u''(\cdot)\|_{L^{p(\cdot)}(a,\infty)}, \qquad u \in C_0^2(a,\infty),$$

where the constant $\overline{C}(p,a,R)$ *is defined by* (1.25).

1.2 Two-Weight Hardy Inequality in $VELS$s and Applications

In this section, using the quantitative estimates of the norm of the two-weighted Hardy operator

$$H_{v,w}f(x) = v(x)\int_0^x f(t)w(t)dt,$$

we obtain two-sided estimates for the distance between the operator $H_{v,w}$ and the class of compact linear operators acting from $L^{p(\cdot)}(\mathbb{R}_+)$ to $L^{q(\cdot)}(\mathbb{R}_+)$ denoted by $\mathcal{K}\big(L^{p(\cdot)}(\mathbb{R}_+), L^{q(\cdot)}(\mathbb{R}_+)\big)$. In particular, we show that if $p(\cdot)$ and $q(\cdot)$ satisfy a log-Hölder continuity condition and $H_{v,w}$ is bounded from $L^{p(\cdot)}(\mathbb{R}_+)$ to $L^{q(\cdot)}(\mathbb{R}_+)$, then the estimate

$$\frac{1}{2}G \leqslant \|H_{v,w}\|_{\mathcal{K}\left(L^{p(\cdot)}(\mathbb{R}_+),L^{q(\cdot)}(\mathbb{R}_+)\right)} \leqslant DG, \tag{1.26}$$

holds for a certain constant D depending on p and q, where

$$G = \lim_{a\to 0} A_a + \lim_{b\to\infty} A_\infty^{(b)} \tag{1.27}$$

with

$$A_a := \sup_{t<a} \|v\chi_{(t,a)}\|_{q(\cdot)} \|w\chi_{(0,t)}\|_{p'(\cdot)} \tag{1.28}$$

and

$$A_\infty^{(b)} := \sup_{t>b} \|v\chi_{(t,\infty)}\|_{q(\cdot)} \|w\chi_{(b,t)}\|_{p'(\cdot)}. \tag{1.29}$$

As a corollary, we have criteria for the compactness of $H_{v,w}$ from $L^{p(\cdot)}(I)$ to $L^{q(\cdot)}(I)$. In particular, if p and q are log-Hölder continuous, then $H_{v,w}$ is compact from $L^{p(\cdot)}(I)$ to $L^{q(\cdot)}(I)$ if and only if

$$\lim_{a\to 0} A_a + \lim_{b\to\infty} A_\infty^{(b)} = 0, \tag{1.30}$$

where A_a and $A^{(b)}$ are defined by (1.28) and (1.29), respectively (cf. Edmunds et al. [91]).

Further, by applying the derived results about the norm of $H_{v,w}$, we get two-weighted variant of the Rellich inequality in $VELS$ defined on an interval I:

$$\|v(\cdot)u(\cdot)\delta^{-2}(\cdot)\|_{q(\cdot),I} \leqslant B_{p,q}\|w(\cdot)u''(\cdot)\|_{p(\cdot),I}, \qquad u \in C_0^2(I), \tag{1.31}$$

with a certain constant $B_{p,q}$, where $I = (a, b)$ and $\delta(x) = \min\{x - a, b - x\}$. Here $C_0^2(I)$ is the class of compactly supported functions defined on I with continuous second-order derivatives.

Such an inequality in the higher-dimensional case was first derived by Rellich [285] in the unweighted space $L^2(\mathbb{R}^n)$.

1.2.1 Preliminaries

We start this section with necessary properties and definitions regarding variable exponent spaces, which slightly differ from the ones in the previous section.

Let E be a measurable set in $\mathbb{R}$ with positive measure. We denote:

$$p_-(E) := \inf_E p, \qquad p_+(E) := \sup_E p$$

for a measurable real-valued function p on E.

By $\mathcal{P}(E)$ we denote the class of measurable functions p for which $1 < p_-(E) \leqslant p_+(E) < \infty$.

If $p : I \to \mathbb{R}_+$ and $p \in \mathcal{P}(I)$, where I is an interval in $\mathbb{R}$, then we will use the notation $p_- := p_-(I)$; $\quad p_+ := p_+(I)$.

We say that a measurable function f on a set E belongs to $L^{p(\cdot)}(E)$ (or to $L^{p(x)}(E)$) if $S_{p(\cdot)}(f) = \int_E |f(x)|^{p(x)} dx < \infty$. It is a Banach space with respect to the norm (see, e.g., Kováčik and Rákosník [198], Samko [292], Sharapudinov [311]):

$$\|f\|_{p(\cdot),E} = \inf\{\lambda > 0 : S_{p(\cdot)}(f/\lambda) \leqslant 1\}.$$

To prove the main statements, we need to recall some known results.

Proposition 1.16 (Kováčik and Rákosník [198], Samko [292], Sharapudinov [311]) *Let I be an interval in $\mathbb{R}$. Suppose that $p \in \mathcal{P}(I)$. Then*

(i)

$$\|f\|_{p(\cdot),I}^{p_+(I)} \leqslant S_{p(\cdot)}(f\chi_I) \leqslant \|f\|_{p(\cdot),I}^{p_-(I)}, \quad \|f\|_{p(\cdot),I} \leqslant 1;$$

$$\|f\|_{p(\cdot),I}^{p_-(I)} \leqslant S_{p(\cdot)}(f\chi_I) \leqslant \|f\|_{p(\cdot),I}^{p_+(I)}, \quad \|f\|_{p(\cdot),I} \geqslant 1;$$

(ii) Hölder's inequality

$$\left|\int_I f(x)g(x)dx\right| \leqslant A(p)\, \|f\|_{p(\cdot),I}\, \|g\|_{p'(\cdot),I}$$

holds, where $f \in L^{p(\cdot)}(I)$, $g \in L^{p'(\cdot)}(I)$ and

$$A(p) := A(p, I) = \frac{1}{p_-(I)} + \frac{1}{(p_+(I))'}. \tag{1.32}$$

Further, let $p \in \mathcal{P}(I)$. It is possible to introduce another norm in $L^{p(\cdot)}$ as follows:

$$\|f\|_{p,I}^* := \sup_{\|\varphi\|_{p'(\cdot),I} \leqslant 1} \left|\int_I f(x)\varphi(x)dx\right|.$$

Then, it is known that (see, e.g., Samko [292])

$$\|f\|_{p(\cdot),I} \leqslant \|f\|_{p(\cdot),I}^* \leqslant A(p)\|f\|_{p(\cdot),I}, \tag{1.33}$$

where the constant $A(p)$ is given by (1.32).

Proposition 1.17 (Kováčik and Rákosník [198], Samko [292], Sharapudinov [311]) *Let E be a set of finite measure and let $r(\cdot)$, $p(\cdot) \in \mathcal{P}(E)$. Suppose that $1 \leqslant r(x) \leqslant p(x)$, $x \in E$. Then the following inequality*

$$\|f\|_{r(\cdot),E} \leqslant (|E| + 1)\|f\|_{p(\cdot),E}$$

holds.

Definition 1.18 Let E be a measurable set in $\mathbb{R}$ and let $p(\cdot) \in \mathcal{P}(E)$. We say that p satisfies a log-Hölder continuity (weak Lipschitz) condition on E ($p \in \mathcal{P}^{\log}(E)$), if there is a positive constant A such that for all x and y in E with $0 < |x - y| < 1/2$, the inequality

$$|p(x) - p(y)| \leqslant \frac{A}{-\log|x-y|}$$

holds.

Definition 1.19 Let $p(\cdot) \in \mathcal{P}(E)$, where E is an unbounded subset of $\mathbb{R}$. We say that p satisfies the decay condition on E at infinity ($p \in \mathcal{P}_\infty(E)$), if there are constants $A_\infty \geqslant 0$ and $p_\infty \in (1, \infty)$ such that for all x in E the inequality

$$|p(x) - p_\infty| \leqslant \frac{A_\infty}{\log(e + |x|)}$$

holds.

It is known (see Diening [71] for bounded E, and Cruz-Uribe et al. [62] for unbounded E) that if E is a bounded open subset of $\mathbb{R}$, then the Hardy–Littlewood maximal operator M_E defined on E is bounded in $L^{p(\cdot)}$ if $p \in \mathcal{P}^{\log}(E)$, while the condition $p \in \mathcal{P}^{\log}_\infty(E) := \mathcal{P}^{\log}(E) \cap \mathcal{P}^{\log}_\infty(E)$ implies the boundedness of M_E in $L^{p(x)}$ when E is unbounded.

The following lemma is from Diening [71], but since we are interested in a bound for the constant, we give the proof.

Lemma 1.20 *Let I be an interval in $\mathbb{R}$ and let $s \in \mathcal{P}^{\log}(I)$. Then*

$$|J|^{s_-(J)-s_+(J)} \leqslant C_s$$

for all intervals $J \subset I$ with $|J| > 0$, where

$$C_s = \max\left\{e^A, 2^{s_+-s_-}\right\}, \tag{1.34}$$

where the constant A is from Definition 1.18.

Proof If $|J| \geqslant 1/2$, then

$$|J|^{s_-(J)-s_+(J)} \leqslant 2^{s_+(J)-s_-(J)} \leqslant 2^{s_+-s_-}.$$

Let now $|J| < 1/2$. Then there are $x, y \in \overline{J}$ such that

$$|J|^{s_-(J)-s_+(J)} = |J|^{-|s(x)-s(y)|} \leqslant \left(\frac{1}{|J|}\right)^{\frac{A}{-\log|y-z|}} \leqslant e^A,$$

from which the result follows. □

Remark 1.21 If $p \in \mathcal{P}^{\log}_{\infty}(\mathbb{R})$, then the following conditions are satisfied at 0 and ∞:

$$|p(x) - p(0)| \leqslant \frac{A_0}{|\log |x||}, \qquad 0 < |x| \leqslant 1, \tag{1.35}$$

$$|p(x) - p_\infty| \leqslant \frac{A_\infty}{\log |x|}, \qquad |x| > 1. \tag{1.36}$$

Remark 1.22 Let $I = \mathbb{R}$- It is known (see, e.g., Kokilashvili et al. [182]) that $\|\chi_{(-r,r)}\|_{p(\cdot),I} \approx r^{1/p(0)}$ as $r \to 0+$ if $p(x)$ satisfies a local log-Hölder continuity condition, and $\|\chi_{(-r,r)}\|_{p(\cdot),I} \approx r^{1/p_\infty}$ as $r \to \infty$, if $p \in \mathcal{P}^{\log}_{\infty}(I)$.

Let v and w be a.e. positive measurable functions on (a, ∞) and let

$$(H^{(a)}_{v,w} f)(x) = v(x) \int_a^x f(t)w(t)dt, \qquad x > a,$$

where a is a real number. Let

$$H_{v,w} := H^{(0)}_{v,w}.$$

The following statements was proved in Kopaliani [194] and Kopaliani [195] (see also Meskhi and Zaighum [247]) for certain values of a and b, but the proof for arbitrary a and b is the same.

Theorem 1.23 *Let $I = (a, b)$, where $-\infty < a < b \leqslant \infty$. Suppose that $p(\cdot)$ and $q(\cdot)$ are exponents on I satisfying the condition $1 < p_-(I) \leqslant p(x) \leqslant q(x) \leqslant q_+(I) < \infty$. Suppose also that $p, q \in \mathcal{P}^{\log}(I)$ for $b < \infty$, and $p, q \in \mathcal{P}^{\log}_{\infty}(I)$ for $b = \infty$. Then $H^{(a)}_{v,w}$ is bounded from $L^{p(\cdot)}(I)$ to $L^{q(\cdot)}(I)$ if and only if*

$$A_{a,b} := \sup_{a<t<b} \|\chi_{(t,b)} v\|_{q(\cdot),I} \|\chi_{(a,t)} w\|_{p'(\cdot),I} < \infty. \tag{1.37}$$

To get the main statements, we will need more precise estimates of the norms of the Hardy operators which will be discussed below.

We will also need the next statement.

Theorem 1.24 (Edmunds and Meskhi [87]) *Let $p(x)$ and $q(x)$ be measurable functions on an interval $I \subseteq \mathbb{R}$. Suppose that $1 < p_-(I) \leqslant p_+(I) < \infty$ and $1 < q_-(I) \leqslant q_+(I) < \infty$. If*

$$\Big\| \|k(x, y)\|_{p'(y),I} \Big\|_{q(x),I} < \infty,$$

where k is a non-negative measurable kernel, then the operator

$$Kf(x) = \int_I k(x, y) f(y) dy$$

is compact from $L^{p(\cdot)}(I)$ to $L^{q(\cdot)}(I)$.

Let X and Y be Banach spaces. Suppose that $\mathcal{K}(X, Y)$ (resp. $F_R(X, Y)$) denotes the class of compact linear operators (resp. finite rank operators) acting from X to Y. For every bounded linear operator $T : X \mapsto Y$, let

$$\|T\|_{\mathcal{K}(X,Y)} := \operatorname{dist}\{T, \mathcal{K}(X, Y)\}; \quad \overline{\alpha}(T) := \operatorname{dist}\{T, F_R(X, Y)\}.$$

Theorem 1.25 (Meskhi [240] p. 80) *Let $I := (0, a)$, where $0 < a \leqslant \infty$. Let $q \in \mathcal{P}_\infty^{\log}(I)$, assume that X is a Banach space and suppose that $1 < q_-(I) \leqslant q_+(I) < \infty$. Then for every bounded linear operator $T : X \mapsto L^{q(\cdot)}(I)$:*

$$\|T\|_{\mathcal{K}\left(X, L^{q(\cdot)}(I)\right)} = \overline{\alpha}(T).$$

1.2.2 *Bounds of Some $L^{p(\cdot)}$ Norm Inequalities*

In this section we give bounds of some $L^{p(\cdot)}$ norm inequalities.

Definition 1.26 Let I be an interval in $\mathbb{R}$. We say that the pair of exponents $\big(p(\cdot), q(\cdot)\big) \in \mathcal{P}(I)$ belong to the class $\mathbb{G}$ if there is a constant C such that for all $f \in L^{q(\cdot)}(I)$ and $g \in L^{p'(\cdot)}(I)$ and all sequences of disjoint intervals $I_k := [x_k, x_{k+1})$ satisfying the condition $\cup_k I_k = I$, the inequality

$$\sum_k \|f\chi_{S_k}\|_{p(\cdot),I} \|g\chi_{S_k}\|_{q'(\cdot),I} \leqslant C\|f\|_{p(\cdot),I}\|g\|_{q'(\cdot),I} \tag{1.38}$$

holds, where

$$S_k := [x_{k-1}, x_{k+1}). \tag{1.39}$$

We denote by C_Σ the best possible constant C in (1.38).

This definition for Banach function spaces was introduced in Berezhnoi [26]. It is known (see Kopaliani [194], Kopaliani [195] and also Meskhi [240], Chapter 1, Meskhi and Zaighum [247] for slight modifications) that if $1 < p_-(I) \leqslant p(x) \leqslant q(x) \leqslant q_+ < \infty$ and $p(\cdot), q(\cdot) \in \mathcal{P}_\infty^{\log}(I)$, then $\big(p(\cdot), q(\cdot)\big) \in \mathbb{G}$.

To formulate the next statements, we need to introduce some notation:

$$C_p := \max\left\{e^{A/(p_-)^2}, 2^{1/p_- - 1/p_+}\right\}; \tag{1.40}$$

$$C_q := \max\left\{e^{B/(q_-)^2}, 2^{1/q_- - 1/q_+}\right\}; \tag{1.41}$$

$$C_{q'} := \max\left\{e^{B/[(q_- - 1)^2((q')_-)^2]}, 2^{1/(q')_- - 1/(q')_+}\right\}, \tag{1.42}$$

$$A(q) = \frac{1}{q_-} + \frac{1}{(q_+)'}. \tag{1.43}$$

Lemma 1.27 *Let p and q be measurable functions on $I := (0, 1)$ satisfying the condition $1 < p_-(I) \leqslant p(x) \leqslant q(x) < q_+(I) < \infty$, $x \in I$. Let p and q belong to the class $\mathcal{P}^{\log}(I)$ with log-Hölder constants A and B, respectively. Then there is a positive constant $C_{p,q}$ depending only on p and q such that for all $f \in L^{p(\cdot)}(I)$, $g \in L^{q'(\cdot)}(I)$ and all sequences of disjoint intervals $I_k := [x_k, x_{k+1})$ satisfying the condition $\cup_k I_k = I$, the inequality*

$$\sum_k \|f\chi_{S_k}\|_{p(\cdot),I}\|g\chi_{S_k}\|_{q'(\cdot),I} \leqslant C_{p,q}\|f\|_{p(\cdot),I}\|g\|_{q'(\cdot),I}$$

holds, where

$$C_{p,q} = 8(C_p)^{p_- + p_+}(C_q)^{q_+ + 1}A(q)(C_{q'})^{(q')_+}, \tag{1.44}$$

with constants C_p, C_q, $C_{q'}$, and $A(q)$ defined by (1.40), (1.41), (1.42), (1.43), *respectively, and S_k is the interval given by* (1.39). *Consequently,*

$$C(\Sigma) \leqslant C_{p,q},$$

where $C(\Sigma)$ is the best possible constant in (1.38).

Proof We follow Kopaliani [194] (see also Meskhi [240], pp. 18-20). Let $\|f\|_{p(\cdot)} \leqslant 1$, $\|g\|_{p(\cdot)} \leqslant 1$. First observe that by Lemma 1.20 and the fact that $|S_k| \leqslant 1$,

$$\|\chi_{S_k}\|_{q(\cdot)}\|\chi_{S_k}\|_{q'(\cdot)} \leqslant |S_k|^{1/q_+(S_k)+1/(q')_+(S_k)}$$
$$\leqslant C_q|S_k|^{1/q_-(S_k)+1/(q_-)'_+(S_k)} = C_q|S_k|.$$

Further,

$$J := \sum_k \|f\chi_{S_k}\|_{p(\cdot),I}\|g\chi_{S_k}\|_{q'(\cdot),I}$$

$$
\begin{aligned}
&\leqslant C_q \sum_k |S_k| \frac{\|\chi_{S_k} f\|_{q(\cdot)} \|\chi_{S_k} g\|_{q'(\cdot)}}{\|\chi_{S_k}\|_{q(\cdot)} \|\chi_{S_k}\|_{q'(\cdot)}} \\
&\leqslant C_q \sum_k \int_0^1 \frac{\|\chi_{S_k} f\|_{p(\cdot)} \|\chi_{S_k} g\|_{q'(\cdot)}}{\|\chi_{S_k}\|_{q(\cdot)} \|\chi_{S_k}\|_{q'(\cdot)}} \chi_{S_k} dx \\
&\leqslant C_q \sum_k \int_0^1 \frac{\|\chi_{S_k} f\|_{p(\cdot)} \|\chi_{S_k} g\|_{q'(\cdot)}}{\|\chi_{S_k}\|_{q(\cdot)} \|\chi_{S_k}\|_{q'(\cdot)}} \chi_{I_{k-1}} dx \\
&\quad + C_q \sum_k \int_0^1 \frac{\|\chi_{S_k} f\|_{p(\cdot)} \|\chi_{S_k} g\|_{q'(\cdot)}}{\|\chi_{S_k}\|_{q(\cdot)} \|\chi_{S_k}\|_{p'(\cdot)}} \chi_{I_k} dx \\
&:= J_1 + J_2.
\end{aligned}
$$

First we estimate J_1. The bound for J_2 is the same. Applying Hölder's inequality we find that

$$
\begin{aligned}
J_1 &\leqslant C_q A(q) \left\| \sum_k \frac{\|f\|_{p(\cdot),S_k}}{\|\chi_{S_k}\|_{q(\cdot)}} \chi_{I_{k-1}} \right\|_{q(\cdot),I} \left\| \sum_k \frac{\|g\|_{p'(\cdot),S_k}}{\|\chi_{S_k}\|_{q'(\cdot)}} \chi_{I_{k-1}} \right\|_{q'(\cdot),I} \\
&=: C_q A(q) I(q) \cdot I(q'),
\end{aligned}
$$

where

$$
I(q) := \left\| \sum_k \frac{\|f\|_{p(\cdot),S_k}}{\|\chi_{S_k}\|_{q(\cdot)}} \chi_{I_{k-1}} \right\|_{q(\cdot),I}, \quad I(q') := \left\| \sum_k \frac{\|g\|_{p'(\cdot),S_k}}{\|\chi_{S_k}\|_{q'(\cdot)}} \chi_{I_{k-1}} \right\|_{q'(\cdot),I}.
$$

Now we claim that

$$
I(q) \leqslant (C_q)^{q_+} (C_p)^{p_-} I(p), \tag{1.45}
$$

where

$$
I(p) := \left\| \sum_k \frac{\|f\|_{p(\cdot),S_k}}{\|\chi_{S_k}\|_{p(\cdot)}} \chi_{I_{k-1}} \right\|_{p(\cdot),I}.
$$

Indeed, let $I(p) \leqslant 1$. Then

$$
\sum_k \frac{1}{|S_k|} \int_{I_{k-1}} \|f \chi_{S_k}\|_{p(\cdot)}^{p(x)} dx \leqslant (C_p)^{p_-} \sum_k \int_{I_{k-1}} \frac{\|f \chi_{S_k}\|_{p(\cdot)}^{p(x)}}{\|\chi_{S_k}\|_{p(\cdot)}^{p(x)}} dx
$$

$$\leqslant (C_p)^{p_-} \int_0^1 \left[\sum_k \frac{\|f\chi_{S_k}\|_{p(\cdot)}}{\|\chi_{S_k}\|_{p(\cdot)}} \right]^{p(x)} \chi_{I_{k-1}}(x)dx$$

$$\leqslant (C_p)^{p_-}.$$

Here we used the inequality that

$$\|\chi_{S_k}\|_{L^{p(\cdot)}} \leqslant (C_p)^{p_-}|S_k|.$$

Hence, since $\|f\|_{L^{p(\cdot)}} \leqslant 1$

$$\int_0^1 \sum_k \left[\frac{\|f\chi_{S_k}\|_{p(\cdot)}^{q(x)}}{\|\chi_{S_k}\|_{q(\cdot)}^{q(x)}} \right] \chi_{I_{k-1}}dx \leqslant (C_q)^{q_+} \sum_k \frac{1}{|S_k|} \int_0^1 \|f\chi_{S_k}\|_{p(\cdot)}^{q(x)} \chi_{I_{k-1}}(x)dx$$

$$\leqslant (C_q)^{q_+} \sum_k \frac{1}{|S_k|} \int_0^1 \|f\chi_{S_k}\|_{p(\cdot)}^{p(x)} \chi_{I_{k-1}}(x)dx$$

$$\leqslant (C_q)^{q_+}(C_p)^{p_-}.$$

Thus, (1.45) holds. Consequently,

$$I(q) \leqslant (C_q)^{q_+}(C_p)^{p_-}A(q)I(p)I(q').$$

Now we estimate $I(p)$. Let us introduce the exponent $\mathbb{P}(t)$ defined by

$$\mathbb{P}(t) := \sum_k p_+(S_k)\chi_{I_{k-1}}(t).$$

It is obvious that $p(t) \leqslant \mathbb{P}(t)$. Hence, by Proposition 1.17 we find that

$$I(p) \leqslant 2\left\| \sum_k \frac{\|f\chi_{S_k}\|_{p(\cdot)}}{\|\chi_{S_k}\|_{p(\cdot)}} \chi_{I_{k-1}} \right\|_{\mathbb{P}(\cdot),I}.$$

Passing to the modular, we find that

$$\int_0^1 \left(\sum_k \frac{\|f\chi_{I_k}\|_{p(\cdot),I}}{\|\chi_{S_k}\|_{p(\cdot)}} \chi_{I_{k-1}} \right)^{\mathbb{P}(x)} dx = \int_0^1 \sum_k \frac{\|f\chi_{S_k}\|_{p(\cdot),I}^{p_+(S_k)}}{\|\chi_{S_k}\|_{p(\cdot)}^{p_+(S_k)}} \chi_{I_{k-1}}dx$$

$$= \sum_k \int_{I_{k-1}} \frac{\|f\chi_{S_k}\|_{p(\cdot),I}^{p_+(S_k)}}{\|\chi_{S_k}\|_{p(\cdot)}^{p_+(S_k)}} dx$$

$$\leqslant (C_p)^{p_+} \sum_k \int\limits_{I_{k-1}} \frac{\|f\chi_{S_k}\|_{p(\cdot),I}^{p_+(S_k)}}{|S_k|} dx$$

$$\leqslant (C_p)^{p_+}.$$

Consequently,

$$I(p) \leqslant 2(C_p)^{p_+}.$$

Now we pass to $I(q')$. In this case we introduce the function:

$$Q'(t) = \sum_k (q')_+(S_k)\chi_{I_{k-1}}(t).$$

Then

$$I(q') \leqslant 2\left\| \sum_k \frac{\|g\chi_{S_k}\|_{q'(\cdot),I}}{\|\chi_{S_k}\|_{q'(\cdot)}} \chi_{I_{k-1}} \right\|_{Q'(\cdot),I}.$$

Further, arguing as in the case of $I(p)$, we find that

$$I(q') \leqslant 2(C_{q'})^{(q')_+}.$$

Summarizing these estimates we get:

$$J_1 \leqslant 4(C_q)^{q_+ +1}(C_p)^{p_- + p_-} A(q)(C_{q'})^{(q')_+}.$$

In a similar manner, we find that

$$J_2 \leqslant 4(C_q)^{q_+ +1}(C_p)^{p_- + p_-} A(q)(C_{q'})^{(q')_+}.$$

Summarizing the estimates for J_1 and J_2, we get the desired result. □

In the next statement, we will use the notation:

$$\overline{C}_p := \max\left\{e^{\overline{A}/(p_-)^2}, 2^{1/p_- - 1/p_+}\right\};$$

$$\overline{C}_q := \max\left\{e^{\overline{B}/(q_-)^2}, 2^{1/q_- - 1/q_+}\right\};$$

$$\overline{C}_{q'} := \max\left\{e^{\overline{B}/[(q_- -1)^2((q')_-)^2]}, 2^{1/(q')_- - 1/(q')_+}\right\}$$

with

$$\overline{A} := \max\left\{\frac{2A\log(b-a)}{\log 2},\ \ 2A,\ \ (p_+ - p_-)\log 2(b-a)\right\}, \tag{1.46}$$

$$\overline{B} := \max\left\{\frac{2B\log(b-a)}{\log 2},\ \ 2B,\ \ (p_+ - p_-)\log 2(b-a)\right\}. \tag{1.47}$$

Lemma 1.28 *Let $p(\cdot), q(\cdot) \in \mathcal{P}^{\log}(I)$ ($I = (a,b)$, $-\infty < a < b < \infty$) have log-Hölder constants A and B, respectively. Suppose that $1 < p_-(I) \leqslant p(x) \leqslant q(x) < q_+(I) < \infty$, $x \in I$. Then there is a positive constant $C_{p,q,a,b}$ depending only on p and q such that for all $f \in L^{p(\cdot)}(I)$, $g \in L^{q'(\cdot)}(I)$ and all sequences of disjoint intervals $I_k := [x_k, x_{k+1})$ satisfying the condition $\cup_k I_k = I$, the inequality*

$$\sum_k \|f\chi_{S_k}\|_{L^{p(\cdot)}(I)}\|g\chi_{S_k}\|_{L^{q'(\cdot)}(I)} \leqslant C_{p,q,a,b}\|f\|_{L^{p(\cdot)}(I)}\|g\|_{L^{q'(\cdot)}(I)}$$

holds, where $C_{p,q,a,b}$ is defined by

$$C_{p,q,a,b} = 8(\widetilde{C}_p)^{p_-+p_+}(\widetilde{C}_q)^{q_++1}A(q)(\widetilde{C}_{q'})^{(q')_+} + 4 \tag{1.48}$$

with constants $\widetilde{C}_p$, $\widetilde{C}_q$, $\widetilde{C}_{q'}$ equal to C_p, C_q and $C_{q'}$ (see (1.40), (1.41), (1.42)*) respectively, for $b - a \leqslant 1$, and to $\overline{C}_p$, $\overline{C}_q$, $\overline{C}_{q'}$ respectively, for $b - a > 1$. Consequently,*

$$C(\Sigma) \leqslant C_{p,q,a,b},$$

where $C(\Sigma)$ is the best possible constant in (1.38).

Proof Let $p, q \in \mathcal{P}^{\log}(I)$. Introduce the function:

$$s(t) := (b-a)t + a, \quad t \in (0,1).$$

Then s maps $(0,1)$ to (a,b). Observe that if $p \in \mathcal{P}^{\log}(I)$ with the constant A, then $p \circ s \in \mathcal{P}^{\log}_{\infty}((0,1))$ with the constant $\overline{A}$, where $\overline{A}$ is defined by (1.46).

(i) Let $b - a \leqslant 1$ and let $|t - \tau| \leqslant 1/2$. Then

$$|s(t) - s(\tau)| = (b-a)|t-\tau| \leqslant 1/2.$$

Consequently,

$$|(p\circ s)(t) - (p\circ s)(\tau)| \leqslant \frac{A}{-\log|(s(t)-s(\tau)|} \leqslant \frac{A}{-\log|t-\tau|}.$$

Thus, the result follows for the same constant A:

(ii) Let now $b - a > 1$. Suppose that $|t - \tau| < 1/2$.

(a) First consider the case $(b - a)^2 < 2$. Then $|t - \tau| < \frac{1}{2} < \frac{1}{(b-a)^2}$. Consequently,

$$|p(s(t)) - p(s(\tau))| \leqslant \frac{A}{-\log(b-a)|t-\tau|} \leqslant \frac{2A}{-\log|t-\tau|}. \tag{1.49}$$

(b) Suppose now that $(b-a)^2 > 2$. Then we consider the cases: $|t-\tau| < \frac{1}{(b-a)^2}$ and $\frac{1}{(b-a)^2} \leqslant |t-\tau| < \frac{1}{2}$ separately. In the first case, we have that (1.49) holds.

For the second case, we consider again two cases: $(b-a)|t-\tau| < 1/2$ and $(b-a)|t-\tau| \leqslant 1/2$ separately. In the first case, we have that

$$|p(s(t)) - p(s(\tau))| \leqslant \frac{2A\log(b-a)}{\log 2}\frac{1}{-\log|t-\tau|},$$

while in the second, one we find that

$$|p(s(t)) - p(s(\tau))| \leqslant (p_+ - p_-)\log 2(b-a)\frac{1}{-\log|t-\tau|}.$$

Summarizing these estimates we have the desired result.

Analogously, if $q \in \mathcal{P}^{\log}(I)$ with the constant B, then $q \circ s \in \mathcal{P}^{\log}((0,1))$, with the constant $\overline{B}$, where $\overline{B}$ is defined by (1.47). Now the result follows from Lemma 1.27. □

In the following statement, we will assume that

$$\overline{c}_p := \max\left\{e^{\widetilde{A}/(p_-)^2}, 2^{1/p_- - 1/p_+}\right\};$$

$$\overline{c}_q := \max\left\{e^{\widetilde{B}/(q_-)^2}, 2^{1/q_- - 1/q_+}\right\};$$

$$\overline{c}_{q'} := \max\left\{e^{\widetilde{B}/[(q_- - 1)^2((q')_-)^2]}, 2^{1/(q')_- - 1/(q')_+}\right\}$$

with

$$\widetilde{A} := \max\left\{\frac{2A\log(2R)}{\log 2},\ 2A,\ (p_+ - p_-)\log(2R)\right\}, \tag{1.50}$$

$$\widetilde{B} := \max\left\{\frac{2B\log(2R)}{\log 2},\ 2B,\ (p_+ - p_-)\log(2R)\right\}. \tag{1.51}$$

Lemma 1.29 *Let $p(\cdot), q(\cdot) \in \mathcal{P}^{\log}(I)$ ($I = (0, \infty)$) have log-Hölder constants A and B, respectively- Suppose that $1 < p_-(I) \leqslant p(x) \leqslant q(x) < q_+(I) < \infty$, $x \in I$. Suppose also that p and q are constants outside some large interval* $(0, R)$. *Then there is a positive constant $C_{p,q,R}$ depending only on p, q and R such that for all $f \in L^{p(\cdot)}(I)$, $g \in L^{q'(\cdot)}(I)$ and all sequences of disjoint intervals $I_k := [x_k, x_{k+1})$ satisfying the condition $\cup_k I_k = I$, the inequality*

$$\sum_k \|f\chi_{S_k}\|_{p(\cdot),I}\|g\chi_{S_k}\|_{q'(\cdot),I} \leqslant C_{p,q,R}\|f\|_{p(\cdot),I}\|g\|_{q'(\cdot),I}$$

holds, where S_k are defined by (1.39) *and $C_{p,q,R}$ is given by*

$$C_{p,q,R} = 8(\widetilde{C}_p)^{p_-+p_+}(\widetilde{C}_q)^{q_++1}A(q)(\widetilde{C}_{q'})^{(q')_+} + 4 \tag{1.52}$$

with constants $\widetilde{C}_p$, $\widetilde{C}_q$, $\widetilde{C}_{q'}$ equal to C_p, C_q, $C_{q'}$ (see (1.40), (1.41), (1.42)*) respectively, for $R \leqslant 1/2$, and to $\overline{c}_p$, $\overline{c}_q$ and $\overline{c}_{q'}$ respectively, for $R > 1/2$. Consequently,*

$$C(\Sigma) \leqslant C_{p,q,R},$$

where $C(\Sigma)$ is the best possible constant in (1.38).

Proof Without loss of generality, assume that and $R < x_m < x_{m+1} \leqslant 2R$ for some $m \in \mathbb{Z}$:

$$\begin{aligned} J &:= \sum_k \|f\chi_{S_k}\|_{p(\cdot),I}\|g\chi_{S_k}\|_{q'(\cdot),I} \\ &\leqslant \sum_{k\leqslant m+1} \|f\chi_{S_k}\|_{p(\cdot),I}\|g\chi_{S_k}\|_{q'(\cdot),I} + \sum_{k>m+1} \|f\chi_{S_k}\|_{p(\cdot),I}\|g\chi_{S_k}\|_{q'(\cdot),I} \\ &:= J_1 + J_2. \end{aligned}$$

By Lemma 1.28 we have that

$$J_1 \leqslant C_1(p, q, R)\|f\|_{p(\cdot),I}\|g\|_{q'(\cdot),I},$$

where $C_1(p, q, R)$ is defined by

$$C_1(p, q, R) = 8(C_p)^{p_-+p_+}(C_q)^{q_++1}A(q)(C_{q'})^{(q')_+}. \tag{1.53}$$

Further, observe that since in the case of J_2, we deal with constant p and q, and $\sum_k \chi_{S_k} \leqslant 2$. We have that

$$J_2 \leqslant 4\|f\|_{p(\cdot),I}\|g\|_{q'(\cdot),I}.$$

Now the result follows. □

Lemma 1.30 *Let $p(\cdot), q(\cdot) \in \mathcal{P}^{\log}(I)$ $(I = (a, \infty),\ -\infty < a < \infty)$ have log-Hölder constants A and B, respectively. Suppose that $1 < p_-(I) \leqslant p(x) \leqslant q(x) < q_+(I) < \infty$, $x \in I$. Suppose also that p and q are constants outside some large interval $(a, a + R)$. Then there is a positive constant $C_{p,q,R}$ such that for all $f \in L^{p(\cdot)}(I)$, $g \in L^{q'(\cdot)}(I)$ and all sequences of disjoint intervals $I_k := [x_k, x_{k+1})$ satisfying the condition $\cup_k I_k = I$, the inequality*

$$\sum_k \|f\chi_{S_k}\|_{p(\cdot),I}\|g\chi_{S_k}\|_{q'(\cdot),I} \leqslant C_{p,q,R}\|f\|_{p(\cdot),I}\|g\|_{L^{q'(\cdot)}(I)}$$

holds, where $C_{p,q,R}$ is defined by (1.52). *Consequently,*

$$C(\Sigma) \leqslant C_{p,q,R},$$

where $C(\Sigma)$ is the best possible constant in (1.38).

The proof of this lemma follows from Lemma 1.29 by using the change of variable $t \mapsto t + a$, $t \in (0, \infty)$.

Theorem 1.31 *Let $I = (a, b)$, where $-\infty < a < b \leqslant \infty$. Suppose that $p(\cdot)$ and $q(\cdot)$ are exponents on I satisfying the condition $1 < p_-(I) \leqslant p(x) \leqslant q(x) \leqslant q_+(I) < \infty$. Suppose also that $p, q \in \mathcal{P}^{\log}_\infty(I)$ if $b = \infty$, and $p, q \in \mathcal{P}^{\log}(I)$ when $b < \infty$. Then the following estimate holds for $H^{(a)}_{v,w}$:*

$$A_{a,b} \leqslant \|H^{(a)}_{v,w}\|_{L^{p(\cdot)}(I)\to L^{q(\cdot)}(I)} \leqslant 4A(p)A(q)C(\Sigma)A_{a,b},$$

where $A(p)$, $A(q)$, $C(\Sigma)$, and $A_{a,b}$ are determined by (1.32), (1.43), (1.38), *and* (1.37), *respectively.*

Proof (Upper estimate) Let $f \geqslant 0$, and $\int_I f(t)dt = \infty$. We construct a sequence $\{x_k\}$ so that

$$\int_a^{x_k} fw = \int_{x_k}^{x_{k+1}} fw = 2^k.$$

It is easy to check that $I = \cup_k[x_k, x_{k+1})$. Let g be a function satisfying the condition, $\|g\|_{q'(\cdot),I} \leqslant 1$. By applying Hölder's inequality for variable exponent Lebesgue spaces and Lemma 1.27, we have that

$$\begin{aligned}\int_I (H^{(a)}_{v,w} f)g &\leqslant \sum_k \Big(\int_{x_k}^{x_{k+1}} gv\Big)\Big(\int_a^{x_{k+1}} fw\Big)\\ &= 4\sum_k \Big(\int_{x_k}^{x_{k+1}} gv\Big)\Big(\int_{x_{k-1}}^{x_k} fw\Big)\end{aligned}$$

$$\leqslant 4\sum_k \|\chi_{(x_k,x_{k+1})}(\cdot)g(\cdot)\|_{q'(\cdot),I}\|\chi_{(x_k,x_{k+1})}(\cdot)v(\cdot)\|_{q(\cdot),I}$$

$$\times \|\chi_{(x_{k-1},x_k)}(\cdot)f(\cdot)\|_{p(\cdot),I}\|\chi_{(x_{k-1},x_k)}(\cdot)w(\cdot)\|_{p'(\cdot),I}$$

$$\leqslant 4A(p)A(q)\sum_k \|\chi_{(x_k,x_{k+1})}(\cdot)g(\cdot)\|_{q'(\cdot),I}\|\chi_{(x_{k-1},x_k)}(\cdot)f(\cdot)\|_{p(\cdot),I}$$

$$\leqslant 4A(p)A(q)C(\Sigma)\|f(\cdot)\|_{p(\cdot),I}\|g(\cdot)\|_{q'(\cdot),I}.$$

Taking the supremum over all g satisfying $|g|_{q'(\cdot),I} \leqslant 1$ establishes sufficiency.

The lower estimate follows by the standard way taking the test function f supported in (a,t) with $\|f\|_{L^{p(\cdot)}} \leqslant 1$. Details are omitted. □

1.2.3 Measure of Noncompactness Estimates: Compactness Criteria

It was shown in Edmunds et al. [91] that if X and Y are Banach function spaces (see, e.g., Bennett and Sharpley [25, p. 2], for the definition of a Banach function space), then there is a positive constant c such that the following estimate holds for the two-weighted Hardy operator $H_{v,w}$ bounded from X to Y:

$$\frac{1}{4}\overline{G} \leqslant \overline{\alpha}(H_{v,w}) \leqslant C\overline{G}, \tag{1.54}$$

where

$$\overline{G} = \lim_{a\to 0} B_a + \lim_{b\to\infty} B_\infty^{(b)},$$

with

$$B_\infty^{(b)} := \sup_{t>b} \|\chi_{(t,\infty)}v\|_Y \|\chi_{(b,t)}w\|_{X'},$$

$$B_a := \sup_{0<t<a} \|\chi_{(t,a)}v\|_Y \|\chi_{(0,t)}w\|_{X'}.$$

Here $\|\cdot\|_X$ and $\|\cdot\|_Y$ are norms in X and Y, respectively; X' is an associate space of X defined with respect to the norm (see Bennett and Sharpley [25, p. 8]):

$$\|f\|_{X'} = \sup\left\{\int_I |f(x)g(x)|dx : \|g\|_X \leqslant 1\right\}.$$

Theorem 1.32 *Let $I := \mathbb{R}_+$. Suppose that $1 < p_-(I) \leqslant p(x) \leqslant q(x) \leqslant q_+(I) < \infty$. Let $p, q \in \mathcal{P}_\infty^{\log}(I)$ and let $A_\infty^{(0)} < \infty$ (see (1.37) with $a = 0$ and $b = \infty$). Then the following estimate holds:*

$$\frac{1}{2}G \leqslant \|H_{v,w}\|_{\mathcal{K}(L^{p(\cdot)}(I),L^{q(\cdot)}(I))} \leqslant D_{p,q}G, \tag{1.55}$$

where

$$G = \lim_{a\to 0} A_a + \lim_{b\to\infty} A_\infty^{(b)}$$

(see (1.28), (1.29)),

$$D_{p,q} = 4A(p)A(q)C(\Sigma),$$

and the constants $A(p)$, $A(q)$ and $C(\Sigma)$ are defined by (1.32), (1.43) and (1.38), respectively.

Proof (Upper estimate) Following classical arguments (see, e.g., Edmunds et al. [91]), we represent $H_{v,w}$ as follows:

$$\begin{aligned} H_{v,w}f(x) &= \chi_{(0,a)}(H_{v,w}f\chi_{(0,a)})(x) + \chi_{(0,a)}(H_{v,w}f\chi_{(a,b)})(x) \\ &\quad + \chi_{(0,a)}(H_{v,w}f\chi_{(b,\infty)})(x) + \chi_{(a,b)}(H_{v,w}f\chi_{(0,a)})(x) \\ &\quad + \chi_{(a,b)}(H_{v,w}f\chi_{(a,b)})(x) + \chi_{(a,b)}(H_{v,w}f\chi_{(b,\infty)})(x) \\ &\quad + \chi_{(b,\infty)}(H_{v,w}f\chi_{(0,a)})(x) + \chi_{(b,\infty)}(H_{v,w}f\chi_{(a,b)})(x) \\ &\quad + \chi_{(b,\infty)}(H_{v,w}f\chi_{(b,\infty)})(x) \\ &=: \sum_{j=1}^{9} H_{v,w}^{(j)}f(x). \end{aligned}$$

It is easy to check that

$$H_{v,w}^{(2)}f \equiv 0; \quad H_{v,w}^{(3)}f \equiv 0, \quad H_{v,w}^{(6)}f \equiv 0.$$

Observe now that by Proposition 1.24, we find that $H_{v,w}^{(4)}$, $H_{v,w}^{(5)}$, $H_{v,w}^{(7)}$, $H_{v,w}^{(8)}$ are compact from $L^{p(\cdot)}(I)$ to $L^{q(\cdot)}(I)$. Consequently, by Theorems 1.31,

$$\begin{aligned} &\|H - H_{v,w}^{(4)} - H_{v,w}^{(5)} - H_{v,w}^{(7)} - H_{v,w}^{(8)}\|_{L^{p(\cdot)}\to L^{q(\cdot)}} \\ &\qquad \leqslant \|H_{v,w}^{(8)}\|_{L^{p(\cdot)}\to L^{q(\cdot)}} + \|H_{v,w}^{(9)}\|_{L^{p(\cdot)}\to L^{q(\cdot)}} \\ &\qquad \leqslant 4A(p)A(q)C(\Sigma)(A_a + A_\infty^{(b)}) \\ &\qquad \leqslant D_2 G, \end{aligned}$$

where A_a and $A_\infty^{(b)}$ are defined by

$$A_a := \sup_{t<a} \|v\chi_{(t,a)}\|_{q(\cdot),I} \|w(\cdot)\chi_{(0,t)}\|_{p'(\cdot),I};$$

and

$$A_\infty^{(b)} := \sup_{t>b} \|v\chi_{(t,\infty)}\|_{q(\cdot),I} \|w(\cdot)\chi_{(b,t)}\|_{p'(\cdot),I}.$$

Now passing to the limit when $a \to 0$ and $b \to \infty$, the upper estimate follows.

(LOWER ESTIMATE) We use the arguments of Edmunds et al. [91] (see also Meskhi and Zaighum [247]). Let us take a positive number λ such that $\lambda > \|H_{v,w}\|_{\mathcal{K}(L^{p(\cdot)}(I),L^{q(\cdot)}(I))}$. Consequently, by Theorem 1.25, we have that $\lambda > \overline{\alpha}(H_{v,w})$. Hence, there exist functions $g_1, \ldots, g_N$ from $L^{q(\cdot)}(I)$ such that

$$\overline{\alpha}(H_{v,w}) \leqslant \|H_{v,w} - F\| < \lambda,$$

where $Ff(x) = \sum_{j=1}^{N} \alpha_j(f) g_j(x)$. Here α_j are linear bounded functionals in $L^{p(\cdot)}(I)$, and g_i are linearly independent. Further, there exist $\bar{g}_1, \ldots, \bar{g}_N$ such that supports of $\bar{g}_i$ are in $[\sigma_i, \eta_i]$, $0 < \sigma_i < \eta_i < \infty$, and

$$\|H_{v,w} - F_0\| < \lambda,$$

where $F_0 f(x) = \sum_{j=1}^{N} \alpha_j(f)\bar{g}_j(x)$. Suppose that $\sigma = \min\{\sigma_j\}$, $\eta = \max\{\eta_j\}$. Then obviously, supp $F_0 f \subset [\sigma, \eta]$.

Let $\|f\|_{L^{p(\cdot)}(I)} \leqslant 1$. Taking $a < \sigma$ and $t \in (0, a)$, and observing that $(F_0 f)(x) = 0$ for $x \in (t, a)$, we find that

$$\begin{aligned} \lambda \geqslant \lambda \cdot \|f\chi_{(0,t)}\|_{L^{p(\cdot)}(I)} &\geqslant \|\chi_{(t,a)}(x)(H_{v,w} f\chi_{(0,t)} - F_0(f\chi_{(0,t)}))\|_{L^{q(\cdot)}(I)} \\ &= \|\chi_{(t,a)}(H_{v,w} f)\|_{L^{q(\cdot)}(I)} \\ &\geqslant \left(\int_0^t f(\tau)w(\tau)d\tau\right) \left\|\chi_{(t,a)} v\right\|_{L^{q(\cdot)}(I)}. \end{aligned}$$

Using relation (1.33) and taking the supremum over all such f and t, we obtain that

$$\lambda \geqslant A_a.$$

Since the left-hand side of this inequality does not depend on a, we find that

$$\lambda \geqslant \lim_{a \to 0} A_a.$$

Similarly, taking $t > b > \eta$ and arguing as above, we have that

$$\begin{aligned}\lambda \geqslant \lambda \cdot \|f\chi_{(b,t)}\|_{L^{p(\cdot)}(I)} &\geqslant \|\chi_{(t,\infty)}(x)(H_{v,w}f\chi_{(b,t)} - F_0(f\chi_{(b,t)}))\|_{L^{q(\cdot)}(I)} \\ &= \|\chi_{(t,\infty)}(H_{v,w}f)\|_{L^{q(\cdot)}(I)} \\ &\geqslant \left(\int_b^t f(\tau)w(\tau)d\tau\right)\left\|\chi_{(t,\infty)}v\right\|_{L^{q(\cdot)}(I)}.\end{aligned}$$

Again taking the relation (1.33) into account, we have that

$$\lambda \geqslant \lim_{b\to\infty} A_\infty^{(b)}.$$

Summarizing these estimates we get the desired result. □

Theorem 1.32 and Lemma 1.29 immediately imply the next statement.

Corollary 1.33 *Let $I := \mathbb{R}_+$. Suppose that $1 < p_-(I) \leqslant p(x) \leqslant q(x) \leqslant q_+(I) < \infty$. Let $p, q \in \mathcal{P}_\infty^{\log}(I)$ and let $A_\infty^{(0)} < \infty$ (see* (1.37) *for $a = 0$ and $b = \infty$). Suppose that p and q are constants outside some large interval* $(0, R)$. *Then, the following estimate holds:*

$$\frac{1}{2}G \leqslant \|H_{v,w}\|_{\mathcal{K}(L^{p(\cdot)}(I),L^{q(\cdot)}(I))} \leqslant D_{p,q,R}G, \tag{1.56}$$

where

$$G = \lim_{a\to 0} A_a + \lim_{b\to\infty} A_\infty^{(b)}$$

(see (1.28), (1.29)*),*

$$D_{p,q,R} = 4A(p)A(q)C_{p,q,R},$$

and the constants $A(p)$, $A(q)$, and $C_{p,q,R}$ are defined by (1.32), (1.43), *and* (1.52), *respectively.*

Corollary 1.34 (cf. Edmunds et al. [91]) *Let $I = \mathbb{R}_+$. Let $1 < p_-(I) \leqslant p(x) \leqslant q(x) \leqslant q_+(I) < \infty$. Suppose that $p, q \in \mathcal{P}_\infty^{\log}(I)$. The operator $H_{v,w}$ is compact from $L^{p(\cdot)}(I)$ to $L^{q(\cdot)}(I)$ if and only if:*

(i)

$$A_\infty^{(0)} < \infty,$$

where

$$A_\infty^{(0)} := \sup_{t>0} \|\chi_{(t,\infty)}(\cdot)v(\cdot)\|_{q(\cdot),I}\|\chi_{(0,t)}(\cdot)w(\cdot)\|_{p'(\cdot),I};$$

(ii)

$$\lim_{a\to 0} A_a + \lim_{b\to\infty} A_\infty^{(b)} = 0,$$

where A_a *and* $A_\infty^{(b)}$ *are defined by* (1.28) *and* (1.29).

1.2.4 Two-Weighted Rellich Inequality

Our result regarding the two-weighted version of the Rellich inequality reads as follows.

Theorem 1.35 *Let* $I := (a, b)$, *where* $-\infty < a < b < \infty$ *and let* $p(\cdot)$ *and* $q(\cdot)$ *be exponents on* $I := (a, b)$ *such that* $p, q \in \mathcal{P}^{\log}(I)$. *Suppose that the weight pair* (v, w) *satisfies the conditions:*

$$\sup_{a<t<b} \|v(\cdot)/(\cdot - t)\chi_{(t,b)}(\cdot)\|_{q(\cdot),I}\|w\chi_{(a,t)}\|_{p'(\cdot),I} < \infty,$$

$$\sup_{a<t<b} \|v(\cdot)/(b - \cdot)\chi_{(a,t)}(\cdot)\|_{q(\cdot),I}\|w\chi_{(t,b)}\|_{p'(\cdot),I} < \infty.$$

Then, the following inequality holds for all $u \in C_0^2(I)$:

$$\|v(\cdot)u(\cdot)\delta^{-2}(\cdot)\|_{p(\cdot),I} \leqslant \widetilde{C}(p, q)\|w(\cdot)u''(\cdot)\|_{p(\cdot),I}, \tag{1.57}$$

where $\widetilde{C}(p, q)$ *is defined by the formula:*

$$\widetilde{C}(p, q) = 4A(p)A(q)C(\Sigma)\Big(A(p) + 1\Big)$$

and $A(p)$, $A(q)$, *and* $C(\Sigma)$ *are constants given by* (1.32), (1.43), *and* (1.38), *respectively.*

Proof The result is a consequence of Theorems 1.23 and the following observation:

$$\|v(\cdot)u(\cdot)\delta^{-2}(\cdot)\|_{p(\cdot),I} \leqslant \|v(\cdot)u(\cdot)(x - a)^{-2}(\cdot)\|_{p(\cdot),(a,c)} + \|v(\cdot)u(\cdot)(b - \cdot)^{-2}(\cdot)\|_{p(\cdot),(c,b)}$$

$$\leqslant \left\| v(x)(x-a)^{-2}\left(\int_a^x (x-t)u''(t)dt\right)\right\|_{p(\cdot),(a,c)}$$

$$+\left\| v(x)(b-x)^{-2}\left(\int_x^b (t-x)u''(t)dt\right)\right\|_{p(\cdot),(c,b)}$$

$$\leqslant \left\| v(x)(x-a)^{-1}\left(\int_a^x |u''(t)|dt\right)\right\|_{p(\cdot),I}+\left\| v(x)(b-x)^{-1}\left(\int_x^b |u''(t)|dt\right)\right\|_{p(\cdot),I}$$

$$\leqslant \left(\|H^{(a)}_{\overline{v},w}\|_{L^{p(\cdot)}(I)\mapsto L^{q(\cdot)}(I)}+\|H^{(a)}_{w,\widetilde{v}}\|_{L^{q'(\cdot)}(I)\mapsto L^{p'(\cdot)}(I)}\right)\|u''\|_{L^{p(\cdot)}(I)}$$

$$\leqslant \left(\overline{c}(p,a,b)+\overline{c}'(p,a,b)\right)\|u''\|_{L^{p(\cdot)}(I)},$$

where

$$\overline{v}(x)=\frac{v(x)}{x-a},\qquad \widetilde{v}(x)=\frac{v(x)}{b-x},$$

$$H^{(a)}_{\overline{v},w}=\overline{v}(x)\int_a^x f(t)w(t)dt,\quad H^{(a)}_{w,\widetilde{v}}f(x)=w(x)\int_a^x f(t)\widetilde{v}(t)dt,$$

and

$$\overline{c}(p,a,b)=4A(p)C(\Sigma),\quad \overline{c}'(p,a,b))=4A(p)^2A(q)C(\Sigma).$$

Here we used the fact that

$$\|(H')^{(b)}_{\widetilde{v},w}\|_{L^{p(\cdot)}(I)\mapsto L^{q(\cdot)}(I)}\leqslant A(p)\|H^{(a)}_{w,\widetilde{v}}\|_{L^{q'(\cdot)}(I)\mapsto L^{p'(\cdot)}(I)},$$

where

$$(H')^{(b)}_{\widetilde{v},w}g(x)=(b-x)^{-1}v(x)\int_x^b g(t)w(t)dt.$$

□

When $b=\infty$, we have as follows.

Theorem 1.36 *Let* $-\infty<a<\infty$ *and let exponents* $p(\cdot)$ *and* $q(\cdot)$ *belong to* $\mathcal{P}^{\log}_{\infty}(I)$, *where* $I=(a,\infty)$. *Suppose that the weight pair* (v,w) *satisfies the conditions:*

$$\sup_{t>a}\|v(\cdot)/(\cdot-t)\chi_{(t,\infty)}(\cdot)\|_{L^{q(\cdot)}(I)}\|w\chi_{(a,t)}\|_{L^{p'(\cdot)}(I)}<\infty.$$

Then for all $u \in C_0^2(I)$

$$\|v(\cdot)u(\cdot)\delta^{-2}(\cdot)\|_{p(\cdot),I} \leqslant \overline{C}(p,q)\|w(\cdot)u''(\cdot)\|_{p(\cdot),I}, \tag{1.58}$$

where the constant $\overline{C}(p,q)$ *is defined by the formula*

$$\overline{C}(p,q) := 4A(p)A(q)C(\Sigma).$$

Remark 1.37 Since for classical Lebesgue spaces Hölder's inequality holds with constant equal to 1, and $C(\Sigma) = 1$, we have that in this case the constants $\widetilde{C}(p,q)$ and $\overline{C}(p,q)$ in (1.57) and (1.58) are equal to 8 and 4, respectively.

From Theorems 1.35, 1.36, Lemmas 1.28, 1.29 follow the next statements:

Corollary 1.38 *Let* a, b, $p(\cdot)$, $q(\cdot)$, v, *and* w *satisfy the conditions of Theorem 1.35. Then inequality* (1.57) *holds for all* $u \in C_0^2(I)$, *where* $\widetilde{C}(p,q)$ *is defined by the formula:*

$$\widetilde{C}(p,q) = 4A(p)A(q)C_{p,q,a,b}\Big(A(p)+1\Big)$$

and $A(p)$, $A(q)$, *and* $C_{p,q,a,b}$ *are constants given by* (1.32), (1.43), *and* (1.48), *respectively.*

Corollary 1.39 *Let* a, b, $p(\cdot)$, $q(\cdot)$, v, *and* w *satisfy the conditions of Theorem 1.36. Then inequality* (1.58) *holds for all* $u \in C_0^2(I)$, *where* $\overline{c}(p,q)$ *is defined by the formula*

$$\overline{c}(p,q) := 4A(p)A(q)C_{p,q,R}$$

with constants $A(p)$, $A(q)$, *and* $C_{p,q,R}$ *given by* (1.32), (1.43), *and* (1.52), *respectively.*

1.3 Multilinear Rellich Inequality

Our aim is to establish an m-linear weighted Rellich inequality:

$$\Big\|\prod_{j=1}^{m} u_j\Big\|_{L^q_{w(\delta(\cdot))}(I)} \leqslant C\prod_{j=1}^{m}\Big\|u_j''\Big\|_{L^{p_j}(I)}, \quad I := (a,b), \quad -\infty \leqslant a < b \leqslant +\infty, \tag{1.59}$$

with a certain positive constant c independent of $u_k \in C_0^\infty(I)$, $k = 1, \dots, m$, where $\delta(x)$ is the distance function on I given by the formula

$$\delta(x) = \min\{x - a, b - x\}, \tag{1.60}$$

and

Throughout this section we shall assume that m is a positive integer, and p is determined by

$$\frac{1}{p} := \sum_{k=1}^{m} \frac{1}{p_k}. \tag{1.61}$$

Observe that in this case, $0 < p < \infty$.

We establish (1.59) by using appropriate multilinear weighted Hardy inequalities:

$$\left\| \prod_{j=1}^{m} \int_a^x f_j(t)dt \right\|_{L_v^q(a,b)} \leqslant c \prod_{j=1}^{m} \|f_j\|_{L^{p_j}(a,b)}, \tag{1.62}$$

$$\left\| \prod_{j=1}^{m} \int_x^b f_j(t)dt \right\|_{L_v^q(a,b)} \leqslant c \prod_{j=1}^{m} \|f_j\|_{L^{p}(a,b)}, \tag{1.63}$$

which are also proved in this section. It should be noticed that necessary and sufficient conditions governing the two-weight bilinear Hardy inequality

$$\left(\int_a^b \left(\int_a^b f \right)^q \left(\int_a^x g \right)^q w(x)dx \right)^{1/q} \leqslant C \left(\int_a^b f^{p_1} w_1 \right)^{1/p_1} \left(\int_a^b g^{p_2} w_1 \right)^{1/p_2},$$

for non-negative f and g were found in Aguilar Cañestro et al. [9] under different conditions on weights for various ranges of p_1, p_2, and q, with $q > 1$ (see also Křepela [329] for different proof). The proofs used in this section enable us to get appropriate two-weight bilinear Rellich inequalities based on the results of Aguilar Cañestro et al. [9], but we do not consider this case here.

1.3.1 Preliminaries

We now recall and fix some notations used in the remaining of the section.

Let v be an a.e. positive function (i.e., a weight) on the interval $I := (a, b)$, $-\infty \leqslant a < b \leqslant \infty$. We denote by $L^r_v(I)$ (or by $L^r_v(a, b)$), $0 < r < \infty$, the Lebesgue space defined by the norm for $r \geqslant 1$ (quasi-norm if $0 < r < 1$):

$$\|g\|_{L^r_v(I)} = \left(\int_a^b |g(x)|^r v(x)dx \right)^{1/r}.$$

If $v \equiv \mathit{const}$, then $L^r_v(I)$ will be denoted by $L^r(I)$ (or by $L^r(a, b)$).

In the sequel we will denote by H_a and H'_b the Hardy-type operators of the form:

$$H_a f(x) = \frac{1}{x-a} \int_a^x f(t)dt, \quad x \in (a, b), \quad -\infty < a < b \leqslant \infty;$$

$$H'_b f(x) = \frac{1}{b-x} \int_x^b f(t)dt; \quad x \in (a, b), \quad -\infty \leqslant a < b < \infty.$$

It is known that (e.g., it follows from the general two-weight theory for the Hardy operator; see, e.g., introduction of Kufner and Persson [202] and references therein) operators H_a and H'_b are bounded in $L^r(a, b)$ for $1 < r < \infty$; moreover,

$$\|H_a\|_{L^r} := \|H_a\|_{L^r(a,b)\to L^r(a,b)} \leqslant r', \quad \text{and} \quad \|H'_b\|_{L^r} := \|H'_b\|_{L^r(a,b)\to L^r(a,b)} \leqslant r', \tag{1.64}$$

where $-\infty < a < b \leqslant \infty$, and

$$r' = \frac{r}{r-1}.$$

To get the main statements of this section, we will derive some weighted multilinear Hardy inequalities. In particular, we prove the following propositions.

Proposition 1.40 *Let* $-\infty < a < b \leqslant \infty$, v *be a weight function on* (a, b). *Then inequality* (1.62) *with a positive constant* c *independent of* f_j, $f_j \in L^{p_j}(a, b)$, $j = 1, \ldots, m$, *holds if and only if*

$$A_{a,b} := \sup_{a<t<b} \left(\int_t^b v(x)dx \right)^{1/q} \left(t-a\right)^{m-1/p} < \infty.$$

Moreover, if c *is the best possible constant in* (1.62), *then*

$$A_{a,b} \leqslant c \leqslant C A_{a,b},$$

where

$$C = \begin{cases} 2^{m-1/p}\left(1 + \prod_{i=1}^{m} \|H_a\|_{L^{p_i}}^{q}\right)^{1/q}, & \text{if } b < \infty, \\ 2^{m-1/p} \prod_{j=1}^{m} \|H_a\|_{L^{p_i}(a,\infty)} & \text{if } \ b = \infty. \end{cases}$$

Proof *Necessity*, and consequently, the lower estimate $c \geqslant A_{a,b}$ follows by the standard way taking the test functions $f_j = \chi_{(a,t)}$ in inequality (1.62). Now we show sufficiency, and hence, the upper estimate for the best constant c in (1.62):
(i) $b < \infty$.

Observe that

$$\left(\int_t^b v(x)dx\right)^{1/q} (t-a)^{m-1/p} = \left(\int_t^b v(x)dx\right)^{1/q} \prod_{j=1}^{m} (t-a)^{1/p'_j}$$

for all $a < t < b$. That is why the condition $A_{a,b} < \infty$, Hölder's inequality and the representation $(a,b) = \cup_{k\geqslant 0}[a + \frac{b-a}{2^{k+1}}, a + \frac{b-a}{2^k})$ yield (we assume that f_j are non-negative and $f_j \in L^{p_j}(I)$, $j = 1, \dots, m$):

$$\left\| \prod_{j=1}^{m} \int_a^x f_j(t)dt \right\|_{L_v^q(a,b)}^q = \sum_{k\geqslant 0} \int_{a+\frac{b-a}{2^{k+1}}}^{a+\frac{b-a}{2^k}} \prod_{j=1}^{m} \left(\int_a^x f_j(t)dt\right)^q v(x)dx$$

$$\leqslant \sum_{k\geqslant 0} \left(\int_{a+\frac{b-a}{2^{k+1}}}^{a+\frac{b-a}{2^k}} v(x)dx \right) \prod_{j=1}^{m} \left(\int_a^{a+\frac{b-a}{2^k}} f_j(t)dt \right)^q$$

$$\leqslant A_{a,b}^q \sum_{k\geqslant 0} \prod_{j=1}^{m} \left(\frac{b-a}{2^{k+1}}\right)^{-q/p'_j} \left(\int_a^{a+\frac{b-a}{2^k}} f_j(t)dt \right)^q$$

$$= A_{a,b}^q \prod_{j=1}^{m} \left(\frac{b-a}{2}\right)^{-q/p'_j} \left(\int_a^b f_j(t)dt \right)^q$$

$$+ A_{a,b}^q \sum_{k\geqslant 1} \prod_{j=1}^{m} \left(\frac{b-a}{2^{k+1}}\right)^{-q/p'_j} \left(\int_a^{a+\frac{b-a}{2^k}} f_j(t)dt \right)^q$$

$$\leqslant A_{a,b}^q \prod_{j=1}^{m} \left(\frac{b-a}{2}\right)^{-q/p'_j} \left(\int_a^b f_j^{p_j}(t)dt \right)^{q/p_j} (b-a)^{q/p'_j}$$

$$+ A^q_{a,b} \prod_{j=1}^m \Big[\sum_{k\geqslant 1} \Big(\frac{b-a}{2^{k+1}}\Big)^{-p_j/p'_j} \Big(\int_a^{a+\frac{b-a}{2^k}} f_j(t)dt \Big)^{p_j} \Big]^{q/p_j}$$

$$\leqslant 2^{q(m-1/p)} A^q_{a,b} \prod_{i=1}^m \|f_j\|^q_{L^{p_j}}$$

$$+ 2^{q(m-1/p)} A^q_{a,b} \prod_{i=1}^m \Big[\sum_{k\geqslant 1} \Big(\frac{b-a}{2^k}\Big)^{-p_j/p'_j} \Big(\frac{(b-a)}{2^k}\Big)^{-1} \frac{(b-a)}{2^k}$$

$$\times \Big(\int_a^{a+(b-a)/2^{k-1}} f_j(t)dt \Big)^{p_j} dx \Big]^{q/p_j} \leqslant 2^{q(m-1/p)} A^q_{a,b} \prod_{i=1}^m \|f_j\|^q_{L^{p_j}}$$

$$+ 2^{q(m-1/p)} A^q_{a,b} \prod_{i=1}^m \Big[\int_a^b \Big(\int_a^x f_j(t)dt \Big)^{p_j} (x-a)^{-p_i} dx \Big]^{q/p_j}$$

$$\leqslant 2^{q(m-1/p)} A^q_{a,b} \prod_{i=1}^m \|f_j\|^q_{L^{p_j}} + 2^{mp-1} A^q_{a,b} \prod_{i=1}^m \|H_a\|^q_{L^{p_i}} \Big(\int_a^b (f_j(t))^{p_j} dt \Big)^{q/p_j}$$

$$= \Big(2^{q(m-1/p)} A^q_{a,b} + 2^{q(m-1/p)} A^q_{a,b} \prod_{i=1}^m \|H_a\|^q_{L^{p_i}} \Big) \|f_j\|^q_{L^{p_j}}.$$

Hence,

$$\Big\| \prod_{i=1}^m \Big(\int_a^x f_i(t)dt \Big) \Big\|_{L^q_v} \leqslant A_{a,b} 2^{m-1/p} \Big(1 + \prod_{i=1}^m \|H_a\|^q_{L^{p_i}} \Big)^{1/q} \prod_{i=1}^m \|f_i\|^q_{L^{p_i}}.$$

(ii) $b = \infty$. In this case, we represent the interval (a, ∞) as the union of intervals $(a + 2^k, a + 2^{k+1}]$, where $k \in \mathbb{Z}$. Consequently, arguing as above we find that

$$\Big\| \prod_{j=1}^m \int_a^x f_j dt \Big\|^q_{L^q_v(a,b)} = \sum_{k\in\mathbb{Z}} \Big(\int_{a+2^k}^{a+2^{k+1}} v(x)dx \Big) \prod_{j=1}^m \Big(\int_a^{a+2^{k+1}} f_j(t)dt \Big)^q$$

$$\leqslant A^q_{a,\infty} \sum_{k\in\mathbb{Z}} \prod_{j=1}^m 2^{-kq/p'_j} \Big(\int_a^{a+2^{k+1}} f_j(t)dt \Big)^q$$

$$\leqslant A^q_{a,\infty}\prod_{j=1}^{m}\Big[\sum_{k\in\mathbb{Z}}2^{-kp_j/p'_j}\Big(\int\limits_a^{a+2^{k+1}} f_j(t)dt\Big)^{p_j}\Big]^{q/p_j}$$

$$\leqslant 2^{q(m-1/p)}A^q_{a,\infty}\prod_{j=1}^{m}\Big[\int\limits_a^{\infty}(x-a)^{-p_j}\Big(\int\limits_a^{x} f_j(t)dt\Big)^{p_j}\Big]^{q/p_j}$$

$$\leqslant 2^{q(m-1/p)}A^q_{a,\infty}\prod_{j=1}^{m}\|H_a\|^q_{L^{p_j}}\prod_{j=1}^{m}\|f_j\|^q_{L^{p_j}}.$$

The proposition has been proved. □

Proposition 1.41 *Let* $-\infty\leqslant a<b<\infty$, v *be a weight function on* (a,b). *Then inequality* (1.63) *with a positive constant* c *independent of* f_j, $f_j\in L^{p_j}(a,b)$, $j=1,\dots,m$, *holds if and only if*

$$B_{a,b}:=\sup_{a<t<b}\Big(\int\limits_a^t v(x)dx\Big)^{1/q}\Big(b-t\Big)^{m-1/p}<\infty.$$

Moreover, if c *is the best possible constant in* (1.63), *then*

$$B_{a,b}\leqslant c\leqslant CB_{a,b},$$

where

$$C=\begin{cases}2^{m-1/p}\Big(1+\prod_{i=1}^m\|H'_b\|^q_{L^{p_i}(a,b)}\Big)^{1/q}, & \text{if } a>-\infty,\\ 2^{m-1/p}\prod_{j=1}^m\|H'_b\|_{L^{p_i}(-\infty,b)} & \text{if } a=-\infty.\end{cases}$$

Proof The proof of this statement is similar to that of Proposition 1.40. We only point out that in this case we use the representations:

$$(a,b)=\cup_{j=0}^{\infty}\Big(b-(b-a)/2^k,b-(b-a)/2^{k+1}\Big],$$

$$(-\infty,b)=\cup_{j\in\mathbb{Z}}\Big(b-2^{k+1},b-2^k].$$

□

1.3.2 Weighted Rellich Inequality

We discuss the case when a weight function is a function of the distance function (1.60) defined on (a, b).

Theorem 1.42 *Let $-\infty < a < b < \infty$, $I := (a, b)$, and let w be a weight function on the interval $\big(0, (b-a)\big)$. Suppose that $q \geqslant \max\{p_1, \ldots, p_m\}$. If*

$$\widetilde{D}_{a,b} := \sup_{0<\tau<b-a} \left(\int_{\tau}^{b-a} w(x) x^{mq} dx \right)^{1/q} \tau^{m-1/p} < \infty,$$

then for all $u_j \in C_0^2(I)$, $j = 1, \ldots, m$, inequality (1.59) holds with the constant C given by the formula:

$$C = 2^{m-1/p+2^{1/q}} \widetilde{D}_{a,b} \left[1 + \prod_{i=1}^{m} (p_i')^q \right]^{1/q}. \tag{1.65}$$

Proof Let d be the midpoint of the interval (a, b). Then by using Propositions 1.40 and 1.41, we have that

$$\begin{aligned}
\left\| \prod_{j=1}^{m} u_j \right\|^q_{L^q_{w(\delta(\cdot))}(a,b)} &\leqslant \left\| \prod_{j=1}^{m} u_j \right\|^q_{L^q_{w(\delta(\cdot))}(a,d)} + \left\| \prod_{j=1}^{m} u_j \right\|^q_{L^q_{w(\delta(\cdot))}(d,b)} \\
&\leqslant \int_a^d w(x-a) \left| \prod_{j=1}^{m} u_j(x) \right|^q dx + \int_d^b w(b-x) \left| \prod_{j=1}^{m} u_j(x) \right|^q dx \\
&\leqslant \int_a^b w(x-a) \left| \prod_{j=1}^{m} \int_a^x (x-t) u_j''(t) dt \right|^q dx \\
&+ \int_a^b w(b-x) \left| \prod_{j=1}^{m} \int_x^b (t-x) u_j''(t) dt \right|^q dx \\
&\leqslant \int_a^b w(x-a)(x-a)^{mq} \left(\prod_{j=1}^{m} \int_a^x |u_j''(t) dt| \right)^q dx \\
&+ \int_a^b w(b-x)(b-x)^{mq} \left(\prod_{j=1}^{m} \int_x^b |u_j''(t)| dt \right)^q dx \\
&\leqslant \widetilde{C} \prod_{j=1}^{m} \|u_j''\|^q_{L^{p_j}(a,b)},
\end{aligned}$$

where $\widetilde{C}$ is defined by

$$\widetilde{C} = 2^{q(m-\frac{1}{p})}\left(1+\prod_{i=1}^{m}\|H_a\|^q_{L^{p_i}(I)}\right)\widetilde{A}^q_{a,b} + 2^{q(m-\frac{1}{p})}\left(1+\prod_{i=1}^{m}\|H'_b\|^q_{L^{p_i}(I)}\right)\widetilde{B}^q_{a,b} \tag{1.66}$$

with

$$\widetilde{A}_{a,b} := \sup_{a<t<b}\left(\int_t^b w(x-a)(x-a)^{mq}dx\right)^{1/q}\left(t-a\right)^{m-1/p},$$

and

$$\widetilde{B}_{a,b} := \sup_{a<t<b}\left(\int_a^t w(b-x)(b-x)^{mq}dx\right)^{1/q}\left(b-t\right)^{m-1/p}.$$

Consequently,

$$\left\|\prod_{j=1}^{m}u_j\right\|_{L^q_{w(\delta(\cdot))}(a,b)} \leqslant \widetilde{C}^{1/q}\prod_{j=1}^{m}\|u''_j\|_{L^{p_j}(a,b)},$$

with the constant $\widetilde{C}$ defined by (1.66).

Observe now that an appropriate change of variables in integrals yields that

$$\widetilde{A}_{a,b} = \widetilde{B}_{a,b} = \widetilde{D}_{a,b} < \infty.$$

Consequently, taking also (1.64) into account, we find that

$$\widetilde{C}^{1/q} = 2^{m-1/p}\widetilde{D}_{a,b}\left[2+\prod_{i=1}^{m}\|H_a\|^q_{L^{p_i}(I)}+\prod_{i=1}^{m}\|H'_b\|^q_{L^{p_i}(I)}\right]^{1/q} \leqslant C,$$

where C is defined by (1.65). □

The next statement deals with the cases $b=\infty$ and $a=-\infty$, respectively.

Theorem 1.43 *Let* $-\infty < a < \infty$. *Suppose that* $q \geqslant \max\{p_1,\dots,p_m\}$, $I := (a,\infty)$. *Let* w *be a weight function on* $(0,\infty)$. *If*

$$\widetilde{D}_{a,\infty} := \sup_{t>0}\left(\int_t^\infty w(x)x^{mq}dx\right)^{1/q} t^{m-1/p} < \infty, \tag{1.67}$$

then for all $u_j \in C_0^2(I)$, $j = 1, \ldots, m$, inequality (1.59) *holds, where*

$$C = 2^{m-1/p} \widetilde{D}_{a,\infty} \prod_{i=1}^{m} p_i'. \tag{1.68}$$

Proof Suppose that $I = (a, \infty)$. Observe that if $x \in (a, \infty)$, we have that $\delta(x) = x - a$. Arguing as in the proof of Theorem 1.42, applying Proposition 1.40 and representations $u_j(x) = \int_a^x (x-t)u_j''(t)dt$ for $u_j \in C_0^2(I)$, $j = 1, \ldots, m$, we get:

$$\left\| \prod_{j=1}^{m} u_j \right\|_{L^q_{w(\delta(\cdot))}(I)} \leqslant 2^{m-1/p} \widetilde{A}_{a,\infty} \left(\prod_{j=1}^{m} \|H_a\|_{L^{p_i}(I)} \right) \prod_{j=1}^{m} \|u_j''\|_{L^{p_j}(I)},$$

where

$$\widetilde{A}_{a,\infty} := \sup_{t>a} \left(\int_t^\infty w(x-a)(x-a)^{mq} dx \right)^{1/q} \left(t - a\right)^{m-1/p}.$$

Since

$$\widetilde{A}_{a,\infty} = \widetilde{D}_{a,\infty} < \infty,$$

and (1.64) holds, we are done. □

Theorem 1.44 *Let* $-\infty < b < \infty$ *and let* $I := (-\infty, b)$. *Suppose that* $q \geqslant \max\{p_1, \ldots, p_m\}$. *Let w be a positive function on* $(0, \infty)$. *If condition* (1.67) *is satisfied, then for all $u_j \in C_0^2(I)$, $j = 1, \ldots, m$, inequality* (1.59) *holds, where*

$$C = 2^{m-1/p} \widetilde{D}_{-\infty,b} \prod_{i=1}^{m} p_i', \tag{1.69}$$

and $\widetilde{D}_{-\infty,b} = \widetilde{D}_{a,\infty}$ is given by (1.67).

Proof Let $I = (-\infty, b)$. The proof of this case is similar to that of the previous one. In this case we use representations $u_j(x) = \int_x^b (t-x)u_j''(t)dt$ which hold for $u_j \in C_0^2(I)$, $j = 1, \ldots, m$, and Proposition 1.41. Hence, we find that

$$\left\| \prod_{j=1}^{m} u_j \right\|_{L^q_{w(\delta(\cdot))}(I)} \leqslant 2^{m-1/p} \widetilde{B}_{-\infty,b} \left(\prod_{j=1}^{m} \|H_b'\|_{L^{p_i}(I)} \right) \prod_{j=1}^{m} \|u_j''\|_{L^{p_j}(I)},$$

where

$$\widetilde{B}_{-\infty,b} := \sup_{t<b} \left(\int_{-\infty}^{t} w(b-x)(b-x)^{mq} dx \right)^{1/q} \left(b-t\right)^{m-1/p}.$$

Again, it can be checked that

$$\widetilde{B}_{-\infty,b} = \widetilde{D}_{-\infty,b} = \widetilde{D}_{a,\infty} < \infty,$$

where $\widetilde{D}_{a,\infty}$ is defined by (1.67).

Since (1.64) holds for the norms $\|H_b'\|_{L^{p_i}(I)}$, $i = 1 \ldots, m$, the result follows. □

By applying Theorems 1.42, 1.43, and 1.44, we can easily deduce the following statements.

Corollary 1.45 *Let* $-\infty < a < b < \infty$ *and let* $I := (a, b)$. *Suppose that* $q \geqslant \max\{p_1, \ldots, p_m\}$. *Then, the following inequality holds for all* $u_j \in C_0^2(I)$, $j = 1, \ldots, m$:

$$\left(\int_I \left| \prod_{j=1}^{m} u_j(x) \right|^q \delta(x)^{-2mq+q/p-1} dx \right)^{1/q} \leqslant C \prod_{j=1}^{m} \|u_j''\|_{L^{p_j}(I)}, \tag{1.70}$$

where

$$C = (mq - q/p)^{-1/q} 2^{m-1/p+1/q} \left[1 + \prod_{i=1}^{m} (p_i')^q \right]^{1/q}.$$

Corollary 1.46 *Let* $-\infty < a < \infty$ *and let* $I := (a, \infty)$. *Suppose that* $q \geqslant \max\{p_1, \ldots, p_m\}$. *Then inequality* (1.70) *holds for all* $u_j \in C_0^2(I)$, $j = 1, \ldots, m$, *where*

$$C = 2^{m-1/p} (mq - q/p)^{-1/q} \prod_{j=1}^{m} p_j'.$$

Corollary 1.47 *Let* $-\infty < b < \infty$. *Suppose that* $I := (-\infty, b)$. *Suppose that* $q \geqslant \max\{p_1, \ldots, p_m\}$. *Then inequality* (1.70) *holds for all* $u_j \in C_0^2(I)$, $j = 1, \ldots, m$, *where* C *is defined by*

$$C = 2^{m-1/p} (mq - q/p)^{-1/q} \prod_{j=1}^{m} p_j'.$$

1.4 Higher-Dimensional Rellich–Sobolev and Hardy–Sobolev Inequalities

One of our aims in this section is to derive Rellich-type inequalities in variable exponent Lebesgue spaces (briefly $VELS$) $L^{p(\cdot)}$ in the higher-dimensional case. In particular, the problem is studied in homogeneous groups $\mathbb{G}$, but the results are new even for the Abelian (Euclidean) case $\mathbb{G} = \left(\mathbb{R}^d, +\right)$, and the Heisenberg groups $\mathbb{G} = \mathbb{H}^n$. The proofs rely on weighted estimates for Riesz potentials with power-type weights in $VELS$ which are also derived in this section. Another consequence of the latter inequalities are higher dimensional Hardy-type inequalities in $VELS$s. These results are obtained under the condition that the Hardy–Littlewood maximal operator is bounded in appropriate unweighted $VELS$, which, for example, is guaranteed if the variable exponent satisfies a log-Hölder continuity condition and a decay condition at infinity. Hardy-type estimates are also investigated in the variable exponent setting. Similar results on domains in $\mathbb{R}^d$ were derived in Harjulehto et al. [133] and Rafeiro and Samko [278] under different conditions on variable exponents.

In the Rellich and Hardy inequalities, the bounds of constants depend on weights, as well as on the norms of the Hardy–Littlewood maximal operator in $VELSs$. That is why we also investigated quantitative estimates for the norm of the maximal operator in these spaces (see Sect. 1.4.3).

1.4.1 Preliminaries

Homogeneous Groups

In this section, we recall definitions and some of the main properties of (stratified) homogeneous groups based on Bonfiglioli et al. [34], Fischer and Ruzhansky [104], Folland and Stein [105], Ruzhansky and Suragan [288, 289].

Definition 1.48 A Lie group (on $\mathbb{R}^d$) $\mathbb{G}$ is said to be homogeneous if there is a dilation $D_\lambda(x)$ such that

$$D_\lambda(x) := \left(\lambda^{\nu_1}x_1, \ldots, \lambda^{\nu_d}x_d\right), \quad \nu_1, \ldots, \nu_d > 0, \quad D_\lambda : \mathbb{R}^d \to \mathbb{R}^d,$$

which is an automorphism of the group $\mathbb{G}$ for each $\lambda > 0$. Throughout this section, we use the notation λx for the dilation $D_\lambda(x)$. The number $Q := \nu_1 + \cdots + \nu_d$ is called the homogeneous dimension of $\mathbb{G}$. A homogeneous quasi-norm on $\mathbb{G}$ is a continuous non-negative function $r : \mathbb{G} \mapsto [0, \infty)$ such that:

(i) $r(x) = r\left(x^{-1}\right)$ for all $x \in \mathbb{G}$,
(ii) $r(\lambda x) = \lambda r\,(x)$ for all $x \in \mathbb{G}$ and $\lambda > 0$,
(iii) $r(x) = 0$ if and only $x = 0$.

The quasi-ball centered at $x \in \mathbb{G}$ with radius $R > 0$ is defined by

$$B(x, R) := \left\{ y \in \mathbb{G} : r\left(x^{-1}y\right) < R \right\}.$$

A homogeneous group is necessarily nilpotent, and the Haar measure on $\mathbb{G}$ coincides with the Lebesgue measure; we will denote it by dx. If $|E|$ denotes the measure of a measurable set $E \subset \mathbb{G}$, then

$$|D_\lambda(E)| = \lambda^Q |E| \quad \text{and} \quad \int_{\mathbb{G}} f(\lambda x)dx = \lambda^{-Q} \int_{\mathbb{G}} f(x)dx.$$

Hence we have that the Haar measure of the quasi-ball has the following property: there is a constant $A \geqslant 1$ such that

$$A^{-1}R^Q \leqslant |B(x, R)| \leqslant AR^Q. \tag{1.71}$$

Throughout this section, by A we denote the constant from (1.71).

In a homogeneous group $\mathbb{G}$, there is a polarization formula in the following form: there is a (unique) positive Borel measure σ on the unit quasi-sphere $\sigma_{\mathbb{G}} := \{x \in \mathbb{G} : r(x) = 1\}$ so that for every $f \in L^1(\mathbb{G})$, we have:

$$\int_{\mathbb{G}} f(x)dx = \int_0^\infty \int_{\sigma_{\mathbb{G}}} f(ty)t^{Q-1}d\sigma(y)dt.$$

In the sequel, by the symbol $B_{\mathbb{G}}$, we mean the surface measure of $\sigma_{\mathbb{G}}$.

The following property for a homogeneous quasi-norm on $\mathbb{G}$ holds.

Proposition 1.49 *Let $\mathbb{G}$ be a homogeneous group. Then there exists a homogeneous quasi-norm on $\mathbb{G}$ which is a norm, that is, a homogeneous quasi-norm $\cdot \mapsto r(\cdot)$ which satisfies the triangle inequality:*

$$r(xy) \leqslant r(x) + r(y), \quad \textit{for all } x, y \in \mathbb{G}.$$

Furthermore, all homogeneous quasi-norms on $\mathbb{G}$ are equivalent.

Taking Proposition 1.49 into account, we will assume that the function $r(\cdot)$ is a norm in $\mathbb{G}$.

A homogeneous group with the quasi-norm $r(\cdot)$ and Haar measure dx is an example of a quasi-metric measure space with a doubling measure, which is also called a space of homogeneous type (briefly SHT) (see Sect. 2.1.1 for the definition of an SHT).

Let us now recall the definition of a homogeneous stratified group (or homogeneous Carnot group). They form an important class of homogeneous groups. We refer, e.g., to Bonfiglioli et al. [34] Fischer and Ruzhansky [104] Ruzhansky and Suragan [288].

Definition 1.50 A Lie group $\mathbb{G} = \left(\mathbb{R}^d, \circ\right)$ is called a homogeneous stratified group if the following conditions hold:

(a) the decomposition $\mathbb{R}^d = \mathbb{R}^{d_1} \times \cdots \times \mathbb{R}^{d_s}$ is valid for some natural numbers $d_1, \ldots, d_s$ with $d_1 + \cdots + d_s = d$; the dilation $\delta_\lambda : \mathbb{R}^d \longrightarrow \mathbb{R}^d$ given by

$$\delta_\lambda(x) \equiv \delta_\lambda\left(x^{(1)}, \ldots, x^{(s)}\right) := \left(\lambda x^{(1)}, \ldots, \lambda^s x^{(s)}\right),$$

where $x^{(k)} \in \mathbb{R}^{d_k}$, $k = 1, \ldots, s$, is an automorphism of the group $\mathbb{G}$ for every $\lambda > 0$.

(b) If d_1 is as in (a) and $X_1, \ldots, X_{d_1}$ are the left-invariant vector fields on $\mathbb{G}$ such that $X_k(0) = \frac{\partial}{\partial x_k}\Big|_0$ for $k = 1, \ldots, d_1$, then

$$\operatorname{rank}\left(\operatorname{Lie}\left\{X_1, \ldots, X_{d_1}\right\}\right) = d,$$

for every $x \in \mathbb{R}^d$. In another words, the iterated commutators of $X_1, \ldots, X_{d_1}$ span the Lie algebra of $\mathbb{G}$.

In the sequel, by the symbol

$$\nabla_{\mathbb{G}} := \left(X_1, \ldots, X_{d_1}\right) \tag{1.72}$$

we denote the horizontal gradient on $\mathbb{G}$. Hence the sub-Laplacian on (homogeneous) stratified groups is determined by the formula:

$$\Delta_{\mathbb{G}} := \nabla_{\mathbb{G}} \cdot \nabla_{\mathbb{G}}. \tag{1.73}$$

In general, a Lie group is called stratified if it is a connected and simply connected Lie group whose Lie algebra is stratified. Any (abstract) stratified group is isomorphic to a homogeneous one.

When we deal with the inequalities involving $\nabla_{\mathbb{G}}$ and $\Delta_{\mathbb{G}}$, we will assume that $\mathbb{G}$ is a stratified homogeneous group.

Variable Exponent Lebesgue Spaces

Keeping some notation the notation of Sect. 1.2.1, for a positive continuous exponent function $p(\cdot)$ defined on a Haar-measurable set $E \subset \mathbb{G}$, we set:

$$p_-(E) := \inf_E p(\cdot); \quad p_+(E) := \sup_E p(\cdot).$$

We will write p_- and p_+ for $p_-(E)$ and $p_+(E)$, respectively, if $E := \mathbb{G}$.

Let $\mathcal{P}(E)$ be a class of continuous exponents $p(\cdot)$ defined on E, $E \subset \mathbb{G}$, such that

$$1 < p_-(E) \leqslant p_+(E) < \infty.$$

Let $p(\cdot) \in \mathcal{P}(E)$. A $VELS$, denoted by $L^{p(\cdot)}(E)$, is the linear space of all Haar-measurable functions f on E for which

$$S_{p(\cdot)}(E, f) := \int\limits_E |f(x)|^{p(x)} dx < \infty.$$

By the symbol $\int\limits_E g(x)dx$, we mean $\int\limits_{\mathbb{G}} g(x)\chi_E(x)dx$.

The norm in $L^{p(\cdot)}(E)$ is defined as follows:

$$\|f\|_{L^{p(\cdot)}(E)} = \inf\left\{\lambda > 0 : \; S_{p(\cdot)}(E, f/\lambda) \leqslant 1\right\}.$$

If $p(\cdot) \equiv p_c \equiv \text{const}$, then $L^{p(\cdot)}(E)$ coincides with the classical Lebesgue space $L^{p_c}(E)$.

It is known (see, e.g., Kováčik and Rákosník [198]) that $L^{p(\cdot)}(E)$ is a Banach space.

Hölder's inequality in $L^{p(\cdot)}(E)$ spaces has the following form (see, e.g., Kováčik and Rákosník [198]):

$$\left|\int\limits_E f(x)g(x)dx\right| \leqslant \kappa \|f\|_{L^{p(\cdot)}(E)} \|g\|_{L^{p'(\cdot)}(E)}, \quad p'(\cdot) = \frac{p(\cdot)}{p(\cdot) - 1}, \tag{1.74}$$

where $\kappa := \frac{1}{p_-} + \frac{1}{(p')_-}$.

Further, for $p(\cdot), q(\cdot) \in \mathcal{P}(\mathbb{G})$, $p(\cdot) \leqslant q(\cdot)$, the following useful estimate is valid:

$$\|f\|_{L^{p(\cdot)}(\mathbb{G})} \leqslant 2^{1/p_-} \|f\|_{L^{q(\cdot)}(\mathbb{G})} \|1\|_{L^{s(\cdot)}(\mathbb{G})}, \quad s(\cdot) = \frac{p(\cdot)q(\cdot)}{q(\cdot) - p(\cdot)}. \tag{1.75}$$

Definition 1.51 Let E be a measurable subset of $\mathbb{G}$. We denote by $\mathcal{P}_0^{\log}(E)$ the class of all positive continuous exponents $s(\cdot)$ on E satisfying the log-Hölder continuity condition on E, i.e., there is a positive constant L such that for all $x, y \in E$, $0 < r\left(xy^{-1}\right) \leqslant \frac{1}{2}$,

$$|s(x) - s(y)| \leqslant \frac{L}{\ln \frac{1}{r(xy^{-1})}}. \tag{1.76}$$

The best possible constant in (1.76) is called the log-Hölder constant.

If L is the log-Hölder constant for $s(\cdot)$, then we sometimes write $s(\cdot) \in \mathcal{P}_0^{\log}(E, L)$.

Remark 1.52 Let E be an open set in $\mathbb{G}$. Then:

(a) if $s(\cdot) \in \mathcal{P}(E) \cap \mathcal{P}_0^{\log}(E, L)$, then $s'(\cdot) = \frac{s(\cdot)}{s(\cdot)-1} \in \mathcal{P}(E) \cap \mathcal{P}_0^{\log}(E, L/(s_- - 1)^2)$;
(b) if $s(\cdot) \in \mathcal{P}(E) \cap \mathcal{P}_0^{\log}(E, L)$, then $1/s(\cdot) \in \mathcal{P}(E) \cap \mathcal{P}_0^{\log}(E, L/(s_-)^2)$;
(c) if $s(\cdot) \in \mathcal{P}_0^{\log}(E, L)$, then $s(\cdot)/s_- \in \mathcal{P}_0^{\log}(E, L/(s_-)^2)$.

Lemma 1.53 *Let* $1 \leqslant s_- \leqslant s_+ < \infty$ *and let* $s(\cdot) \in \mathcal{P}_0^{log}(\mathbb{G}, L)$. *Then for all balls* $B \subset \mathbb{G}$

$$|B|^{s_-(B)-s_+(B)} \leqslant (A\mathcal{R}_0)^{s_- - s_+},$$

where

$$\mathcal{R}_0 := \min\left\{\frac{e^{(2L)/A(s_- - s_+)}}{A}, 1/2\right\}. \tag{1.77}$$

Proof This statement is proved in Diening [71] for Euclidean spaces, but the same proof works for homogeneous groups (see also Kokilashvili et al. [182, p. 834] for an SHT). We give the proof to clarify bounds of constants there.

Let $B := B(x_0, R)$, $x_0 \in \mathbb{G}$, $R \leqslant \mathcal{R}_0$. Then for $x, y \in B$, we have $|B| \leqslant 1$, and consequently,

$$|B|^{s_-(B)-s_+(B)} \leqslant |B|^{\frac{L}{-\ln(2R)}} \leqslant e^{2LQ}.$$

The latter inequality follows from the fact that we can assume $\mathcal{R}_0$ so small that $Q/2 \leqslant \frac{\ln|B|}{\ln(2R)} \leqslant \frac{\ln(AR^Q)}{\ln(2R)} \leqslant 2Q$.

Suppose now that $R \geqslant \mathcal{R}_0$. Then $|B|^{s_-(B)-s_+(B)} \leqslant (A\mathcal{R}_0)^{s_- - s_+}$. Since $\mathcal{R}_0$ might be arbitrarily small, we conclude that the desired estimate follows. □

In the sequel, we denote:

$$\widetilde{C}(s(\cdot)) := (A\mathcal{R}_0)^{s_- - s_+}, \tag{1.78}$$

where $\mathcal{R}_0$ is the constant defined by (1.77).

Definition 1.54 We say that an exponent function $s(\cdot) \in \mathcal{P}(\mathbb{G})$ satisfies the decay condition $\left(s(\cdot) \in \mathcal{P}_\infty^{\log}(\mathbb{G})\right)$ if there exists the limit $s(\infty) = \lim_{s(x)\to\infty} s(x)$ and a positive constant L_∞ such that

$$|s(x) - s(\infty)| \leqslant \frac{L_\infty}{\ln(e + r(x))}, \quad x \in \mathbb{G}. \tag{1.79}$$

Let us call "decay constant" the best possible constant in (1.79).

It should be noted that condition (1.79) is equivalent to the condition:

$$|s(x) - s(y)| \leqslant \frac{L_\infty}{\ln(e + \min\{r(x), r(y)\})}.$$

We denote $\mathcal{P}^{\log}(E) := \mathcal{P}_0^{\log}(E) \cap \mathcal{P}_\infty^{\log}(\mathbb{G})$.

It is known that the condition $p \in \mathcal{P}(E) \cap \mathcal{P}^{\log}(E)$ (resp. $p \in \mathcal{P}(E) \cap \mathcal{P}_0^{\log}(E)$) guarantees the boundedness of various important operators of harmonic analysis in $VELS$ defined on an unbounded set E (resp. bounded set E). The same is true for an SHT with finite measure (see Cruz-Uribe and Fiorenza [58], Diening [71], Diening et al. [74], Kokilashvili et al. [182], Kokilashvili et al. [183]).

If L and L_∞ are log-Hölder and decay constants, respectively, for $s(\cdot) \in \mathcal{P}^{\log}(E)$, then we write $s(\cdot) \in \mathcal{P}^{\log}(E, L, L_\infty)$.

Remark 1.55 Let E be an open set in $\mathbb{G}$. Then:

(i) if $s \in \mathcal{P}(E) \cap \mathcal{P}^{\log}(E, L, L_\infty)$, then $s' = \frac{s(\cdot)}{s(\cdot)-1} \in \mathcal{P}(E) \cap \mathcal{P}^{\log}(E, L/(s_- - 1)^2, L_\infty/(s_- - 1)^2)$;
(ii) if $s \in \mathcal{P}(E) \cap \mathcal{P}^{\log}(E, L, L_\infty)$, then $1/s \in \mathcal{P}(E) \cap \mathcal{P}^{\log}(E, L/(s_-)^2, L_\infty/(s_-)^2)$;
(iii) if $s \in \mathcal{P}^{\log}(E, L, L_\infty)$, then $s/s_- \in \mathcal{P}^{\log}(E, L/(s_-)^2, L_\infty/(s_-)^2)$.

Remark 1.56 Let a positive function $\varphi(\cdot)$ satisfy the decay condition (1.79). Then the weight $(1+r(x))^{-\varphi(x)}$ is equivalent to the weight $(1+r(x))^{-\varphi(\infty)}$, i.e., there is a constant $C \geqslant 1$ such that $\frac{1}{C}(1+r(x))^{-\varphi(x)} \leqslant (1+r(x))^{-\varphi(\infty)} \leqslant C(1+r(x))^{-\varphi(x)}$, $x \in \mathbb{G}$.

The next estimate is well-known (see, e.g., Diening et al. [74, Corollary 4.5.9] for Euclidean case), but we give the proof for completeness.

Lemma 1.57 *Let $p(\cdot) \in \mathcal{P}(\mathbb{G}) \cap \mathcal{P}^{\log}(\mathbb{G})$. Then there is a positive constant C such that*

$$\|\chi_{B(x,R)}\|_{L^{p(\cdot)}(\mathbb{G})} \leqslant C R^{\frac{Q}{p(x,R)}},$$

where

$$p(x,R) := \begin{cases} p(x) & \text{if } 0 < R < 1, \\ p(\infty) & \text{if } R \geqslant 1. \end{cases}$$

Proof The estimate for $R \leqslant 1$ is well-known (see Diening et al. [74, Corollary 4.5.9] for Euclidean spaces and Kokilashvili et al. [182, Lemma 2.57] for an SHT).

Suppose that $R \geqslant 1$. We follow the proof of Lemma 1.4 of Kokilashvili et al. [182] (see p. 6).

We will show that there exists a constant $C > 0$ such that

$$\int\limits_{B(x,R)} \frac{dy}{R^{Q\frac{p(y)}{p(\infty)}}} \leqslant C \quad \text{for all} \;\; x \in \mathbb{G}.$$

For this, we will estimate the function:

$$F(x, R) := \int\limits_{B(x,R)\cap(\mathbb{G}\backslash B(0,1))} \frac{dy}{R^{Q\frac{p(y)}{p(\infty)}}}.$$

Let $r(x) \geqslant 2R$. Then $r(y) > r(x) - r\left(xy^{-1}\right) > R$. We have:

$$|p(y) - p(\infty)| \ln R \leqslant |p(y) - p(\infty)| \ln r(y) \leqslant C$$

by the decay condition which, on the other hand, yields

$$\frac{1}{R^{Q\frac{p(y)}{p(\infty)}}} \leqslant \frac{C}{R^Q}.$$

Consequently,

$$F(x, R) \leqslant \frac{C}{R^Q}|B(x, R)| \leqslant C.$$

Now let $r(x) < 2R$. Then $B(x, R) \subset B(0, 3R)$, so that

$$F(x, R) \leqslant \int\limits_{1<r(y)<3r} \frac{dy}{R^{Q\frac{p(y)}{p(\infty)}}}.$$

Now we proceed as follows:

$$F(x, R) \leqslant \sum_{k=0}^{N} \int\limits_{m_k(R)<r(y)<2^{-k}3R} \frac{dy}{R^{Q\frac{p(y)}{p(\infty)}}}$$

with the notation $m_k(R) := \max\left(1, 2^{-k-1}3R\right)$ and $N = \left[\log_2(3R)\right] + 1$. Hence

$$\begin{aligned} F(x, R) &\leqslant \sum_{k=0}^{N} \int\limits_{2m_k(R)<r(y)<2^{-k}3R} \frac{dy}{\left(2^k r(y)/3\right)^{Q\frac{p(y)}{p(\infty)}}} \\ &\leqslant C\sum_{k=0}^{N} 2^{-k\frac{p}{p(\infty)}} \int\limits_{m_k(R)<r(y)<2^{-k}3R} \frac{dy}{r(y)^{Q\frac{p(y)}{p(\infty)}}}. \end{aligned}$$

Therefore, in view of the decay condition, we obtain:

$$\begin{aligned} F(x,R) &\leqslant C\sum_{k=0}^{N} 2^{-k\frac{p}{p(\infty)}} \int\limits_{m_k(R)<r(y)<2^{-k}3R} \frac{dy}{r(y)^Q} \\ &= C\sum_{k=0}^{N} 2^{-k\frac{p}{p(\infty)}} \int\limits_{m_k(R)}^{2^{-k}3R} \frac{d\rho}{\rho} \\ &\leqslant C < \infty, \end{aligned}$$

which ends the proof. □

The next lemma will be useful for us (see Kokilashvili and Samko [176] for Euclidean spaces, but the proof is the same for the case of homogeneous groups).

Lemma 1.58 *Suppose that* $p(\cdot) \in \mathcal{P}(\mathbb{G}) \cap \mathcal{P}^{log}(\mathbb{G})$ *and let the following conditions hold:*

$$\sup_{x\in\mathbb{G}} |\beta(x)| < \infty; \quad \sup_{x\in\mathbb{G}}(Q+\beta(x)p(x)) < 0, \quad \sup_{x\in\mathbb{G}}(Q+\beta(\infty)p(x)) < 0.$$

Then

$$\left\| r\left(xx_0^{-1}\right)^{\beta(x_0)} \right\|_{L^{p(\cdot)}(\mathbb{G}\setminus B(x_0,R))} \leqslant CR^{\beta(x_0)+\frac{Q}{p(x_0)}}, \quad R \leqslant 1, \tag{1.80}$$

and

$$\left\| r\left(xx_0^{-1}\right)^{\beta(x_0)} \right\|_{L^{p(\cdot)}(\mathbb{G}\setminus B(x_0,R))} \leqslant CK(x_0)R^{\beta(x_0)+\frac{Q}{p(\infty)}}, \quad R > 1,$$

where $C > 0$ *is an absolute constant not depending on* R *and* x_0 *and*

$$K(x_0) = (1 + r(x_0))^{L_\infty|\beta(x_0)|/p(\infty)}$$

with the decay constant L_∞ *for* $p(\cdot)$. *Moreover, estimate* (1.80) *is valid for all* $R \in (0,\infty)$ *if* $p(x) \leqslant p(\infty)$, $x \in \mathbb{G}$.

Hardy–Littlewood Maximal Operator

Let $Mf(x)$ be the Hardy–Littlewood maximal function on $\mathbb{G}$ defined by the formula

$$Mf(x) := \sup_{B\ni x} \frac{1}{|B|} \int\limits_B |f(y)|dy, \quad f \in L_{\mathrm{loc}}(\mathbb{G}),$$

where the supremum is taken over all balls B containing $x \in \mathbb{G}$.

It is known that the condition $p(\cdot) \in \mathcal{P}^{\mathrm{loc}}(\mathbb{G})$ guarantees the boundedness of the operator M in $L^{p(\cdot)}(\mathbb{G})$ (see Cruz-Uribe et al. [62], Diening [71] for Euclidean spaces and Cruz-Uribe and Shukla [67] for an SHT). In particular, the following statement holds.

Theorem 1.59 *Let* $p(\cdot) \in \mathcal{P}(\mathbb{G})$ *and let conditions* (1.76), (1.79) *hold for* $p(\cdot)$. *Then* M *is bounded in* $L^{p(\cdot)}(\mathbb{G})$.

In the sequel, we say that $p(\cdot) \in (\mathbb{G})$ if the operator M is bounded in space $L^{p(\cdot)}(\mathbb{G})$.

In Sect. 1.4.3, we derive the upper bound of $\|M\|_{L^{p(\cdot)}(\mathbb{G})}$ in the case when $p(\cdot) \in \mathcal{P}(\mathbb{G}) \cap \mathcal{P}_0^{\mathrm{loc}}(\mathbb{G})$ and $p(\cdot)$ is constant outside some large ball.

Hardy Operator on $\mathbb{G}$

Let us recall the two-weight criteria for the Hardy operator

$$Hf(x) = \int\limits_{B(0,r(x))} f(y)dy, \quad x \in \mathbb{G}.$$

For the next statement in a more general setting, we refer, e.g., to Edmunds et al. [92, Ch. 1]).

Proposition 1.60 *Let* $1 < p_0 \leqslant q_0 < \infty$. *Let* v *and* w *be a.e. positive functions on* $\mathbb{G}$. *Then the operator* H *is bounded from* $L_w^{p_0}(\mathbb{G})$ *to* $L_v^{q_0}(\mathbb{G})$ *if and only if*

$$\mathcal{H} := \sup_{t>0} \left(\int\limits_{r(x)>t} v(x)dx \right)^{\frac{1}{q_0}} \left(\int\limits_{r(x)<t} w^{1-p_0'}(x)dx \right)^{\frac{1}{p_0'}} < \infty, \quad p_0' = \frac{p_0}{p_0 - 1}.$$

Moreover, $\mathcal{H} \leqslant \|H\| \leqslant 4\mathcal{H}$.

The proof of this statement is given, e.g., in Edmunds et al. [92, Ch. 1] for a more general Hardy operator, in particular, for the operator:

$$H_\mu f(x) = \int\limits_{d(x_0,y)<d(x_0,x)} f(y)d\mu, \quad x \in X,$$

defined on an SHT (X, d, μ), where x_0 is a fixed point in X and d is a quasi-metric on X (see Sect. 2.1.1 for the definition of a quasi-metric).

Hardy inequalities play an important role in harmonic analysis and its applications. They are very useful tool to study, for example, weighted inequalities for integral operators such as the Riesz potential operator:

$$I_\alpha f(x) = \int\limits_{\mathbb{G}} r\left(y^{-1}x\right)^{\alpha - Q} f(y)dy, \quad 0 < \alpha < Q.$$

In recent years, it became evident that it is also important to study mapping properties of the variable-parameter Riesz potential operator:

$$I_{\alpha(\cdot)} f(x) = \int_{\mathbb{G}} r\left(y^{-1}x\right)^{\alpha(x)-Q} f(y)dy, \quad 0 < \alpha_- \leqslant \alpha_+ < Q.$$

Muckenhoupt's Weights and Weighted Inequalities

Definition 1.61 We say that a weight function $\rho \in A_{p_0}$ if

$$[\rho]_{p_0} := \sup_B \left(\frac{1}{|B|}\int_B \rho(x)dx\right)\left(\frac{1}{|B|}\int_B \rho^{1-p_0'}(x)dx\right)^{p_0-1} < \infty$$

for $1 < p_0 < \infty$, and

$$(M\rho)(x) \leqslant c\rho(x), \quad x \in \mathbb{G}, \tag{1.81}$$

for $p = 1$, where M is the Hardy–Littlewood maximal operator on $\mathbb{G}$ and B is a ball in $\mathbb{G}$.

It is easy to check that condition (1.81) is equivalent to the following one: there is a positive constant c such that for all balls B

$$\frac{1}{|B|}\int_B \rho(y)dy \leqslant c \operatorname*{ess\,inf}_B \rho \tag{1.82}$$

(see also Duoandikoetxea [81] for Euclidean spaces and Strömberg and Torchinsky [316] for an SHT). The best possible constant C in (1.82) is the A_1-characteristic of ρ denoted by $[\rho]_1$.

It is easy to check that

$$[\rho]_{p_0} \leqslant [\rho]_1, \quad p_0 > 1.$$

Muckenhoupt A_{p_0} weights guarantee one-weight characterization for operators of harmonic analysis. Moreover, there are known the following sharp forms of the Muckenhoupt maximal theorem:

$$\|M\|_{L^{p_0}_w \to L^{p_0}_w} \leqslant C_{p_0}[w]_{p_0}^{p_0'-1}, \quad p_0 > 1, \tag{1.83}$$

and the Muckenhoupt–Wheeden theorem for the constant parameter Riesz potential operator I_α:

$$\|I_\alpha\|_{L^{p_0}_{w^{p_0}} \to L^{q_0}_{w^{q_0}}} \leqslant C_{p_0,\alpha}[w]_{p_0,q_0}^{\left(1-\frac{\alpha}{Q}\right)\max\{1,\frac{p_0'}{q_0}\}}, \tag{1.84}$$

where

$$1 < p_0 < Q/\alpha, \quad q_0 = \frac{p_0 Q}{Q - \alpha p_0}$$

and

$$[w]_{A_{p_0,q_0}} := \sup_B \left(\frac{1}{|B|}\int_B w^{q_0}(x)dx\right)\left(\frac{1}{|B|}\int_B w^{-p_0'}(x)dx\right)^{p-1}.$$

In these inequalities sharpness means that $[w]_{p_0}^{p_0'-1}$ (resp. $[w]_{p_0,q_0}^{(1-\frac{\alpha}{Q})\max\{1,\frac{p_0'}{q_0}\}}$) cannot be replaced by $\psi([w]_{p_0})$ (resp. by $\varphi([w]_{p_0,q_0})$) for any positive nondecreasing function $\psi(s)$ (resp. $\varphi(s)$) growing slower than $s^{p'-1}$ $\left(\text{resp. } s^{(1-\frac{\alpha}{Q})\max\{1,p_0'/q_0\}}\right)$.

Inequality (1.83) was established in Buckley [37] for Euclidean spaces (see Hytönen et al. [144] for an SHT), while estimate (1.84) was proved in Lacey et al. [204] in the Euclidean setting, and in Kairema [159] for the case of an SHT (see also Kokilashvili et al. [186]).

Theorem 1.62 *Let $0 < \alpha < Q$ and let $1 < p_0 < \frac{Q}{\alpha}$. We set $\frac{1}{p_0} - \frac{1}{q_0} = \frac{\alpha}{Q}$. Suppose that $\rho \in A_{1+q_0/p_0}(\mathbb{G})$. Then the inequality*

$$\left\|I_\alpha\left(f\rho^{\frac{\alpha}{Q}}\right)\right\|_{L^{q_0}_\rho(\mathbb{G})} \leqslant C[\rho]_{1+q_0/p_0'}^{(1-\frac{\alpha}{Q})\max\{1,p_0'/q_0\}} \|f\|_{L^{p_0}_\rho(\mathbb{G})} \tag{1.85}$$

holds with the positive constant $C \equiv C_{p_0,\alpha,Q}$, where $\left(1-\frac{\alpha}{Q}\right)\max\{1, p'/q\}$ is sharp in the sense that it cannot be replaced by the smaller one.

Although the two-weight problem for the Riesz potentials has been solved in terms of various conditions for weight pairs (see, e.g., Kokilashvili and Krbec [170] and the references therein), it is useful to have easily verifiable two-weight criteria for special type of weights. One of the such results was derived in Meskhi [238]. We formulate that statement in a slightly more general form.

Theorem 1.63 *Let $0 < \alpha < Q$ and let $1 < p_0 < \frac{Q}{\alpha}$. We set $\frac{1}{p_0} - \frac{1}{q_0} = \frac{\alpha}{Q}$. Suppose that $\rho \in A_{1+q_0/p_0'}(\mathbb{G})$. We set $v(x) = \sigma(r(x))\rho(x)$, $w(x) = u(r(x))\rho(x)$, where ρ and u are positive increasing functions on $(0, \infty)$. If $B_\rho < \infty$, where*

$$B_\rho := \sup_{t>0}\left(\int_{r(x)>t} v(x)r(x)^{(\alpha-Q)p_0}dx\right)^{1/q_0}\left(\int_{r(x)<t}\left(\rho^{-\frac{\alpha p_0}{\alpha}}(x)w(x)\right)^{1-p_0'}dx\right)^{\frac{1}{p_0'}},$$

then there is a positive constant $C := C_{\alpha,p_0,Q,v,w}$ *such that*

$$\left\| I_\alpha\left(f \cdot \rho^{\frac{\alpha}{Q}}\right)\right\|_{L^{q_0}_v(\mathbb{G})} \leqslant C\|f\|_{L^{p_0}_w(\mathbb{G})}. \tag{1.86}$$

Moreover, if C *is the best possible constant in* (1.85), *then*

$$C \leqslant C_{\alpha,p_0,Q} B_\rho^{1/p_0}\left([\rho]_{1+q_0/p_0'}^{(1-\frac{\alpha}{Q})\max\{1,p_0'/q_0\}+\frac{1}{q_0}} + 1\right). \tag{1.87}$$

The proof of this statement is the same as for the radial weight ρ given in Meskhi [238] (see also Guliyev [125, 126] for similar results). In this case, instead of the classical Hardy inequality, it is used the two-weight criterion for the Hardy-type transform defined on $\mathbb{G}$ given by Proposition 1.60. Here we point out the following easily verifiable estimates:

$$B(x,2R) \leqslant C_{A,p_0,q_0,Q}[\rho]_{p_0} B(x,R), \quad \rho \in A_{p_0},$$

$$\left[\rho^{-p_0'/q_0}\right]_{1+p_0'/q_0} = [\rho]_{1+q_0/p_0'}^{p_0'/q_0}, \quad \rho \in A_{1+q_0/p_0'}.$$

In Theorem 1.63, we are also interested in bounds of the best possible constants C given by (1.87).

Lemma 1.64 *Let conditions of Theorem 1.63 be fulfilled. Suppose that* $\rho \in A_1(\mathbb{G})$. *If*

$$\overline{B} := \sup_{t>0}\left(\int_{r(x)>t} \rho(r(x))r(x)^{(\alpha-Q)q_0}dx\right)^{\frac{1}{q_0}}\left(\int_{r(x)<t} u^{1-p'}(r(x))dx\right)^{\frac{1}{p_0}} < \infty,$$

then inequality (1.86) *holds.*

Proof It is enough to check that

$$B_\rho \leqslant c\overline{B}.$$

Indeed, we have:

$$\int_{r(x)>t} \sigma(r(x))\rho(x)r(x)^{(\alpha-Q)q_0}dx\left(\int_{r(x)<t}\left(\rho^{-\frac{\alpha p}{Q}}(x)w(x)\right)^{1-p_0'}\right)^{\frac{q_0}{p_0'}}$$

$$=\left(\int_{r(x)<t}\left(\rho^{-\frac{\alpha p}{Q}}(x)w(x)\right)^{1-p'}dx\right)^{\frac{q_0}{p_0'}}$$

$$\times \sum_{k=0}^{\infty} \int_{2^k t < r(x) < 2^{k+1} t} \sigma(r(x))\rho(x) r(x)^{(\alpha-Q)q_0} dx$$

$$\leqslant \left(\int_{B(0,t)} \rho^{\left(-\frac{\alpha p}{Q}+1\right)(1-p_0')} u(r(x))^{1-p_0'} dx \right)^{q_0/p_0'}$$

$$\times \sum_{k=0}^{\infty} \sigma\left(2^{k+1} t\right) \left(2^k t\right)^{(\alpha-Q)q_0} \left(\int_{B(0,2^{k+1}t)} \rho(x) dx \right)$$

$$\leqslant c_{\alpha,Q,p_0} [\rho]_1 \overline{B},$$

which ends the proof. □

For the next statement, we refer to Cruz-Uribe et al. [63] for Euclidean spaces and Kokilashvili et al. [183, p. 845] for an SHT.

Proposition 1.65 *Let $\mathcal{F}$ be a family of pairs of non-negative functions (f, g) and let $0 < p_0 \leqslant q_0 < \infty$. Suppose that for every weight $w \in A_1(\mathbb{G})$:*

$$\int_{\mathbb{G}} (f(x))^{p_0} w(x) dx \leqslant C_{p_0,A,Q}([w]_1) \left(\int_{\mathbb{G}} (g(x))^{p_0} w^{p_0/q_0}(x) dx \right)^{\frac{q_0}{p_0}},$$

where $C_{p_0,A,Q}([w]_1)$ is the positive constant depending on p_0, A, Q and $[w]_{A_1}$ so that the mapping $\cdot \to C_{p_0,A,Q}(\cdot)$ is nondecreasing. Suppose also that $p(\cdot)$ is a variable exponent such that $p_0 < p_- \leqslant p_+ < \frac{p_0 q_0}{q_0 - p_0}$. We set:

$$\frac{1}{p(\cdot)} - \frac{1}{q(\cdot)} = \frac{1}{p_0} - \frac{1}{q_0}. \tag{1.88}$$

If the maximal operator M is bounded in $L^{\overline{q}'(\cdot)}(\mathbb{G})$, where $\overline{q}(\cdot) = \frac{q(\cdot)}{q_0}$, then for all $(f, g) \in \mathcal{F}$,

$$\|f\|_{L^{q(\cdot)}(\mathbb{G})} \leqslant 2c_0 \left(2^Q A^2 \|M\|_{L^{\overline{q}'(\cdot)}(\mathbb{G})}\right) \|g\|_{L^{q(\cdot)}(\mathbb{G})}.$$

Remark 1.66 Observe that the doubling constant C_{DC} in $\mathbb{G}$ (i.e., the best possible constant in $|B(x, 2R)| \leqslant C_{DC} B(x, R))$ is less than or equal to $2^Q A^2$.

Lemma 1.64 implies the following.

Lemma 1.67 *Let $\sigma(t) = t^{\gamma}$, $u(t) = t^{\gamma p_0/q_0}$, where $0 < \gamma < \frac{Q q_0}{p_0}$. Then the following inequality holds:*

$$\left\| I_\alpha f \right\|_{L^{q_0}_{\sigma(r(\cdot))}(\mathbb{G})} \leqslant C \|f\|_{L^{p_0}_{u(r(\cdot))}(\mathbb{G})}.$$

Denote by $K_{\alpha,\eta} f(x)$ the weighted fractional integral operator

$$K_{\alpha,\eta} f(x) = r^{\eta}(x) I_\alpha \left(f(\cdot) r^{-\eta}(\cdot)\right)(x).$$

Then, we have:

Proposition 1.68 *Let* $1 < p_0 < \infty$ *and* $0 < \alpha < Q/p_0$. *Let* $-\frac{Q}{q_0} < \eta < \frac{Q}{p_0'}$. *We set* $\frac{1}{p_0} - \frac{1}{q_0} = \frac{\alpha}{Q}$. *Then there is a positive constant* $C \equiv C_{p(\cdot),\alpha,\eta,Q}$ *such that*

$$\left\| K_{\alpha,\eta} f \right\|_{L^{q_0}(\mathbb{G})} \leqslant C \| f \|_{L^{p_0}(\mathbb{G})}.$$

Proof The case $0 \leqslant \eta < Q/p_0'$ is a consequence of Lemma 1.67, while the statement for negative η follows from duality arguments. □

Lemma 1.64 and Propositions 1.65 and 1.68 imply the next statement.

Lemma 1.69 *Let* α, p_0, *and* q_0 *satisfy the conditions of Proposition 1.68. Suppose that* $0 < \eta < Q/p_0'$ *and let* $\rho \in A_1(\mathbb{G})$. *Then the inequality*

$$\left\| K_{\alpha,\eta} f \right\|_{L^{q_0}_{\rho}(\mathbb{G})} \leqslant C \| f \|_{L^{p_0}_{\rho^{p_0/q_0}}(\mathbb{G})}$$

holds for positive constant C *independent of* f.

Further Extrapolation Results

Definition 1.70 Given an exponent $p(\cdot) \in \mathcal{P}(\mathbb{G})$ and a weight w such that $0 < w(x) < \infty$ almost everywhere on $\mathbb{G}$, we say that $w \in A_{p(\cdot)}$ if

$$[w]_{A_{p(\cdot)}} = \sup_B |B|^{-1} \| w \chi_B \|_{L^{p(\cdot)}(\mathbb{G})} \left\| w^{-1} \chi_B \right\|_{L^{p'(\cdot)}(\mathbb{G})} < \infty,$$

where the supremum is taken over all balls $B \subset \mathbb{R}^n$.

Observe that if $w \in A_{p(\cdot)}$, then $w \in L^{p(\cdot)}_{\text{loc}}(\mathbb{G})$ and $w^{-1} \in L^{p'(\cdot)}_{\text{loc}}(\mathbb{G})$. Further, if $w \in A_{p(\cdot)}$, then $w^{-1} \in A_{p'(\cdot)}$.

We say that $f \in L^{p(\cdot)}(w, \mathbb{G})$ if $\| f \|_{L^{p(\cdot)}(w,\mathbb{G})} := \| f w \|_{L^{p(\cdot)}} < \infty$.

The following result is from Cruz-Uribe and Wang [61] (see Cruz-Uribe and Cummings [57] for an SHT).

Theorem 1.71 *Suppose that* $p(\cdot) \in \mathcal{P}(\mathbb{G}) \cap \mathcal{P}^{\log}(\mathbb{G})$. *Then for every* $w \in A_{p(\cdot)}$

$$\| w M f \|_{L^{p(\cdot)}(\mathbb{G})} \leqslant C \| w f \|_{L^{p(\cdot)}(\mathbb{G})}. \tag{1.89}$$

Conversely, given any $p(\cdot)$ *and* w*, if* (1.89) *holds for* $f \in L^{p(\cdot)}(\mathbb{G})$*, then* $p_- > 1$ *and* $w \in A_{p(\cdot)}$.

For given $p(\cdot) \in \mathcal{P}(\mathbb{G})$ and a weight w, we will say $(p(\cdot), w)$ is an M-pair if the maximal operator is bounded on $L^{p(\cdot)}(w, \mathbb{G})$ and $L^{p'(\cdot)}\left(w^{-1}, \mathbb{G}\right)$. By Theorem 1.71 we necessarily have $w \in A_{p(\cdot)}$ (equivalently, $w^{-1} \in A_{p'(\cdot)}$) and $p_- > 1$. Conversely, if $p(\cdot) \in \mathcal{P}^{\log}(\mathbb{G})$, with $p_- > 1$, then for any $w \in A_{p(\cdot)}$, $(p(\cdot), w)$ is an M-pair.

For the next result, see Cruz-Uribe and Wang [61] in the case of Euclidean spaces, but the proof of that result enables us to formulate it for $\mathbb{G}$ (we also refer to Cruz-Uribe and Cummings [57] for the boundedness of the maximal operator in weighted variable exponent Lebesgue spaces defined on an SHT with Muckenhoupt $A_{p(\cdot)}$ weights).

Theorem 1.72 *Suppose that for some* p_0, $1 < p_0 < \infty$, *and every* $w_0 \in A_{p_0}$,

$$\int_{\mathbb{G}} f(x)^{p_0} w_0(x)dx \leqslant C \int_{\mathbb{G}} g(x)^{p_0} w_0(x)dx, \quad (f, g) \in \mathcal{F}. \tag{1.90}$$

Then for any M*-pair* $(p(\cdot), w)$,

$$\|f\|_{L^{p(\cdot)}(w,\mathbb{G})} \leqslant C\|g\|_{L^{p(\cdot)}(w,\mathbb{G})}, \quad (f, g) \in \mathcal{F}. \tag{1.91}$$

The theorem remains valid in case $p_0 = 1$ *if we assume only that the maximal operator is bounded on* $L^{p'(\cdot)}\left(w^{-1}\right)$.

Definition 1.73 Given $1 < p_0 \leqslant q_0 < \infty$, we say that $w \in A_{p,q}$ if

$$\sup_B \left(\frac{1}{|B|}\int_B w(x)^q dx\right)^{1/q} \left(\frac{1}{|B|}\int_B w(x)^{-p'} dx\right)^{1/p'} < \infty,$$

where the supremum is taken over all balls $B \subset \mathbb{G}$. If $p = 1$, then $w \in A_{1,q}$ if

$$\sup_B \frac{\int_B w(x)^q dx}{\operatorname{ess\,inf}_{x\in B} w(x)^q} < \infty.$$

Definition 1.74 Let $p(\cdot), q(\cdot) \in \mathcal{P}(\mathbb{G})$ be such that for some γ: $0 < \gamma < 1$,

$$\frac{1}{p(x)} - \frac{1}{q(x)} = \gamma.$$

Given w such that $0 < w(x) < \infty$ almost everywhere, we say that $w \in A_{p(\cdot),q(\cdot)}$ if

$$\sup_B |B|^{\gamma-1} \|w\chi_B\|_{L^{q(\cdot)}(\mathbb{G})} \left\|w^{-1}\chi_B\right\|_{L^{p'(\cdot)},(\mathbb{G})} < \infty.$$

Theorem 1.75 *Suppose that for some p_0, q_0, $1 < p_0 \leqslant q_0 < \infty$, and every $w_0 \in A_{p_0,q_0}$,*

$$\left(\int_{\mathbb{G}} f(x)^{q_0} w_0(x)^{q_0} dx\right)^{1/q_0} \leqslant C\left(\int_{\mathbb{G}} g(x)^{p_0} w_0(x)^{p_0} dx\right)^{1/p_0}, \quad (f, g) \in \mathcal{F}. \tag{1.92}$$

Given $p(\cdot), q(\cdot) \in \mathcal{P}(\mathbb{G})$, suppose:

$$\frac{1}{p(x)} - \frac{1}{q(x)} = \frac{1}{p_0} - \frac{1}{q_0}.$$

Define $\sigma \geqslant 1$ by $1/\sigma' = 1/p_0 - 1/q_0$. If $w \in A_{p(\cdot),q(\cdot)}$ and $(q(\cdot)/\sigma, w^\sigma)$ is an M-pair, then

$$\|f\|_{L^{q(\cdot)}(w,\mathbb{G})} \leqslant C\|g\|_{L^{p(\cdot)}(w,\mathbb{G})}, \quad (f, g) \in \mathcal{F}. \tag{1.93}$$

The theorem remains valid in case $p_0 = 1$ if we assume only that the maximal operator is bounded on $L^{(q(\cdot)/q_0)'}\left(w^{-q_0,\mathbb{G}}, \mathbb{G}\right)$.

Remark 1.76 When $\sigma = 1$, Theorem 1.75 is reduced to Theorem 1.72.

Proposition 1.77 *Given constants p_0, q_0, $1 \leqslant p_0 < q_0 < \infty$, suppose $w \in A_{p_0,q_0}$. Then $w^{q_0} \in A_{r_0}$ when $r_0 = 1 + q_0/p_0'$.*

The next result is the variable exponent variant of Proposition 1.77.

Proposition 1.78 *Given $p(\cdot), q(\cdot) \in \mathcal{P}(\mathbb{G})$ with $1 < p(x) \leqslant q(x) < \infty$, suppose there exists $\sigma > 1$ such that $\frac{1}{p(x)} - \frac{1}{q(x)} = \frac{1}{\sigma'}$. Then $w \in A_{p(\cdot),q(\cdot)}$ if and only if $w^\sigma \in A_{q(\cdot)/\sigma}$.*

1.4.2 Weighted Sobolev and Rellich Inequalities

Now we formulate the main statements of this section. We will use the notation:

$$\overline{q}(\cdot) := q(\cdot)/q_0, \quad q_0 < q_-.$$

Observe that taking into account Remark 1.55, we have that, for example, the condition $p(\cdot) \in \mathcal{P}^{\log}(\mathbb{G})$ together with the relations $\frac{1}{p(\cdot)} - \frac{1}{q(\cdot)} = \text{conts}$ and $p(\cdot) \leqslant q(\cdot)$ guarantees the boundedness of M in $L^{\overline{q}'(\cdot)}(\mathbb{G})$.

Estimates with Power-Type Weights

Theorem 1.79 (Weighted Sobolev Inequality) *Let $\mathbb{G}$ be a homogeneous group with homogeneous dimension Q. Let $p \in \mathcal{P}(\mathbb{G})$ and $0 < \alpha < \frac{Q}{p_-}$. We set $\frac{1}{p(\cdot)} - \frac{1}{q(\cdot)} = \frac{\alpha}{Q}$. Suppose that $-\frac{Q}{q_-} < \eta < \frac{Q}{(p_-)'}$ and $\overline{q}'(\cdot) \in \mathcal{B}(\mathbb{G})$ for some $q_0 < q_-$. Then the operator $K_{\alpha,\eta}$ is bounded from the space $L^{p(\cdot)}(\mathbb{G})$ to $L^{q(\cdot)}(\mathbb{G})$. Moreover, the norm $\|K_{\alpha,\eta}\|_{L^{p(\cdot)}(\mathbb{G})\to L^{q(\cdot)}(\mathbb{G})}$ is bounded from above by the constant $C_{p(\cdot),\alpha,\eta,A,Q}\big(\|M\|_{L^{\overline{q}'(\cdot)}(\mathbb{G})}\big)$ depending on $p(\cdot), \alpha, \eta, A, Q$ and $\|M\|_{L^{\overline{q}'(\cdot)}(\mathbb{G})}$ such that the mapping $\cdot \to C_{p(\cdot),\alpha,\eta,A,Q}(\cdot)$ is nondecreasing.*

Proof Let us take a constant p_0, $p_0 < p_-$ sufficiently close to p_-. Then, we can assume that $\alpha < \frac{Q}{p_0}$ and q_0 is also close to q_-, with $q_0 < q_-$, so that identity (1.88) holds. Now Lemma 1.69 and Proposition 1.65 imply the validity of the case $0 < \eta < Q/(p_-)'$, while the case of negative η follows from the fact that $K_{\alpha,\eta}$ is bounded from $L^{p(\cdot)}(\mathbb{G})$ to $L^{q(\cdot)}(\mathbb{G})$ if and only if $K_{\alpha,-\eta}$ is bounded from $L^{q'(\cdot)}(\mathbb{G})$ to $L^{p'(\cdot)}(\mathbb{G})$. □

Theorem 1.80 (Weighted Rellich–Sobolev Inequality) *Let $\mathbb{G}$ be a stratified homogeneous group with homogeneous dimension $Q > 2$. Suppose that $1 < p_- \leqslant p_+ < \frac{Q}{2}$ and $q(\cdot) = \frac{p(\cdot)Q}{Q-2p(\cdot)}$. Let $-\frac{Q}{q_-} < \eta < \frac{Q}{(p_-)'}$. If $\overline{q}'(\cdot) \in \mathcal{B}(\mathbb{G})$ for some $q_0 < q_-$, then for $u \in C_0^\infty(\mathbb{R}^d, \mathbb{R})$ the following inequality holds:*

$$\left\|r^\eta(\cdot)u(\cdot)\right\|_{L^{q(\cdot)}(\mathbb{G})} \leqslant C\left\|r^\eta(\cdot)\Delta_{\mathbb{G}}u(\cdot)\right\|_{L^{p(\cdot)}(G)} \tag{1.94}$$

with the positive constant $C \equiv C_{p(\cdot),\eta,A,Q}\big(\|M\|_{L^{\overline{q}'(\cdot)}(\mathbb{G})}\big)$. Moreover, the constant C is such that the mapping $\cdot \longmapsto C_{p(\cdot),\eta,A,Q}(\cdot)$ is nondecreasing.

Proof The result follows from the estimate:

$$|u(x)| \leqslant c \int_{\mathbb{G}} \frac{|\Delta_{\mathbb{G}}u(y)|}{r\left(y^{-1}x\right)^{Q-2}}dy \tag{1.95}$$

(see, e.g., Bonfiglioli et al. [34, p. 280]) and Theorem 1.79, where $\Delta_{\mathbb{G}}u$ is defined by (1.73). □

Remark 1.81 By Theorem 1.59 and Remark 1.55, Theorems 1.79 and 1.80 remain valid if we replace the condition $\overline{q}'(\cdot) \in \mathcal{B}(\mathbb{G})$ by the following one: $p(\cdot) \in \mathcal{P}^{\log}(\mathbb{G})$.

Theorem 1.82 (Weighted Rellich Inequality) *Let $\mathbb{G}$ be a stratified homogeneous group with homogeneous dimension $Q > 2$. Let $1 < p_- \leqslant p_+ < \frac{Q}{2}$. Suppose that $\frac{2p_- - Q}{p_-} < \eta < \frac{Q}{(p_-)'}$. Let β be a constant such that $\beta < -2$. If $p(\cdot) \in \mathcal{P}^{\log}(\mathbb{G})$, then for $u \in C_0^\infty(\mathbb{R}^d, \mathbb{R})$ the following estimate holds:*

$$\left\|(1+r(\cdot))^\beta r^\eta(\cdot)u(\cdot)\right\|_{L^{p(\cdot)}(\mathbb{G})} \leqslant C_{p(\cdot),\eta,A,Q}\big(\|M\|_{L^{\overline{q}}(\mathbb{G})}\big)\|r(\cdot)^\eta\Delta_{\mathbb{G}}u(\cdot)\|_{L^{p(\cdot)}(\mathbb{G})}$$

with the positive constant $C_{p(\cdot),\eta,A,Q}\left(\|M\|_{L^{\overline{q}'(\cdot)}(\mathbb{G})}\right)$ *independent of* u *and depending on* $p(\cdot)$, η, A, Q, *and* $\|M\|_{L^{\overline{q}'(\cdot)}(\mathbb{G})}$. *Moreover, the constant* $C_{p(\cdot),\eta,A,Q}\left(\|M\|_{L^{\overline{q}'(\cdot)}(\mathbb{G})}\right)$ *is such that the mapping* $\cdot \longmapsto C_{p(\cdot),\eta,A,Q}(\cdot)$ *is nondecreasing.*

Proof Since $p(\cdot) \in \mathcal{P}^{\log}(\mathbb{G})$, by Theorem 1.59 and Remark 1.55, we have that $\overline{q}'(\cdot) \in \mathcal{B}(\mathbb{G})$. Now applying Theorem 1.80, Lemma 1.57, and Hölder's inequality, we have

$$\begin{aligned}\left\| (1+r(\cdot))^{\beta} r^{\eta}(\cdot)u(\cdot)\right\|_{L^{p(\cdot)}(\mathbb{G})} &\leqslant C\left\|(1+r(\cdot))^{\beta}\right\|_{L^{\frac{Q}{2}}(\mathbb{G})}\left\|r^{\eta}(\cdot)u(\cdot)\right\|_{L^{q(\cdot)}(\mathbb{G})}\\ &\leqslant C\left\|r^{\eta}(\cdot)u(\cdot)\right\|_{L^{q(\cdot)}(\mathbb{G})} \leqslant C\left\|r^{\eta}(\cdot)\Delta_{\mathbb{G}}u(\cdot)\right\|_{L^{p(\cdot)}(\mathbb{G})},\end{aligned}$$

where $q(\cdot) = \frac{p(\cdot)Q}{Q-2}$. □

It is known (see Bonfiglioli et al. [34, pp. 237–239]) that

$$u(x) = -C_Q \int_{\mathbb{G}} r\left(y^{-1}x\right)^{1-Q} \nabla_{\mathbb{G}} u(y)dy,$$

where $\nabla_{\mathbb{G}}$ is defined by (1.72). Hence

$$|u(x)| \leqslant C_Q \int_{\mathbb{G}} r\left(y^{-1}x\right)^{1-Q} \left|\nabla_{\mathbb{G}} u(y)\right|dy.$$

By the latter estimate and Theorem 1.79, we also have Hardy-type inequality for $\mathbb{G}$.

Theorem 1.83 (Weighted Hardy–Sobolev (Weighted Sobolev–Stein) Embedding) *Let* $\mathbb{G}$ *be a stratified homogeneous group. Let* $1 < p_- \leqslant p_+ < Q$ *and let* $q(\cdot) = \frac{p(\cdot)Q}{Q-p(\cdot)}$. *Suppose that* $-\frac{Q}{q_-} < \eta < \frac{Q}{(p_-)'}$. *If* $\overline{q}'(\cdot) \in \mathcal{B}(\mathbb{G})$ *for some* $q_0 < q_-$, *then there is a positive constant* $C \equiv C_{p,\eta,A,Q}\left(\|M\|_{L^{\overline{q}'(\cdot)}(\mathbb{G})}\right)$ *such that the inequality*

$$\left\|u(\cdot)r^{\eta}(\cdot)\right\|_{L^{q(\cdot)}(\mathbb{G})} \leqslant C\left\|\nabla_{\mathbb{G}} u(\cdot)r^{\eta}(\cdot)\right\|_{L^{p(\cdot)}(\mathbb{G})}$$

holds for all $u \in C_0^{\infty}\left(\mathbb{R}^d, \mathbb{R}\right)$. *Moreover, the constant* $C_{p(\cdot),\eta,A,Q}\left(\|M\|_{L^{\overline{q}'(\cdot)}(\mathbb{G})}\right)$ *is such that the mapping* $\cdot \longmapsto C_{p(\cdot),\eta,A,Q}(\cdot)$ *is nondecreasing.*

Remark 1.84 By Theorem 1.59, Theorem 1.83 remains valid if we replace the condition $\overline{q}'(\cdot) \in \mathcal{B}(\mathbb{G})$ by $p(\cdot) \in \mathcal{P}^{\log}(\mathbb{G})$.

Theorem 1.85 (Weighted Hardy Inequality) *Let $1 < p_- \leqslant p_+ < Q$. Suppose that $\beta < -1$ and $\frac{p_- - Q}{p_-} < \eta < \frac{Q}{(p_-)'}$. If $p(\cdot) \in \mathcal{P}^{log}(\mathbb{G})$, then there is a positive constant $C \equiv C_{p,\eta,A,Q}\left(\|M\|_{L^{\overline{q}'(\cdot)}(\mathbb{G})}\right)$ such that*

$$\left\|u(\cdot)(1+r(\cdot))^{\beta} r^{\eta}(\cdot)\right\|_{L^{p(\cdot)}(\mathbb{G})} \leqslant C\left\|\nabla_{\mathbb{G}} u(\cdot) r^{\eta}(\cdot)\right\|_{L^{p(\cdot)}(\mathbb{G})}$$

holds for $u \in C_0^{\infty}\left(\mathbb{R}^d, \mathbb{R}\right)$. Moreover, the constant $C_{p(\cdot),\alpha,\eta,A,Q}\left(\|M\|_{L^{\overline{q}'(\cdot)}(\mathbb{G})}\right)$ is such that the mapping $\cdot \longmapsto C_{p(\cdot),\eta,A,Q}(\cdot)$ is nondecreasing.

The proof is similar to that of Theorem 1.82.

Estimates with General Weights

Now we show that the weighted Rellich–Sobolev inequalities with general weights hold under certain conditions on the weights. These results imply the inequalities with power-type weights under stronger condition ($p(\cdot) \in \mathcal{P}^{\log}(\mathbb{G})$) on exponents than in the statements of the previous subsection.

Theorem 1.86 *Let $Q > 2$, $1 < p_- \leqslant p_+ < \frac{Q}{2}$ and let $q(\cdot) = \frac{2p(\cdot)}{2-Qp(\cdot)}$. If $w \in A_{p(\cdot),q(\cdot)}(\mathbb{G})$ and $\left(q(\cdot)\frac{Q-2}{Q}, w^{\frac{Q}{Q-2}}\right)$ is an M-pair, then for all $u \in C^0_{\infty}\left(\mathbb{R}^d, \mathbb{R}\right)$,*

$$\|wu\|_{L^{q(\cdot)}(\mathbb{G})} \leqslant C\|w\Delta u\|_{L^{p(\cdot)}(\mathbb{G})} \tag{1.96}$$

for some positive constant C.

The proof of this statement follows immediately from Proposition 1.77, (1.95) and the fact that the condition $w \in A_{p_0,q_0}(\mathbb{G})$ implies the estimate:

$$\|I_2 f\|_{L^{q_0}_{w^{q_0}}(\mathbb{G})} \leqslant C\|f\|_{L^{p_0}_{w^{p_0}}(\mathbb{G})},$$

where $\frac{1}{p_0} - \frac{1}{q_0} = \frac{2}{Q}$.

Corollary 1.87 *Let $p(\cdot)$ and $q(\cdot)$ be such that $1 < p_- \leqslant p_+ < \frac{Q}{2}$, $q(\cdot) = \frac{2p(\cdot)}{2-Qp(\cdot)}$, where $Q > 2$. Let $p(\cdot) \in \mathcal{P}^{log}(\mathbb{G})$ and $w \in A_{p(\cdot),q(\cdot)}(\mathbb{G})$. Then (1.96) holds for all $u \in C^0_{\infty}\left(\mathbb{R}^d, \mathbb{R}\right)$.*

Proof It is easy to see that the condition $w \in A_{p(\cdot),q(\cdot)}(\mathbb{G})$ is equivalent to the condition $w^{\frac{Q}{Q-2}} \in A_{q(\cdot)(Q-2)/Q}(\mathbb{G})$. The latter condition is equivalent to $w^{-\frac{Q}{Q-2}} \in A_{\left(\frac{q(\cdot)(Q-2)}{Q}\right)'}(\mathbb{G})$. Since $p(\cdot) \in \mathcal{P}^{\log}(\mathbb{G})$, we have that $w \in A_{p(\cdot),q(\cdot)}(\mathbb{G})$ is equivalent to the condition that $\left(\frac{q(\cdot)(Q-2)}{Q}, w^{\frac{Q}{Q-2}}\right)$ is an M-pair (see Cruz-Uribe and Wang [61] for Euclidean spaces and Cruz-Uribe and Cummings [57] for a space of homogeneous type). Now the conclusion follows from Theorem 1.86. □

Corollary 1.88 *Let $p(\cdot)$ and $q(\cdot)$ satisfy the conditions of Corollary 1.87. Let η be a constant such that $-\frac{Q}{q_+} < \eta < \frac{Q}{(p_+)'}$. Then inequality (1.94) holds for all $u \in C^0_\infty(\mathbb{G})$.*

Proof The proof is a consequence of the fact that the weight function $w(x) = r(x)^{-a}$ belongs to $A_{p(\cdot)}(\mathbb{G})$ for $-\frac{Q}{p_+} < -a \leqslant 0$ (see Cruz-Uribe and Wang [61] for Euclidean spaces, but the proof is the same for homogeneous groups). □

1.4.3 Further Results

Norm of the Hardy–Littlewood Maximal Operator in $VELS$s

Since in the main statements of this section we derive weighted estimates involving the norm of the maximal operator M in $VELS$s, it is important to have upper bounds for that norm $\|M\|_{L^{p(\cdot)}(\mathbb{G})}$.

Proposition 1.89 *Let $1 < p_- \leqslant p_+ < \infty$ and let $p(\cdot) \in \mathcal{P}^{log}(\mathbb{G}, L)$. Then the following estimate holds for a.e. $x \in \mathbb{G}$ and f with $\|f\|_{L^{p(\cdot)}(\mathbb{G})} \leqslant 1$:*

$$(Mf(x))^{p(x)/p_-} \leqslant \overline{C}(p(\cdot), A, Q)\left[M\left(|f|^{p(\cdot)/p_-}\right)(x) + 1\right],$$

where

$$\overline{C}(p(\cdot), A, Q) := \max\left\{3^{p_+/p_- -1}\widetilde{C}\,(p(\cdot)/p_-),\ \left(2^Q A^{Q+1} + 1\right)^{p_+/p_-}\right\} \tag{1.97}$$

and $\widetilde{C}\,(p(\cdot)/p_-)$ is defined by (1.78) taking $s(\cdot) = p(\cdot)/p_-$ there.

Proposition 1.90 *Let $1 < p_- \leqslant p_+ < \infty$ and let $p(\cdot) \in \mathcal{P}^{log}_0(\mathbb{G}, L)$. Suppose that $p(\cdot)$ is constant outside some large ball $B(0, R_0) := B_{R_0}$. Then the estimate*

$$(Mf(x))^{p(x)/p_-} \leqslant C(p(\cdot), A, Q) M\left(f^{p/p_-}\right)(x) + h(x),$$

holds for a.e. $x \in \mathbb{G}$ and f with $\|f\|_{L^{p(\cdot)}(\mathbb{G})}$, where

$$C(p(\cdot), A, Q) = \max\left\{\overline{C}(p(\cdot), A, Q),\ 2^{p_+/p_- -1}\right\}, \tag{1.98}$$

$$h(x) = \chi_{2R_0}(x) \cdot \overline{C}(p(\cdot), A, Q) + \chi_{\mathbb{G}\setminus B_{2R_0}}\overline{\overline{C}}(p(\cdot), A, Q) r(x)^{-Q},$$

$$\overline{\overline{C}}(p(\cdot), A, Q) = 2^{p_+/p_- -1}\frac{1 + AR_0^Q}{C_A}$$

and $\overline{C}(p(\cdot), A, Q)$ is defined by (1.97).

To prove Propositions 1.89 and 1.90, we need some lemmas.

Lemma 1.91 *Let $1 < p_- \leqslant p_+ < \infty$ and let $q(\cdot) := \frac{p(\cdot)}{p_-}$. We set $h(x) := C_1\chi_{B_{2R_0}}(x) + C_2\chi_{\mathbb{G}\setminus B_{2R_0}} r(x)^{-Q}$, where $B_{2R_0} := B(0, 2R_0)$, C_1 and C_2 are positive constants. Then, the following estimate holds:*

$$\|h\|_{L^{p_-}} \leqslant \left[C_1^{p_-} A(2R_0)^Q + C_2^{p_-}(2R_0)^{Q-Qp_-}\frac{B_{\mathbb{G}}}{Qp_- - Q}\right]^{1/p_-}.$$

Proof We have:

$$\|h\|_{L^{p_-}} = \left[\int_{B_{2R_0}} C_1^{p_-}dx + \int_{\mathbb{G}\setminus B_{2R_0}} C_2^{p_-} r(x)^{-Qp_-}dx\right]^{1/p_-}$$

$$\leqslant \left[C_1^{p_-} A(2R_0)^Q + C_2^{p_-}\left(\int_{2R_0}^{\infty} t^{Q-1-Qp_-}dt\right)B_{\mathbb{G}}\right]^{1/p_-}$$

$$= \left[C_1^{p_-} A(2R_0)^Q + C_2^{p_-}\frac{(2R_0)^{Q-Qp_-}}{Qp_- - Q}\cdot B_{\mathbb{G}}\right]^{1/p_-}.$$

□

Lemma 1.92 *Let $1 < p_- \leqslant p_+ < \infty$. If $S_{p(\cdot)}(Mf) \leqslant C$ for some positive constant C, then*

$$\|Mf\|_{L^{p(\cdot)}} \leqslant C^{1/p_-}.$$

As the proof is straightforward, the details are omitted.

Proof of Proposition 1.89 Let us denote $q := p/p_-$. By Remark 1.52 (c), we have that $q(\cdot) \in \mathcal{P}^{\log}((E, L/(p_-)^2)$. Let $\|f\|_{L^{p(\cdot)}} \leqslant 1$. Then $S_{p(\cdot)}(f) \leqslant 1$. We consider two cases: (i) $R \geqslant \frac{1}{2A}$; (ii) $0 < R < \frac{1}{2A}$, where R is the radius of a ball B. We will estimate:

$$M_B f := \frac{1}{|B|}\int_B |f(y)|dy.$$

Case (i): Observe that

$$(M_B f)^{q(x)} \leqslant \frac{1}{|B|^{q(x)}}\left(\int_{B\cap\{f\geqslant 1\}} |f(y)|dy + \int_{B\cap\{f<1\}} |f(y)|dy\right)^{q(x)}$$

$$\leqslant \frac{1}{|B|^{q(x)}}\left(\int_B |f(y)|^{p(y)}dy + |B|\right)^{q(x)}$$

$$\leqslant \left(\frac{1}{|B|}\int_B |f(y)|^{p(y)}dy + 1\right)^{q(x)}$$

$$\leqslant \left(1 + 2^Q A^{Q+1}\right)^{q_+}.$$

Case (ii): Let $0 < R \leqslant \frac{1}{2A}$. Then $|B| < 1$. Therefore, by Jensen's inequality and simple estimates, we have:

$$(M_B f)^{q(x)}\left(\frac{1}{|B|}\int_B |f(y)|^{q_-(B)}dy\right)^{\frac{q(x)}{q_-(B)}}$$

$$\leqslant \left(\frac{1}{|B|}\int_{B\cap\{f>1\}} |f(y)|^{q_-(B)}dy + \frac{1}{|B|}\int_{B\cap\{f\leqslant 1\}} |f(y)|^{q_-(B)}dy\right)^{\frac{q(x)}{q_-(B)}}$$

$$\leqslant \left(\frac{1}{|B|}\int_B |f(y)|^{q(y)}dy + 1\right)^{\frac{q(x)}{q_-(B)}}$$

$$\leqslant 3^{q_+}|B|^{-\frac{q(x)}{q_-(B)}}\left[\frac{1}{3}\int_B \left(|f(y)|^{q(y)}dy + 1\right)dy\right]^{q(x)/q_-(B)}.$$

Observe now that since

$$\frac{1}{3}\int_B \left(|f(y)|^{q(y)}dy + 1\right)dy \leqslant \frac{1}{3}\int_B |f(y)|^{\frac{p(y)}{p_-}}dy + \frac{|B|}{3}$$

$$\leqslant \frac{1}{3}\left(\int_B |f(y)|^{p(y)}dy\right)^{\frac{1}{p_-}}|B|^{1/(p_-)'} + \frac{|B|}{3}$$

$$\leqslant \frac{1}{3} + \frac{|B|}{3} \leqslant \frac{2}{3} < 1,$$

we have:

$$(M_B f)^{q(x)} \leqslant 3^{q_+}|B|^{-\frac{q(x)}{q_-(B)}}\left(\frac{1}{3}\int_B |f(y)|^{q(y)}dy + \frac{1}{3}|B|\right)$$

$$\leqslant 3^{q_+-1}|B|^{-\frac{q_-(B)-q_+(B)}{q_-(B)}}\left[\frac{1}{|B|}\int_B |f(y)|^{q(y)}dy + 1\right]$$

$$\leqslant 3^{q_+-1}\left[\widetilde{C}(p(\cdot)/p_-)\right]^{1/q_-}\left[\frac{1}{|B|}\int_B |f(y)|^{q(y)}dy+1\right]$$

$$=3^{\frac{p_+}{p_-}-1}\widetilde{C}(p(\cdot)/p_-)\left[\frac{1}{|B|}\int_B |f(y)|^{q(y)}dy+1\right].$$

□

Proof of Proposition 1.90 Let $\|f\|_{L^{p(\cdot)}}\leqslant 1$, which implies that $S_{p(\cdot)}(f)\leqslant 1$. Split f into $f_0 := \chi_{B_{R_0}}f$ and $f_1 := \chi_{\mathbb{G}\backslash B_{R_0}}f$. We denote by p_∞ the value of $p(\cdot)$ on $\mathbb{G}\backslash B_{R_0}$. Here, $B_{R_0} := B(0, R_0)$. We denote $q(\cdot) = p(\cdot)/p_-$ and $q_\infty = p_\infty/p_-$. By Proposition 1.89, we have that for all $x \in B_{2R_0} := B(0, 2R_0)$:

$$(Mf(x))^{q(x)} \leqslant \overline{C}(p(\cdot), A, Q)\left[M\left(|f(\cdot)|^{q(\cdot)}\right)+1\right],$$

where $\overline{C}(p(\cdot), A, Q)$ is defined by (1.97). Denote $B_R := B$. Let $x \in \mathbb{G}\backslash B_{2R_0}$. Then $r(x) - R_0 \geqslant \frac{1}{2}r(x)$ and, consequently, $\left|B_{r(x)-R_0}\right| \geqslant C_A r(x)^Q$ with the positive constant C_A depending on A.

Thus, for $x \in \mathbb{G}\backslash B_{R_0}$, we find that

$$\begin{aligned}
\left(Mf_0(x)\right)^{q(x)} &\leqslant \left(\sup_{R>r(x)-R_0}\frac{1}{|B_R|}\int_{B_R}|f_0(y)|dy\right)^{q(x)}\\
&\leqslant \left(\frac{1}{|B_{r(x)-R_0}|}\int_{B_R\cap B_{R_0}}|f(y)|dy\right)^{q(x)}\\
&\leqslant \left(\frac{1}{C_A r(x)^Q}\int_{B_{R_0}}|f(y)|dy\right)^{q(x)}\\
&\leqslant \left[\frac{1}{C_A r(x)^Q}\left(\int_{B_{R_0}}|f(y)|^{p(y)}dy+|B_{R_0}|\right)\right]^{q(x)}\\
&\leqslant \left(\frac{1}{C_A r(x)^Q}\left[1+AR_0^Q\right]\right)^{q(x)}\\
&\leqslant \left[\frac{1}{C_A r(x)^Q}+\frac{AR_0^Q}{C_A r(x)^Q}\right]^{q(x)}\\
&\leqslant \frac{C_{p(\cdot),R_0,A}}{C_A r(x)^Q},
\end{aligned}$$

where $C_{p(\cdot),R_0,A} = 1 + AR_0^Q$. Here we used the fact that $\frac{1}{C_A r(x)^Q} + \frac{AR_0^Q}{C_A r(x)^Q} \leqslant 1$ if R_0 is sufficiently large.

Further, let $x \in \mathbb{G}\backslash B_{2R_0}$. Then, by Jensen's inequality,

$$(Mf_1(x))^{q(x)} = (Mf_1(x))^{q_\infty} \leqslant \left(M|f_1|^{q_\infty}\right)(x) \leqslant \left(M|f|^{q_\cdot}\right)(x).$$

Summarizing these estimates with Proposition 1.89, we find that

$$\begin{aligned}
&(Mf(x))^{q(x)}\\
&\leqslant \chi_{B_{2R_0}}(Mf(x))^{q(x)} + \chi_{\mathbb{G}\backslash B_{2R_0}}[Mf_0(x) + Mf_1(x)]^{q(x)}\chi_{B_{2R_0}}(Mf(x))^{q(x)}\\
&+ 2^{q_+-1}\chi_{\mathbb{G}\backslash B_{2R_0}}\left[Mf_0(x)^{q(x)} + Mf_1(x)^{q(x)}\right]\\
&\leqslant \overline{C}(p(\cdot), A, Q)\chi_{B_{2R_0}}(x)M\left(|f|^{q(\cdot)}\right)(x) + \overline{C}(p(\cdot), A, Q)\chi_{B_{2R_0}}(x)\\
&+ \frac{C_{p(\cdot),R_0,A}}{C_A r(x)^Q}2^{q_+-1}\chi_{\mathbb{G}\backslash B_{2R_0}}(x) + 2^{q_+-1}\chi_{\mathbb{G}\backslash B_{2R_0}}(x)M\left(|f|^{q(\cdot)}\right)(x)\\
&\leqslant \max\left\{\overline{C}(p(\cdot), A, Q),\ 2^{q_+-1}\right\}M\left(|f|^{q(\cdot)}\right)(x) + \overline{C}(p(\cdot), Q)\chi_{B_{2R_0}}(x)\\
&+ \frac{2^{q_+-1}C_{p(\cdot),R_0,A}}{C_A r(x)^Q}\chi_{\mathbb{G}\backslash B_{2R_0}}(x)\\
&= C(p(\cdot), R_0, A)M\left(|f|^{p(\cdot)/p_-}\right)(x) + h(x),
\end{aligned}$$

where $C_{p(\cdot),R_0,A}$, and $\overline{C}(p(\cdot), A, Q)$ are defined by (1.98) and (1.97), respectively, and $h(x) = \overline{C}(p(\cdot), A, Q)\chi_{B_{2R_0}}(x) + \frac{2^{q_+-1}C(p(\cdot),R_0,A)}{C_A r(x)^Q}\chi_{\mathbb{G}\backslash B_{2R_0}}(x)$. □

The next statement gives an estimate of the norm of $\|M\|_{L^{p(\cdot)}(\mathbb{G})}$ in the case when $p(\cdot)$ is constant outside some large ball.

Proposition 1.93 *Let* $1 < p_- \leqslant p_+ < \infty$ *and* $p \in \mathcal{P}^{\log}(\mathbb{G})$. *Let* $p(\cdot)$ *be constant outside some large ball* $B_{R_0} := B(0, R_0)$. *Then*

$$\begin{aligned}
&\|M\|_{L^{p(\cdot)}\to L^{p(\cdot)}}\\
&\leqslant 2^{1/(p_-)'}\Bigg[C(p(\cdot), A, Q)^{p_-}\|M\|_{L^{p_-}\to L^{p_-}}^{p_-} + \overline{C}(p(\cdot), A, Q)^{p_-}(2R_0)^Q A\\
&\qquad + \left(\overline{\overline{C}}(p(\cdot), A, Q)\right)^{p_-}(2R_0)^{Q-Qp_-}\cdot\frac{B_{\mathbb{G}}}{Qp_- - Q}\Bigg]^{1/p_-},
\end{aligned}$$

where $C(p(\cdot), A, Q)$, $\overline{C}(p(\cdot), Q)$ *and* $\overline{\overline{C}}(p(\cdot), A, Q)$ *are defined by* (1.98), (1.97) *and* (1.90), *respectively.*

Proof Let $\|f\|_{L^{p(\cdot)}} \leqslant 1$. Then $S_{p(\cdot)}(f) \leqslant 1$. Further, as before, we denote $q(\cdot) := p(\cdot)/p_-$. By Proposition 1.90, Lemma 1.91, and the boundedness of M in $L^{p_-}(\mathbb{G})$, we find that

$$
\begin{aligned}
&S_{p(\cdot)}(Mf)\\
&= \|(Mf)^{q(\cdot)}\|_{L_{p_-}}^{p_-} \leqslant \left(C(p(\cdot), A, Q)\left\|M\left(|f|^{q(\cdot)}\right)\right\|_{L^{p_-}}\right)^{p_-}\\
&\leqslant \left(C\left(p(\cdot), A, Q\right)\|M\|_{L^{p_-}\to L^{p_-}}\||f|^q\|_{L^{p_-}} + \|h\|_{L^{p_-}}\right)^{p_-}\\
&= \left[C(p(\cdot), A, Q)\|M\|_{L^{p_-}\to L^{p_-}} + \|h\|_{L^{p_-}}\right]^{p_-}\\
&\leqslant \left[C(p(\cdot), A, Q)\|M\|_{L^{p_-}\to L^{p_-}} + \|h\|_{L^{p_-}}\right]^{p_-} \leqslant\\
&\leqslant \Bigg[C(p(\cdot), A, Q)\|M\|_{L^{p_-}\to L^{p_-}}\\
&+\left(\overline{C}(p(\cdot), Q)^{p_-}(2R_0)^Q A + \left(\overline{\overline{C}}(p(\cdot), A, Q)\right)^{p_-}(2R_0)^{Q-Qp_-}\cdot\frac{B_{\mathbb{G}}}{Qp_- - Q}\right)^{1/p_-}\Bigg]^{p_-}\\
&\leqslant 2^{p_- -1}C(p(\cdot), A, Q)^{p_-}\|M\|_{L^{p_-}\to L^{p_-}}^{p_-} + 2^{p_- -1}\overline{C}(p(\cdot), Q)^{p_-}(2R_0)^Q A\\
&+ 2^{p_- -1}\left(\overline{\overline{C}}(p(\cdot), A, Q)\right)^{p_-}(2R_0)^{Q-Qp_-}\frac{B_{\mathbb{G}}}{Qp_- - Q}.
\end{aligned}
$$

Now the result follows since $S_{p(\cdot)}(Mf) \leqslant C \Rightarrow \|Mf\|_{L^{p(\cdot)}(\mathbb{G})} \leqslant C^{1/p_-}$. □

Since (see, e.g., Hytönen et al. [144] or Kokilashvili et al. [183, pp. 877–879])

$$
\|Mf\|_{L^{p_0}\to L^{p_0}} \leqslant C_S(p_0')^{1/p_0}, \quad 1 < p_0 < \infty,
$$

where C_S is a structural constant, Proposition 1.93 yields the next statement.

Corollary 1.94 *Let* $1 < p_- \leqslant p_+ < \infty$ *and* $p(\cdot) \in \mathcal{P}_0^{log}(\mathbb{G}, L)$. *Let* $p(\cdot)$ *be constant outside some large ball* $B_{R_0} := B(0, R_0)$. *Then*

$$
\begin{aligned}
&\|Mf\|_{L^{p(\cdot)}\to L^{p(\cdot)}}\\
&\quad\leqslant 2^{1/(p_-)'}\Bigg[C(p(\cdot), A, Q)^{p_-}C_S[(p_-)']^{1/p_-} + \overline{C}(p(\cdot), A, Q)^{p_-}L(2R_0)^Q A\\
&\quad+\left(\overline{\overline{C}}(p(\cdot), A, Q)\right)^{p_-}(2R_0)^{Q-Q_{p_-}}\cdot\frac{B_{\mathbb{G}}}{Q_{p_-} - Q}\Bigg]^{1/p_-}.
\end{aligned}
$$

Variable Parameter Riesz Potential in $VELS$s

The next statement concerns the Sobolev-type inequality for $I_{\alpha(\cdot)}$ in $VELS$s defined on homogeneous groups. To formulate it, we need some notation:

$$\alpha_0 := \inf_{x\in\mathbb{G}} \alpha(x) > 0.$$

Proposition 1.95 *Let $p(\cdot)$ satisfy the condition:*

$$1 < p(\infty) \leqslant p(x) \leqslant p_+ < \infty$$

and let $p(\cdot) \in \mathcal{P}(\mathbb{G})$. Suppose that $p(\cdot) \in \mathcal{P}$ and

$$\sup_{x\in\mathbb{G}} p(x)\alpha(x) < Q, \quad \sup_{x\in\mathbb{G}} p(\infty)\alpha(x) < Q.$$

Then the following weighted estimate holds:

$$\left\| (1+r(x))^{-\gamma(x)} I_{\alpha(\cdot)} f \right\|_{L^{q(\cdot)}(\mathbb{G})} \leqslant c\|f\|_{L^{p(\cdot)}(\mathbb{G})},$$

where

$$q(\cdot) = \frac{p(\cdot)Q}{Q-\alpha(\cdot)p(\cdot)}$$

is the variable Sobolev exponent and

$$\gamma(x) = \frac{L_\infty}{(p_- - 1)^2}\alpha(x)\left[1 - \frac{\alpha(x)}{Q}\right]$$

with the decay constant v from (1.79).

Proof We follow the proof of Theorem A in Kokilashvili and Samko [176]. Using the standard arguments of Hedberg [135], we represent $I_{\alpha(\cdot)} f(x)$ as the sum:

$$I_{\alpha(\cdot)} f(x) = \int\limits_{B(x,R)} \frac{f(y)}{r\left(y^{-1}x\right)^{Q-\alpha(x)}} dy + \int\limits_{\mathbb{G}\backslash B(x,R)} \frac{f(y)}{r\left(y^{-1}x\right)^{Q-\alpha(x)}}$$
$$:= I_{1,R}(x) + I_{2,R}(x).$$

Further, we have:

$$|I_{1,R}(x)| \leqslant A2^{Q-\alpha(x)} \frac{2^{\alpha(x)} R^{\alpha(x)}}{2^{\alpha(x)} - 1} Mf(x).$$

Now we estimate $I_{2,R}(x)$. Suppose that $\|f\|_{L^{p(\cdot)}(\mathbb{G})} \leqslant 1$. Due to Hölder's inequality, we find that

$$I_{2,R}(x) \leqslant \kappa \nu_{\alpha,Q,p(\cdot)}(x, R)\|f\|_{L^{p(\cdot)}(\mathbb{G})} \leqslant \kappa \nu_{\alpha,Q,p(\cdot)}(x, R),$$

where κ is the constant from (1.74), and

$$\nu_{\alpha,Q,p(\cdot)}(x, R) := \left\| r\left(x^{-1}\right)^{\alpha(x)-Q} \chi_{\mathbb{G}\setminus B(x,R)} \right\|_{L^{p'(\cdot)}(\mathbb{G})}.$$

Using Lemma 1.58 for $\beta(x) = \alpha(x) - Q$, we have that there is a constant $C_{\alpha,Q,p(\cdot)}$ such that

$$|B(x, R)| \leqslant C_{\alpha,Q,p(\cdot)}(1 + r(\cdot))^{\frac{[Q-\alpha(\cdot)]\overline{L}_\infty}{p(\infty)}} R^{-\frac{Q}{q(x)}}, \quad x \in \mathbb{G},$$

where $\overline{L} := \frac{L_\infty}{(p_- -1)^2}$. Then, summing up the above inequalities, we conclude that for $0 < R < \infty$ and $x \in \mathbb{G}$,

$$\left|I_{\alpha(\cdot)} f(x)\right| \leqslant C\left[R^{\alpha(x)} Mf(x) + (1 + r(\cdot))^{\frac{[Q-\alpha(x)]\overline{L}_\infty}{p(\infty)}} R^{-\frac{Q}{q(x)}}\right]. \tag{1.99}$$

Taking now $R = R_{\min} = \left[\frac{\alpha(x)q(x)}{Q(1+r(\cdot))^{\frac{[Q-\alpha(\cdot)]\overline{L}_\infty}{p(\infty)}}} Mf(x)\right]^{-\frac{p(x)}{Q}}$ in (1.99) we find that

$$\left|I_{\alpha(\cdot)} f(x)\right| \leqslant C\left[(1 + r(\cdot))^{\frac{[Q-\alpha(\cdot)]\overline{L}_\infty}{p(\infty)}}\right]^{\frac{\alpha(x)p(x)}{Q}} [Mf(x)]^{\frac{p(x)}{q(x)}}.$$

Since $p(x)$ satisfies the decay condition at infinity, we may replace $p(x)$ in $\left[(1+r(\cdot))^{\frac{[Q-\alpha(\cdot)]\overline{L}_\infty}{p(\infty)}}\right]^{\frac{\alpha(x)p(x)}{Q}}$ by $p(\infty)$. Then

$$\left|I_{\alpha(\cdot)} f(x)\right| \leqslant C(1+r(x))^{\alpha(x)\left(1-\frac{\alpha(x)}{Q}\right)\overline{L}_\infty} [Mf(x)]^{\frac{p(x)}{q(x)}} = C(1+r(x))^{\gamma(x)} [Mf(x)]^{\frac{p(x)}{q(x)}}.$$

Finally,

$$\left((1 + r(x))^{-\gamma(x)} I^{\alpha(\cdot)} f(x)\right) \leqslant C \int_{\mathbb{G}} |Mf(x)|^{p(x)} dx \leqslant C,$$

by Corollary 1.94. □

1.5 Comments to Chap. 1

The classical Rellich inequalities (1.1) and (1.2) are already well-known in the setting of L^2. The first step toward an L^p version appears to have been taken by Davies and Hinz [69], most notably when $\Omega = \mathbb{R}^n \backslash \{0\}$. In Edmunds and Evans [86] an L^p form of the Rellich inequality was obtained in general open subsets Ω of $\mathbb{R}^n$, together with a somewhat similar inequality involving the p-Laplacian. We summarize the main results of Edmunds and Evans [86] as follows.

Theorem 1.96 (one-dimensional case) *Suppose that* $-\infty < a < b \leqslant \infty$ *and let* $r \in (1, \infty)$; *put* $\delta(t) = \min\{t - a, b - t\}$. *Then for all* $u \in C_0^2(a, b)$,

$$\int_a^b \frac{|u(t)|^r}{\delta(t)^{2r}} dt \leqslant \left(\frac{r}{2r-1}\right)^r \left(\frac{r}{r-1}\right)^r \int_a^b |u''(t)|^r \, dt.$$

Theorem 1.97 (multidimensional case) *Let* Ω *be a non-empty, proper open subset of* $\mathbb{R}^n$ *and let* $r \in (1, \infty)$; *suppose that* $u \in C_0^2(\Omega)$. *If* $r = 2$, *then*

$$\int_\Omega \frac{|u(x)|^2}{\delta_{M,4}(x)^4} dx \leqslant \frac{16}{9} \int_\Omega |\Delta u(x)|^2 \, dx,$$

while if $r \in (1, \infty) \backslash \{2\}$, *then for some explicit constant* $K(r, n)$,

$$\int_\Omega \frac{|u(x)|^r}{\delta_{M,2r}(x)^{2r}} dx \leqslant K(r, n) \int_\Omega |\Delta u(x)|^r \, dx.$$

Here $\delta_{M,4}$ *and* $\delta_{M,2r}$ *are mean distance functions obtained by averaging, in a certain sense, the distance to the boundary of* Ω *in all possible directions.*

The boundedness/compactness of the two-weighted Hardy operator

$$H_{v,w} f(x) = v(x) \int_0^x f(t) w(t) dt$$

acting between $VELS$s was investigated for the first time in Edmunds et al. [94], where the authors established necessary conditions and sufficient conditions for the operator $H_{v,w}$ to be bounded/compact under modular-type conditions on weights. In that paper the exponent of spaces are constants outside some large interval and are free from log-type condition. Later, Kopaliani [195] established criteria for the validity of the two-weighted Hardy inequality, assuming log-Hölder continuity of the variable exponents and conditions on the weights expressed in terms of $L^{p(\cdot)}-$ norms. We refer also to Diening and Samko [72], Cruz-Uribe and Mamedov [59],

Mamedov and Mammadli [221], Mamedov and Zeren [225], and Mamedov and Harman [223] for related topics. We recall the papers Ashraf et al. [22], Kokilashvili et al. [178], Meskhi and Zaighum [247] for boundedness/compactness criteria regarding weighted integral operators with more general kernels in $L^{p(\cdot)}$ spaces.

Compactness criteria for the two-weight Hardy operator $H_{v,w}$ acting between Banach function spaces X and Y were established in Edmunds et al. [91]. In the same paper, two-sided estimates for the measure of noncompactness of $H_{v,w}$ were also given. Further, we refer also to the papers Edmunds and Meskhi [87], Edmunds et al. [94], Mamedov and Zeren [225], Mamedov et al. [226], Rafeiro and Samko [277] for the compactness of Hardy-type operators in $VELS$s under various conditions on weights and exponents. In the papers Ashraf et al. [22], Kokilashvili et al. [178], Meskhi and Zaighum [247] and the monograph Meskhi [240] (see also Kokilashvili et al. [182]), the author established two-sided estimates of the measure of noncompactness for Hardy-type operators (even for more general positive kernel operators) in the variable exponent setting.

Finally, we mention the paper Cruz-Uribe OFS and Suragan [68], where the authors established two-weighted Rellich-Sobolev and Hardy–Sobolev inequalities in $\mathbb{R}^n$.

Section 1.1 is based on the paper Edmunds and Meskhi [82]. Section 1.2 covers the results published in Edmunds and Meskhi [88]. Section 1.3 presents the results published in Edmunds and Meskhi [89], and Sect. 1.4 is based on Edmunds and Meskhi [84] (see also Edmunds and Meskhi [83]).

Chapter 2
Trace Inequalities for Fractional Integrals

We establish necessary and sufficient condition on a non-negative locally integrable function v guaranteeing the (trace) inequality:

$$\|I_\alpha f\|_{L^p_v(\mathbb{R}^n)} \leqslant C\|f\|_{L^{p,1}(\mathbb{R}^n)}, \quad f \in L^{p,1}(\mathbb{R}^n),$$

for the Riesz potential I_α, where $L^{p,1}(\mathbb{R}^n)$ is the Lorentz space. The condition on v is of D. Adams type. The same problem is studied for potentials defined on spaces of homogeneous type. We also find necessary and sufficient conditions on a weight v for the boundedness of multilinear Riemann–Liouville operators from $\prod_{j=1}^m L^{p_j}$ to L^q_v.

2.1 Trace Inequality Criteria for Riesz Potentials

The Riesz potential operator

$$I_\alpha f(x) = \int\limits_{\mathbb{R}^n} \frac{f(y)}{|x-y|^{n-\alpha}} dy, \quad 0 < \alpha < n, \quad x \in \mathbb{R}^n,$$

plays an important role in the theory of partial differential equations. It is worth mentioning its role in the theory of Sobolev's embeddings (see, e.g., Maz'ya [235]).

The trace inequality for I_α deals with non-negative measures ν such that

$$\left(\int\limits_{\mathbb{R}^n} |I_\alpha f(x)|^q d\nu(x) \right)^{1/q} \lesssim \left(\int\limits_{\mathbb{R}^n} |f(x)|^p dx \right)^{1/p}. \tag{2.1}$$

A. Meskhi et al., *Integral Operators in Non-Standard Function Spaces*,
Operator Theory: Advances and Applications 310,
https://doi.org/10.1007/978-3-032-03691-9_2

Adams [3] proved necessary and sufficient condition on ν guaranteeing (2.1) for $1 < p < q < \infty$ and $0 < \alpha < n/p$ is that measure ν satisfies the condition: there is a positive constant C such that for all balls $B \subset \mathbb{R}^n$:

$$\nu(B) \lesssim |B|^{(1/p-\alpha/n)q}.$$

The appropriate fractional maximal operator is given by the formula:

$$M_\alpha f(x) = \sup_{B \ni x} \frac{1}{|B|^{1-\alpha/n}} \int_B |f(y)| dy, \quad 0 \leqslant \alpha < n, \quad x \in \mathbb{R}^n,$$

where the convention $M_0 f = Mf$ refers to the Hardy–Littlewood maximal function, which is significant in harmonic analysis, particularly in the theory of singular integrals.

Let v be a non-negative locally integrable function on $\mathbb{R}^n$. We are interested in the inequality (2.1) for $d\nu = v(x)dx$, i.e.,

$$\|I_\alpha f\|_{L^q_v(\mathbb{R}^n)} \lesssim \|f\|_{L^p(\mathbb{R}^n)}. \tag{2.2}$$

In this case, by the result of Adams [3], the condition

$$[v]_{p,q,\alpha} := \sup_B \Big(v(B)\Big)^{1/q} |B|^{\alpha/n-1/p} < \infty, \tag{2.3}$$

where the supremum is taken over all balls $B \subset \mathbb{R}^n$, is simultaneously necessary and sufficient whenever $1 < p < q < \infty$ and $0 < \alpha < n/p$. In the case $p = q$, the implication (2.2) $\Rightarrow$ (2.3) can be checked easily by considering the test functions χ_B, but the fact that (2.3) $\Rightarrow$ (2.2) fails (see Adams [5], Kerman and Sawyer [164], Lemarié-Rieusset [207]).

Our aim is to find a Lorentz space $L^{p,s}$, which is narrower than the class $L^p(\mathbb{R}^n)$ (i.e., $s < p$) and for which the inequality

$$\|I_\alpha f\|_{L^p_v(\mathbb{R}^n)} \lesssim \|f\|_{L^{p,s}(\mathbb{R}^n)} \tag{2.4}$$

holds if and only if (2.3) is satisfied for $p = q$. In particular, we show that (2.4) is equivalent to the condition (2.3) for $s = 1$. The question for $1 < s < p$ remains open.

It should be mentioned that there are known various different criteria for (2.2) with $p = q$ (see Adams [5], Maz'ya [233, 234], Kerman and Sawyer [164], Maz'ya and Verbitsky [236]). For the solution of the two-weight problem for Riesz potential operators I_α, we refer to Gabidzashvili and Kokilashvili [108], Sawyer [309] (see also the monograph Kokilashvili and Krbec [170]).

Inequality (2.2) for $p = q$ implies the estimate:

$$\|f\|_{L^q_v(\mathbb{R}^n)} \lesssim \|\nabla f\|_{L^p(\mathbb{R}^n)}, \quad f \in C_0^\infty, \tag{2.5}$$

which follows from the estimate

$$|f(x)| \lesssim I_1(|\nabla f|)(x).$$

The following Fefferman–Phong-type theorem (Fefferman [100]) holds.

Theorem 2.1 *Let* $1 < p < \infty$ *and let* $0 < \alpha < n/p$. *Then, for some* $p < r$, *the following inequality holds:*

$$\|I_\alpha f\|_{L^p_v} \lesssim [v]^*_{p,r,\alpha}\|f\|_{L^p},$$

where

$$[v]^*_{p,r,\alpha} := \sup_B |B|^{\alpha/n-1/r}\left(\int_B v^{r/p}(x)dx\right)^{1/r} < \infty. \tag{2.6}$$

Remark 2.2 By Hölder's inequality, it is easy to see that condition (2.6) is stronger than (2.3) for $p = q$. In particular, $[v]_{p,\alpha} \leqslant [v]^*_{p,r,\alpha}$ for $r > p$, where $[v]_{p,\alpha} = [v]_{p,p,\alpha}$.

Let f be a measurable function on $\mathbb{R}^n$ and let $1 \leqslant p < \infty$, $1 \leqslant s \leqslant \infty$. We say that f belongs to the Lorentz space $L^{p,s}$ if

$$\|f\|_{L^{p,s}} = \begin{cases} \left(s\int\limits_0^\infty \left(|\{x \in \mathbb{R}^n : |f(x)| > \tau\}|\right)^{s/p}\tau^{s-1}d\tau\right)^{1/s}, & \text{if } 1 \leqslant s < \infty,\\ \sup\limits_{\tau>0} \tau\left(|\{x \in \mathbb{R}^n : |f(x)| > \tau\}|\right)^{1/p}, & \text{if } s = \infty,\end{cases}$$

is finite.

If $p = s$, then $L^{p,s}$ coincides with the Lebesgue space L^p.

Denote by f^* a nonincreasing rearrangement of f. Then by integration by parts, it can be checked that (see also Hunt [141])

$$\|f\|_{L^{p,s}} = \begin{cases} \left(\frac{s}{p}\int\limits_0^\infty \left(t^{1/p}f^*(t)\right)^s \frac{dt}{t}\right)^{1/s}, & \text{if } 1 \leqslant s < \infty,\\ \sup\limits_{t>0}\{t^{1/p}f^*(t)\}, & \text{if } s = \infty.\end{cases}$$

Now we list some useful properties of Lorentz spaces (see, e.g., Hunt [141]):

(i) $\|\chi_E\|_{L^{p,s}} = |E|^{1/p}$;
(ii) If $1 \leqslant p < \infty$, $s_2 \leqslant s_1$, then $L^{p,s_2} \hookrightarrow L^{p,s_1}$ with the embedding constant C_{p,s_1,s_2} depending only on p, s_1, and s_2;

(iii) positive constant $C_{p,s}$ such that

$$C_{p,s}^{-1}\|f\|_{L^{p,s}} \leqslant \sup_{\|h\|_{L^{p',s'}}\leqslant 1}\left|\int_X f(x)h(x)d\mu(x)\right| \lesssim_{p,s} \|f\|_{L^{p,s}}$$

for every $f \in L_w^{p,s}$, where $p' = p/(p-1)$, $s' = s/(s-1)$.

(iv) (Hölder's inequality) Let $\frac{1}{p} = \frac{1}{p_1} + \frac{1}{p_2}$, $\frac{1}{s} = \frac{1}{s_1} + \frac{1}{s_2}$. Then

$$\|f_1 f_2\|_{L^{p,s}} \leqslant C\|f_1\|_{L^{p_1,s_1}}\|f_2\|_{L^{p_2,s_2}}$$

for all $f_1 \in L^{p_1,s_1}$ and $f_2 \in L^{p_2,s_2}$, where $C = C_{p,s,p_1,p_2,s_1,s_2}$.

In the sequel for a non-negative function v and a set E by the symbol $v(E)$ will denote the integral:

$$v(E) := \int_E v(x)dx.$$

Now we formulate the main statement of this section.

Theorem 2.3 *Let $1 < p < \infty$ and let $0 < \alpha < n/p$. Suppose that v is a non-negative locally integrable function on $\mathbb{R}^n$. Then, the following statements are equivalent:*

(i) there is a positive constant C such that for all $f \in L^{p,1}(\mathbb{R}^n)$

$$\|I_\alpha f\|_{L_v^p(\mathbb{R}^n)} \lesssim \|f\|_{L^{p,1}(\mathbb{R}^n)} \tag{2.7}$$

(ii) there is a positive constant c such that for all $f \in L^{p,1}(\mathbb{R}^n)$

$$\|M_\alpha f\|_{L_v^p(\mathbb{R}^n)} \lesssim \|f\|_{L^{p,1}(\mathbb{R}^n)} \tag{2.8}$$

(iii)

$$[v]_{p,\alpha} = \sup_B \big(v(B)\big)^{1/p}|B|^{\alpha/n-1/p} < \infty.$$

Moreover, if C and c are best constant in (2.7) *and* (2.8) *respectively, then*

$$C \approx c \approx [v]_{p,\alpha}.$$

To prove this theorem, we need some lemmas.

For a non-negative a.e. finite function w, we denote $E_w := \{x \in \mathbb{R} : w(x) \neq 0\}$. We will also denote $p' = \frac{p}{p-1}$ for $1 \leqslant p \leqslant \infty$.

Lemma 2.4 *Let $1 < p < \infty$ and $0 < \alpha < n/p'$. Then the condition*

$$[w]'_{p,\alpha} \equiv \sup_B \left(\int_B w^{1-p'}(x)\chi_{E_w}(x)dx \right)^{\frac{1}{p'}} |B|^{\frac{\alpha}{n}-\frac{1}{p'}} < \infty \tag{2.9}$$

guarantees that there is a positive constant $C = C_{\alpha,p}$ such that for all $f \in L^p_w(E_w)$,

$$\|M_\alpha(f\chi_{E_w})\|_{L^{p,\infty}(\mathbb{R}^n)} \leqslant C\|f\chi_{E_w}\|_{L^p_w(\mathbb{R}^n)}.$$

Proof Observe that, by Hölder's inequality, we find that

$$\begin{aligned}
\frac{1}{|B|^{1-\frac{\alpha}{n}}} \int_B |f(y)|\chi_{E_w}(y)dy &= \frac{1}{|B|^{1-\frac{\alpha}{n}}} \int_B |f(y)|\chi_{E_w}(y)w^{\frac{1}{p}}(y)w^{-\frac{1}{p}}(y)dy \\
&\leqslant \frac{1}{|B|^{1-\frac{\alpha}{n}}} \left(\int_B |f(y)|^p w(y)\chi_{E_w}(y)dy \right)^{\frac{1}{p}} \left(\int_B w^{1-p'}(y)\chi_{E_w}(y)dy \right)^{\frac{1}{p'}} \\
&\leqslant [w]'_{p,\alpha} \left(\frac{1}{|B|} \int_B |f(y)|^p \chi_{E_w} w(y)dy \right)^{\frac{1}{p}} \\
&= [w]'_{p,\alpha} \big(M(|f|^p \chi_{E_w} w) \big)^{\frac{1}{p}},
\end{aligned}$$

where M is the Hardy–Littlewood maximal operator.

Hence,

$$M_\alpha(f\chi_{E_w})(x) \leqslant [w]'_{p,\alpha}(M(|f|^p\chi_{E_w}w))^{\frac{1}{p}}.$$

Consequently,

$$\begin{aligned}
|\{x \in \mathbb{R}^n : M_\alpha(f\chi_{E_w})(x) > \lambda\}| &\leqslant |\{x \in \mathbb{R}^n : C[w]'_{p,\alpha}(M(|f\chi_{E_w}|^p w)(x))^{\frac{1}{p}} > \lambda\}| \\
&\leqslant \frac{C[w]'_{p,\alpha}}{\lambda^p} \||f|^p \chi_{E_w} w\|_{L^1} \\
&= \frac{C[w]'_{p,\alpha}}{\lambda^p} \|f\|^p_{L^p_w(E_w)},
\end{aligned}$$

which ends the proof. □

Lemma 2.5 *Let $1 < p < \infty$ and let $0 < \alpha < n/p'$. If $[w]'_{p,\alpha} < \infty$ (see Lemma 2.4), then there is a positive constant $C = C_{p,\alpha}$ such that for all $f \in L^p_w(E_w)$*

$$\|I_\alpha(f\chi_{E_w})\|_{L^{p,\infty}(\mathbb{R}^n)} \leqslant C\|f\chi_{E_w}\|_{L^p_w(\mathbb{R}^n)}.$$

Proof It is known (see, e.g., Muckenhoupt and Wheeden [256], Th. 1 for more general setting) that

$$\|I_\alpha g\|_{L^{p,\infty}(\mathbb{R}^n)} \approx \|M_\alpha g\|_{L^{p,\infty}(\mathbb{R}^n)}$$

for a measurable function g. Thus, by Lemma 2.4, the result follows. □

Lemma 2.6 *Let $1 < p < \infty$ and $0 < \alpha < n/p'$. If $[w]'_{p,\alpha} < \infty$, then there is a positive constant $C = C_{p,\alpha}$ such that for all $f \in L^{p',1}(E_w)$,*

$$\|I_\alpha f\|_{L^{p'}_{w^{1-p'}}(E_w)} \leqslant C[w]'_{p,\alpha}\|f\|_{L^{p',1}(E_w)}.$$

Proof For $f \geqslant 0$, we have:

$$\begin{aligned}
\|I_\alpha f\|_{L^{p'}_{w^{1-p'}}(E_w)} &= \sup_{\|g\|_{L^p_w(E_w)}\leqslant 1} \left| \int_{\mathbb{R}^n} (I_\alpha f)(x)g(x)w^{-\frac{1}{p}}(x)w^{\frac{1}{p}}(x)\chi_{E_w}(x)dx \right| \\
&\leqslant \sup_{\|g\|_{L^p_w(E_w)}\leqslant 1} \left(\int_{\mathbb{R}^n} (I_\alpha f)(x)|g(x)|\chi_{E_w}(x)dx \right) \\
&= \sup_{\|g\|_{L^p_w(E_w)}\leqslant 1} \int_{E_w} I_\alpha(|g|\chi_{E_w})(y)f(y)dy \\
&\leqslant \sup_{\|g\|_{L^p_w(E_w)}\leqslant 1} \|I_\alpha|g|\|_{L^{p,\infty}(E_w)}\|f\|_{L^{p',1}(E_w)} \\
&\lesssim_{p,\alpha} [w]'_{p,\alpha}\|f\|_{L^{p',1}(E_w)},
\end{aligned}$$

thereby completing the proof. □

Proof of Theorem 2.3 $(iii) \Rightarrow (i)$: Observe, that if $1 < p < \infty, 0 < \alpha < \frac{n}{p}$, and v is a non-negative function satisfying $[v]_{p,\alpha} < \infty$, then $|\{x \in \mathbb{R}^n : v(x) = \infty\}| = 0$. Hence, $|\{x \in \mathbb{R}^n : v^{1-p}(x) = 0\}| = 0$. Consequently, since

$$[v]_{p,\alpha} = \sup_B \left(\int_B v^{(1-p)(1-p')}(x)dx \right)^{\frac{1}{p}} |B|^{\frac{\alpha}{n}-\frac{1}{p}} < \infty,$$

from Lemma 2.6, we have:

$$\|I_\alpha f\|_{L^p_v(\mathbb{R}^n)} \lesssim_{p,\alpha} [v]_{p,\alpha} \|f\|_{L^{p,1}(\mathbb{R}^n)}.$$

The pointwise estimate

$$M_\alpha g(x) \leqslant I_\alpha g(x) \quad g \geqslant 0.$$

entails the implication $(i) \Rightarrow (ii)$.

Taking the test functions

$$f_{B_0} = \chi_{B_0},$$

in (2.8), where B_0 is a ball in $\mathbb{R}^n$, using the estimate

$$M_\alpha f_{B_0}(x) \geqslant C_{n,\alpha} |B_0|^{\frac{\alpha}{n}}, \quad x \in B_0,$$

and observing that

$$\|f_{B_0}\|_{L^{p,1}} = c|B_0|^{\frac{1}{p}}$$

we can easily see that (ii) implies

$$v(B_0)^{1/p} \lesssim |B_0|^{1/p-\alpha/n}$$

with a positive constant C independent of B_0. The theorem has been proved. □

2.1.1 The Case of Spaces of Homogeneous Type

Let (X, d, μ) be a space of homogeneous type (SHT), i.e., X is an abstract set, and $d : X \times X \to \mathbb{R}_+$ is a quasi-metric satisfying the following conditions:

(i) $d(x, y) = 0$ if and only if $x = y$;
(ii) $d(x, y) = d(y, x)$ for all $x, y \in X$;
(iii) There is a constant $\kappa > 1$ such that $d(x, y) \leqslant \kappa\,(d(x, z) + d(z, y))$ for all $x, y, z \in X$,

and μ is a complete measure defined on a σ-algebra over X, such that the balls $B(x, r) := \{y \in X;\ d(x, y) < r\}$ are measurable with positive and finite μ-measure for all $x \in X$ and $r > 0$. Moreover, the doubling condition is satisfied:

$$\mu(B(x, 2r)) \leqslant C_{DC}\,\mu(B(x, r)), \tag{2.10}$$

with a positive constant C_{DC} independent of x and r.

The best possible constant C_{DC} in (2.10) is called the *doubling constant*.

Definition 2.7 We say that a measure satisfies the *reverse doubling condition*, or the triple (X, d, μ) is called an RD-space, if it is an SHT and μ satisfies the reverse doubling condition: there exist constants $a, b > 1$ such that for all $x \in X$ and $0 < r < d_X/a$

$$b\mu(B(x,r)) \leqslant \mu B(x, ar).$$

In this case we write $\mu \in RD(X)$.

Remark 2.8 (i) It is known that (X, d, μ) is an RD-space if and only if it is an SHT and there is a constant $\overline{c}$ such that for all $x \in X$ and $0 < r < \frac{d_X}{\overline{c}}$

$$B(x, \overline{c}r)\backslash B(x,r) \neq 0, \quad x \in X.$$

For the proof we refer to, e.g., Strömberg and Torchinsky [316, p. 11, Lemma 20], Han et al. [131, Remark 1.2].

For a given SHT, (X, d, μ), and q satisfying $1 \leqslant q \leqslant \infty$, as usual, we denote by $L^q = L^q(X, \mu)$ the Lebesgue space equipped with the standard norm. Let $L^{p,s}(X, \mu)$ be the Lorentz space defined on (X, d, μ). If ν is another measure on X, then we denote the Lebesgue and Lorentz spaces defined with respect to ν by $L^q(X, \nu)$ and $L^{p,s}(X, \nu)$, respectively. If ν is absolutely continuous with respect to μ, i.e., $d\nu(x) = \upsilon(x)d\mu(x)$ for some μ-a.e. positive locally integrable function on X (i.e., υ is a weight function on (X, d, μ)), then we use the symbols $L^q_\upsilon(X, \mu)$ and $L^{p,s}_\upsilon(X, \mu)$ for $L^q(X, \nu)$ and $L^{p,s}(X, \nu)$, respectively.

Define by $I_{\alpha,X} f$ the Riesz potential of a μ-measurable function f given by the formula:

$$I_{\alpha,X} f(x) = \int_X \mu(B_{xy})^{\alpha-1} f(y)\, d\mu(y), \quad x \in X,$$

where $0 < \alpha < 1$ and $B_{xy} := B(x, d(x, y))$.

The appropriate fractional maximal function has the form:

$$M_{\alpha,X} f(x) = \sup_{B \ni x} \frac{1}{\mu(B)^{1-\alpha}} \int_B |f(y)|\, d\mu(y), \quad x \in X,$$

The following trace inequality for an SHT was proved by Gabidzashvili [109].

Theorem 2.9 *Let $1 < p < q < \infty$ and let $0 < \alpha < 1/p$. Suppose that (X, d, μ) is an SHT and ν is another measure on X. Then the inequality*

$$\|I_{\alpha,X} f\|_{L^q(X,\nu)} \lesssim \|f\|_{L^p(X,\mu)}$$

holds if and only if

$$\sup_B \nu(B)^{1/q} \mu(B)^{\alpha-1/p} < \infty,$$

where the supremum is taken over all balls B in X.

Analyzing the proof of Theorem 2.3, we can formulate the same result for an SHT. In particular, the following theorem holds.

Theorem 2.10 *Let $1 < p < \infty$ and let $0 < \alpha < 1/p$. Suppose that (X, d, μ) be an SHT. Assume that v is a weight function on (X, d, μ). Then, the following statements are equivalent:*

(i) there is a positive constant C such that for all $f \in L^{p,1}(X, \mu)$,

$$\|I_{\alpha,X} f\|_{L^p_v(X,\mu)} \leqslant C \|f\|_{L^{p,1}(X,\mu)}; \tag{2.11}$$

(ii) there is a positive constant c such that for all $f \in L^{p,1}(X, \mu)$,

$$\|M_{\alpha,X} f\|_{L^p_v(X,\mu)} \leqslant c \|f\|_{L^{p,1}(X,\mu)}; \tag{2.12}$$

(iii)

$$[v]_{p,\alpha,X,\mu} := \sup_B \left(\int_B v(x) d\mu(x) \right)^{1/p} \mu(B)^{\alpha-1/p} < \infty.$$

Moreover, if C and c are best constants in (2.11) *and* (2.12) *respectively, then*

$$C \approx c \approx [v]_{p,\alpha,X,\mu}.$$

2.2 Trace Inequality Characterization for Multilinear Riemann–Liouville Fractional Integrals

In this section we establish necessary and sufficient condition on a weight function v which guarantees the weighted inequality:

$$\|R_\alpha(\vec{f})\|_{L^q_v(\mathbb{R}_+)} \leqslant C \prod_{k=1}^m \|f_k\|_{L^{p_k}(\mathbb{R}_+)}, \quad f_k \in L^{p_k}(\mathbb{R}_+), \ k = 1, \ldots, m, \tag{2.13}$$

for the one-sided multilinear fractional integral operator

$$R_\alpha(\vec{f})(x) = \underbrace{\int_0^x \cdots \int_0^x}_{m} \frac{f_1(x-t_1)\cdots f_m(x-t_m)}{(t_1+\cdots+t_m)^{m-\alpha}} dt_1 \cdots dt_m,$$

where $\vec{f} := (f_1, \ldots, f_m)$.

Similar results for the operator

$$\mathcal{R}_{\vec{\alpha}}(\vec{f})(x) = \prod_{k=1}^{m} \int_0^x \frac{f_k(t_k)}{(x-t_k)^{1-\alpha_k}} dt_k$$

are also obtained, where $\vec{\alpha} := (\alpha_1, \ldots, \alpha_m)$.

Taking $\alpha = m$ in R_α, or $\alpha_1 = \cdots = \alpha_m = 1$ in $\mathcal{R}_{\vec{\alpha}}$, we get the multilinear Hardy operator:

$$H_m(\vec{f})(x) = \prod_{k=1}^{m} \int_0^x f_k(t_k) dt_k, \quad x \in (0, \infty).$$

Two-weight criteria for H_2 were found in Aguilar Cañestro et al. [9] (we refer also to Křepela [329] for different approach).

The operator H_m plays an important role, for example, in the derivation of multilinear Rellich inequalities (see, e.g., Edmunds and Meskhi [89], or Sect. 1.3).

In the linear case (i.e., when $m = 1$), the operators R_α and $\mathcal{R}_{\vec{\alpha}}$ give the classical Riemann–Liouville operator :

$$(RL)_\alpha \varphi(x) = \int_0^x \frac{\varphi(t)}{(x-t)^{1-\alpha}} dt, \quad x > 0$$

defined on $(0, \infty)$.

Two-weight criteria (L^p_w to L^q_v boundedness) for $(RL)_\alpha$, in the case $\alpha > 1$, are well-known (see Stepanov [315], Opic and Kufner [268]), but for the singular case, i.e., when $0 < \alpha < 1$, it is known the $(L^p \to L^q_v)$-boundedness (see Meskhi [239], Prokhorov [275], Edmunds et al. [92, Chapter 2]).

The multilinear operator R_α was introduced in the paper Fu et al. [106]. It is the one-sided variant of the multilinear fractional integral operator:

$$\mathcal{I}_\gamma(f_1, \ldots, f_m)(x) = \int_{(\mathbb{R}^n)^m} \frac{f_1(y_1)\cdots f_m(y_m)}{(|x-y_1|+\cdots+|x-y_m|)^{mn-\gamma}} dy_1 \cdots dy_m, \quad x \in \mathbb{R}^n,$$

where $0 < \gamma < nm$.

The operator $\mathcal{I}_\gamma$ is a very natural intermediate operator (written in m-linear form) between $(I_{\alpha_1} f_1)(I_{\alpha_2} f_2)$ and $B_{\alpha_1+\alpha_2}(f_1, f_2)$ (see Kenig and Stein [163]), where I_{α_1} and I_{α_2} are Riesz potential operators:

$$I_{\alpha_i} f(x) = \int\limits_{\mathbb{R}^n} \frac{f(y)}{|x-y|^{n-\alpha_i}} dy, \quad x \in \mathbb{R}^n, \ 0 < \alpha_i < n, \ i = 1, 2,$$

and

$$B_\alpha(f_1, f_2)(x) = \int\limits_{\mathbb{R}^n} \frac{f_1(x+t) f_2(x-t)}{|t|^{n-\alpha}} dt, \quad 0 < \alpha < n,$$

is the bilinear fractional integral operator introduced and studied in Grafakos [118]. For the one-weight and trace inequalities regarding $\mathcal{I}_\gamma$, we refer to Moen [251], Kokilashvili et al. [179], respectively.

The operator $\mathcal{R}_{\vec{\alpha}} \vec{f}$ is the fractional variant of the m-linear Hardy operator H_m.

Observe that the operator R_α can be rewritten as follows:

$$R_\alpha(\vec{f})(x) = \underbrace{\int\limits_0^x \cdots \int\limits_0^x}_{m} \frac{f_1(t_1) \cdots f_m(t_m)}{[(x-t_1) + \cdots + (x-t_m)]^{m-\alpha}} dt_1 \cdots dt_m.$$

It is easy to see that the following estimates hold for non-negative f_j, $j = 1, \ldots, m$:

$$R_\alpha(\vec{f})(x) \geqslant m^{\alpha-m} M_\alpha(\vec{f})(x), \ \ 0 < \alpha < m, \tag{2.14}$$

and

$$\mathcal{R}_{\vec{\alpha}}(\vec{f})(x) \geqslant \mathcal{M}_{\vec{\alpha}}(\vec{f})(x), \ \ 0 < \alpha_k < 1, \ k = 1, \ldots, m, \ \ \vec{\alpha} = (\alpha_1, \ldots, \alpha_m), \tag{2.15}$$

where

$$M_\alpha(\vec{f})(x) = \sup_{0<h<x} \prod_{k=1}^m \frac{1}{h^{1-\alpha/m}} \int\limits_{x-h}^x f_k(t_k) dt_k,$$

and

$$\mathcal{M}_{\vec{\alpha}}(\vec{f})(x) = \sup_{0<h<x} \prod_{k=1}^m \frac{1}{h^{1-\alpha_k}} \int\limits_{x-h}^x f_k(t_k) dt_k.$$

Let $1 < s < \infty$, Ω be an open set in $\mathbb{R}^n$, and v be an a.e. positive locally integrable function (i.e., a weight) on Ω. Denote by L^s_v the weighted Lebesgue space defined with respect to the norm

$$\|f\|_{L^s_v(\Omega)} := \left(\int_\Omega |f(x)|^s v(x) dx \right)^{1/s}.$$

If $v \equiv 1$, then we use the symbol L^s for L^s_v.

Let us recall the following statements regarding classical (linear) Hardy and Riemann–Liouville operators.

Theorem 2.11 (Aguilar Cañestro et al. [9]) *Let $1 < p \leqslant q < \infty$. Then, the two-weight inequality*

$$\|Hf\|_{L^q_v(\mathbb{R}_+)} \leqslant C\|f\|_{L^p_w(\mathbb{R}_+)}, \quad f \in L^p_w,$$

for the linear Hardy operator $H := H_1$ holds if and only if

$$\sup_{t>0} \left(\int_t^\infty v(x)\, dx \right)^{\frac{1}{q}} \left(\int_0^t w^{1-p'}(x)\, dx \right)^{\frac{1}{p'}} < \infty,$$

where $p' = \frac{p}{p-1}$.

The following statement is also known.

Theorem 2.12 (Meskhi [239]) *Let $1 < p \leqslant q < \infty$ and let $\gamma > \frac{1}{p}$. Then the following are equivalent:*

(i) $(RL)_\gamma$ *is bounded from $L^p(\mathbb{R}_+)$ to $L^q_v(\mathbb{R}_+)$;*

(ii)

$$\sup_{t>0} \left(\int_t^\infty v(x) x^{(\gamma-1)q} dx \right)^{\frac{1}{q}} t^{\frac{1}{p'}} < \infty;$$

(iii)

$$\sup_{k\in\mathbb{Z}} \left(\int_{2^k}^{2^{k+1}} v(x) dx \right)^{\frac{1}{q}} 2^{k\left(\gamma-\frac{1}{p}\right)} < \infty.$$

Two-weight criteria of Sawyer and integral type for one-sided potentials defined on the whole real line $\mathbb{R}$ were given in Lorente [213], Edmunds et al. [92, Chapter 2].

For weighted inequalities regarding one-sided multilinear fractional integrals and maximal functions, we refer to Imerlishvili and Meskhi [151].

2.2.1 Preliminaries

In the sequel we assume that p is defined by the formula:

$$\frac{1}{p} := \sum_{j=1}^{m} \frac{1}{p_j}. \tag{2.16}$$

A weight function defined on $\mathbb{R}_+$ will be denoted by v.

To prove the main theorems, we need some Lemmas.

Lemma 2.13 *Let* $1 < p_1, \ldots, p_m, q < \infty$ *and let* p *be defined by* (2.16). *Then, the following conditions are equivalent:*

(i)

$$B_1 := \sup_{t>0} B_1(t) := \sup_{t>0} \left(\int_t^{\infty} v(x) x^{(\alpha-m)q} dx \right)^{\frac{1}{q}} t^{m-\frac{1}{p}} < \infty; \tag{2.17}$$

(ii)

$$B_2 := \sup_{t>0} B_2(t) := \sup_{t>0} \left(\int_t^{\infty} v(x) x^{q\left(\alpha - m + \sum_{k\neq i_0} \frac{1}{p_k'}\right)} dx \right)^{\frac{1}{q}} t^{\frac{1}{p'_{i_0}}} < \infty$$

for any index i_0, $1 \leqslant i_0 \leqslant m$.

(iii)

$$B_3 := \sup_{k\in\mathbb{Z}} B_3(t) := \sup_{k\in\mathbb{Z}} \left(\int_{2^k}^{2^{k+1}} v(x) dx \right)^{\frac{1}{q}} 2^{k(\alpha-\frac{1}{p})} < \infty. \tag{2.18}$$

Proof For simplicity we assume that $m = 2$. The implication (i) $\Rightarrow$ (iii) follows immediately. Let us show that (iii) $\Rightarrow$ (i). Let $t > 0$. Then $t \in \left[2^{m_0}, 2^{m_0+1}\right)$ for some $m_0 \in \mathbb{Z}$. Then, we have:

$$
\begin{aligned}
B_1^q(t) &= \left(\int_t^{\infty} v(x) x^{(\alpha-2)q} dx \right) t^{\left(2-\frac{1}{p}\right)q} \\
&\leqslant \left(\int_{2^{m_0}}^{\infty} v(x) x^{(\alpha-2)q} dx \right) 2^{(m_0+1)\left(2-\frac{1}{p}\right)q} \\
&\lesssim \sum_{k=m_0}^{\infty} \left(\int_{2^k}^{2^{k+1}} v(x) dx \right) 2^{k(\alpha-2)q} 2^{(m_0+1)\left(2-\frac{1}{p}\right)q} \\
&\lesssim \left[\sum_{k=m_0}^{\infty} 2^{k\left(\frac{1}{p}-\alpha\right)q} 2^{k(\alpha-2)q} \right] 2^{m_0\left(2-\frac{1}{p}\right)q} \\
&\lesssim \left[\sum_{k=m_0}^{\infty} 2^{k\left(\frac{1}{p}-2\right)q} \right] 2^{m_0\left(2-\frac{1}{p}\right)q} \\
&\leqslant C,
\end{aligned}
$$

because $\frac{1}{p} < 2$.

Now we will show that (ii) $\Leftrightarrow$ (iii). The fact that (ii) $\Rightarrow$ (iii) is straightforward. We show that (iii) $\Rightarrow$ (ii). Let $t > 0$. Then $t \in \left[2^{m_0}, 2^{m_0+1}\right)$ for some $m_0 \in \mathbb{Z}$. Suppose that $i_0 = 2$. Then

$$
\begin{aligned}
B_2^q(t) &= \left(\int_t^{\infty} v(x) x^{(\alpha-2)q+\frac{q}{p_1'}} dx \right) t^{\frac{q}{p_2'}} \\
&\leqslant 2^{(m_0+1)\frac{q}{p_2'}} \sum_{k=m_0}^{\infty} \int_{2^k}^{2^{k+1}} v(x) x^{(\alpha-2)q+\frac{q}{p_1'}} dx \\
&\lesssim 2^{(m_0+1)\frac{q}{p_2'}} \sum_{k=m_0}^{\infty} \left[\left(\int_{2^k}^{2^{k+1}} v(x) dx \right) 2^{k\left[(\alpha-2)q+\frac{q}{p_2'}\right]} \right] \\
&\lesssim B_3^q \left[\sum_{k=m_0}^{\infty} 2^{-\frac{kq}{p_2'}} \right] 2^{(m_0+1)\frac{q}{p_2'}} \\
&\lesssim B_3^q.
\end{aligned}
$$

Thus, $B_2 \lesssim B_3$. □

Lemma 2.14 *Let* $1 < p_1, \ldots, p_m < \infty$ *and let* $0 < \alpha < m$. *Then for all non-negative* $f_k \in L^{p_k}$, $k = 1, 2, 3, \ldots, m$, *the estimate*

$$R_\alpha(\vec{f})(x) \leqslant \left(\int_0^x (x - t_1)^{\alpha - m} f_1(t_1)\, dt_1 \right) x^{\frac{1}{p_2'} + \cdots + \frac{1}{p_m'}} \prod_{k=2}^{m} \| f_k \|_{L^{p_k}} \tag{2.19}$$

holds.

Proof The inequality (2.19) follows easily by using the fact that $\alpha < m$ and Hölder's inequality for the exponents $p_2, \ldots, p_m$. Indeed, since $\alpha < m$, we have: $\left(\sum_k (x - t_k)\right)^{\alpha - m} \leqslant (x - t_1)^{\alpha - m}$. Consequently,

$$\begin{aligned} R_\alpha(\vec{f})(x) &= \int_0^x f_1(t_1) \int_0^x f_m(t_m) \left[\sum_{k=1}^{m} (x - t_k) \right]^{\alpha - m} dt_m \cdots dt_1 \\ &\leqslant \left(\int_0^x (x - t_1)^{\alpha - m} f_1(t_1)\, dt_1 \right) \prod_{k=2}^{m} \int_0^x f_k(t_k)\, dt_k \\ &\leqslant \left(\int_0^x (x - t_1)^{\alpha - m} f_1(t_1)\, dt_1 \right) \left(\prod_{k=2}^{m} \| f_k \|_{L^{p_k}} \right) x^{\frac{1}{p_2'} + \cdots + \frac{1}{p_m'}}, \end{aligned}$$

which ends the proof. □

Lemma 2.15 *Let* $1 < p_1, \ldots, p_m < \infty$ *and let* $m - \frac{1}{p_m} < \alpha < m$. *Then for all non-negative* $f_1, \ldots, f_m$ *with* $f_k \in L^{p_k}$, *and* $x > 0$:

$$\begin{aligned} I := \int_{x/2}^{x} f_1(t_1) \Bigg(\int_0^x f_2(t_2) \cdots \int_0^x f_{m-1}(t_{m-1}) \int_0^x f_m(t_m) \\ \times \left[\sum_{k=1}^{m} (x - t_k) \right]^{\alpha - m} dt_m \cdots dt_2 \Bigg) dt_1 \\ \lesssim \left(\int_{x/2}^{x} f_1(u_1)\, (mx - u_1)^{\alpha - m}\, du_1 \right) \prod_{k=2}^{m} \| f_k \|_{L^{p_k}}\, x^{\frac{1}{p_2'} + \cdots + \frac{1}{p_m'}}, \end{aligned}$$

where the positive constant C is independent of x and $\vec{f}$.

Proof By Hölder's inequality with respect to p_k, $k = m, \ldots, 2$, and change of variables, we find that

$$
\begin{aligned}
I &\leqslant \left(\int_{\frac{x}{2}}^{x} f_1(t_1) \cdots \int_0^x f_{m-1}(t_{m-1}) \right. \\
&\qquad \left. \times \left[\int_0^x \left[\sum_{k=1}^{m} (x - t_m) \right]^{(\alpha - m)p'_m} dt_m \right]^{\frac{1}{p'_m}} dt_{m-1} \cdots dt_1 \right) \times \| f_m \|_{L^{p_m}} \\
&= \left(\int_0^{\frac{x}{2}} f_1(x - u_1) \cdots \int_0^x f_{m-1}(x - u_{m-1}) \right. \\
&\qquad \left. \times \left[\int_0^x (u_1 + \cdots + u_m)^{(\alpha - m)p'_m} du_m \right]^{\frac{1}{p_{m'}}} du_{m-1} \cdots du_1 \right) \| f_m \|_{L^{p_m}} \\
&= \left(\int_0^{\frac{x}{2}} f_1(x - u_1) \cdots \int_0^x f_{m-1}(x - u_{m-1}) \right. \\
&\qquad \left. \times \left(\int_{u_1 + \cdots + u_{m-1}}^{u_1 + \cdots + u_{m-1} + x} \tau^{(\alpha - m)p'_m} d\tau \right)^{\frac{1}{p_m}} du_{m-1} \cdots du_1 \right) \| f_m \|_{L^{p_m}} \\
&\lesssim \left(\int_0^{\frac{x}{2}} f_1(x - u_1) \cdots \right. \\
&\quad \times \left. \left(\int_0^x f_{m-1}(x - u_{m-1}) (u_1 + \cdots + u_{m-1} + x)^{\alpha - m + \frac{1}{p'_m}} du_{m-1} \right) \cdots du_1 \right) \\
&\qquad \times \| f_m \|_{L^{p_m}} \, x^{\frac{1}{p_m}} \\
&\lesssim \left(\int_0^{\frac{x}{2}} f_1(x - u_1) \cdots \left(\int_0^x (u_1 + \cdots + u_{m-1} + x)^{(\alpha - m)p'_{m-1}} du_{m-1} \right)^{\frac{1}{p'_{m-1}}} \cdots du_1 \right) \\
&\qquad \times \| f_m \|_{L^{p_m}} \, \| f_{m-1} \|_{L^{p_{m-1}}} \, x^{\frac{1}{p'_m}} \\
&= c \left(\int_0^{x/2} f_1(x - u_1) \cdots \left(\int_{u_1 + \cdots + u_{m-2} + x}^{u_1 + \cdots + u_{m-2} + 2x} \tau^{(\alpha - m)p'_{m-1}} d\tau \right)^{\frac{1}{p'_{m-1}}} \cdots du_1 \right)
\end{aligned}
$$

$$\times \left\|f_m\right\|_{L^{p_m}} \left\|f_{m-1}\right\|_{L^{p_{m-1}}} \leqslant (\text{since } \alpha < m)$$

$$\lesssim \left(\int_0^{x/2} f_1(x-u_1) \cdots \right.$$

$$\left. \times \left(\int_0^x (u_1 + \cdots + u_{m-2} + 2x)^{\alpha - m} f_{m-2}(x-u_{m-2})\, du_{m-2} \right) \cdots du_1 \right)$$

$$\times \left\|f_m\right\|_{L^{p_m}} \left\|f_{m-1}\right\|_{L^{p_{m-1}}} x^{\frac{1}{p'_m} + \frac{1}{p'_{m-1}}}$$

$$\leqslant \cdots \lesssim \left(\int_0^{\frac{x}{2}} f_1(x-u_1)(u_1 + (m-1)x)^{\alpha - m}\, du_1 \right) \prod_{k=2}^m \left\|f_k\right\|_{L^{p_k}} x^{\frac{1}{p'_2} + \cdots + \frac{1}{p'_m}}$$

$$\lesssim \left(\int_{x/2}^{x} f_1(u_1)(mx - u_1)^{\alpha - m}\, du_1 \right) \prod_{k=2}^m \left\|f_k\right\|_{L^{p_k}} x^{\frac{1}{p'_2} + \cdots + \frac{1}{p'_m}},$$

which ends the proof. □

Lemma 2.16 *Let* $1 < p_1, \ldots, p_m < \infty$. *Suppose that* $0 < \alpha < m$. *Then for all non-negative* $f_1, \ldots, f_m$ *with* $f_k \in L^{p_k}$, *and* $x > 0$*:*

$$I \lesssim \left(\int_{\frac{x}{2}}^{x} f_1(t_1)(x - t_1)^{\alpha - m}\, dt_1 \right) \prod_{k=2}^m \left\|f_k\right\|_{L^{p_k}} x^{\frac{1}{p'_2} + \cdots + \frac{1}{p'_m}},$$

where I *is defined in Lemma 2.15.*

Proof By Hölder's inequality with respect to p_k, $k = m, \ldots, 2$ and change of variable, we find that

$$I \leqslant \left(\int_{\frac{x}{2}}^{x} f_1(t_1) \cdots \int_0^x f_{m-1}(t_{m-1}) \right.$$

$$\left. \times \left[\int_0^x \left[\sum_{k=1}^m (x - t_m) \right]^{(\alpha - m)p'_m} dt_m \right]^{\frac{1}{p'_m}} dt_{m-1} \cdots dt_1 \right) \left\|f_m\right\|_{L^{p_m}}$$

$$= \left(\int_0^{\frac{x}{2}} f_1(x - u_1) \cdots \int_0^x f_{m-1}(x - u_{m-1}) \right.$$

$$\times \left[\int_0^x (u_1+\cdots+u_m)^{(\alpha-m)p'_m}\,du_m\right]^{\frac{1}{p_{m'}}} du_{m-1}\cdots du_1\Bigg)\,\|f_m\|_{L^{p_m}}$$

$$= \left(\int_0^{\frac{x}{2}} f_1(x-u_1)\cdots\int_0^x f_{m-1}(x-u_{m-1})\right.$$

$$\times \left(\int_{u_1+\cdots+u_{m-1}}^{u_1+\cdots+u_{m-1}+x} \tau^{(\alpha-m)p'_m}d\tau\right)^{\frac{1}{p_m}} du_{m-1}\cdots du_1\Bigg)$$

$$\times \|f_m\|_{L^{p_m}} \quad \text{(by the condition } \alpha < m)$$

$$\lesssim \left(\int_0^{\frac{x}{2}} f_1(x-u_1)\cdots\right.$$

$$\times \left(\int_0^x f_{m-1}(x-u_{m-1})(u_1+\cdots+u_{m-1})^{\alpha-m+\frac{1}{p'_m}} du_{m-1}\right)\cdots du_1\Bigg)$$

$$\times \|f_m\|_{L^{p_m}}\, x^{\frac{1}{p_m}}$$

$$\leqslant \cdots \lesssim \left(\int_0^{\frac{x}{2}} f_1(x-u_1)\cdots\right.$$

$$\times \left(\int_0^x (u_1+\cdots+u_{m-1})^{(\alpha-m)p'_{m-1}}\,du_{m-1}\right)^{\frac{1}{p'_{m-1}}}\cdots du_1\Bigg)$$

$$\times \|f_m\|_{L^{p_m}}\,\|f_{m-1}\|_{L^{p_{m-1}}}\, x^{\frac{1}{p'_m}}$$

$$= c\left(\int_{\frac{x}{2}}^x f_1(x-u_1)\cdots\left(\int_{u_1+\cdots+u_{m-2}}^{u_1+\cdots+u_{m-2}+x} \tau^{(\alpha-m)p'_{m-1}}d\tau\right)^{\frac{1}{p'_{m-1}}}\cdots du_1\right)$$

$$\times \|f_m\|_{L^{p_m}}\,\|f_{m-1}\|_{L^{p_{m-1}}} \leqslant \text{(since } \alpha < m)$$

$$\lesssim \left(\int_{\frac{x}{2}}^x f_1(u_1)\cdots\left(\int_0^x (u_1+\cdots+u_{m-2})^{\alpha-m} f_{m-2}(x-u_{m-2})\,du_{m-2}\right)\cdots du_1\right)$$

$$\times \|f_m\|_{L^{p_m}} \|f_{m-1}\|_{L^{p_{m-1}}} x^{\frac{1}{p'_m}+\frac{1}{p'_{m-1}}}$$

$$\leqslant \cdots \lesssim \left(\int_0^{\frac{x}{2}} f_1(x-u_1)\, u_1^{\alpha-m} du_1 \right) \prod_{k=2}^{m} \|f_k\|_{L^{p_k}} x^{\frac{1}{p'_2}+\cdots+\frac{1}{p'_m}}$$

$$= c \left(\int_{x/2}^{x} f_1(t_1)(x-t_1)^{\alpha-m} du_1 \right) \prod_{k=2}^{m} \|f_k\|_{L^{p_k}} x^{\frac{1}{p'_2}+\cdots+\frac{1}{p'_m}}.$$

□

The two-weight characterizations for bilinear Hardy operator H_2 is already known (see Aguilar Cañestro et al. [9], Křepela [329]), but we the give the statement for the m-linear case, i.e., for H_m.

Lemma 2.17 *Let* $1 \leqslant \min\{p_1, \ldots, p_m\} \leqslant q < \infty$ *and let* p *be an exponent defined by* (2.16). *Then, the following statements are equivalent:*

(i) $\left\| H_m(\vec{f}) \right\|_{L^q_v} \lesssim \prod_{k=1}^{m} \|f_k\|_{L^{p_k}}$;

(ii) $A := \sup_{t>0} A(t) := \sup_{t>0} \left(\int_t^{\infty} v(x)dx \right)^{\frac{1}{q}} t^{m-\frac{1}{p}} < \infty.$

Proof Let $f_1 \geqslant 0, \ldots, f_m \geqslant 0$. We show that (ii) $\Rightarrow$ (i). Let $f_k \in L^{p_k}$, $k = 1, \ldots, m$.

Without loss of generality, assume that $p_1 \leqslant q$. By using Hölder's inequality step by step with respect to the exponents $p_m, \ldots, p_2$, we have that

$$\left\| H_m(\vec{f}) \right\|_{L^q_v} = \left(\int_0^{\infty} v(x) \left(\int_0^x f_1(t_1)\, dt_1 \right)^q \cdots \left(\int_0^x f_m(t_m)\, dt_m \right)^q dx \right)^{1/q}$$

$$\leqslant \left(\int_0^{\infty} v(x) \left(\int_0^x f_1(t_1)\, dt_1 \right)^q x^{q\left(\frac{1}{p'_2}+\cdots+\frac{1}{p'_m}\right)} dx \right)^{\frac{1}{q}} \prod_{k=2}^{m} \|f_k\|_{L^{p_k}}.$$

Due to Theorem 2.11, the proposition will be proved if we show that

$$\bar{A} := \sup_{t>0} \bar{A}(t) := \sup_{t>0} \left(\int_t^{\infty} v(x) x^{q\left(\frac{1}{p_2'}+\cdots+\frac{1}{p_m'}\right)} dx \right)^{1/q} t^{\frac{1}{p'_1}} < \infty.$$

Let $t \in \left[2^{m_0}, 2^{m_0+1}\right)$ for some integer m_0.Then

$$
\begin{aligned}
(\bar{A}(t))^q &\lesssim \left(\int_{2^{m_0}}^{\infty} v(x) x^{q\left(\frac{1}{p_2'}+\cdots+\frac{1}{p_m'}\right)} dx \right) 2^{m_0\left(\frac{q}{p_1'}\right)} \\
&= c \left[\sum_{k=m_0}^{\infty} \left(\int_{2^k}^{2^{k+1}} v(x) dx \right) 2^{kq\left(\frac{1}{p_2'}+\cdots+\frac{1}{p_m'}\right)} \right] 2^{m_0\left(\frac{q}{p_1'}\right)} \\
&\lesssim A^q \left(\sum_{k=m_0}^{\infty} 2^{-\frac{kq}{p_1'}} \right) 2^{m_0 \frac{q}{p_1'}} \lesssim A^q.
\end{aligned}
$$

Hence $\bar{A} \lesssim A$. The implication (i) $\Rightarrow$ (ii) follows by choosing the test functions $f_1 = \cdots = f_m = \chi_{(0,t)}$ in the weighted inequality. □

2.2.2 *Main Results*

Now we formulate and prove the main results of this section.

Theorem 2.18 *Let* $1 < \min\{p_1, \ldots, p_m\} \leqslant q < \infty$ *and let* $\alpha > m - \frac{1}{p_{i_0}'}$ *for some* i_0, $1 \leqslant i_0 \leqslant m$, *where* $p_{i_0}' = \frac{p_{i_0}}{p_{i_0}-1}$. *Then the following statements are equivalent:*

(i)

$$
\left\| R_\alpha(\vec{f}) \right\|_{L_v^q} \lesssim \prod_{k=1}^{m} \|f_k\|_{L^{p_k}}, \quad f_k \in L^{p_k}, \ k = 1, \ldots, m;
$$

(ii)

$$
B_1 < \infty,
$$

where B_1 *is defined by* (2.17);

(iii)

$$
B_3 < \infty,
$$

where B_3 *is defined by* (2.18).

Proof Let $f_1, \ldots, f_m$ be non-negative functions such that $f_k \in L^{p_k}, k = 1, \ldots, m$. If $\alpha > m$, then

$$
R_\alpha(\vec{f})(x) \leqslant (mx)^{\alpha-m} H_m(\vec{f})(x). \tag{2.20}
$$

Hence, applying Proposition 2.17, we have that (ii) $\Rightarrow$ (i) for this case.

Now suppose that $m - \frac{1}{p'_{i_0}} < \alpha < m$ for some index i_0, $1 \leqslant i_0 \leqslant m$. Represent $R_\alpha(\vec{f})$ as follows:

$$R_\alpha(\vec{f})(x) = R_\alpha^{(1)}(\vec{f})(x) + R_\alpha^{(2)}(\vec{f})(x), \tag{2.21}$$

where

$$R_\alpha^{(1)}(\vec{f})(x) = \int_0^{x/2}\int_0^x \cdots \int_0^x \frac{f_1(t_1)\cdots f_m(t_m)}{\left[\sum_{k=1}^m (x-t_k)\right]^{m-\alpha}} dt_1 \cdots dt_m;$$

and

$$R_\alpha^{(2)}(\vec{f})(x) = \int_{x/2}^{x}\int_0^x \cdots \int_0^x \frac{f_1(t_1)\cdots f_m(t_m)}{\left[\sum_{k=1}^m (x-t_k)\right]^{m-\alpha}} dt_1 \cdots dt_m.$$

Observe that if $t_k < x/2$, $k = 1, \dots, m$, then $\sum_{k=1}^m (x-t_k)^{\alpha-m} \leqslant \left(\frac{x}{2}\right)^{\alpha-m}$. Hence,

$$R_\alpha^{(1)}(\vec{f})(x) \leqslant 2^{m-\alpha} x^{\alpha-m} H_m(\vec{f})(x).$$

Then Proposition 2.17 yields:

$$\left\| R_\alpha^{(1)}(\vec{f}) \right\|_{L_v^q} \lesssim \prod_{k=1}^m \|f_k\|_{L^{p_k}}.$$

Suppose now that $p'_{i_0} \leqslant q$ for the index i_0. Without loss of generality, we assume that $i_0 = 1$. By Lemma 2.16 we find that

$$\left\| R_\alpha^{(2)}(\vec{f}) \right\|_{L_v^q}^q \lesssim \left(\int_0^\infty \left(\int_{x/2}^x f_1(t_1)(x-t_1)^{\alpha-m} dt_1 \right)^q x^{q\left(\frac{1}{p'_2}+\cdots+\frac{1}{p'_m}\right)} dx \right) \times \prod_{k=2}^m \|f_k\|_{L^{p_k}}^q.$$

Now Hölder's inequality and the condition $p_1 \leqslant q$ give the estimates:

$$\left\| R_\alpha^{(2)}(\vec{f}) \right\|_{L_v^q}^q \lesssim \Bigg(\sum_{k\in\mathbb{Z}} \int_{2^k}^{2^{k+1}} v(x) \Big(\int_{x/2}^{x} f_1^{p_1}(t_1)dt_1 \Big)^{\frac{q}{p_1}}$$

$$\times \left(\int_{x/2}^{x} (x-t_1)^{(\alpha-m)p_1'} dt_1 \right)^{\frac{q}{p_1'}} x^{q\left(\frac{1}{p_2'}+\cdots+\frac{1}{p_m'}\right)} dx \Bigg) \prod_{k=2}^{m} \| f_k \|_{L^{p_k}}^q$$

$$\lesssim \Bigg[\sum_{k\in\mathbb{Z}} \Big(\int_{2^k}^{2^{k+1}} v(x) x^{(\alpha-m)q} dx \Big) 2^{kq\sum_{k=1}^m \frac{1}{p_k'}} \Big(\int_{2^{k-1}}^{2^{k+1}} f_1^{p_1}(t_1)dt_1 \Big)^{\frac{q}{p_1}}$$

$$\times \prod_{k=2}^{m} \| f_k \|_{L^{p_k}}^q \leqslant c B_1^q \sum_{k\in\mathbb{Z}} \Big(\int_{2^{k-1}}^{2^{k+1}} f_1^{p_1}(t_1)dt_1 \Big)^{\frac{q}{p_1}} \prod_{k=2}^{m} \| f_k \|_{L^{p_k}}^q$$

$$\lesssim B_1^q \prod_{k=1}^{m} \| f_k \|_{L^{p_k}}^q .$$

Let now $m - \frac{1}{p_{i_0}'} < \alpha < m$ hold for the index i_0 for which $q < p_{i_0}'$. Without loss of generality, assume that $i_0 = m$ and $p_1 \leqslant q$. Then applying Lemma 2.15, we have that

$$\left\| R_\alpha^{(2)} f \right\|_{L_v^q}^q \lesssim \left(\int_0^\infty v(x) \left(\int_{x/2}^{x} f_1(t_1)(mx-t_1)^{\alpha-m} dt_1 \right)^q x^{\sum_{k=2}^m 1/p_k'} dx \right)$$

$$\times \prod_{k=2}^{m} \| f_k \|_{L^{p_k}}^q$$

$$\lesssim \Bigg[\sum_{k\in\mathbb{Z}} \Big(\int_{2^k}^{2^{k+1}} v(x) \left(\int_{(m-1/2)x}^{(m-1)x} \tau^{(\alpha-m)p_1'} d\tau \right)^{\frac{q}{p_1'}} x^{\sum_{k=2}^m \frac{1}{p_k'}} dx \Big)$$

$$\times \Big(\int_{2^{k-1}}^{2^{k+2}} f_1^{p_1}(\tau)d\tau \Big)^{\frac{q}{p_1}} \prod_{k=2}^{m} \| f_k \|_{L^{p_k}}^q$$

$$\lesssim B_1^q \prod_{k=2}^{m} \| f_k \|_{L^{p_k}}^q .$$

Here we used again the fact that $p_1 \leqslant q$. Finally we derive the implication (ii) $\Rightarrow$ (i). Taking the test functions $f_1 = \cdots = f_m = \chi_{(2^k, 2^{k+1})}$ in the weighted inequality (i) we have that (i) $\Rightarrow$ (iii).

The equivalence of (ii) and (iii) follows from Lemma 2.13. □

Now we investigate the boundedness of the operator $\mathcal{R}_{\vec{\alpha}}$.

Lemma 2.19 *Let $1 < p_1, \dots, p_m < \infty$ and let $\vec{\alpha} = (\alpha_1, \dots, \alpha_m)$ be such a vector that $\alpha_1 > 0$ and $\alpha_k > 1/p_k$, $k = 2, \dots, m$. Then for all $f_1 \geqslant 0, \dots, f_m \geqslant 0$ with $f_k \in L^{p_k}$, $k = 2, \dots, m$, the estimate*

$$\mathcal{R}_{\vec{\alpha}}(\vec{f})(x) \leqslant \left(\int_0^x f_1(t_1)(x-t_1)^{\alpha_1 - 1} dt_1 \right) \prod_{k=2}^m \|f_k\|_{L^{p_k}} \cdot x^{\sum_{k=2}^m (\alpha_k - \frac{1}{p_k})}$$

holds, where the positive constant c is independent of $\vec{f}$ and x.

Proof This statement follows immediately by applying Hölder's inequality for p_k, $k = 2, \dots, m$:

$$\int_0^x f_k(t_k)(x-t_k)^{\alpha_k - 1} dt_k \leqslant \left(\int_0^x f_k^{p_k}(t_k) dt_k \right)^{1/p_k} \left(\int_0^x (x-t_k)^{(\alpha_k - 1)p_k'} dt_k \right)^{1/p_k'}$$

and the condition $\alpha_k > 1/p_k$. □

Theorem 2.20 *Let $1 < \min\{p_1, \dots, p_m\} \leqslant q < \infty$ and let $\alpha_k > 1/p_k$, $k = 1, 2, \dots, m$. Then, the following statements are equivalent:*

(i)

$$\left\| \mathcal{R}_{\vec{\alpha}}(\vec{f}) \right\|_{L^q_v} \lesssim \prod_{k=1}^m \|f_k\|_{L^{p_k}};$$

(ii)

$$B_1 < \infty,$$

where B_1 is defined by (2.17) *and $\alpha := \alpha_1 + \cdots + \alpha_m$;*

(iii)

$$B_3 < \infty,$$

where B_3 is defined by (2.18) *and $\alpha := \alpha_1 + \cdots + \alpha_m$.*

Proof The implication (ii) $\Rightarrow$ (i) is a consequence of Lemma 2.19, Theorem 2.12 (for $\gamma = \alpha_1$) and the fact that $B_1 < \infty$ implies the condition $B_2 < \infty$ (for $i_0 = 1$), where B_1 and B_2 are defined in Lemma 2.13. The implication (i) $\Rightarrow$ (iii) is a consequence of choosing test functions $f_j, k = \chi_{[2^k, 2^{k+1})}$, $j = 1, \ldots, m$, in the weighted inequality. The implication (iii) $\Rightarrow$ (ii) follows from Lemma 2.13. □

Finally we note that Theorems 2.18 and 2.20 yield the following statements.

Corollary 2.21 *Let* $1 < \min\{p_1, \ldots, p_m\} \leqslant q < \infty$ *and let there exist an index* $1 \leqslant i_0 \leqslant m$ *such that* $m > \alpha > m - \frac{1}{p'_{i_0}}$, *where* $p'_{i_0} = \frac{p_{i_0}}{p'_{i_0} - 1}$.

Then, the following statements are equivalent:

(i)

$$\left\| R_\alpha(\vec{f}) \right\|_{L^q_v} \lesssim \prod_{k=1}^{m} \|f_k\|_{L^{p_k}},$$

(ii)

$$\left\| M_\alpha(\vec{f}) \right\|_{L^q_v} \lesssim \prod_{k=1}^{m} \|f_k\|_{L^{p_k}},$$

(iii)

$$B_1 < \infty;$$

(iv)

$$B_3 < \infty,$$

where B_1 *and* B_3 *are defined in Lemma 2.13.*

Proof The implication (i) $\Rightarrow$ (ii) follows from the pointwise inequality (2.14). Taking the test functions $f_1 = \cdots = f_m = \chi_{(2^k, 2^{k+1}]}$, $k \in \mathbb{Z}$, we obtain that (ii) $\Rightarrow$ (iv). The implications (iv) $\Rightarrow$ (iii) and (iii) $\Rightarrow$ (i) are due to Theorem 2.18. □

The next statement follows in a similar manner as Corollary 2.21 was derived by applying the estimate (2.15) and Theorem 2.20.

Corollary 2.22 *Let* $1 < \min\{p_1, \ldots, p_m\} \leqslant q < \infty$ *and let* $1 > \alpha_k > 1/p_k$, $k = 1, 2, \ldots, m$. *We set* $\alpha := \alpha_1 + \cdots + \alpha_m$. *Then, the following statements are equivalent:*

(i)

$$\left\| \mathcal{R}_{\vec{\alpha}}(\vec{f}) \right\|_{L^q_v} \lesssim \prod_{k=1}^{m} \|f_k\|_{L^{p_k}};$$

(ii)

$$\left\|\mathcal{M}_{\vec{\alpha}}(\vec{f})\right\|_{L_v^q} \lesssim \prod_{k=1}^{m} \|f_k\|_{L^{p_k}};$$

(iii)

$$B_1 < \infty,$$

where B_1 *is defined by* (2.17)*;*

(iv)

$$B_3 < \infty,$$

where B_3 *is defined by* (2.18).

2.3 Comments to Chap. 2

The results of Sect. 2.1 are taken from Imerlishvili and Meskhi [150] (see also Korobkov and Kristensen [197] for different approach). We refer also to de Almeida and Lima [70], Liu and Xiao [212], Xiao [337], and Bhat and Kosuru [27] for related topics.

Section 2.2 is based on the paper of Edmunds et al. [95] (see also Meskhi [244]).

Chapter 3
Integral Transforms Associated with Measures

From the beginning of this century, a considerable attention of researchers was attracted to the study of the mapping properties of integral operators defined on metric measure spaces with non-doubling measure (see, e.g., the monographs Volberg [330], Tolsa [322], Edmunds et al. [92], Yang et al. [340] and references cited therein).

In this chapter we give a complete characterization of a measure μ governing the boundedness of fractional integral operator:

$$J_{\gamma,\mu} f(x) = \int_X \frac{f(y)}{d(x,y)^{1-\gamma}} d\mu(y), \quad 0 < \gamma < 1,$$

defined on a quasi-metric measure space (X, d, μ) (nonhomogeneous space) from one grand Lebesgue spaces $L^{p),\theta_1}_{\mu}(X)$ into another $L^{q),\theta_2}_{\mu}(X)$. As a corollary, we have a generalization of the Sobolev inequality for $J_{\alpha,\mu}$. D. Adams trace inequality (i.e., $L^{p),\theta_1}_{\mu}(X) \to L^{q),\theta_2}_{\nu}(X)$-boundedness for $J_{\alpha,\mu}$) is also derived for these operators in grand Lebesgue/Morrey spaces. In the case of Morrey spaces, we assume that the underlying sets of spaces might be of infinite measure. Under some additional conditions on a measure, we investigate the sharpness of the second parameter θ_2 in the target space.

Necessary and sufficient conditions on a measure μ guaranteeing the boundedness of the multilinear fractional integral operator $T^{(m)}_{\gamma,\mu}$ (defined with respect to a measure μ) from the product of Lorentz spaces $\prod_{k=1}^{m} L^{r_k,s_k}_{\mu}(X)$ to the Lorentz space $L^{p,q}_{\mu}(X)$ are established. From these results we have the similar results for linear fractional integrals $J_{\gamma,\mu}$ (i.e., for $m = 1$). From the general results, we have a criterion for the validity of Sobolev-type inequality in Lorentz spaces defined for non-doubling measures. Finally, we investigate the same problem for Morrey–Lorentz spaces.

A. Meskhi et al., *Integral Operators in Non-Standard Function Spaces*,
Operator Theory: Advances and Applications 310,
https://doi.org/10.1007/978-3-032-03691-9_3

Furthermore, we study the generalized fractional integral transform T_φ associated with a measure on a quasi-metric space. We give a characterization of those measures for which these operators are bounded between L_p-spaces defined on nonhomogeneous spaces. The key in the proof of one of the main theorems is the boundedness of the modified sublinear Hardy–Littlewood maximal operator in the classical Lebesgue space with general measure. We also provide necessary and sufficient conditions for some classes of integral operators to be bounded from Lorentz to Marcinkiewicz spaces.

We also establish necessary and sufficient conditions for the compactness of fractional integral operators from $L^p_\mu(X)$ to $L^q_\mu(X)$ with $1 < p < q < \infty$, where μ is a measure on a quasi-metric measure space X. As an application, we obtain criteria for the compactness of fractional integral operators defined in weighted Lebesgue spaces over bounded domains of the Euclidean space $\mathbb{R}^n$ with the Lebesgue measure, and also for the fractional integral operator associated with rectifiable curves of the complex plane.

Moreover, necessary and sufficient condition for exponent q governing the compactness of K_α (a fractional type integral operator defined on a space of homogeneous type (X, d, μ)) from $L^p_\mu(X)$ to $L^q_\mu(X)$ is established.

Additionally, we investigate the multilinear variants of the quantities which measure the noncompactness of multilinear operators taking values in Banach spaces with the uniform approximation property. We show applications to multilinear variant of the Hilbert and Riesz transform on rearrangement invariant spaces. We derive lower estimates of the essential norm of these transforms. Along the way we obtain also, as a by-product, the lower estimates of the measure noncompactness of these transforms. As a consequence we conclude that the Hilbert and Riesz transforms are not compact multilinear transforms.

We end the chapter deriving lower two-weight estimates for the essential norm (measure of noncompactness) for multilinear Hilbert and Riesz transforms, and Riesz potential operators in $BFLs$. As a corollary we have appropriate results in weighted Lebesgue spaces. From these statements we conclude that there is no $m+1$ tuple of weights $(v, w_1, \dots, w_m)$ for which these operators are compact from $L^{p_1}_{w_1} \times \cdots \times L^{p_m}_{w_m}$ to L^q_v.

3.1 Fractional Integrals in Grand Lebesgue and Morrey Spaces

Let (X, d, μ) be a quasi-metric measure space ($QMMS$ briefly), i.e., it is a topological space endowed with a locally finite complete measure μ and quasi-metric $d : X \times X \to \mathbb{R}_+$ satisfying the following conditions:

(i) $d(x, y) = 0$ if and only if $x = y$;
(ii) $d(x, y) = d(y, x)$ for all $x, y \in X$;

(iii) There is a constant $\kappa > 1$ such that $d(x,y) \leqslant \kappa\,(d(x,z) + d(z,y))$ for all $x, y, z \in X$.

In what follows, we will assume that the balls $B(x,r) := \{y \in X;\ d(x,y) < r\}$ are measurable with positive and finite μ measure for all $x \in X$ and $r > 0$. It is also assumed that for any neighborhood V of every $x \in X$, there exists $r > 0$ such that the ball $B(x,r)$ is contained in V.

We denote by d_X the diameter of X, i.e., $d_X := \sup\{d(x,y) : x, y \in X\}$.

Let β be a positive constant. We say that a measure is *upper Ahlfors β-regular* if there is a positive constant C_0 such that for all $x \in X$ and $r \in (0, d_X)$:

$$\mu B(x,r) \leqslant C_0 r^{\beta}. \tag{3.1}$$

holds. It is easy to see that if a measure μ satisfies the condition (3.1), then (X, d, μ) does not contain any atoms. Furthermore, if X is bounded, then the condition (3.1) for some positive β guarantees that $\mu(X) < \infty$.

Let $A(\cdot)$ be a function on $(0, p-1]$, $1 < p < \infty$, such that it is positive bounded on $(0, p-1]$ and $\lim_{x\to 0} A(x) = 0$. The class of such functions will be denoted by $\mathcal{A}^{(p)}$.

Let $\mu(X) < \infty$. Denote by $L^{p),A(\cdot)}_{\mu}(X)$ (or simply $L^{p),A(\cdot)}_{\mu}$) the grand Lebesgue space defined by the norm:

$$\|f\|_{L^{p),A(\cdot)}_{\mu}} = \sup_{0<\eta\leqslant p-1} A(\eta)^{\frac{1}{p-\eta}} \|f\|_{L^{p-\eta}_{\mu}(X)},$$

where $L^{r}_{\mu}(X)$, $1 \leqslant r < \infty$, is the classical Lebesgue space with respect to a measure μ, and defined by the standard norm

$$\|f\|_{L^{r}_{\mu}(X)} = \left(\int_X |f(x)|^r d\mu(x)\right)^{1/r}.$$

If $A(x) = \frac{x^{\theta}}{\mu(X)}$, where $\theta > 0$, then we denote:

$$\|\cdot\|_{L^{p),A(\cdot)}_{\mu}(X)} := \|\cdot\|_{L^{p),\theta}_{\mu}(X)}.$$

We will need the following notation:

$$\beta := \frac{pq(1-\gamma)}{pq + p - q}, \quad 1 < p < q < \infty; \tag{3.2}$$

and

$$\alpha := 1 - \frac{1-\gamma}{\beta}. \tag{3.3}$$

It can be checked that $0 < \alpha < 1$.

One of our main aims in this section is to give a complete characterization of measures guaranteeing the boundedness of $J_{\gamma,\mu}$ from the grand Lebesgue space $L_{\mu}^{p),\theta}(X)$ into another one $L_{\mu}^{q),\theta q/p}(X)$. It should be noticed that the second parameter in the target space cannot be less than $q\theta/p$ (see Proposition 3.5). From the general statement, we have Sobolev-type inequality for I_{γ}.

The following statement will be useful for us.

Theorem 3.1 (Kokilashvili and Meskhi [171]) *Let (X, d, μ) be a $QMMS$ and let $1 < p < q < \infty$, $0 < \gamma < 1$. Then $J_{\gamma,\mu}$ is bounded from $L_{\mu}^{p}(X)$ to $L_{\mu}^{q}(X)$ if and only if μ is upper Ahlfors β regular, where*

$$\beta = \frac{(1-\gamma)pq}{pq+p-q}. \tag{3.4}$$

Theorem 3.1 implies the next statement (see García-Cuerva and Gatto [110], Kokilashvili and Meskhi [171]):

Corollary 3.2 *If $1 < p < \infty$, $0 < \gamma < \frac{1}{p}$ and $q = \frac{p}{1-\gamma p}$, then $J_{\gamma,\mu}$ is bounded from $L_{\mu}^{p}(X)$ to $L_{\mu}^{q}(X)$ if and only if μ is upper Ahlfors 1-regular, i.e.,*

$$\mu B(x,r) \leqslant cr \tag{3.5}$$

with the positive constant c independent of $x \in X$ and $r > 0$.

Adams [3]-type trace inequalities are also established in grand Lebesgue spaces for fractional integral operator $J_{\gamma,\mu}$ and also for the following potential operator:

$$I_{\gamma,X} f(x) = \int\limits_{X} \frac{f(y)}{\mu B(x, d(x,y))^{1-\gamma}} d\mu(y), \quad 0 < \gamma < 1.$$

Finally, we derive the boundedness of fractional integrals with measure in grand Morrey spaces. The results are obtained in the case when μ is finite measure; however, some statements are true for the case when μ is σ-finite. All the results of this section are proved in the case when μ is a non-doubling measure.

3.1.1 Boundedness of $J_{\gamma,\mu}$ in Grand Lebesgue Spaces

Let $\theta > 0$. We set:

$$\varphi(x) := \left[\frac{x-q}{1-\alpha(x-q)} + p\right]^{1-(x-q)\alpha}, \quad 1 < p < q < \infty, \tag{3.6}$$

$$\psi(x) := \big(\varphi(x)\big)^{\theta}. \tag{3.7}$$

It can be checked that (see also Meskhi [242, Lemma 3.1])

$$\varphi(x) \sim x^{1+\alpha q}, \quad \text{as } x \to 0. \tag{3.8}$$

Consequently,

$$\psi(x) \sim x^{\theta(1+\alpha q)}, \quad \text{as } x \to 0. \tag{3.9}$$

Theorem 3.3 *Let $\mu(X) < \infty$, $1 < p < q < \infty$ and Let $0 < \gamma < 1$. Suppose that $\theta > 0$. Then $J_{\gamma,\mu}$ is bounded from $L^{p),\theta}_{\mu}(X)$ to $L^{q),\frac{q\theta}{p}}_{\mu}(X)$ if and only if μ is upper Ahlfors β-regular (i.e.,* (3.1) *holds) with β defined by* (3.2).

Proof Observe that the $L^{p),\theta}_{\mu}(X) \mapsto L^{q),\frac{q\theta}{p}}_{\mu}(X)$ boundedness is equivalent to the $L^{p),\theta}_{\mu}(X) \mapsto L^{q),\psi(\cdot)}_{\mu}(X)$ one. Indeed, (3.2) and (3.3) yield that $0 < \alpha < 1/p$ and $q = \frac{p}{1-\alpha p}$. Hence, $1 + \alpha p = \frac{p}{q}$.

Necessity. Let $J_{\gamma,\mu}$ be bounded from $L^{p),\theta}_{\mu}(X)$ to $L^{q),\psi(\cdot)}_{\mu}(X)$. Take any ball $B \subset X$ and let $f_B = \chi_B$. Then for $x \in B$

$$J_{\gamma,\mu} f_B(x) \geqslant \int\limits_X \frac{\chi_B(y)}{d(x,y)^{1-\gamma}} d\mu(y) \geqslant c_\gamma r^{\gamma-1}\mu(B)$$

Hence,

$$\|J_{\gamma,\mu} f_B\|_{L^{q),\psi(\cdot)}_{\mu}(X)} \geqslant c_\gamma r^{\gamma-1} \cdot \mu(B) \|\chi_B\|_{L^{q),\psi(\cdot)}_{\mu}(X)}.$$

Taking into account that $J_{\gamma,\mu}$ is bounded from $L^{p),\theta}_{\mu}(X)$ to $L^{q),\frac{q\theta}{p}}_{\mu}(X)$, we have:

$$c_\gamma r^{\gamma-1}\mu(B)\|\chi_B\|_{L^{q),\psi(\cdot)}_{\mu}(X)} \leqslant c\|\chi_B\|_{L^{p),\theta}_{\mu}(X)}.$$

Since the function $R(\varepsilon) := \left(\varepsilon^\theta \mu(B)\right)^{\frac{1}{p-\varepsilon}}$ attains its maximum on the interval $[0, p-1]$ and $R(0) = 0$, we have that there is a number $\eta_B \in (0, p-1]$ such that

$$\|\chi_B\|_{L^{p),\theta}_{\mu}(X)} = \sup_{0<\varepsilon\leqslant p-1} \left(\varepsilon^\theta \mu(B)\right)^{\frac{1}{p-\varepsilon}} = \left(\eta_B^\theta \mu(B)\right)^{\frac{1}{p-\eta_B}}.$$

Hence,

$$c\frac{\mu(B)}{r^{1-\gamma}}\|\chi_B\|_{L^{q),\psi(\cdot)}_{\mu}} \leqslant c\eta_B^{\frac{\theta}{p-\eta_B}} (\mu(B))^{\frac{1}{p-\eta_B}}. \tag{3.10}$$

Let us choose ε_B such that

$$\frac{1}{p-\eta_B} - \frac{1}{q-\varepsilon_B} = \alpha, \tag{3.11}$$

where α is defined by (3.3).

Without loss of generality, we can assume that $\varepsilon_B \in (0, q-1]$. By virtue of (3.10), we get:

$$c_\gamma \frac{\mu(B)}{r^{1-\gamma}} \mu(B)^{\frac{1}{q-\varepsilon_B} - \frac{1}{p-\eta_B}} \eta_B^{-\frac{\theta}{p-\eta_B}} \psi(\varepsilon_B)^{\frac{1}{q-\varepsilon_B}} \leqslant c.$$

By (3.8) we get that $\eta_B^{-\frac{\theta}{p-\eta_B}} \psi(\varepsilon_B)^{\frac{1}{q-\varepsilon_B}} \sim 1$. Consequently, due to (3.11) we find that

$$c_\gamma \frac{\mu(B)}{r^{1-\gamma}} \mu(B)^{-\alpha} \leqslant c.$$

Since $0 < \alpha < 1$, we have that

$$\mu(B) \leqslant c r^{\frac{1-\gamma}{1-\alpha}} = c r^{\beta}.$$

Sufficiency. Since $\mu(X) < \infty$, by Hölder's inequality and (3.8) it is enough to see that

$$\sup_{0<\varepsilon<\varepsilon_0} \psi(\varepsilon)^{\frac{1}{q-\varepsilon}} \|J_{\gamma,\mu} f\|_{L_\mu^{q-\varepsilon}(X)} \leqslant c \|f\|_{L_\mu^{p),\theta}(X)}$$

for some small positive number $\varepsilon_0 \in (0, q-1)$, where ψ is defined by (3.7).

Take ε_0. Choose a number η_0 such that

$$\beta = \frac{(p-\eta_0)(q-\varepsilon_0)(1-\gamma)}{(p-\eta_0)(q-\varepsilon_0) + p - \eta_0 - (q-\varepsilon_0)}.$$

This is equivalent to

$$\frac{1}{p-\eta_0} - \frac{1}{q-\varepsilon_0} = \alpha.$$

Thus

$$\frac{1}{p} - \frac{1}{q} = \frac{1}{p-\eta_0} - \frac{1}{q-\varepsilon_0} = \alpha$$

(recall that $\alpha = 1 - \frac{1-\gamma}{\beta}$).

Let $0 < t < 1$. Define $0 < \varepsilon < \varepsilon_0$ and $0 < \eta < \eta_0$ so that the equalities

$$\frac{1}{p-\eta} = \frac{t}{p} + \frac{1-t}{p-\eta_0}, \qquad \frac{1}{q-\varepsilon} = \frac{1}{q} + \frac{1-t}{q-\varepsilon_0}$$

hold. Then for such ε and η, we have:

$$\begin{aligned}\frac{1}{p-\eta} - \frac{1}{q-\varepsilon} &= \frac{t}{p} - \frac{t}{q} + \frac{1-t}{p-\eta_0} + \frac{1-t}{q-\varepsilon_0} \\ &= t\alpha + (1-t)\alpha = \alpha.\end{aligned}$$

Thus, for the same ε and η, we get:

$$\beta = \frac{(p-\eta)(q-\varepsilon)(1-\gamma)}{(p-\eta)(q-\varepsilon) + p - \eta - (q-\varepsilon)}.$$

By virtue of Theorem 3.1, we have the following two boundedness:

$$J_{\gamma,\mu} : L^p_\mu(X) \longrightarrow L^q_\mu(X)$$

and

$$J_{\gamma,\mu} : L^{p-\eta_0}_\mu(X) \longrightarrow L^{q-\varepsilon_0}_\mu(X).$$

Consequently, the Riesz–Thorin interpolation theorem yields the boundedness:

$$J_{\gamma,\mu} : L^{p-\eta}_\mu(X) \longrightarrow L^{q-\varepsilon}_\mu(X)$$

with

$$\|J_{\gamma,\mu}\|_{L^{p-\eta}_\mu(X) \longrightarrow L^{q-\varepsilon}_\mu(X)} \leqslant \|J_{\gamma,\mu}\|_{L^p_\mu(X) \longrightarrow L^q_\mu(X)} \|J_{\gamma,\mu}\|_{L^{p-\eta_0}_\mu(X) \longrightarrow L^{q-\varepsilon_0}_\mu(X)}.$$

Consequently, for $0 < \eta < \eta_0$,

$$\begin{aligned}\psi(\varepsilon)^{\frac{1}{q-\varepsilon}} \|J_{\gamma,\mu} f\|_{L^{q-\varepsilon}_\mu(X)} &\leqslant c\psi(\varepsilon)^{\frac{1}{q-\varepsilon}} \|f\|_{L^{p-\eta}_\mu(X)} \leqslant c\eta^{\frac{\theta}{p-\eta}} \|f\|_{L^{p-\eta}_\mu(X)} \\ &\leqslant c\|f\|_{L^{p),\theta}_\mu(X)}.\end{aligned}$$

Here we used the fact that, since $\frac{1}{p-\eta} - \frac{1}{q-\varepsilon} = \alpha$

$$\psi(\varepsilon)^{\frac{1}{q-\varepsilon}} \cdot \eta^{-\frac{\theta}{p-\eta}} \sim 1$$

(see also Meskhi [242, Lemma 3.1]).

Taking the supremum with respect to ε in the latter inequality, we conclude that the theorem has been proved. □

We have the following Sobolev-type inequality which is a generalization of Corollary 3.2 in the case when μ is a finite measure.

Corollary 3.4 *Let $\mu(X) < \infty$, $1 < p < q < \infty$ and let $0 < r < \frac{1}{p}$. We set $q = \frac{p}{1-\gamma p}$. Suppose that $\theta > 0$. Then, there is a positive constant C such that for all $f \in L_{\mu}^{p),\theta}(X)$, the inequality*

$$\|J_{\gamma,\mu} f\|_{L_{\mu}^{q),\frac{q\theta}{p}}(X)} \leqslant C \|f\|_{L_{\mu}^{p),\theta}(X)}$$

holds if and only if condition (3.5) *is satisfied for μ.*

The next statement gives an information about the second parameter of grand Lebesgue spaces from the $J_{\gamma,\mu} : L_{\mu}^{p),\theta_1}(X) \to L_{\mu}^{q),\theta_2}(X)$ boundedness viewpoint (see Meskhi [242] for Euclidean spaces).

Proposition 3.5 *Let $1 < p < q < \infty$ and $0 < \gamma < 1$. Suppose that (X, d, μ) be a quasi-metric measure space (see Sect. 2.1.1) without atoms. Let there exist a positive constant b such that for all $x \in X$ and $r \in (0,\ diam\ (X))$:*

$$\mu(B(x,r)) \geqslant b r^{\beta}, \tag{3.12}$$

where β is defined by (3.2)*. Then, the boundedness of $J_{\gamma,\mu}$ from $L_{\mu}^{p),\theta_1}(X)$ to $L_{\mu}^{q),\theta_2}(X)$ implies that $\theta_2 \geqslant \frac{\theta_1 q}{p}$.*

Proof We follow the proof of Theorem 2.1 of Meskhi [242]. Suppose that $J_{\gamma,\mu}$ is bounded from $L_{\mu}^{p),\theta_1}(X)$ to $L_{\mu}^{q),\theta_2}(X)$, i.e., the inequality

$$\|J_{\gamma,\mu} f\|_{L_{\mu}^{q),\theta_2}(X)} \leqslant c \|f\|_{L_{\mu}^{p),\theta_1}(X)} \tag{3.13}$$

holds, where the positive constant c does not depend on f. Let $B := B(a,r)$, where $a \in X$ and let $f = \chi_B$ in (3.13). Observe that

$$\|J_{\gamma,\mu} f\|_{L_{\mu}^{q),\theta_2}(X)} \geqslant c r^{\gamma-1} \mu(B) \|\chi_B\|_{L_{\mu}^{q),\theta_2}(X)}.$$

Applying inequality (3.13) we get:

$$r^{\gamma-1} \mu(B) \|\chi_B\|_{L_{\mu}^{q),\theta_2}(X)} \leqslant c \|\chi_B\|_{L_{\mu}^{p),\theta_1}(X)}, \tag{3.14}$$

where the positive constant c does not depend on B.

Let us define the number $\varepsilon_B \in (0, p-1]$ satisfying the condition:

$$\sup_{0<\varepsilon\leqslant p-1} \left(\varepsilon^{\theta_1}\mu(B)\right)^{\frac{1}{p-\varepsilon}} = \left(\varepsilon_B^{\theta_1}\mu(B)\right)^{\frac{1}{p-\varepsilon_B}}. \tag{3.15}$$

Further, it can be checked that (see also the proof of this fact in Meskhi [242, p.122] for intervals, but the proof for the case of quasi-metric measure space is the same):

$$\lim_{\mu(B)\to 0} \varepsilon_B = 0.$$

Further, we choose η_B so that

$$\alpha = 1 - \frac{1-\gamma}{\beta} = \frac{1}{p} - \frac{1}{q} = \frac{1}{p-\varepsilon_B} - \frac{1}{q-\eta_B}.$$

This is equivalent to say that

$$\eta_B = q - \frac{p-\varepsilon_B}{1-\alpha(p-\varepsilon_B)}. \tag{3.16}$$

By (3.14) and (3.15), we get:

$$r^{\gamma-1}\mu(B)\eta_B^{\frac{\theta_2}{q-\eta_B}}\mu(B)^{\frac{1}{q-\eta_B}} \leqslant c\varepsilon_B^{\frac{\theta_1}{p-\varepsilon_B}}\mu(B)^{\frac{1}{p-\varepsilon_B}}.$$

Hence, (3.1.1) and (3.12) imply:

$$\eta_B^{\frac{\theta_2}{q-\eta_B}}\varepsilon_B^{-\frac{\theta_1}{p-\varepsilon_B}} \leqslant c. \tag{3.17}$$

Further, (3.16) and (3.17) yield:

$$\left(\frac{q-\frac{p-\varepsilon_B}{1-\alpha(p-\varepsilon_B)}}{\varepsilon_B}\right)^{\frac{\theta_2}{p-\varepsilon_B}-\alpha\theta_2} \varepsilon_B^{-\frac{\theta_1}{p-\varepsilon_B}+\frac{\theta_2}{p-\varepsilon_B}-\alpha\theta_2} \leqslant c.$$

Passing now to the limit as $\mu(B) \to 0$, we see that the left-hand side of the latter inequality goes to $+\infty$ unless $\theta_2 \geqslant \frac{\theta_1 q}{p}$.

The statement has been proved. □

Analyzing the proof of Theorem 3.3, we can formulate the next statement which is more general in the sense of the second parameter.

Theorem 3.6 *Let* $1 < p < q < \infty$ *and Let* $0 < \gamma < 1$. *Suppose that* $A(\cdot) \in \mathcal{A}^{(q)}$. *We set:*

$$\overline{A}(x) := \left[A\left(q - \frac{p-x}{1-\alpha(p-x)}\right)\right]^{1-\alpha(p-x)},$$

where α *is defined by* (3.3). *Then* $J_{\gamma,\mu}$ *is bounded from* $L_{\mu}^{p),\overline{A}(\cdot)}(X)$ *to* $L_{\mu}^{q),A(\cdot)}(X)$ *if and only if* μ *is upper Ahlfors* β*-regular (i.e.,* (3.1) *holds) with* β *defined by* (3.2).

3.1.2 D. Adams-Type Trace Inequality for Fractional Integrals

We start this section with the following statement.

Theorem 3.7 (Edmunds et al. [92, p. 375]) *Let* (X, d, μ) *be a quasi-metric measure space (see Sect. 2.1.1 for the definition) and let* $1 < p < q < \infty$. *Suppose that* ν *is another measures on* X *satisfied the condition: there are constants* $c > 0$, $\beta > 0$ *such that for all* $x \in X$ *and* $r > 0$*:*

(a) the inequality (3.1) *holds;*
(b) the inequality

$$\nu B(x,r) \leqslant c r^{(1-\gamma-\beta/p')q} \tag{3.18}$$

holds, where $p' = \frac{p}{p-1}$ *and* $1 - \beta < \gamma < 1 - \frac{\beta}{p'}$.

Then $J_{\gamma,\mu}$ *is bounded from* $L_{\mu}^{p}(X)$ *to* $L_{\nu}^{q}(X)$.

Now we prove Adams [3]-type theorem in grand Lebesgue spaces defined on quasi-metric measure spaces.

Theorem 3.8 *Let* (X, d, μ) *be a quasi-metric measure space (see Sect. 2.1.1 for the definition) with bounded* X. *Suppose that the conditions of Theorem 3.7 are satisfied and* $\theta > 0$. *Then* $J_{\gamma,\mu}$ *is bounded from* $L_{\mu}^{p),\theta}(X)$ *to* $L_{\nu}^{q),\frac{q\theta}{p}}(X)$.

Proof First of all observe that conditions (3.1) and (3.18) imply that μ and ν are finite measures. Let (3.1) and (3.18) hold. Choose small positive numbers ε_0 and η_0 such that

$$\left(\frac{1-\gamma}{\beta} - \frac{1}{(p-\eta_0)'}\right)(q - \varepsilon_0) = A,$$

where

$$A := \left(\frac{1-\gamma}{\beta} - \frac{1}{p'}\right)q.$$

Observe that these equalities are equivalent to

$$\frac{1}{p-\eta_0} - \frac{A}{q-\varepsilon_0} = \alpha \quad \text{and} \quad \frac{1}{p} - \frac{A}{q} = \alpha$$

respectively, where α is defined by (3.3).

Let us choose $0 < \varepsilon < \varepsilon_0$ and $0 < \eta < \eta_0$ so that

$$\frac{t}{p-\eta_0} + \frac{1-t}{p} = \frac{1}{p-\eta}; \quad \frac{t}{q-\varepsilon_0} + \frac{1-t}{q} = \frac{1}{q-\varepsilon}. \tag{3.19}$$

Multiplying the second equality of (3.19) by A and subtract it from the first inequality, we get:

$$\frac{t}{p-\eta_0} - \frac{At}{q-\varepsilon_0} + \frac{1-t}{p} - \frac{A(1-t)}{q} = \frac{1}{p-\eta} - \frac{A}{q-\varepsilon}.$$

The left-hand side of this equality is equal to

$$t\left[\frac{1}{p-\eta_0} - \frac{A}{q-\varepsilon_0}\right] + (1-t)\left[\frac{1}{p} - \frac{A}{q}\right] = t\alpha + (1-t)\alpha = \alpha.$$

Hence,

$$\frac{1}{p-\eta} - \frac{A}{q-\varepsilon} = \alpha \tag{3.20}$$

which is equivalent to

$$\left(\frac{1-\gamma}{\beta} - \frac{1}{(p-\eta)'}\right)(q-\varepsilon) = A.$$

By Theorem 3.7 we have the boundedness:

$$J_{\gamma,\mu} : L^p_\mu(X) \longrightarrow L^q_\nu(X); \quad J_{\gamma,\mu} : L^{p-\eta_0}_\mu(X) \longrightarrow L^{q-\varepsilon_0}_\nu(X).$$

Then, for all $0 < \varepsilon < \varepsilon_0$ and $0 < \eta < \eta_0$, by Riesz–Thorin interpolation theorem, the boundedness

$$J_{\gamma,\mu} : L^{p-\eta}_\mu(X) \longrightarrow L^{q-\varepsilon}_\nu(X),$$

holds, where ε and η satisfy (3.20) and

$$\|J_{\gamma,\mu}\|_{L^{p-\eta}_\mu \to L^{q-\varepsilon}_\nu} \leqslant \|J_{\gamma,\mu}\|_{L^p_\mu \to L^q_\nu} \|J_{\gamma,\mu}\|_{L^{p-\eta_0}_\mu \to L^{q-\varepsilon_0}_\nu}. \tag{3.21}$$

Observe that condition $1-\beta < \gamma < 1 - \frac{\beta}{p'}$ guarantees that $0 < \alpha < 1$.

Introduce the function:

$$D(\varepsilon) := p - \frac{q-\varepsilon}{A+\alpha(q-\varepsilon)},$$

where $\varepsilon \in (0, \varepsilon_0)$ for some small ε_0.

Inverse of $D(\varepsilon)$ is given by the formula:

$$B(\eta) = q - \frac{(p-\eta)A}{1-\alpha(p-\eta)}.$$

It is easy to see that

$$\lim_{\varepsilon\to 0} D(\varepsilon) = \lim_{\eta\to 0} B(\eta) = 0,$$

and

$$D(\varepsilon) \sim \varepsilon \quad \text{as} \quad \varepsilon \longrightarrow 0.$$

Introduce another function:

$$\overline{D}(\varepsilon) = D(\varepsilon)^{A+\alpha(q-\varepsilon)}.$$

It is easy to check that

$$\overline{D}(\varepsilon) \sim \varepsilon^{q/p} \quad \text{as} \quad \varepsilon \to 0.$$

Let us show that the $L^{p),\theta}_{\mu}(X) \longrightarrow L^{q),\theta q/p}_{\nu}$ boundedness holds. For that it is enough to show that

$$\|J_{\gamma,\mu} f\|_{L^{q),\tilde{D}(\cdot)}_{\nu}(X)} \leqslant c\|f\|_{L^{p),\theta}_{\mu}(X)},$$

where $\widetilde{D}(\varepsilon) = \left(\overline{D}(x)\right)^{\theta}$. Since $\widetilde{D}(\varepsilon) \sim x^{\theta q/p}$, we will be done.

By using inequality (3.21), we get:

$$\begin{aligned}
\widetilde{D}(\varepsilon)^{\frac{1}{q-\varepsilon}} \|J_{\gamma,\mu} f\|_{L^{q-\varepsilon}_{\nu}(X)} &\leqslant c\widetilde{D}(\varepsilon)^{\frac{1}{q-\varepsilon}} \|f\|_{L^{p-\eta}_{\mu}(X)} \\
&\leqslant c\eta^{\frac{\theta}{p-\eta}} \|f\|_{L^{p-\eta}_{\mu}(X)} \\
&\leqslant c\|f\|_{L^{p),\theta}_{\mu}(X)},
\end{aligned}$$

where the positive constant c does not depend on f.

Since $\nu(X) < \infty$ and the latter inequality holds for ε less than or equal to ε_0, by virtue of Hölder's inequality, we conclude that the desired result holds (see also sufficiency of the proof of Theorem 3.3 for similar arguments). □

The next statement was proved in Kokilashvili and Meskhi [171] (see also Edmunds et al. [92, p. 372]).

Theorem 3.9 *Let $1 < p < q < \infty$ and let $0 < \gamma < 1/p$. Suppose that $\mu B(x,r)$ is continuous with respect to r and let $\mu\{x\} = 0$ for all $x \in X$. If the pair of measures (μ, ν) satisfy the condition*

$$(\nu B(x, 6\kappa r))^{1/q} \leqslant c\,(\mu B(x,r))^{1/p-\gamma}$$

with the positive constant c independent of $x \in X$ and $r \in (0,\ d_X)$, then $I_{\gamma,X}$ is bounded from $L^p_\mu(X)$ to $L^q_\nu(X)$.

By using the proof of Theorem 3.8 and repeating the interpolation arguments, we have the next statement.

Theorem 3.10 *Suppose that the triple (X, d, μ) is a quasi-metric measure space with bounded X and that the conditions of Theorem 3.9 are satisfied. Let $\theta > 0$. Then $I_{\gamma,X}$ is bounded from $L^{p),\theta}_\mu(X)$ to $L^{q),q\theta/p}_\nu(X)$.*

Remark 3.11 Observe that in Theorems 3.8 and 3.10 the fact that X is bounded guarantees that ν is a finite measure which follows from conditions (3.18) and (3.9), respectively.

3.1.3 Boundedness in Grand Morrey Spaces

Let (X, d, μ) be a quasi-metric measure space and let $M^{p,r}_{\mu,\ell}(X)$ denote the Morrey space defined with respect to a measure μ which is the class at all measurable functions $f : X \to \mathbb{R}$ for which the norm

$$\|f\|_{M^{p,r}_{\mu,\ell}(X)} := \sup_{\substack{a\in X\\ t>0}} \frac{1}{t^{(1/p-1/r)\ell}} \|f\|_{L^p_\mu(B(a,t))}$$

$$:= \sup_{\substack{a\in X\\ t>0}} \frac{1}{t^{(1/p-1/r)\ell}} \left(\int\limits_{B(a,t)} |f(y)|^p d\mu(y) \right)^{1/p}$$

is finite, where $1 < p \leqslant r < \infty$, $\ell > 0$. If $p = r$, then $M^{p,p}_{\mu,\ell}(X)$ coincides with the Lebesgue space $L^p_\mu(X)$. If $\ell = 1$, then $M^{p,r}_{\mu,\ell}(X)$ is denoted by $M^{p,r}_\mu(X)$.

On the base of $M^{p,r}_{\mu,\ell}$, we denote grand Morrey space denoted by $M^{p),r,\varphi(\cdot)}_{\mu,\ell}(X)$ and defined by the norm:

$$\|f\|_{M^{p),r,\varphi(\cdot)}_{\mu,\ell}(X)} = \sup_{0<\varepsilon<p-1} \varphi(\varepsilon)\|f\|_{M^{p-\varepsilon,r}_{\mu,\ell}(X)},$$

where $\varphi \in \mathcal{A}^{(p)}$. If $\varphi(\varepsilon) = \varepsilon^\theta$, then we denote $M^{p),r,\varphi(\cdot)}_{\mu,\ell}(X)$ by $M^{p),r,\theta}_{\mu,\ell}(X)$.

Proposition 3.12 *Let $1 < s < p \leqslant r < \infty$ and let μ be upper Ahlfors $\ell-$ regular, i.e., (3.1) holds for $\beta = \ell$. Then, the following embeddings hold:*

$$L^r_\mu(X) \hookrightarrow M^{p,r}_{\mu,\ell}(X) \hookrightarrow M^{p),r,\varphi(\cdot)}_{\mu,\ell}(X) \hookrightarrow M^{s),r,\varphi(\cdot)}_{\mu,\ell}(X).$$

If $\mu(X) < \infty$, then

$$M^{p),r,\varphi(\cdot)}_{\mu,\ell}(X) \hookrightarrow \mathcal{L}^{p),\varphi(\cdot)}_\mu(X),$$

where $\mathcal{L}^{p),\varphi(\cdot)}_\mu(X)$ is the grand Lebesgue space defined by the norm

$$\|f\|_{\mathcal{L}^{p),\varphi(\cdot)}_\mu(X)} = \sup_{0<\varepsilon<p-1} \varphi(\varepsilon)\|f\|_{L^{p-\varepsilon}_\mu(X)}.$$

Proof The embedding $L^r_\mu(X) \hookrightarrow M^{p,r}_{\mu,\ell}(X)$ is a consequence of Hölder's inequality with respect to the exponent $\frac{r}{p}$ and the upper Ahlfors ℓ-regularity condition for μ. Indeed,

$$\frac{1}{t^{(1/p-1/r)\ell}}\left(\int\limits_{B(x,t)} |f(y)|^p d\mu(y)\right)^{1/p}$$

$$\leqslant \frac{1}{t^{(1/p-1/r)\ell}}\left(\int\limits_{B(x,t)} |f(y)|^r d\mu(y)\right)^{1/r} (\mu B(x,t))^{\frac{r-p}{pr}}$$

$$\leqslant c\frac{t^{(1/p-1/r)\ell}}{t^{(1/p-1/r)\ell}}\left(\int\limits_{B(x,t)} |f(y)|^r d\mu(y)\right)^{1/r} \leqslant c\|f\|_{L^r_\mu(X)}.$$

Taking the supremum over all $x \in X$ and $t > 0$, we have the desired result.

Let us check the embedding $M^{p,r}_{\mu,\ell} \hookrightarrow M^{p),r,\varphi(\cdot)}_{\mu,\ell}$. For that, observe that Hölder's inequality with respect to $\frac{p}{p-\varepsilon}$ yields:

$$\varphi(\varepsilon)t^{\left(\frac{1}{r}-\frac{1}{p-\varepsilon}\right)\ell}\left(\int\limits_{B(x,t)}|f(y)|^{p-\varepsilon}d\mu(y)\right)^{\frac{1}{p-\varepsilon}}$$

$$\leqslant \varphi(\varepsilon)t^{\left(\frac{1}{r}-\frac{1}{p-\varepsilon}\right)\ell}\left(\int\limits_{B(x,t)}|f(y)|^{p}d\mu(y)\right)^{\frac{1}{p}}(\mu(x,t))^{\frac{\varepsilon}{p(p-\varepsilon)}}$$

$$\leqslant \varphi(\varepsilon)t^{\left(\frac{1}{r}-\frac{1}{p-\varepsilon}\right)\ell}t^{\left(-\frac{1}{p}+\frac{1}{p-\varepsilon}\right)\ell}\|f\|_{L^p(B(x,t))}$$

$$\leqslant ct^{\left(\frac{1}{r}-\frac{1}{p}\right)\ell}\|f\|_{L^p(B(x,t))},$$

where $0<\varepsilon<p-1$. Taking the supremum with respect to ε and t on both sides of the latter inequality, we are done.

The embedding $M^{p),r,\varphi(\cdot)}_{\mu,\ell}(X)\hookrightarrow M^{s),r,\varphi(\cdot)}_{\mu,\ell}(X)$ follows similarly.

Let now $\mu(X)<\infty$. Then for $\varepsilon\in(0,p-1)$:

$$\varphi(\varepsilon)\left(\int\limits_{X}|f(y)|^{p-\varepsilon}d\mu(y)\right)^{\frac{1}{p-\varepsilon}}$$

$$\leqslant c\varphi(\varepsilon)\Big(d_X\Big)^{\left(\frac{1}{r}-\frac{1}{p-\varepsilon}\right)\ell}\left(\int\limits_{B(x,d_X}|f(y)|^{p-\varepsilon}d\mu(y)\right)^{\frac{1}{p-\varepsilon}}$$

$$\leqslant c\|f\|_{M^{p),r,\varphi(\cdot)}_{\mu,\ell}}.$$

Hence the result follows. □

The following theorem was proved in Eridani et al. [98, Theorem 3.4].

Theorem 3.13 *Let* $1<p<q<\infty$ *and let* $0<\gamma<1$. *Suppose that* μ *is upper Ahlfors* β*-regular, where* β *is defined by* (3.2). *Suppose that* $1<r,s<\infty$ *and let*

$$\frac{1}{p}-\frac{1}{r}=\frac{1}{q}-\frac{1}{s}. \tag{3.22}$$

Then there is a positive constant $\overline{C}:=\overline{C}(p,q,r,s,\beta,\gamma)$ *such that for all* $f\in M^{p,r}_{\mu}(X)$, *the inequality*

$$\|J_{\gamma,\mu}f\|_{M^{q,s}_{\mu,\beta}(X)}\leqslant\overline{C}\|f\|_{M^{p,r}_{\mu}(X)} \tag{3.23}$$

holds.

Moreover, observing the proof of Theorem 3.13, we conclude that the constant $\overline{C}$ in (3.23) is such that

$$\sup_{\substack{0<\eta<\eta_0 \\ 0<\varepsilon<\varepsilon_0}} \overline{C}(p-\eta, q-\varepsilon, r, s, \beta, \gamma) < \infty \tag{3.24}$$

for small positive numbers $0 < \eta_0 < p-1$, $0 < \varepsilon_0 < q-1$, where in (3.24) ε and η satisfy the condition

$$\frac{(p-\eta)(q-\varepsilon)(1-\gamma)}{(p-\eta)(q-\varepsilon)+(p-\eta)-(q-\varepsilon)} = \beta.$$

Our aim is to prove the following statement.

Theorem 3.14 *Let p, q, r, s, γ, and μ satisfy the conditions of Theorem 3.13. Let $\theta > 0$. Then $J_{\gamma,\mu}$ is bounded from $M_\mu^{p),r,\theta}(X)$ to $M_{\mu,\beta}^{q),s,\theta}(X)$.*

Proof Let a function $\overline{\varphi}$ defined on $[0, q-1)$ be given by the formula:

$$\overline{\varphi}(x) := \varphi(x)^{1/[1-(x-q)\alpha]},$$

where φ is defined by (3.6). We set $\overline{\psi}(x) := \left(\overline{\varphi}(x)\right)^\theta$. Then, it is easy to see that $\overline{\psi}(x) \sim x^\theta$, as $x \to 0$. We will show that the inequality

$$\|J_{\gamma,\mu} f\|_{L_{\mu,\beta}^{q),s,\overline{\psi}(\cdot)}(X)} \leqslant c\|f\|_{L_\mu^{p),r,\theta}(X)}$$

holds. We have:

$$\begin{aligned}
\|J_{\gamma,\mu} f\|_{L_{\mu,\beta}^{q),s,\overline{\psi}(\cdot)}(X)} &= \sup_{0<\varepsilon<q-1} \overline{\psi}(\varepsilon)\|J_{\gamma,\mu} f\|_{M_{\mu,\beta}^{q-\varepsilon,s}(X)} \\
&= \max\left\{ \sup_{0<\varepsilon\leqslant\sigma} (\cdots), \sup_{\sigma\leqslant\varepsilon<q-1} (\cdots) \right\} = \max\{S_1, S_2\},
\end{aligned}$$

where $\sigma \in (0, q-1)$.

Observe that Hölder's inequality and the upper Ahlfors β-regularity condition for μ give the estimates:

$$\begin{aligned}
S_2 &\leqslant \Big(\sup_{\sigma\leqslant\varepsilon<q-1} \overline{\psi}(\varepsilon)\Big) \sup_{\substack{x\in X,\\ t>0}} t^{\left(\frac{1}{s}-\frac{1}{q-\varepsilon}\right)\beta} \|J_{\gamma,\mu} f\|_{L_\mu^{q-\varepsilon}(B(x,t))} \\
&\leqslant \Big(\sup_{\sigma\leqslant\varepsilon<q-1} \overline{\psi}(\varepsilon)\Big) \sup_{\substack{x\in X,\\ t>0}} t^{\left(\frac{1}{s}-\frac{1}{q-\varepsilon}\right)\beta} \left(\int_{B(x,t)} |J_{\gamma,\mu} f(x)|^{q-\sigma} d\mu(x)\right)^{\frac{1}{q-\sigma}}
\end{aligned}$$

$$\times\,(\mu B(x,t))^{\frac{\varepsilon-\sigma}{(q-\sigma)(q-\varepsilon)}}$$

$$\leqslant c^{\frac{\varepsilon-\sigma}{(q-\sigma)(q-\varepsilon)}}\left(\sup_{\sigma\leqslant\varepsilon<q-1}\overline{\psi}(\varepsilon)\right)\sup_{\substack{x\in X,\\ t>0}} t^{\left(\frac{1}{s}-\frac{1}{q-\sigma}\right)\beta}\|J_{\gamma,\mu}f\|_{L^{q-\sigma}_{\mu}(X)}$$

$$\leqslant C_{\sigma,q}\sup_{0<\varepsilon\leqslant\sigma}\overline{\psi}(\varepsilon)\|J_{\gamma,\mu}f\|_{M^{q-\varepsilon,s}_{\mu,\beta}}.$$

Thus, for small positive σ

$$\|J_{\gamma,\mu}f\|_{L^{q),s,\overline{\psi}(\cdot)}_{\mu,\beta}(X)}\leqslant C_{\sigma,q}\sup_{0<\varepsilon\leqslant\sigma}\overline{\psi}(\varepsilon)\|J_{\gamma,\mu}f\|_{M^{q-\varepsilon,s}_{\mu,\beta}}.$$

By Theorem 3.13 and condition (3.22), we find that, for $0<\varepsilon\leqslant\sigma$ and $0<\eta\leqslant\delta$, satisfying the condition:

$$\frac{1}{p-\eta}-\frac{1}{q-\varepsilon}=\frac{1}{p}-\frac{1}{q}=\alpha$$

with α defined by (3.3)

$$\begin{aligned}
&\overline{\psi}(\varepsilon)\|J_{\gamma,\mu}f\|_{M^{q-\varepsilon,s}_{\mu,\beta}(X)}\\
&\quad\leqslant\overline{C}(p-\eta,q-\varepsilon,r,s,\gamma)\overline{\psi}(\varepsilon)\|f\|_{M^{p-\eta,r}_{\mu}(X)}\\
&\quad\leqslant\overline{C}(p-\eta,q-\varepsilon,r,s,\gamma)\overline{\psi}\left(\frac{\eta-p}{1-(p-\eta)\alpha}+q\right)\|f\|_{M^{p-\eta,r}_{\mu}(X)}\\
&\quad=\overline{C}(p-\eta,q-\varepsilon,r,s,\gamma)\eta^{\theta}\|f\|_{M^{p-\eta,r}_{\mu}(X)}\leqslant C\|f\|_{M^{p),r,\theta}_{\mu}(X)}.
\end{aligned}$$

Since $\varphi(x)\sim x$ near 0, we are done. □

Analyzing the proof of Theorem 3.14, we can formulate the next statement which is more general than the previous one from the second parameter viewpoint.

Theorem 3.15 *Suppose that p,q,r,s,γ, and μ satisfy conditions of Theorem 3.13. Suppose that $\overline{\psi}(\cdot)$ is an increasing function on $(0,q-1]$ such that $\lim_{x\to0+}\overline{\psi}(x)=0$. Then $J_{\gamma,\mu}$ is bounded from $M^{p),r,\overline{\psi}(\cdot)}_{\mu}(X)$ to $M^{q),s,\widetilde{\psi}(\cdot)}_{\mu,\beta}(X)$, where*

$$\widetilde{\psi}(\varepsilon)=\overline{\psi}\left(\frac{\eta-p}{1-(p-\eta)\alpha}+q\right).$$

3.2 Sobolev Inequality with Non-doubling Measure in Lorentz Spaces

In this section, criteria for a measure μ governing the boundedness of the multilinear fractional integral operator: $T^{(m)}_{\gamma,\mu}$ (defined with respect to a measure μ) from the product of Lorentz spaces $\prod_{k=1}^{m} L^{r_k,s_k}_{\mu}(X)$ to the Lorentz space $L^{p,q}_{\mu}(X)$ are established. The results regarding the boundedness of such operators in function spaces were mainly obtained under the growth condition on a measure. We investigate similar problem for Morrey–Lorentz spaces.

The brief history of multilinear fractional integral operators was discussed in Sect. 2.2.

Let (X, d, μ) be a quasi-metric measure space $(QMMS)$. Our aim is to characterize measures ensuring the boundedness of the fractional integral operator

$$T^{(m)}_{\gamma,\mu}(\vec{f})(x) = \int\limits_{X^m} \frac{f_1(y_1)\cdots f_m(y_m)d\mu(\vec{y})}{\big(d(x, y_1) + \cdots + d(x, y_m)\big)^{m-\gamma}}, \quad 0 < \gamma < m, \quad x \in X,$$

from the space $\prod_{j=1}^{m} L^{r_j,s_j}_{\mu}(X)$ to $L^{p,q}_{\mu}(X)$, where $\vec{f} := (f_1, \ldots, f_m)$, $d\mu(\vec{y}) := d\mu(y_1)\cdots d\mu(y_m)$, $L^{r_j,s_j}_{\mu}(X)$ and $L^{p,q}_{\mu}(X)$ are Lorentz spaces defined on an (X, d, μ).

This result is new even for the linear case $m = 1$. In particular, as a corollary, we have a complete characterization of a measure μ guaranteeing the boundedness of the fractional integral operator:

$$J_{\gamma,\mu}g(x) = \int\limits_{X} \frac{g(y)}{d(x, y)^{1-\gamma}} d\mu(y) \quad 0 < \gamma < 1, \ x \in X,$$

from $L^{r,s}_{\mu}(X)$ to $L^{p,q}_{\mu}(X)$. As a consequence of the main result, we have a generalization of the Sobolev-type inequality in Lorentz spaces. In particular, we give necessary and sufficient condition on a measure μ for which the inequality

$$\|J_{\gamma,\mu}g\|_{L^{p,q}_{\mu}(X)} \leqslant C\|g\|_{L^{r,s}_{\mu}(X)}, \quad p = \frac{r}{1-\gamma r}, \tag{3.25}$$

holds.

Similar results for Lebesgue spaces in the multilinear setting were derived in Kokilashvili et al. [189], and for the linear case go back to Kokilashvili and Meskhi [171] (see also Edmunds et al. [92], Ch. 6) for quasi-metric measure spaces. We refer also to García-Cuerva and Martell [111], García-Cuerva and Gatto [110] for the Sobolev-type inequalities in the classical Lebesgue spaces for non-doubling measure (see also Tao and He [318] for related topics in the multilinear setting).

3.2.1 Preliminaries

Let (X, d, μ) be a $QMMS$ (see Sect. 3.1 for the definition of a $QMMS$).

Recall that a measure μ is said to be Ahlfors upper β-regular if there is a positive constant C_0 such that condition (3.1) is satisfied for all $x \in X$ and $r \in (0, d_X)$, where d_X is the diameter of X. In this section we assume that might be of infinite measure.

Let f be a μ-measurable function on X and let $1 \leqslant p, s \leqslant \infty$. We say that f belongs to the Lorentz space $L^{p,s}_\mu(X)$ (or simply $L^{p,s}_\mu$) if

$$\|f\|_{L^{p,s}_\mu} = \begin{cases} \left(s \int\limits_0^\infty \left(\mu\{x \in X : |f(x)| > \tau\}\right)^{s/p} \tau^{s-1} d\tau \right)^{1/s}, & \text{if } s < \infty, \\ \sup\limits_{\tau>0} \tau \Big(\mu(\{x \in X : |f(x)| > \tau\}\Big)^{1/p}, & \text{if } s = \infty \end{cases}$$

is finite. It is easy to see that $L^{p,p}_\mu$ coincides with the Lebesgue space L^p_μ with measure μ.

Denote by f^*_μ a weighted nonincreasing rearrangement of f with respect to the measure μ. Then, by integration by parts, it can be checked that (see also Hunt [141])

$$\|f\|_{L^{p,s}_\mu} = \begin{cases} \left(\frac{s}{p} \int\limits_0^\infty \left(t^{1/p} f^*_\mu(t)\right)^s \frac{dt}{t} \right)^{1/s}, & \text{if } s < \infty, \\ \sup\limits_{t>0}\{t^{1/p} f^*_\mu(t)\}, & \text{if } s = \infty. \end{cases}$$

Hölder's inequality in Lorentz spaces has the form:

$$\left| \int\limits_X fg d\mu \right| \leqslant \|f\|_{L^{r,s}_\mu} \|g\|_{L^{r',s'}_\mu}, \quad r' = \frac{r}{r-1}, \quad s' = \frac{s}{s-1}. \tag{3.26}$$

The following inequality will be useful for us:

$$\|f\|_{L^{r,s}_\mu} \leqslant \prod_{j=1}^m \|f_j\|_{L^{r_j,s_j}_\mu}, \quad \frac{1}{r} = \sum_{j=1}^m \frac{1}{r_j}, \quad \frac{1}{s} = \sum_{j=1}^m \frac{1}{s_j}. \tag{3.27}$$

Let us denote by $\widetilde{\mathcal{M}}$ the modified maximal operator defined on X (see Kokilashvili and Meskhi [171], Edmunds et al. [92, p. 368]):

$$\widetilde{\mathcal{M}}g(x) = \sup_{R>0} \frac{1}{B(x, 3\kappa R)} \int\limits_{B(x,R)} |g(y)| d\mu(y),$$

where g is the locally integrable function on X and κ is the quasi-metric constant.

We will need some auxiliary results.

Lemma 3.16 *Let* $1 < r, s < \infty$. *Then* $\widetilde{\mathcal{M}}$ *is bounded in* $L^{r,s}_{\mu}$.

Proof We know that $\widetilde{\mathcal{M}}$ is of strong (∞, ∞) type and weak $(1, 1)$ type (see Kokilashvili and Meskhi [171]). Then, by the Stein–Weiss interpolation theorem (see Stein and Weiss [314, Theorem 3.15]), $\widetilde{\mathcal{M}}$ is bounded in $L^{r,s}_{\mu}$. □

Lemma 3.17 *Let* $1 < r_j, s_j < \infty$, $j = 1, \ldots, m$. *Then the modified multisublinear maximal operator*

$$\widetilde{\mathcal{M}}^{(m)}(\vec{f})(x) = \sup_{R>0} \prod_{k=1}^{m} \frac{1}{\mu(B(x, 3\kappa R))} \int_{B(x,R)} |f_k(y)| d\mu(y), \quad x \in X, \tag{3.28}$$

is bounded from $\prod_{j=1}^{m} L^{r_j,s_j}_{\mu}$ *to* $L^{r,s}_{\mu}$, *where* $\frac{1}{r} = \sum_{k=1}^{m} \frac{1}{r_k}$, *and* $\frac{1}{s} = \sum_{k=1}^{m} \frac{1}{s_k}$.

Proof By inequality (3.27) we have that

$$\left\| \widetilde{\mathcal{M}}^{(m)}(\vec{f}) \right\|_{L^{r,s}_{\mu}} \leqslant \left\| \prod_{j=1}^{m} \widetilde{\mathcal{M}} f_j \right\|_{L^{r,s}_{\mu}} \leqslant \prod_{j=1}^{m} \left\| \widetilde{\mathcal{M}} f_j \right\|_{L^{r_j,s_j}_{\mu}} \leqslant c \prod_{j=1}^{m} \|f_j\|_{L^{r_j,s_j}_{\mu}},$$

which ends the proof. □

3.2.2 Main Results

Now we formulate and prove our main statements of this section.

Theorem 3.18 *Let* $m \in \mathbb{N}$ *and let* (X, d, μ) *be a quasi-metric measure space. Suppose that* $0 < \gamma < m$, $1 < r_j, s_j < \infty$, $j = 1, \ldots, m$. *We set* $\frac{1}{r} = \sum_{j=1}^{m} \frac{1}{r_j}$ *and* $\frac{1}{s} = \sum_{j=1}^{m} \frac{1}{s_j}$. *Suppose that* $1 < r < p < \infty$ *and* q *be such that* $\frac{r}{p} = \frac{s}{q}$. *Then the inequality*

$$\|T^{(m)}_{\gamma,\mu}(\vec{f})\|_{L^{p,q}_{\mu}(X)} \leqslant C \prod_{k=1}^{m} \|f_k\|_{L^{r_k,s_k}_{\mu}(X)} \tag{3.29}$$

with the positive constant C *independent of* $\vec{f} = (f_1, \ldots, f_m)$, *holds if and only if* μ *is Ahlfors upper* β*-regular with*

$$\beta = \frac{rp(m - \gamma)}{rpm + r - p}. \tag{3.30}$$

Proof We will assume that $\mu(X) = \infty$. The proof for the case $\mu(X) < \infty$ is similar and therefore it is omitted.

Necessity. Let (3.29) hold. Taking the test functions $f_1 = f_2 = \cdots = f_m = \chi_{B(a,R)}$ in (3.29), where $a \in X$ and $R > 0$, we find that the right-hand side of (3.29) is equal to

$$C\prod_{k=1}^{m}\big(\mu B(a,R)\big)^{1/r_k} = C\big(\mu B(a,R)\big)^{1/r}.$$

Here we used the equality $\|\chi_E\|_{L^{\ell,q}_\mu} = \mu(E)^{1/\ell}$ which holds for a measurable set E. Further, the left-hand side of equation (3.29) is greater than or equal to $(mR)^{\gamma-m}\big(\mu(B(a,R))\big)^{m+1/p}$. Now, inequality (3.29) yields that

$$\mu B(a,R) \leqslant bR^{\beta},$$

where the positive constant b is equal to $(Cm^{m-\gamma})^{\frac{pr}{mpr+r-p}}$.

Sufficiency. Let $f_j \geqslant 0$, $j = 1,\ldots,m$. Let $f_j \in L^{r_j}_\mu$, $j = 1,\ldots,m$. By Tonelli's theorem, we find that

$$
\begin{aligned}
T^{(m)}_{\gamma,\mu}(\vec{f})(x) &= \int\limits_{X^m} \frac{f_1(y_1)\cdots f_m(y_m)}{\Big[d(x,y_1)+\cdots+d(x,y_m)\Big]^{m-\gamma}} d\mu(y_1)\cdots d\mu(y_m)\\
&= c\int\limits_{X^m} f_1(y_1)\cdots f_m(y_m)\left(\int\limits_{d(x,y_1)+\cdots+d(x,y_m)}^{\infty} t^{\gamma-m-1}dt\right)d\mu(y_1)\cdots d\mu(y_m)\\
&= c\int\limits_0^{\infty} t^{\gamma-m-1}\\
&\quad\times\left(\int\limits_{\{(y_1,\cdots,y_m)\in X^m:\, d(x,y_1)+\cdots+d(x,y_m)<t\}} f_1(y_1)\cdots f_m(y_m)d\mu(y_1)\cdots d\mu(y_m)\right)dt\\
&\leqslant c\int\limits_0^{\infty} t^{\gamma-m-1}\left(\prod_{j=1}^{m}\int\limits_{B(x,t)} f_j(y_j)d\mu(y_j)\right)dt\\
&= c\left[\int\limits_0^{R}(\cdots)+\int\limits_R^{\infty}(\cdots)\right] =: c\left(I_1(x,R)+I_2(x,R)\right),
\end{aligned}
$$

where c is the positive constant depending only on γ and m and R is a positive number which will be chosen later. Observe that

$$I_1(x,R) \leqslant c \int_0^R t^{\gamma-m-1+m\beta} \prod_{j=1}^m \left(\frac{1}{t^\beta} \int_{B(x,t)} f_j(y_j) d\mu(y_j) \right) dt$$

$$\leqslant C\Omega^m(x) \int_0^R t^{\gamma-m-1+m\beta} \prod_{j=1}^m \left(\frac{1}{\mu B(x,3\kappa t)} \int_{B(x,t)} f_j(y_j) d\mu(y_j) \right) dt$$

$$\leqslant C\Omega^m(x) \widetilde{\mathcal{M}}^{(m)}(\vec{f})(x) R^{\gamma-m+m\beta},$$

where $\widetilde{\mathcal{M}}^{(m)}$ is the multisublinear modified maximal operator defined by (3.28) and $\Omega(x)$ is a function given by the formula

$$\Omega(x) := \sup_{R>0} \frac{\mu B(x,R)}{R^\beta}. \tag{3.31}$$

Here we used the fact that $\gamma - m + m\beta > 0$.

Applying Hölder's inequality for Lorentz spaces (see (3.26)), we have:

$$I_2(x,R) \leqslant c \int_R^\infty t^{\gamma-m-1} \prod_{j=1}^m \|f_j\|_{L_\mu^{r_j,s_j}} \mu B(x,t)^{\frac{1}{r_j'}} dt$$

$$\leqslant c(\Omega(x))^{m-\frac{1}{r}} \left(\int_R^\infty t^{\gamma-m-1+\beta\left(m-\frac{1}{r}\right)} dt \right) \prod_{j=1}^m \|f_j\|_{L_\mu^{r_j,s_j}}$$

$$\leqslant c(\Omega(x))^{m-\frac{1}{r}} R^{\gamma-m+\beta\left(m-\frac{1}{r}\right)} \prod_{j=1}^m \|f_j\|_{L_\mu^{r_j,s_j}}.$$

Summarizing the estimates for $I_1(x,R)$ and $I_2(x,R)$ we find that

$$T_{\gamma,\mu}^{(m)} \vec{f}(x) \leqslant c \Bigg(\widetilde{\mathcal{M}}^{(m)}(\vec{f})(x) \Omega^m(x) R^{\gamma-m+m\beta} + (\Omega(x))^{m-\frac{1}{r}} R^{\gamma-m+\beta\left(m-\frac{1}{r}\right)} \prod_{j=1}^m \|f_j\|_{L_\mu^{r_j,s_j}} \Bigg).$$

Let us now choose:

$$R = \left(\widetilde{\mathcal{M}}^{(m)}(\vec{f})(x) \right)^{-r/\beta} \left(\prod_{j=1}^m \|f_j\|_{L_\mu^{r_j,s_j}} \right)^{\frac{r}{\beta}}. \tag{3.32}$$

Then the latter inequality yields:

$$T^{(m)}_{\gamma,\mu}(\vec{f})(x) \leqslant c\left(\widetilde{\mathcal{M}}^{(m)}(\vec{f})(x)\right)^{1-\frac{r(\gamma-m+\beta m)}{\beta}}\left(\prod_{j=1}^{m}\|f_j\|_{L^{r_j,s_j}_{\mu}}\right)^{\frac{r(\gamma-m+\beta m)}{\beta}}.$$

Applying now the maximal theorem (see Lemma 3.16), we have:

$$\begin{aligned}\left\|T^{(m)}_{\gamma,\mu}(\vec{f})\right\|_{L^{p,q}_{\mu}} &\leqslant c\left\|\widetilde{\mathcal{M}}^{(m)}(\vec{f})^{1-\frac{r(\gamma-m+\beta m)}{\beta}}\right\|_{L^{p,q}_{\mu}}\prod_{j=1}^{m}\|f_j\|^{\frac{r(\gamma-m+\beta m)}{\beta}}_{L^{r_j,s_j}_{\mu}}\\ &= c\left\|\widetilde{\mathcal{M}}^{(m)}(\vec{f})\right\|^{1-\frac{r(\gamma-m+\beta m)}{\beta}}_{L^{r,s}_{\mu}}\prod_{j=1}^{m}\|f_j\|^{\frac{r(\gamma-m+\beta m)}{\beta}}_{L^{r_j,s_j}_{\mu}}\\ &\leqslant c\prod_{j=1}^{m}\|f_j\|_{L^{r_j,s_j}_{\mu}}.\end{aligned}$$

Here we used the identities:

$$p\left(1-\frac{\gamma-m+\beta m}{\beta}r\right)=r;\quad q\left(1-\frac{\gamma-m+\beta m}{\beta}r\right)=s$$

which holds because of the condition $r/p = s/q$. □

Theorem 3.19 *Let p, q, r_i, s_i, $i = 1, \ldots, m$, satisfy the conditions of Theorem 3.18. Then, the following inequality holds:*

$$\|T^{(m)}_{\gamma,\mu}(\vec{f})\Omega(\cdot)^{(\gamma-m)/\beta}\|_{L^{p,q}_{\mu}(X)} \leqslant C\prod_{k=1}^{m}\|f_k\|_{L^{r_k,s_k}_{\mu}(X)},$$

where $\Omega(\cdot)$ is the function defined by (3.31).

Proof The statement follows in the same way as Theorem 3.18 was proved. Here we mention that in this case we choose

$$R = \left(\widetilde{\mathcal{M}}^{(m)}(\vec{f})(x)\right)^{-r/\beta}\left(\prod_{j=1}^{m}\|f_j\|_{L^{r_j,s_j}_{\mu}}\right)^{\frac{r}{\beta}}\Omega(x)^{-1/\beta}$$

instead of (3.32).

Indeed, this choice gives:

$$T^{(m)}_{\gamma,\mu}(\vec{f})(x) \leqslant c(\Omega(x))^{\frac{m-\gamma}{\beta}} \left(\widetilde{\mathcal{M}}^{(m)}(\vec{f})(x)\right)^{1-\frac{r(\gamma-m+\beta m)}{\beta}} \left(\prod_{j=1}^{m} \|f_j\|_{L^{r_j,s_j}_{\mu}}\right)^{\frac{r(\gamma-m+\beta)}{\beta}}.$$

Consequently, by using the boundedness of $\widetilde{\mathcal{M}}^{(m)}$ in Lorentz spaces, we have the desired result. □

Theorem 3.18 yields the next statement.

Theorem 3.20 *Let (X, d, μ) be a quasi-metric measure space, $0 < \gamma < 1$, and let $1 < r_j, s_j < \infty$, $j = 1, \ldots, m$. We set $\frac{1}{r} = \sum_{j=1}^{m} \frac{1}{r_j}$ and $\frac{1}{s} = \sum_{j=1}^{m} \frac{1}{s_j}$. Suppose that $1 < r < \frac{1}{\gamma}$, and $\frac{1}{p} = \frac{1}{r} - \gamma$. Let q be such that $\frac{r}{p} = \frac{s}{q}$. Then the inequality*

$$\|T^{(m)}_{\gamma,\mu}(\vec{f})\|_{L^{p,q}_{\mu}(X)} \leqslant C \prod_{k=1}^{m} \|f_k\|_{L^{r_k,s_k}_{\mu}(X)},$$

with the positive constant C independent of $\vec{f}$ holds if and only if μ is Ahlfors upper 1-regular.

In the linear case ($m = 1$), we have the following statements.

Theorem 3.21 *Let (X, d, μ) be a quasi-metric measure space, $0 < \gamma < 1$, $1 < r < p < \infty$. Let s and q be such that $\frac{r}{p} = \frac{s}{q}$. Then, the inequality* (3.25) *with the positive constant C independent of g holds if and only if μ is Ahlfors upper λ-regular with $\lambda = \frac{rp(1-\gamma)}{rp+r-p}$.*

Theorem 3.22 *Let (X, d, μ) be a quasi-metric measure space, $0 < \gamma < 1$. Let $1 < r < \frac{1}{\gamma}$, and $\frac{1}{p} = \frac{1}{r} - \gamma$. Let s and q be such that $\frac{r}{p} = \frac{s}{q}$. Then, inequality* (3.25) *with the positive constant C independent of g holds if and only if μ is Ahlfors upper 1-regular.*

3.2.3 Boundedness in Morrey–Lorentz Spaces

In this section, we establish the boundedness of the operator $T^{(m)}_{\gamma,\mu}$ in Morrey–Lorentz spaces.

Historically, Morrey spaces $L^{p,\lambda}$ were introduced in 1938 by C. Morrey [252] to study regularity problems for partial differential equations, and provided a useful tool in the regularity theory of PDEs. For properties of Morrey-type spaces and related topics, we refer to the recent two-volume monograph Sawano et al. [305, 306]. Mapping properties of fractional integrals in Morrey spaces with non-doubling measure were studied in Eridani et al. [98], Sawano [299], Sawano and Tanaka [300].

Let k be a constant such that $k > \kappa$, where κ is the quasi-metric constant. Suppose that $0 < \lambda < 1$, $1 < p, q < \infty$. We say that a locally integrable function f on X belongs to the modified Morrey–Lorentz space $L^{p,q,\lambda}_{\mu,k}$ defined with respect to the measure μ if

$$\|f\|_{L^{p,q,\lambda}_{\mu,k}} := \sup_{B} \frac{1}{\|\chi_{kB}\|^{\lambda}_{L^{p,q}_{\mu}}} \|f\chi_B\|_{L^{p,q}_{\mu}} < \infty,$$

where $kB := B(a, kr)$ for $B := B(a, r)$.

By using the definition of the Lorentz norm, we can easily see that

$$\|f\|_{L^{p,q,\lambda}_{\mu,k}} = \sup_{B} \frac{1}{\mu(kB)^{\lambda/p}} \|f\chi_B\|_{L^{p,q}_{\mu}}.$$

If $\lambda = 0$, then we have Lorentz space defined with respect to measure μ.

Definition 3.23 The quasi-metric space (X, ρ) is called geometrically doubling if there is a positive integer N_0 such that any ball $B(x, r) \subset X$ can be covered by at most N_0 balls $B(x_i, r/2)$ with $x_i \in B(x, r)$.

Lemma 3.24 (Hytönen [142]) *If the quasi-metric space (X, d) is geometrically doubling, then for any constant $\delta \in (0, 1)$, any ball $B(x, r)$ can be covered by at most $N_0\delta^{-n}$ balls $B(x_i, \delta r)$ with $x_i \in B(x, r)$, where $n = \log_2 N_0$.*

Lemma 3.25 *Let $k_1, k_2 > \kappa$ and let (X, d) is geometrically doubling quasi-metric space. Then the spaces $L^{p,q,\lambda}_{\mu,k_1}$ and $L^{p,q,\lambda}_{\mu,k_2}$ coincide.*

Proof Let $k_1 \leqslant k_2$. Since the relation $L^{p,q,\lambda}_{\mu,k_1} \subset L^{p,q,\lambda}_{\mu,k_2}$ is clear, we will show that $L^{p,q,\lambda}_{\mu,k_2} \subset L^{p,q,\lambda}_{\mu,k_1}$. We use the arguments of Sawano [299], Thiele [321]. The geometrically doubling condition implies that the ball $B := B(x, r)$ can be covered by at most $N = N_0\delta^{-n}$ balls $B(x_i, \delta r)$ with $x_i \in B(x, r)$ for any $\delta \in (0, 1)$. Moreover, the triangle inequality yields that $B(x_i, k_2\delta r) \subset B(x, k_1 r)$ for δ less than $(k_1 - \kappa)/(\kappa k_2)$. Consequently,

$$\frac{1}{\mu(B(x, k_1 r)^{\frac{\lambda}{p}}} \|f\chi_B\|_{L^{p,q}_{\mu}}$$

$$\leqslant \sum_{i=1}^{N} \frac{1}{\mu(B(x_i, k_2\delta r))^{\frac{\lambda}{p}}} \|f\chi_{B(x_i,\delta r)}\|_{L^{p,q}_{\mu}} \leqslant C_N \|f\|_{L^{p,q,\lambda}_{\mu,k_2 r}}.$$

which implies the desired result. □

We are now ready to prove the main result of this section.

Theorem 3.26 *Let (X, d, μ) be a quasi-metric measure space satisfying geometrically doubling condition, $0 < \gamma < m$, $1 < r_j, s_j < \infty$, $j = 1, \dots, m$. We set $\frac{1}{r} = \sum_{j=1}^{m} \frac{1}{r_j}$ and $\frac{1}{s} = \sum_{j=1}^{m} \frac{1}{s_j}$. Suppose that $1 < r < p < \infty$ and q be such that $\frac{r}{p} = \frac{s}{q}$. Suppose that σ_j are numbers such that $0 < \sigma_j < \frac{r_j}{m}\left[\frac{1}{p} - \frac{1}{r}\right] + 1$ for each $j \in \{1, \dots, m\}$. Let k be the constant greater than 1. Then the condition that the measure μ is upper Ahlfors β-regular, where β is defined by (3.30), guarantees the validity of the inequality:*

$$\|T_{\gamma,\mu}^{(m)}(\vec{f})\|_{L_{\mu,k}^{p,q,\lambda}} \leqslant C \prod_{k=1}^{m} \|f_k\|_{L_{\mu,k}^{r_k,s_k,\sigma_i}}, \tag{3.33}$$

where $\lambda = \eta p$ with $\eta = \sum_{j=1}^{m} \sigma_j / r_j$.

Proof We use the arguments of Iida et al. [149], Kokilashvili et al. [189], Tao and He [318]. Suppose that f_j are non-negative functions for all $j \in \{1, \dots, m\}$. Let us denote $B := B(x_0, R)$. We set $B^* := B(x_0, 2\kappa r)$. We represent $f_j \in L_{\mu,\kappa}^{r_j,s_j}$ as the sum $f_j = f_j^0 + f_j^\infty$, where $f_j^0 = f_j \chi_{B^*}$, $f_j^\infty = f - f_j^0$. We have:

$$T_{\gamma,\mu}^{(m)}(\vec{f})(x) = T_{\gamma,\mu}^{(m)}(f_1^0, \dots, f_m^0)(x) + \sum T_{\gamma,\mu}^{(m)}(f_1^{\tau_1}, \dots, f_m^{\tau_m})(x),$$

where each term in the last sum contains at least one $\tau_i = \infty$. Suppose that k is the constant, $k > \kappa$. Taking Theorem 3.18, Lemma 3.25, and the condition $\lambda = \eta p$ into account, find that

$$\begin{aligned}
&\frac{1}{\mu(kB^*)^{\lambda/p}} \|\chi_B\big(T_{\gamma,\mu}^{(m)}(f_1^0, \dots f_m^0\big)\|_{L_\mu^{p,q}} \\
&\leqslant C \frac{1}{\mu(kB^*)^{\lambda/p}} \prod_{j=1}^{m} \|f_j\|_{L_\mu^{r_j,s_j}} \leqslant C \frac{1}{\mu(kB^*)^{\lambda/p}} \prod_{j=1}^{m} \mu(kB^*)^{\sigma_j/r_j} \|f_j\|_{L_{\mu,k}^{r_j,s_j,\sigma_j}} \\
&\leqslant C \prod_{j=1}^{m} \|f_j\|_{L_{\mu,k}^{r_j,s_j\sigma_j}}.
\end{aligned}$$

Let us now discuss the case when $\tau_1 = \dots = \tau_m = \infty$. We let

$$R^* := \kappa^{-1}(C_0)^{-1/\beta} \mu(kB^*)^{1/\beta},$$

where the constant C_0 is defined by (3.1) and κ is the quasi-metric constant. Then it is easy to see that

$$(X \setminus B^*)^m \subseteq \prod_{i=1}^{m} \{y_i : d(x, y_i) \geqslant R^*\} \subset \left\{(y_1, \dots, y_m) : \sum_{i=1}^{m} d(x, y_i) \geqslant R^*\right\}.$$

Applying Hölder's inequality and the relation $\sum\prod \leqslant \prod\sum$, we conclude that for $x \in B^*$

$$
\begin{aligned}
&T^{(m)}_{\gamma,\mu}(f_1^\infty,\dots,f_m^\infty)(x)\\
&= \int\limits_{(X\setminus B^*)^m} \frac{f_1(y_1),\dots,f_m(y_m)}{\big(d(x,y_1)+\cdots+d(x,y_m)\big)^{m-\gamma}}\, d\mu(y_1)\cdots d\mu(y_m)\\
&\leqslant \sum_{j=0}^{\infty} \int\limits_{\{\vec{y}:2^j R^* \leqslant \sum_{i=1}^m d(x,y_i) \leqslant 2^{j+1}R^*\}} \frac{f_1(y_1),\dots,f_m(y_m)}{\big(d(x,y_1)+\cdots+d(x,y_m)\big)^{m-\gamma}}\\
&\quad d\mu(y_1)\cdots d\mu(y_m)\\
&\leqslant \sum_{j=0}^{\infty} \frac{1}{(2^j R^*)^{m-\gamma}} \prod_{i=1}^{m} \|\chi_{B(x,2^{j+1}R^*)} f_j\|_{L^{r_i,s_i}_\mu} \big(B(x,2^{j+1}R^*)\big)^{1/r_i'}\\
&\leqslant C\sum_{j=0}^{\infty} \frac{1}{(2^j R^*)^{m-\gamma}} \prod_{i=1}^{m} \mu\big(B(x,2^{j+1}R^*)\big)^{\frac{\sigma_i}{r_i}+\frac{1}{r_i'}} \|f_i\|_{L^{r_i,s_i,\sigma_i}_{\mu,k}}\\
&= C\prod_{i=1}^{m}\sum_{j=0}^{\infty} (2^j R^*)^{\frac{\gamma-m}{m}} 2^{(j+1)\beta\left(\frac{\sigma_i}{r_i}+\frac{1}{r_i'}\right)} \mu(kB^*)^{\frac{\sigma_i}{r_i}+\frac{1}{r_i'}} \|f_i\|_{L^{r_i,s_i,\sigma_i}_{\mu,k}}\\
&= C\prod_{i=1}^{m} \mu(kB^*)^{\frac{\sigma_i}{r_i}+\frac{1}{r_i'}+\frac{\gamma-m}{\beta m}} \|f_i\|_{L^{r_i,s_i,\sigma_i}_{\mu,k}} \sum_{j=0}^{\infty} 2^{j\left(\frac{\gamma-m}{m}+\frac{\sigma_i\beta}{r_i}+\frac{\beta}{r_i'}\right)}\\
&= C\prod_{i=1}^{m} \mu(kB^*)^{\frac{\sigma_i}{r_i}+\frac{1}{r_i'}+\frac{\gamma-m}{\beta m}} \|f_i\|_{L^{r_i,s_i,\sigma_i}_{\mu,k}}\\
&= C\mu(kB^*)^{\eta+m-\frac{1}{r}+\frac{\gamma-m}{\beta}} \prod_{i=1}^{m} \|f_i\|_{L^{p_i,s_i,\sigma_i}_{\mu,k}}.
\end{aligned}
$$

Here we used the convergence of the series $\sum_{j=0}^{\infty} 2^{j\left(\frac{\gamma-m}{m}+\frac{\beta\sigma_i}{r_i}+\frac{\beta}{r_i'}\right)}$. Hence, we conclude that

$$
\begin{aligned}
&\frac{1}{\mu(kB^*)^{\lambda/p}} \|\chi_{B^*} T^{(m)}_{\gamma,\mu}(f_1^\infty,\dots,f_m^\infty)\|_{L^{p,q}_\mu}\\
&\qquad \leqslant C\mu(kB^*)^{\eta+m-\frac{1}{r}+\frac{\gamma-m}{\beta}-\frac{\lambda}{p}+\frac{1}{p}} \prod_{i=1}^{m} \|f_i\|_{L^{r_i,s_i,\sigma_i}_{\mu,k}} = C\prod_{i=1}^{m} \|f_i\|_{L^{r_i,s_i,\sigma_i}_{\mu,k}}.
\end{aligned}
$$

In the last equality, we used the fact that the condition $\eta = \frac{\lambda}{p}$ implies that $\eta + m - 1/r + (\gamma - m)/\beta - \lambda/p + 1/p = 0$.

Let now $\tau_{i_1} = \cdots = \tau_{i_\ell} = 0$ for some $\{i_1, \ldots, i_\ell\} \subset \{1, \ldots, m\}$ with $1 \leqslant \ell < m$. Observe that in this case for $x \in B^*$

$$T^{(m)}_{\gamma,\mu}(f_1^{\tau_1}, \ldots, f_m^{\tau_m})(x) \leqslant \int\limits_{X^m} \frac{f_1(y_1)\cdots f_m(y_m)}{(d(x,y_1)+\cdots+d(x,y_m))^{m-\gamma}}\, d\mu(y_1)\cdots d\mu(y_m)$$

$$\leqslant \Bigg(\int\limits_{(B^*)^\ell} \prod_{i\in\{i_1,\ldots,i_\ell\}} f_i(y_i) d\mu(y_i) \Bigg)$$

$$\times \Bigg(\int\limits_{(X\setminus B^*)^{m-\ell}} \frac{\prod_{i\notin\{i_1,\ldots,i_m\}} f_i(y_i)}{\big(\sum_{i\notin\{i_1,\ldots,i_\ell\}} d(x,y_i)\big)^{m-\gamma}}\, d\mu(y_i) \Bigg)$$

$$:= I_1(x)\cdot I_2(x).$$

By using Hölder's inequality, we get:

$$I_1(x) \leqslant C \prod_{i\in\{i_1,\ldots,i_\ell\}} \|\chi_{B^*} f_i\|_{L^{r_i,s_i}_\mu} \mu(kB^*)^{1-1/r_i}$$

$$\leqslant C \prod_{i\in\{i_1,\ldots,i_\ell\}} \mu(kB^*)^{1+\frac{\sigma_i}{r_i}-\frac{1}{r_i}} \|f_i\|_{L^{r_i,s_i,\sigma_i}_{\mu,k}}.$$

Due to the same arguments as in the case all $\tau_i = \infty$, we see that

$$I_2(x) \leqslant \sum_{j=0}^{\infty} \prod_{i\notin\{i_1,\ldots,i_\ell\}} 2^{j\frac{\gamma-m}{m-\ell}} \mu(kB^*)^{\frac{\gamma-m}{\beta(m-\ell)}}$$

$$\times \prod_{i\notin\{i_1,\ldots,i_\ell\}} 2^{j\beta\left(\frac{\sigma_i}{r_i}+\frac{1}{r_i'}\right)} \Big(\mu(kB^*)\Big)^{\frac{\sigma_i}{r_i}+\frac{1}{r_i'}} \|f_i\|_{L^{r_i,s_i,\sigma_i}_{\mu,k}}.$$

Combining the estimates for $I_1(x)$ and $I_2(x)$ and taking into account the condition $\sigma_j < \frac{r_j}{m}\left[\frac{1}{p} - \frac{1}{r}\right] + 1$, $i \in \{1, \ldots, m\}$, we deduce that

$$T^{(m)}_{\gamma,\mu}(f_1^{\tau_1}, \ldots, f_m^{\tau_m})(x) \leqslant C\mu(kB^*)^{\frac{\gamma-m}{\beta}+\eta+m-\frac{1}{r}} \prod_{i=1}^{m} \|f_i\|_{L^{r_i,s_i,\sigma_i}_{\mu,k}}.$$

Arguing as in the case $\tau_i = \infty$, $i \in \{1, \ldots, m\}$, we obtain the desired result. Summarizing all these estimates, we conclude that the theorem is proved. □

Remark 3.27 We do not know if the Ahlfors upper β-regularity condition for μ, where β is given by (3.30), is simultaneously necessary and sufficient for the inequality (3.33). However, it can be checked that the upper Ahlfors β-regularity condition is also necessary for (3.33) (under the conditions of parameters of spaces given in Theorem 3.26) when μ satisfies the doubling condition: $\mu B(x, 2R) \leqslant C_{DC}\ \mu B(x, R)$ with the positive constant C_{DC} independent of $x \in X$ and $R > 0$. This follows by taking the test functions $f_1 = f_2 = \cdots = f_m = \chi_{B(a,R)}$, $a \in X$, $R > 0$, in (3.33), and taking into account the following observations which hold for doubling measure μ:

(i) the norms $\|\cdot\|_{L^{p,q,\lambda}_{\mu,k}}$ and $\|\cdot\|_{L^{r_j,s_j,\sigma_j}_{\mu,k}}$ are equivalent to $\|\cdot\|_{L^{p,q,\lambda}_{\mu}}$ and $\|\cdot\|_{L^{r_j,s_j,\sigma_j}_{\mu}}$, respectively;

(ii)

$$\|\chi_{B(a,R)}\|_{L^{r_j,s_j,\sigma_j}_{\mu}} = \sup_{B}\left(\frac{\mu\big(B(a, R) \cap B\big)}{(\mu B)^{\sigma_j}}\right)^{\frac{1}{r_j}} \leqslant C\mu(B(a, R))^{\frac{1-\sigma_j}{r_j}},$$

for $j = 1, \ldots, m$.

We refer to Eridani et al. [98] (see Lemma 3.25) for the latter inequality.

3.3 Fractional Integral Operators Between Banach Function Lattices

The aim of this section is to establish boundedness criteria for generalized fractional integral operators between Banach function lattices (BFLs briefly). It is motivated by applications of classical fractional integrals in the theory of Sobolev embeddings. We recall that in the Euclidean space $\mathbb{R}^n$, the fractional integral operator I_α of order α, $0 < \alpha < n$, is defined by

$$I_\alpha f(x) = \int_{\mathbb{R}^n} \frac{f(y)}{|x-y|^{n-\alpha}}\, dy, \qquad f \in L^1_{\mathrm{loc}}(\mathbb{R}^n).$$

By the well-known Hardy–Littlewood–Sobolev theorem, I_α is a bounded operator from $L^p(\mathbb{R}^n)$ to $L^q(\mathbb{R}^n)$ if and only if $p > 1$ and $1/p - 1/q = \alpha/n$.

We recall that by a quasi-metric, we mean a mapping that satisfies the following conditions $d(x, y) = 0$ if and only if $x = y$; $d(x, y) = d(y, x)$ for all $x, y \in X$; $d(x, y) \leqslant \kappa\ (d(x, z) + d(z, y))$ for some constant $\kappa \geqslant 1$ and all $x, y, z \in X$.

As usual, for a ball $B = B(x, r) := \{y \in X;\ d(x, y) < r\}$ with $x \in X, r > 0$ and $a > 0$, the notation $aB := B(x, ar)$ stands for the concentric dilation of B. We assume that all balls are measurable with positive finite measure. This implies that our measures are σ-finite and that the set $\{x \in X : \mu(\{x\}) > 0\}$ of atoms is at most countable.

Recall that say that the measure μ is *lower α-Ahlfors regular* (resp., *upper β-Ahlfors regular*) with $\alpha > 0$ and $\beta > 0$, if there exists $C > 0$ such that $\mu(B(x,r)) \geqslant Cr^{\alpha}$ $\left(\text{resp., } \mu(B(x,r)) \leqslant Cr^{\beta}\right)$, for all $x \in X$ and $0 < r < d_X$. When $\alpha = \beta$, the measure μ is called *α-Ahlfors regular*.

We use standard definitions and notation from the theory of Banach lattices (see, e.g., Kantorovich and Akilov [161]). Let (Ω, Σ, μ) be a complete σ-finite measure space and let $L^0(\mu) = L^0(\Omega, \mu)$ denote the space of all equivalence classes of μ-a.e. finite real-valued measurable functions on Ω with the topology of convergence in measure on μ-finite sets and let $\widetilde{L}^0(\mu, \Omega)$ denote the space of extended real-valued measurable functions on Ω. A linear subspace E of $L^0(\mu)$ is said to be an *ideal* if $f \in E$ and $|g| \leqslant |f|$ in $L^0(\mu)$, then $g \in E$. By E_+ we denote the collection of non-negative functions in E.

Let E and F be two vector subspaces of $L^0(\mathcal{S}, \Sigma_1, \mu)$ and $L^0(\mathcal{T}, \Sigma_2, \nu)$, respectively. A linear operator $T\colon E \to F$ is said to be an *integral* operator if there exists a $\Sigma_1 \times \Sigma_2$ measurable function K (called the *kernel* of the operator) such that we have

$$Tf(x) = \int\limits_{\mathcal{T}} K(x, y) f(y)\, d\nu(y)$$

for every $f \in E$ and for μ-almost all $x \in \mathcal{S}$.

A Banach (function) lattice $(E, \|\cdot\|_E)$ on (Ω, Σ, μ) is an ideal of $L^0(\mu)$, which is complete with respect to a norm $\|\cdot\|_E$. We also assume that the support of the space E is Ω ($\operatorname{supp}(E) = \Omega$), that is, there is an element $u \in E$ with $u > 0$ μ-a.e. on Ω.

It is said that E has the *Fatou property* (or E is *maximal*) if for any $f \in L^0$, $f_n \in E_+$ such that $f_n \uparrow f$ a.e. and $\sup \|f_n\|_E < \infty$, we have that $f \in E$ and $\|f_n\|_E \uparrow \|f\|_E$.

The *Köthe dual* space E' of a Banach lattice E on (Ω, Σ, μ) is defined as the space of all $f \in L^0(\mu)$ such that $\int\limits_{\Omega} |fg|\, d\mu < \infty$ for every $g \in E$. It is a Banach lattice on (Ω, Σ, μ) when equipped with the norm:

$$\|f\|_{E'} = \sup_{\|g\|_E \leqslant 1} \int\limits_{\Omega} |fg|\, d\mu, \qquad f \in E'.$$

Let us remark that E' of E is a maximal Banach lattice on (Ω, μ) as well as a number of classical spaces such as Lebesgue spaces L^p, $1 \leqslant p \leqslant \infty$, Orlicz spaces, or more general Musielak-Orlicz spaces. Notice that E is a maximal Banach lattice if and only if $E = E'' := (E')'$ with equality of norms (see, e.g., Kantorovich and Akilov [161, p. 30]).

Finally we mention that the classical Lebesgue space defined with respect to a measure μ will be denoted by $L^p(\mu)$.

3.3.1 Generalized Fractional Integral Operators

Let (X, d, μ) be a quasi-metric space. We say that an integral operator T_K is an operator of potential type if it is of the form

$$T_K f(x) = \int_X K(x, y) f(y)\, d\mu(y), \qquad x \in X,$$

where the kernel $K\colon X \times X \to [0, \infty]$ is a function which satisfies the following monotonicity conditions: for every $k_2 > 1$, there exists $k_1 > 1$ such that

$$K(x, y) \leqslant k_1 K(x', y) \quad \text{whenever} \quad d(x', y) \leqslant k_2 d(x, y),$$

and

$$K(x, y) \leqslant k_1 K(x, y') \quad \text{whenever} \quad d(x, y') \leqslant k_2\, d(x, y).$$

Important examples of potential type operators are the fractional integrals. We mention that the fractional operators

$$J_{\alpha,\mu,n} f(x) = \int_X \frac{f(y)}{d(x, y)^{n-\alpha}}\, d\mu(y), \qquad x \in X,$$

where $0 < \alpha < n$, have been studied intensively for both doubling and non-doubling settings. Another type of fractional integral is given by

$$I_{\alpha,X} f(x) := \int_X \frac{f(y)}{\mu(B(x, d(x, y)))^{1-\gamma}}\, d\mu(y), \qquad x \in X,$$

where $\gamma \in (0, 1)$. We refer, e.g., to Edmunds et al. [92] (Chapter 6) and also to the papers Gatto [113] Kairema [159] and references therein where this type operator is studied.

Recall that the $\varphi\colon [0, a) \to [0, \infty)$ defined on $[0, a)$ with $0 < a \leqslant \infty$ is called *quasi-concave* if it is nondecreasing on $[0, a)$ with $\varphi(0) = 0$ and $t \mapsto \varphi(t)/t$ is nonincreasing on $[0, a)$. To simplify the presentation, we denote the set of quasi-concave functions on $[0, \infty)$ by $\mathcal{P}$. In the case when $\varphi \in \mathcal{P}$ is concave, we write $\varphi \in \widetilde{\mathcal{P}}$. For any quasi-concave function $\varphi \in \mathcal{P}$, we define $\varphi_*(t) := t/\varphi(t)$ for all $t > 0$.

For any $\varphi \in \mathcal{P}$, we define $\overline{\varphi}$ by

$$\overline{\varphi}(t) := \sup_{s>0} \frac{\varphi(st)}{\varphi(s)}, \qquad t > 0.$$

Since $\overline{\varphi}$ is sub-multiplicative, it follows that there exist the limits (see, e.g., Krein et al. [201, pp. 52–54])

$$\alpha_\varphi := \lim_{t\to 0+} \frac{\ln \overline{\varphi}(t)}{\ln t}, \qquad \beta_\varphi := \lim_{t\to\infty} \frac{\ln \overline{\varphi}(t)}{\ln t},$$

which satisfy $0 \leqslant \alpha_\varphi \leqslant \beta_\varphi \leqslant 1$.

Throughout this section, we consider functions $\varphi \in \mathcal{P}$ which satisfy $0 < \alpha_\varphi \leqslant \beta_\varphi < 1$. For general examples of such functions, we refer to Gustavsson [130]. For the sake of completeness, we include an example; if $0 < \theta < 1$, $0 < \beta, \gamma$ satisfy $\theta + \gamma\beta < 1$, then φ given by the formula

$$\varphi(t) = t^\theta \ln^\beta(1 + t^\gamma), \qquad t > 0,$$

is a quasi-concave function with $\alpha_\varphi = \theta$ and $\beta_\varphi = \theta + \gamma\beta$.

Let (X, d, μ) be a quasi-metric space and let $\varphi \in \mathcal{P}$. In this section, we consider the kernel $K_\varphi \colon X \times X \to [0, \infty)$ defined by

$$K_\varphi(x, y) = \begin{cases} \frac{1}{\varphi_*(d(x,y))}, & \text{if} \quad x \neq y, \\ 0, & \text{if} \quad x = y. \end{cases}$$

The integral operator generated by K_φ is denoted by T_φ.

In the case when $\varphi(t) = t^\alpha$ for some $\alpha \in (0, 1)$ and all $t \geqslant 0$, we write T_α instead of T_φ.

We will need the following modified maximal operator on a quasi-metric space (X, d, μ) given by

$$\widetilde{M} f(x) = \sup_{r>0} \frac{1}{\mu(B(x, N_0 r))} \int\limits_{B(x,r)} |f|\, d\mu.$$

where $N_0 = 3\kappa$ with the quasi-metric constant κ from the definition of a quasi-metric d.

It is well-known (see Edmunds et al. [92, p. 368]) that for any $f \in L^1(\mu)$

$$\mu(\{x \in X;\ \widetilde{M} f(x) > \lambda\}) \leqslant \frac{1}{\lambda} \int\limits_X |f(x)|\, d\mu(x), \qquad \lambda > 0,$$

i.e., $\widetilde{M}$ is of weak type (1, 1). Since $\widetilde{M}$ is bounded in $L^\infty(\mu)$, Marcinkiewicz's interpolation theorem yields that $\widetilde{M}$ is bounded in $L^p(\mu)$ for every $1 < p < \infty$.

We are ready to state and prove a result on the boundedness of T_φ between corresponding $L^p(\mu)$ spaces.

Theorem 3.28 *Let $1 < p < q < \infty$ and $\varphi \in \mathcal{P}$ be such that $0 < \alpha_\varphi \leqslant \beta_\varphi < 1$. Then the operator T_φ is bounded from $L^p(\mu)$ to $L^q(\mu)$ if and only if there exists*

$C > 0$ such that

$$\mu(B(x,r)) \leqslant C\varphi_*(r)^{s'}, \quad x \in X,\ r > 0, \tag{3.34}$$

where $1/s = 1/p - 1/q$ and $1/s + 1/s' = 1$.

Proof Assume that T_φ is bounded from $L^p(\mu) \to L^q(\mu)$. Then there exists $C > 0$ such that

$$\|T_\varphi f\|_{L^q(\mu)} \leqslant C\|f\|_{L^p(\mu)}, \quad f \in L^p(\mu). \tag{3.35}$$

For $f = \chi_{B(a,r)}$, where $a \in X$, $r > 0$, we have:

$$T_\varphi \chi_{B(a,r)}(x) = \int\limits_{B(a,r)} \frac{1}{\varphi_*(d(x,y))}\, d\mu(y), \quad x \in X.$$

Observe that for all $x, y \in B(a,r)$ $d(x,y) \leqslant \kappa(d(x,a)+d(a,y)) \leqslant 2\kappa r$. Denoting $C_0 := 2\kappa$ and combining this estimate with $\varphi_*(C_0 r)/C_0 r \leqslant \varphi_*(r)/r$ for all $r > 0$, we have that

$$\frac{1}{C_0}\frac{\mu(B(a,r))}{\varphi_*(r)} \leqslant \frac{\mu(B(a,r))}{\varphi_*(C_0 r)} \leqslant T_\varphi \chi_{B(a,r)}, \quad x \in B(a,r).$$

This shows that

$$\frac{1}{C_0}\frac{\mu(B(a,r))}{\varphi_*(r)}\chi_{B(a,r)} \leqslant \chi_{B(a,r)}\, T_\varphi \chi_{B(a,r)}.$$

Applying inequality (3.35), we conclude that

$$\frac{1}{C_0}\frac{\mu(B(a,r))}{\varphi_*(r)}\|\chi_{B(a,r)}\|_{L^q(\mu)} \leqslant C\,\|\chi_{B(a,r)}\|_{L^p(\mu)}$$

and whence

$$\frac{1}{C_0}\frac{\mu(B(a,r))}{\varphi_*(r)}^{1+1/q} \leqslant C\mu(B(a,r))^{1/p}$$

gives the required estimate.

For sufficiency we claim that the estimate (3.34) implies that T_φ is bounded operator from $L^p(\mu)$ to $L^q(\mu)$. To see this we define a bounded function $\omega\colon X \to [0,\infty)$:

$$\omega(x) := \sup_{r>0} \frac{\mu(B(x,r))}{\varphi_*(r)^{s'}}, \quad x \in X,$$

which was introduced in Kokilashvili [168] in the case of Euclidean spaces and $\varphi(t) = t^\alpha, \alpha \in (0, 1)$.

Further, for $x \in X$, and $f \in D(T_\varphi)$, we have:

$$T_\varphi f(x) = \int_{B(x,r)} \frac{f(y)}{\varphi_*(d(x,y))} d\mu + \int_{X\setminus B(x,r)} \frac{f(y)}{\varphi_*(d(x,y))} d\mu = I_1(x) + I_2(x).$$

Set $D_k := B(x, 2^{-k}r) \setminus B(x, 2^{-k-1}r)$ for each integer $k \geqslant 0$. Then we have:

$$\begin{aligned}
|I_1(x)| &\leqslant \sum_{k=0}^{\infty} \int_{D_k} \frac{|f(y)|}{\varphi_*(d(x,y))} d\mu(y) \\
&\leqslant \sum_{k=0}^{\infty} \frac{1}{\varphi_*(2^{-k}r)} \frac{\mu(B(x, N_0 2^{-k}r)}{\mu(B(x, N_0 2^{-k}r)} \int_{B(x,2^{-k}r)} |f(y)|\, d\mu(y) \\
&\leqslant C \sum_{k=0}^{\infty} \frac{1}{\varphi_*(2^{-k}r)} \varphi_*(N_0 2^{-k}r)^{s'} \widetilde{M} f(x)\, \omega(x) \\
&\leqslant C \sum_{k=0}^{\infty} \varphi_*(2^{-k}r)^{s'-1} \widetilde{M} f(x)\, \omega(x).
\end{aligned}$$

Here we used the estimate $\varphi_*(N_0 s) \leqslant C' \varphi_*(s)$ for some $C' > 0$ and for all $s > 0$.

Now observe that our hypothesis implies $0 < \beta_{\varphi_*} \leqslant \alpha_{\varphi_*} < 1$ and so

$$\overline{\varphi_*}(t) \leqslant C_2 t^{\alpha_{\varphi_*}+\varepsilon}, \quad t \in (0, 1] \tag{3.36}$$

for some $C_2 > 0$ and $0 < \varepsilon < 1 - \alpha_{\varphi_*}$. This estimate together with the inequality $\varphi_*(uv) \leqslant \varphi_*(u)\overline{\varphi_*}(v)$, $u, v > 0$, implies that for all $r > 0$

$$\sum_{k=0}^{\infty} \varphi_*(2^{-k}r)^{s'-1} \leqslant \sum_{k=0}^{\infty} [\varphi_*(r)\overline{\varphi_*}(2^{-k})]^{s'-1}. \tag{3.37}$$

Since $s' > 1$, the estimates (3.36) and (3.37) imply:

$$\sum_{k=0}^{\infty} \varphi_*(2^{-k}r)^{s'-1} \leqslant (C_2 \varphi_*(r))^{s'-1} \sum_{k=0}^{\infty} 2^{-k(\alpha_{\varphi_*}+\varepsilon)(s'-1)} \leqslant C_3\, \varphi_*(r)^{s'-1},$$

where $C_3 := \sum_{k=0}^{\infty} 2^{-k(\alpha_{\varphi_*}+\varepsilon)} < \infty$. Combining this fact with the above estimate of I_1, we find that

$$|I_1(x)| \leqslant C_3\, \varphi_*(r)^{s'-1}\, \widetilde{M} f(x)\, \omega(x), \quad x \in X, r > 0.$$

To estimate $I_2(x)$, we put $E_k := B(x, 2^{k+1}r) \setminus B(x, 2^k r)$ for each integer $k \geqslant 0$. Then Hölder's inequality gives that for all $f \in L^p(\mu)$

$$|I_2(x)| \leqslant \|f\|_{L^p(\mu)} \left(\int_{X \setminus B(x,r)} \frac{1}{\varphi_*(d(x,y))^{p'}} \, d\mu \right)^{1/p'}.$$

Fix $f \in L^p(\mu)$ with $\|f\|_{L^p(\mu)} \leqslant 1$. Then, we have by (3.34)

$$\begin{aligned}
|I_2(x)| &\leqslant \left(\sum_{k=0}^{\infty} \int_{E_k} \varphi_*(d(x,y))^{-p'} \, d\mu \right)^{1/p'} \\
&\leqslant \left(\sum_{k=0}^{\infty} \varphi_*(2^k r)^{-p'} \mu(B(x, 2^{k+1} r)) \right)^{1/p'} \\
&\leqslant \big(C\omega(x)\big)^{1/p'} \left(\sum_{k=0}^{\infty} \varphi_*(2^k r)^{-p'} \varphi_*(2^{k+1} r)^{s'} \right)^{1/p'} \\
&\leqslant C_4 \, \omega(x)^{1/p'} \left(\sum_{k=0} \varphi_*(2^k r)^{s'-p'} \right)^{1/p'} \\
&= C_4 \, \omega(x)^{1/p'} \left(\sum_{k=0}^{\infty} \left(\frac{\varphi(2^k r)}{2^k r} \right)^{p'-s'} \right)^{1/p'} \\
&\leqslant C_4 \, \omega(x)^{1/p'} \left(\sum_{k=0}^{\infty} \left(\frac{\varphi(r)}{r} \, \frac{\varphi(2^k r)}{2^k \varphi(r)} \right)^{p'-s'} \right)^{1/p'} \\
&\leqslant C_4 \, \omega(x)^{1/p'} \left(\frac{\varphi(r)}{r} \right)^{1-s'/p'} \left(\sum_{k=0}^{\infty} \left(\frac{\overline{\varphi}(2^k)}{2^k} \right)^{p'-s'} \right)^{1/p'}.
\end{aligned}$$

Here we used the fact that $s' < p'$. Our hypothesis on φ implies that

$$\overline{\varphi}(t) \leqslant C_5 t^{\beta_\varphi + \varepsilon}, \quad t \geqslant 1$$

with some $C_5 > 0$ and $0 < \varepsilon < 1 - \beta_\varphi$.

This inequality and the observation $1 - s'/p' = s'/q$ give that there is a constant $C_6 > 0$ such that for all $r > 0$

$$\begin{aligned}
|I_2(x)| &\leqslant C_6 \, \omega(x)^{1/p'} \big(\varphi_*(r)^{s'}\big)^{-1/q} \left(\sum_{k=0}^{\infty} \left(\frac{1}{2^{(1-\beta_\varphi-\varepsilon)k}} \right)^{p'-s'} \right)^{1/p'} \\
&\leqslant C_7 \, \omega(x)^{1/p'} \varphi_*(r)^{-s'/q}.
\end{aligned}$$

The obtained estimates for I_1 and I_2 imply the following pointwise inequality:

$$|T_\varphi f(x)| \leqslant \widetilde{C}\big(\varphi_*(r)^{s'-1}\widetilde{M} f(x)\omega(x) + \omega(x)^{1/p'}\varphi_*(r)^{-s'/q}\big).$$

Since $x \mapsto \omega(x)$ is a bounded function

$$|T_\varphi f(x)| \leqslant \widetilde{C}\big(\varphi_*(r)^{s'-1}\widetilde{M} f(x) + \varphi_*(r)^{-s'/q}\big).$$

Since $0 < \alpha_{\varphi_*} \leqslant \beta_{\varphi_*} < 1$, $\varphi_*(s) \to 0$ as $s \to 0+$ and $\varphi_*(s) \to \infty$ as $s \to \infty$. This implies that the image of φ_* is $[0, \infty)$. Thus taking $r > 0$ such that $\varphi_*(r) = \widetilde{M} f(x)^{-p/s'}$, we obtain the pointwise inequality true for all $f \in L^p(\mu)$ with $\|f\|_{L^p(\mu)} = 1$,

$$|T_\varphi f(x)| \leqslant C\Big(\big(\widetilde{M} f(x)\big)^{1-p(1-1/s')} + \big(\widetilde{M} f(x)\big)^{p/q}\Big) = 2C\big(\widetilde{M} f(x)\big)^{p/q}, \quad x \in X.$$

Combining with the fact that $\widetilde{M}\colon L^p(\mu) \to L^p(\mu)$ is bounded, we conclude that for all $f \in L^p(\mu)$ with $\|f\|_{L^p(\mu)} = 1$

$$\|T_\varphi f\|^q_{L^q(\mu)} \leqslant 2C \int\limits_X \widetilde{M} f(x)^p \, d\mu = 2C\|\widetilde{M} f\|^p_{L^p(\mu)} \leqslant 2C_0\|f\|_{L^p(\mu)} \leqslant C_0.$$

This shows that T_φ is bounded from $L^p(\mu)$ to $L^q(\mu)$, and this completes the proof. □

As an application, we obtain the following version of the Hardy–Littlewood–Sobolev theorem (see García-Cuerva and Gatto [110], Kokilashvili and Meskhi [171]) for the fractional integral $J_{\alpha,\mu,n}$.

Corollary 3.29 *For $1 < p < n/\alpha$ and $1/q = 1/p - \alpha/n$, $J_{\alpha,\mu,n}$ is a bounded operator from $L^p(\mu)$ into $L^q(\mu)$ provided that the measure μ is upper n-Ahlfors regular.*

Before the proof of the next result, we need some definition. Let $\psi\colon (0, \infty) \to (0, \infty)$ be nondecreasing function. The Marcinkiewicz class $M^{(*)}_\psi = M^{(*)}_\psi(\mu)$ on a measure space (Ω, Σ, μ) is defined to be the set of all measurable functions $f \in L^0(\mu)$ such that

$$\|f\|_\psi = \sup_{t>0} \psi(t) f^*(t) < \infty,$$

where f^* denotes the *decreasing rearrangement* of f, i.e., $f^*(t) := \inf\{\lambda \geqslant 0;\ \mu_f(\lambda) \leqslant t\}$ for all $t > 0$, and $\mu_f(\lambda) := \mu\{t \in \Omega;\ |f(t)| > \lambda\}$, $\lambda \geqslant 0$.

In particular, for $\psi(t) = t^{1/q}$, $0 < q < \infty$, we obtain the classical spaces $L_{q,\infty}(\mu)$. One can readily show that $\|\cdot\|_\psi$ is a quasi-norm (i.e.,

$$\|f + g\|_\psi \leqslant C(\|f\|_\psi + \|g\|_\psi), \quad f, g \in M^{(*)}_\psi,$$

for some $C > 0$) if the function ψ satisfies the Δ_2-condition, i.e., there exists $K \geqslant 1$ such that $\psi(2t) \leqslant K\psi(t)$ for all $t > 0$.

It can be easily shown that for any $f \in M_\psi^{(*)}$, we have:

$$\|f\|_\psi = \sup_{\lambda>0} \lambda\psi(\{\omega \in \Omega;\ |f(\omega)| > \lambda\}).$$

Theorem 3.30 *Let $1 \leqslant p < \infty$ and let (X, d, μ) be a quasi-metric space. Let $\varphi\colon [0,\infty) \to [0,\infty)$ be such that the function ϕ given by $\phi(t) = t^{-1/p}\varphi(t)$ for all $t > 0$ be decreasing with $\phi(\mathbb{R}_+) = \mathbb{R}_+$. Assume that K is a positive kernel on X which satisfies the following conditions:*

$$\int_{B(x,r)} K(x,y)\,d\mu(y) \leqslant C_1\varphi(r), \qquad \int_{B(y,r)} K(x,y)\,d\mu(x) \leqslant C_2\varphi(r)$$

and

$$\left\|\chi_{X\setminus B(x,r)} K(x,\cdot)\right\|_{L^{p'}(\mu)} \leqslant C_3\phi(r)$$

for some $C_j > 0$ $(1 \leqslant j \leqslant 3)$ and all $x \in X$ and $r > 0$, where $1/p + 1/p' = 1$. Then the integral operator T_K is bounded from $L^p(\mu)$ into $M_\psi^{(*)}(\mu)$, where $\psi(t) = \frac{1}{\phi(t)}$ for all $t > 0$.

Proof We can, and do, assume that $C_j = 1$ for each $1 \leqslant j \leqslant 3$. Fix $f \in L^p(\mu)$ with $\|f\|_{L^p(\mu)} \leqslant 1$. For any ball $B(x,r)$ in X, we have:

$$|T_K f|(x) \leqslant \int_X K(x,y)|f(y)|\,d\mu(y) = I_1(x) + I_2(x),$$

where

$$I_1(x) := \int_{B(x,r)} K(x,y)|f(y)|\,d\mu(y)$$

and

$$I_2(x) = \int_{X\setminus B(x,r)} K(x,y)|f(y)|\,d\mu(y).$$

We present the proof for $p > 1$. Then by Hölder's inequality, we obtain:

$$|I_2(x)| \leqslant \|f\|_{L^p(\mu)} \left(\int_{X\setminus B(x,r)} K(x,y)^{p'}\,d\mu(y)\right)^{1/p'} \leqslant \frac{1}{\psi(r)}.$$

For $\lambda > 0$, we take $r = \psi^{-1}(2/\lambda)$. Then the above inequality yields $|I_2(x)| \leqslant \lambda/2$ and whence

$$\{x \in X;\ |T_K f|(x) > \lambda\} \subset \{x \in X;\ |I_1(x)| > \lambda/2\} \cup \{x \in X;\ |I_2(x)| > \lambda/2\}$$
$$= \{x \in X;\ |I_1(x)| > \lambda/2\}.$$

Using Hölder's inequality, we get:

$$|I_1(x)| \leqslant \Big(\int_{B(x,r)} K(x,y)|f(y)|^p\,d\mu(y)\Big)^{1/p} \Big(\int_{B(x,r)} K(x,y)\,d\mu(y)\Big)^{1/p'}$$
$$\leqslant \varphi(r)^{1/p'} \Big(\int_{B(x,r)} K(x,y)|f(y)|^p\,d\mu(y)\Big)^{1/p}.$$

Combining with Tchebischev's inequality and Tonelli's theorem, we find that

$$\mu(\{x \in X; |T_K f(x)| > \lambda\}) \leqslant \mu(\{x \in X; |I_1(x)| > \lambda/2\})$$
$$\leqslant \varphi(r)^{p/p'}(2/\lambda)^p \int_X \int_{B(x,r)} K(x,y)|f(y)|^p\,d\mu(y)\,d\mu(x)$$
$$= \varphi(r)^{p/p'}(2/\lambda)^p \int_X \int_{B(y,r)} K(x,y)\,d\mu(x)|f(y)|^p\,d\mu(y) \leqslant \varphi(r)^{p/p'}\varphi(r)(2/\lambda)^p$$
$$= (2\varphi(r)/\lambda)^p = \psi^{-1}(2/\lambda).$$

This shows that $T_K f \in M_\psi^{(*)}$ with

$$\|T_K f\|_\psi = \sup_{\lambda>0} \lambda\psi(\mu\{x \in X; |T_K f(x)| > \lambda\}) \leqslant 2.$$

To finish we note that the proof for $p = 1$ is a simple modification of the proof for $p > 1$. □

The following lemma provides examples of generalized fractional operators which satisfies the integral conditions of the above theorem.

Lemma 3.31 *Let (X, d, μ) be a quasi-metric space and $\varphi \in \mathcal{P}$. Suppose that $\mu(B(x,r)) \leqslant Cr$ for some $C > 0$ and all $0 < r < d_X$, $x \in X$. Then we have:*

(i) If $\alpha_\varphi > 0$, then there exists $C_1 > 0$ such that

$$\int_{B(x,r)} \frac{1}{\varphi_*(d(x,y))}\,d\mu(y) \leqslant C_1\varphi(r).$$

(ii) If φ is sub-multiplicative (i.e., $\varphi(st) \leqslant C\varphi(s)\varphi(t)$ for some $C > 0$ and all $s, t > 0$) such that $1 \leqslant p < 1/\beta_\varphi$, then there exists $C_2 > 0$ for which

$$\left(\int\limits_{X \setminus B(x,r)} \frac{1}{\varphi_*(d(x,y))^{p'}} \, d\mu(y) \right)^{1/p'} \leqslant C_2 \frac{\varphi(r)}{r^{1/p}}.$$

Proof (i). Since $\alpha := \alpha_\varphi > 0$, for every $\varepsilon > 0$, there exists $C(\varepsilon)$ such that $\overline{\varphi}(s) \leqslant C(\varepsilon)s^{\alpha-\varepsilon}$ for all $s \in (0,1)$. Fix $x \in X$ and $r > 0$ and put $D_j = \{y \in X;\ 2^{-j-1}r \leqslant d(x,y) < 2^{-j}r\}$; then our hypothesis $\mu(B(x,r)) \leqslant Cr$, $\alpha_\varphi > 0$ implies:

$$\begin{aligned}
\int\limits_{B(x,r)} \frac{1}{\varphi_*(d(x,y))} \, d\mu(y) &= \sum_{j=0}^{\infty} \int\limits_{D_j} \frac{1}{\varphi_*(d(x,y))} \, d\mu(y) \\
&\leqslant \sum_{j=0}^{\infty} \frac{1}{\varphi_*(2^{-j-1}r)} \mu(B(x, 2^{-j}r)) \\
&\leqslant 2C \sum_{j=0}^{\infty} \overline{\varphi}(2^{-j-1}) \varphi(r) \leqslant C_1 \varphi(r).
\end{aligned}$$

(ii). Since φ is sub-multiplicative, φ_* is super-multiplicative and $\beta_\varphi < 1/p$:

$$\begin{aligned}
\int\limits_{X \setminus B(x,r)} \frac{1}{\varphi_*(d(x,y))^{p'}} d\mu(y) &= \sum_{j=0}^{\infty} \int\limits_{2^j r \leqslant d(x,y) < 2^{j+1} r} \left(\frac{1}{\varphi_*(d(x,y))} \right)^{p'} d\mu(y) \\
&\leqslant \sum_{j=0}^{\infty} \left(\frac{1}{\varphi_*(2^j r)} \right)^{p'} \mu(B(x, 2^{j+1} r)) \\
&\leqslant C \sum_{j=0}^{\infty} \left(\frac{\varphi(2^j r)}{2^j r} \right)^{p'} 2^j r \\
&\leqslant C \varphi(r)^{p'} r^{1-p'} \sum_{j=0}^{\infty} 2^{j(1-p')} \left(\varphi(2^j) \right)^{p'} \\
&\leqslant C_\varphi \varphi(r)^{p'} r^{1-p'}
\end{aligned}$$

and so this gives the required estimate:

$$\left(\int\limits_{X \setminus B(x,r)} \frac{1}{(\varphi_*(d(x,y))^{p'}} \, d\mu(y) \right)^{1/p'} \leqslant C_2 \frac{\varphi(r)}{r^{1/p}},$$

which ends the proof. □

3.3.2 Fractional Integral Operators Between Symmetric Spaces

An important class of Banach lattices on a measure space (Ω, Σ, μ) are *symmetric* quasi-Banach spaces. We recall that a Banach lattice E on (Ω, μ) is said to be a symmetric Banach space if $g \in E$ and $\|f\|_E = \|g\|_E$ whenever $f \in E, g \in L^0(\mu)$ and $\mu_f(\lambda) = \mu_g(\lambda)$ for all $\lambda > 0$, where $\mu_f(\lambda) = \mu(\{\omega \in \Omega;\ |f(\omega)| > \lambda\})$.

Clearly for a symmetric Banach space E, $\|\chi_A\|_E$ depends only on $\mu(A)$ for any measurable set A of finite measure. Thus, for every $t \in \{\mu(A);\ A \in \Sigma\}$, we may define the function φ_E by $\varphi_E(t) = \|\chi_A\|_E$ where A is any measurable set with $\mu(A) = t$. This function is called the *fundamental function* of E.

For the theory of symmetric spaces, we refer to Bennett and Sharpley [25] and Krein et al. [201]. We recall some important properties of symmetric spaces which we will need to use later.

Let E be a symmetric space on nonatomic measure space (Ω, Σ, μ) with $0 < \mu(\Omega)$, and let $\phi := \varphi_E$ be the fundamental function of E. It is well-known that ϕ is a quasi-concave on $[0, a)$, where $a = \mu(\Omega)$. Taking $\widetilde{\phi}(t) = \inf_{s \in (0,a)}(1 + t/s)\phi(s)$ for all $t \in [0, a)$, we obtain a concave function $\widetilde{\phi}$ on $[0, a)$ such that $\phi(t) \leqslant \widetilde{\phi}(t) \leqslant 2\phi(t)$ for all $t \in [0, a)$. In what follows $\widetilde{\phi}$ is called a concave majorant of ϕ.

For any quasi-concave function $\varphi : [0, a) \to [0, \infty)$, the *Marcinkiewicz space* $M_\varphi = M_\varphi(\mu)$ is defined by the norm:

$$\|f\|_{M_\varphi} = \sup_{0<t<a} \frac{1}{\varphi(t)} \int_0^t f^*(s)\, ds,$$

where f^* denotes the decreasing rearrangement of f given by

$$f^*(t) = \inf\{\lambda > 0;\ \mu(\{\omega \in \Omega;\ |f(\omega)| > \lambda\}) \leqslant t\}, \qquad t > 0.$$

We note that the largest symmetric space on (Ω, Σ, μ) with the same fundamental φ is the Marcinkiewicz space M_{φ_*}. We also note that the fundamental function of a symmetric space $E = (E, \|\cdot\|_E)$ is not necessarily concave; however, there exists an equivalent norm under which E is a symmetric space with concave fundamental function.

For any symmetric space E with concave fundamental function $\varphi = \varphi_E$, there is also the smallest symmetric space with the same fundamental function; it is the *Lorentz space* $\Lambda_\varphi = \Lambda_\varphi(\mu)$ given by the norm:

$$\|f\|_{\Lambda_\varphi} = \int_0^a f^*(s)\, d\varphi(s).$$

In what follows, we will use the following easily verified fact that in the case when μ is a nonatomic measure, we have:

$$\|f\|_{M_\varphi} = \sup_{A\in\Sigma,\, 0<\mu(A)<\infty} \frac{1}{\varphi(\mu(A))} \int_A |f|\, d\mu, \qquad f \in M_\varphi.$$

For the theory of symmetric spaces, we refer to Bennett and Sharpley [25] and Krein et al. [201].

To show applications of the Lemma 3.31 to fractional integral operators, we define a special class of kernels. Let (X, d, μ) be a metric space. A kernel $K \in L^0(\mu \times \mu)_+$ is said to be a Borel kernel whenever there is a constant $C \geqslant 1$ such that for every Borel set A in X with $0 < \mu(A) < \infty$ and $x \in X$, there is a ball $B(x, r)$ such that $\mu(B(x, r)) = \mu(A)$ and

$$\int_A K(x, y)\, d\mu(y) \leqslant C \int_{B(x,r)} K(x, y)\, d\mu(y).$$

The following proposition gives examples of Borel kernels.

Proposition 3.32 *Let (X, d, μ) be a nonatomic quasi-metric space. Let $\phi\colon [0, \infty) \times [0, \infty) \to [0, \infty)$ be a nonincreasing function in each variable. Then $K\colon X \times X \to [0, \infty)$ defined by the following formula is a Borel kernel:*

$$K(x, y) = \begin{cases} \phi\big(d(x, y), \mu(B(x, d(x, y)))\big), & \text{if } x \neq y \\ 0, & \text{if } \quad x = y. \end{cases}$$

Proof Let A be an arbitrary Borel set with a finite positive measure. Our hypothesis that μ is a nonatomic measure yields that for any $x \in A$, there exists a ball $B(x, r)$ in X such that $\mu(B(x, r)) = \mu(A)$. Put $B := B(x, r) \cap A$; then $\mu(A \setminus B) = \mu(B(x, r) \setminus B)$. Since $y \in A \setminus B$ implies $d(x, y) \geqslant r$, $\phi\big(d(x, y), \mu(B(x, d(x, y)))\big) \leqslant \phi\big(r, \mu(B(x, r))\big)$ and so we have:

$$\begin{aligned} &\int_{B(x,r)} K(x, y)\, d\mu(y) - \int_A K(x, y)\, d\mu(y) \\ &\qquad = \int_{B(x,r)\setminus B} K(x, y)\, d\mu(y) - \int_{A\setminus B} K(x, y)\, d\mu(y) \\ &\qquad \geqslant \phi(r, \mu(B(x, r)))\, (\mu(B(x, r) \setminus B) - \mu(A \setminus B)) = 0 \end{aligned}$$

and this completes the proof. □

Lemma 3.33 *Let $(\mathcal{S}, \mu)$ and $(\mathcal{T}, \nu)$ be measure spaces, and let T_K be an integral operator generated by the kernel $K \in L^0(\mu \times \nu)_+$. Then T_K is bounded from* *a Banach lattice E on $(\mathcal{T}, \nu)$ to the Marcinkiewicz space $M_\psi(\mu)$ on $(\mathcal{S}, \mu)$ if and only if*

$$C := \sup_{0<\mu(A)<\infty} \frac{1}{\psi(\mu(A))} \Big\| \int_A K(s, \cdot)\, d\mu(s) \Big\|_{E'} < \infty.$$

Proof Combining Tonelli's theorem with the fact that T_K is positive, we obtain:

$$\begin{aligned}
\sup_{\|f\|_E \leqslant 1} \|T_K f\|_{M_\psi(\mu)} &= \sup_{\|f\|_E \leqslant 1} \sup_{\mu(A)>0} \frac{1}{\psi(\mu(A))} \int_A \int_{\mathcal{T}} K(s, t)|f(t)|\, d\nu(t)\, d\mu(s) \\
&= \sup_{\mu(A)>0} \sup_{\|f\|_E \leqslant 1} \frac{1}{\psi(\mu(A))} \int_{\mathcal{T}} |f| \Big(\int_A K(s, \cdot)\, d\mu(s) \Big)\, d\nu \\
&= \sup_{\mu(A)>0} \frac{1}{\psi(\mu(A))} \Big\| \int_A K(s, \cdot)\, d\mu(s) \Big\|_{E'},
\end{aligned}$$

which ends the proof. □

Corollary 3.34 *Let (X, d, μ) be a nonatomic quasi-metric space. If T_K is an integral operator generated by a positive Borel kernel K such that*

$$\int_{B(x,r)} K(x, y)\, d\mu(y) \leqslant \phi(\mu(B(x, r))) \tag{3.38}$$

for all balls $B(x, r)$ in X and some concave function $\phi\colon [0, \mu(X)) \to [0, \infty)$, then T_K is a bounded operator from the Lorentz space $\Lambda_\phi(\mu)$ into $L^\infty(\mu)$.

Proof Since K is a Borel kernel, for every Borel set A with a finite positive measure and every $x \in X$, there exists a ball $B(x, r)$ in X such that $\mu(B(x, r)) = \mu(A)$ and

$$\int_A K(x, y)\, d\mu(y) \leqslant \int_{B(x,r)} K(x, y)\, d\mu(y).$$

Combining the latter with (3.38), we have:

$$T_K \chi_A(x) \leqslant \int_{B(x,r)} K(x, y)\, d\mu(y) \leqslant \phi(\mu(B(x, r)) = \phi(\mu(A)).$$

It is well-known that this inequality (see Krein et al. [201, Corollary 1, p. 112]) for arbitrary characteristic functions χ_A implies that T_K is bounded from $\Lambda_\phi(\mu)$ into $L^\infty(\mu)$. □

As an immediate consequence of the above corollary, we have the following result.

Theorem 3.35 *Let* (X, d, μ) *be a nonatomic quasi-metric space, such that, for any* $t > 0$, $h(t) := \inf\{\mu(B(x,t);\ x \in X\} > 0$. *Assume that* $K\colon X \times X \to [0, \infty)$ *is a Borel kernel for which*

$$\theta(t) := \sup_{x \in X} \int_{B(x,t)} K(x, y)\, d\mu(y) < \infty, \qquad t > 0.$$

If $t \mapsto \theta \circ h^{-1}(t)/t$ *is almost nonincreasing, then the integral operator* T_K *is bounded from the Lorentz space* $\Lambda_\phi(\mu)$ *into* $L^\infty(\mu)$, *where* ϕ *is a concave majorant of* $\theta \circ h^{-1}$.

In what follows we will need a general interpolation theorem for operators from Lorentz into Marcinkiewicz spaces from which it follows (see, e.g., Pustyl'nik [276]): if φ_0, φ_1 and ψ_1, ψ_1 be concave functions on $(0, a)$ with $0 < a \leqslant \infty$ such that

$$\Phi(t) = \varphi_0(t)\theta\Big(\frac{\varphi_1(t)}{\varphi_0(t)}\Big), \qquad \Psi_*(t) = \psi_{0*}(t)\theta\Big(\frac{\psi_{1*}(t)}{\psi_{0*}(t)}\Big), \qquad t \in (0, a)$$

for some $\theta \in \widetilde{\mathcal{P}}$. Then every bounded operator T from Λ_{φ_j} into M_{ψ_j} for $j = 0, 1$ is also bounded from Λ_Φ into M_Ψ.

Theorem 3.36 *Let* (X, d, μ) *be a nonatomic quasi-metric space. Suppose* T_K *is an integral operator generated by a positive and symmetric Borel kernel* K *such that the condition (3.38) holds for some concave function* $\phi\colon [0, \mu(X)) \to [0, \infty)$, *and all balls* $B(x, r)$ *in* X. *Suppose further that* φ *and* ψ *are concave functions such that*

$$\frac{\varphi(t)}{\psi_*(t)} \geqslant C\phi(t), \qquad t \in (0, \mu(X))$$

for some $C > 0$. *Then the integral operator* T_K *is bounded from the Lorentz space* $\Lambda_\varphi(\mu)$ *into the Marcinkiewicz space* $M_\psi(\mu)$.

Proof From Lemma 3.33 and Corollary 3.34, we have that $T_K\colon \Lambda_\phi(\mu) \to L^\infty(\mu)$ and $T_K\colon L^1(\mu) \to M_\phi$. Since $M_{\psi_0}(\mu) = L^\infty(\mu)$ and $\Lambda_{\varphi_1}(\mu) = L^1(\mu)$ with equality of norms whenever $\psi_0(t) = \varphi_1(t) = t$ for every $t \in (0, \mu(X))$, the abovementioned interpolation theorem implies that $T_K\colon \Lambda_\Phi(\mu) \to M_\Psi(\mu)$ is bounded for any quasi-concave function $\widetilde{\rho}$, where

$$\Phi(t) = \phi(t)\widetilde{\rho}\Big(\frac{t}{\phi(t)}\Big), \qquad \Psi_*(t) = \widetilde{\rho}\Big(\frac{t}{\phi(t)}\Big), \qquad t \in (0, \mu(X))$$

and $\widetilde{\rho}$ is a concave majorant of ρ. Now we define $\rho \in \mathcal{P}$ by

$$\rho(t) := \sup_{0<s<\mu(X)} \min\left\{1, \frac{t\,\phi(s)}{s}\right\}\frac{s}{\psi(s)}, \qquad t \geqslant 0.$$

Our hypothesis on φ and ψ implies that $\Psi(t) \leqslant c\psi(t)$ and $\varphi(t) \geqslant c\Phi(t)$ for all $t \in (0, \mu(X))$ and some $c > 0$. Hence

$$\Lambda_\varphi(\mu) \hookrightarrow \Lambda_\Phi(\mu), \qquad M_\Psi(\mu) \hookrightarrow M_\psi(\mu)$$

with continuous inclusions and so the desired statement follows. □

As an application we obtain the following corollary.

Corollary 3.37 *Let (X, d, μ) be a nonatomic quasi-metric space, such that the function $h\colon (0, \infty) \to (0, \mu(X))$ given by $h(t) = \inf\{\mu(B(x, t));\ x \in X\} > 0$ for all $t > 0$ is continuous. If T_K is an integral operator generated by a positive and symmetric Borel kernel K such that*

$$C_1\phi(\mu(B(x, r)) \leqslant \int_{B(x,r)} K(x, y)\, d\mu(y) \leqslant C_2\phi(\mu(B(x, r)))$$

for some concave function $\phi\colon [0, \mu(X)) \to [0, \infty)$, some constants $C_1, C_2 > 0$ and all balls $B(x, r)$ in X, then the integral operator T_K is bounded from the Lorentz space $\Lambda_\varphi(\mu)$ into the Marcinkiewicz space $M_\psi(\mu)$ if and only if there exists $C > 0$ such that

$$\frac{\varphi(t)}{\psi_*(t)} \geqslant C\phi(t), \qquad t \in (0, \mu(X)).$$

Proof It follows from Theorem 3.36 that we need to show only necessity; suppose that T_K is bounded from $\Lambda_\varphi(\mu)$ into $M_\psi(\mu)$. Since $\Lambda_\varphi(\mu)' = M_\varphi(\mu)$ (see Acosta et al. [1]), from Lemma 3.33, we have:

$$\sup_{0<\mu(A)<\infty} \frac{1}{\psi(\mu(A))}\left\| \int_A K(\cdot, y)\, d\mu(y)\right\|_{M_\varphi} < \infty.$$

Combining the latter with the condition $C_1\phi(\mu(B(x, r)) \leqslant \int_{B(x,r)} K(x, y)\, d\mu(y)$ yields that there exists a constant $\widetilde{C} > 0$ such that for $x, y \in X$ and $r > 0$, we have:

$$\widetilde{C} \geqslant \frac{1}{\psi(\mu(B(x, r)))}\frac{1}{\varphi(\mu(B(y, r)))}\int_{B(y,r)} \phi(\mu(B(x, r)))\, d\mu(x)$$

$$\geqslant \frac{1}{\psi(\mu(B(x,r)))}\frac{\mu(B(y,r))}{\varphi(\mu(B(y,r)))}\phi(h(r))$$
$$\geqslant \frac{1}{\psi(\mu(B(x,r)))}\frac{h(r)}{\varphi(h(r))}\phi(h(r)).$$

Consequently, there exits $C > 0$ such that

$$\frac{\varphi(h(r))}{\psi_*(h(r))} \geqslant C\phi(h(r)), \qquad r > 0.$$

and so the required estimate follows by continuity of h. □

We conclude this section with the following result.

Corollary 3.38 *Let (X, d, μ) be a nonatomic quasi-metric space such that μ is 1-Ahlfors regular measure. Let $\phi\colon [0,\infty) \to [0,\infty)$ be a concave function such that $\alpha_\varphi > 0$. Suppose that φ and ψ are concave functions on $[0, \mu(X))$. Then the integral operator T_ϕ is bounded from the Lorentz space $\Lambda_\varphi(\mu)$ into the Marcinkiewicz space $M_\psi(\mu)$ if and only if there exists $C > 0$ such that*

$$\frac{\varphi(t)}{\psi_*(t)} \geqslant C\phi(t), \qquad t \in (0, \mu(X)).$$

Proof For any ball $B(x,r)$ in X, we have the obvious estimate:

$$C_1\phi(r) \leqslant \int\limits_{B(x,r)} \frac{1}{\phi_*(d(x,y))}\,d\mu(y).$$

Since μ is 1-Ahlfors regular, it follows from Lemma 3.31 that there exist constants $C_1, C_2 > 0$ such that

$$C_1\phi(\mu(B(x,r))) \leqslant \int\limits_{B(x,r)} \frac{1}{\phi_*(d(x,y))}\,d\mu(y) \leqslant C_2\phi(\mu(B(x,r))).$$

Now, applying Corollary 3.37 for the symmetric Borel kernel K_ϕ yields the required result. □

3.4 Compactness Criteria for Fractional Integral Operators

In the theory of potential analysis and compact embeddings between function spaces, compact fractional integral operators play a significant role. We start our discussion about these operators with the classical compactness result, which states that if G is a bounded domain in $\mathbb{R}^n$, $p \leqslant q < \frac{pn}{n-\alpha p}$, then the Riesz potential operator

$$I_{\alpha,G} f(x) = \int\limits_G \frac{f(y)}{|x-y|^{n-\alpha}}\, dy, \quad 0 < \alpha < n,$$

is compact from $L^p(G)$ to $L^q(G)$ (see, e.g., Krasnosel'skii et al. [200, p. 146]).

Let (X, d, μ) be a quasi-metric measure space. Our aim is to provide a complete characterization of those measures μ for which the fractional integrals (or Riesz potentials) $J_{\gamma,\mu}$ associated with μ and a parameter $\gamma \in (0, 1)$ given by the formula

$$J_{\gamma,\mu} f(x) := \int\limits_X \frac{f(y)}{d(x, y)^{1-\gamma}}\, d\mu(y), \quad f \in L^p(X, \mu), \quad x \in X$$

are compact operators from $L^p(\mu)$ into $L^q(\mu)$ for $1 < p < q < \infty$.

As an application, we obtain weighted variant of compactness criteria for Riesz potential operator $I_{\alpha,G}$, where G is a bounded domain of $\mathbb{R}^n$ and $0 < \alpha < n$, and for fractional integral operators defined on a rectifiable curve Γ in the complex plane $\mathbb{C}$:

$$I_{\alpha,\Gamma}\varphi(t) = \int\limits_\Gamma \frac{\varphi(\tau)}{|t-\tau|^{1-\alpha}}\, d\nu(\tau), \quad \varphi \in L^p(\Gamma, \nu), \quad t \in \Gamma,$$

where $0 < \alpha < 1$ and ν is the arc-length measure on Γ.

To study mapping properties of fractional and singular integrals defined with respect to a non-doubling measure μ, it is important to have the boundedness of modified Hardy–Littlewood maximal operator:

$$\widetilde{M} f(x) = \sup_{r>0} \frac{1}{\mu(B(x, N_0 r))} \int\limits_{B(x,r)} |f|\, d\mu, \quad f \in L^1_{\mathrm{loc}}(\mu), \quad x \in X,$$

where the constant N_0 depends only on a quasi-metric (see, e.g., Edmunds et al. [92, Chapters 6, 8], and references cited therein).

The above operator is a modification of the Hardy–Littlewood maximal operator:

$$Mf(x) = \sup_{r>0} \frac{1}{\mu(B(x, r))} \int\limits_{B(x,r)} |f|\, d\mu, \quad f \in L^1_{\mathrm{loc}}(\mu), \quad x \in X.$$

3.4.1 Preliminaries

Let (X, d, μ) be a quasi-metric space with the quasi-metric constant κ (see Sect. 3.1).

In what follows, we denote by $L^1_{\text{loc}}(\mu)$ the space of all $f \in L^0(\mu)$ which are μ-integrable on all balls of X.

Recall that a measure μ is said to be upper Ahlfors β-regular with $\beta > 0$ if there exists a constant $C_0 > 0$ such that (3.1) holds for every $x \in X$ and arbitrary $r \in (0, d_X)$, where d_X denotes the diameter of X.

For such measure μ, we define a family $\{\Omega_{\beta,\ell}\}_{\ell \in (0,d_X)}$ of positive and finite functions $\Omega_\ell \colon X \to (0, \infty)$ given by the formula:

$$\Omega_{\beta,\ell}(x) = \sup_{0<r\leqslant \ell} \frac{\mu(B(x,r))}{r^\beta}, \qquad x \in X,$$

which will play an important role in the proofs of the main statements.

Let (Ω, Σ, μ) be a σ-finite measure space and $L^0(\mu) := L^0(\Omega, \mu)$ be the space of (equivalence classes) of all real-valued μ-measurable functions. A linear subspace E of $L^0(\mu)$ is called an (order) ideal of measurable functions, whenever it follows from $f \in E$ and $g \in L^0(\mu)$, with $|g| \leqslant |f|$ that $g \in E$. An order ideal $E \subset L^0(\mu)$ provided with an absolute and monotone norm $\|\cdot\|$ is called a normed function lattice in $L^0(\mu)$ (or on (Ω, Σ, μ)). If the normed function lattice is norm complete and there exists $f \in E$ such that $f > 0$ μ-a.e., then E is called a Banach function lattice in $L^0(\mu)$.

We say that a BFL X in $L^0(\mu)$ satisfies the *Fatou property* (resp., *weak Fatou property*) if for any $f \in L^0(\mu)$, $f_n \in X$ such that $0 \leqslant f_n \leqslant f$, $f_n \uparrow f$ a.e. and $\sup \|f_n\|_E < \infty$, we have that $f \in X$ and $\|f_n\|_X \to \|f\|_X$ (resp., whenever if $f_n, f \in X, 0 \leqslant f_n \leqslant f$, $f_n \uparrow f$ a.e., then $\|f_n\|_X \to \|f\|_X$).

The Köthe dual space X' of X is the collection of all elements $g \in L^0(\mu)$ such that

$$\|g\|_{X'} := \sup\left\{ \int_\Omega |fg|\, d\mu;\ \|f\|_X \leqslant 1\right\} < \infty.$$

The space X' equipped with the norm $\|\cdot\|_{X'}$ is a Banach function lattice in $L^0(\mu)$ lattice with the Fatou property. It is well-known that $X = X''$ with equality of norms whenever X satisfies the Fatou property (see Kantorovich and Akilov [161]).

Let $(\mathcal{S}, \Sigma_1, \mu)$ and $(\mathcal{T}, \Sigma_2, \nu)$ be two σ-finite measure spaces, and let F and E be ideals of measurable functions in $L^0(\mathcal{S}, \mu)$ and $L^0(\mathcal{T}, \nu)$, respectively. Then, a linear mapping $T \colon E \to F$ is called an *integral operator* if there exists a $(\mu \times \nu)$-

measurable function (called a kernel) $K \in L^0(\mathcal{S} \times \mathcal{T}, \mu \times \nu)$ such that, for every $f \in E$,

$$Tf(s) = \int_{\mathcal{T}} K(s,t) f(t)\, d\nu(t)$$

holds for μ-almost all $s \in \mathcal{S}$.

If (Ω, Σ, μ) is a measure space and $1 \leqslant p \leqslant \infty$, the space $L^p(\mu) \equiv L^p(\Omega, \mu)$ is the usual Lebesgue space of p-integrable functions with the usual norm. If $\Omega = G$ and μ is given by $d\mu = v\, d\lambda_n$, where G is a measurable subset in $\mathbb{R}^n$ and v a weight function (i.e., $v\colon G \to (0, \infty)$ is a positive measurable function) and λ_n is the Lebesgue measure, then we denote $L^p(\Omega, \mu)$ by $L^p_v(G)$.

Let $F \subset L^0(\mathcal{S}, \mu)$ and $E \subset L^0(\mathcal{T}, \nu)$ be BFLs. Assume that F has the weak Fatou property. We denote by $E'[F]$ the mixed Banach function lattice of all $K \in L^0(\mathcal{S} \times \mathcal{T}, \mu \times \nu)$ such that the function g_K given by $g_K(t) := \|K(\cdot, t)\|_F$ for all $t \in \mathcal{T}$ belongs to E', equipped with the norm:

$$\|K\|_{E'[F]} := \|g_K\|_{E'}.$$

In what follows for any given $K \in E'[F]$, we let, for all $(s,t) \in \mathcal{S} \times \mathcal{T}$:

$$\widetilde{K}(s,t) := \begin{cases} \frac{K(s,t)}{g_K(t)} & \text{if } g_K(t) > 0 \\ 0 & \text{if } g_K(t) = 0\,. \end{cases}$$

We will need the following factorization result.

Lemma 3.39 *If $K \in E'[F]$, then a linear mapping T_K defined on E by the formula*

$$T_K f(s) := \int_{\mathcal{T}} K(s,t) f(t)\, d\nu(t), \qquad f \in E,\ s \in \mathcal{S}$$

is a bounded integral operator from E into F with $\|T_K\|_{E \to F} \leqslant \|K\|_{E'[F]}$ that admits the following factorization:

$$T_K \colon E \xrightarrow{M} L_1(\nu) \xrightarrow{T_{\widetilde{K}}} F,$$

where M is a pointwise multiplication operator given by $Mf := f g_K$ for all $f \in E$ with $\|M\| = \|K\|_{E'[F]}$ and $T_{\widetilde{K}}$ is an integral operator with $\|T_{\widetilde{K}}\| \leqslant 1$.

Proof Clearly, $g_K \in E'$ implies that the multiplication operator $M\colon E \to L_1(\nu)$ with

$$\|M\| = \sup_{\|f\|_E \leqslant 1} \int_{\mathcal{S}} |f g_K|\, d\nu = \|g_K\|_{E'} = \|K\|_{E'[F]}.$$

Observe that $\widetilde{K} \in L^0(\mu \times \nu)$ with $\|\widetilde{K}(\cdot, t)\|_F \leqslant 1$ for all $t \in \mathcal{T}$. Since F has the Fatou property, it follows from generalized Minkowski's inequality that, for every $h \in L_1(\nu)$, we have:

$$\|T_{\widetilde{K}} h\|_F = \Big\| \int_{\mathcal{S}} \widetilde{K}(\cdot, t) h(t)\, d\nu(t) \Big\|_F \leqslant \int_{\mathcal{S}} \|\widetilde{K}(\cdot, t)\|_F |h(t)|\, d\nu(t) \leqslant \|h\|_{L_1(\nu)}.$$

To conclude it is enough to observe that $T_K = T_{\widetilde{K}} M$. □

To show applications of the above lemma, we recall, following Grothendieck [124], that a Banach space E has the *Dunford–Pettis* property whenever $x_n \to 0$ weakly in X and $x_n^* \to 0$ weakly in E^* imply $x_n^*(x_n) \to 0$. We will use the well-known characterization of the Dunford–Pettis property (see Grothendieck [124]) which states that a Banach space E has the Dunford–Pettis property if and only if every weakly compact operator from E onto an arbitrary Banach space maps weakly compact sets into norm compact sets.

We note that Grothendieck [124] proved that any L^1-space on any measure space has the Dunford–Pettis property. As a consequence of this fact and Lemma 3.39, we have the following result.

Theorem 3.40 *Let E and F be reflexive BFLs in $L^0(\mathcal{T}, \nu)$ and $L^0(\mathcal{S}, \mu)$, respectively. Then, every integral operator T_K generated by a kernel $K \in E'[F]$ is a compact operator from E into F.*

Proof It is well-known that reflexive BFLs are KB-spaces (i.e., every increasing norm-bounded sequence in the "positive cone" is norm-convergent), and so they have the Fatou property (see, e.g., Kantorovich and Akilov [161] or Lindenstrauss and Tzafriri [211]). From Lemma it follows that $T_K : E \to F$ is a bounded operator which admits the following factorization:

$$T_K : E \xrightarrow{M} L_1(\nu) \xrightarrow{T_{\widetilde{K}}} F.$$

Since E and F are reflexive spaces, M and $T_{\widetilde{K}}$ are weakly compact operators. Since $L^1(\nu)$ has the Dunford–Pettis property, the abovementioned facts imply that the image $T_K(B_E)$, of the closed unit ball B_E of E, is a compact set in F, and this completes the proof. □

Since any L^p-space is reflexive for any $1 < p < \infty$, it follows immediately from the above corollary the following result (see Kantorovich and Akilov [161, p. 79] or Kantorovich and Akilov [161, Chapter XI, Section 3] for direct different proofs).

Corollary 3.41 *If $T_K : L^p(\nu) \to L^q(\mu)$ is a bounded integral operator generated by a kernel $K \in L^{p'}[L^q]$ for some $p, q \in (1, \infty)$, then T_K is a compact operator.*

3.4.2 Necessary and Sufficient Conditions for Compactness

In this section we provide necessary and sufficient conditions for compactness of the integral operator $J_{\gamma,\mu}$ from $L^p(X,\mu)$ to $L^q(X,\mu)$, with $1 < p < q < \infty$, associated with a finite quasi-metric space (X, d, μ).

Lebesgue space on an (X, d, μ) will be denoted by $L^p_\mu(X)$ (or by $L^p(X,\mu)$), where $1 \leqslant p < \infty$.

Now we prove the main result of this section.

Theorem 3.42 *Let* $0 < \gamma < 1 < p < q < \infty$ *and let* (X, d, μ) *be a quasi-metric measure space with finite* μ*. Then, the integral operator* $J_{\gamma,\mu}$ *is compact from* $L^p(X,\mu)$ *to* $L^q(X,\mu)$ *if and only if* μ *is upper Ahlfors* β*-regular with* β *given by* $(1-\gamma)/\beta = 1/p' + 1/q$*, and*

$$\lim_{\ell\to 0}\sup_{x\in X}\Omega_{\beta,\ell}(x) = 0.$$

In the proof of the above theorem, we will use the following lemmas.

Lemma 3.43 *Let* μ *be a upper Ahlfors* β*-regular measure on a quasi-metric space* (X, d, μ) *with* $\beta < (1-\gamma)r$ *for some* $\gamma \in (0, 1)$ *and* $r > 0$*. Then, there is a positive constant* $C = C(r, \gamma)$ *such that, for all* $\ell \in (0, d_X)$ *and all* $x \in X$*:*

$$\int_{B(x,\ell)^c} d(x, y)^{(\gamma-1)r}\, d\mu(y) \leqslant C\ell^{\beta+(\gamma-1)r}.$$

Proof For simplicity of notation, we put $\eta := (\gamma - 1)r$. Then, for any fixed $\ell \in (0, d_X)$ and $x \in X$, we have:

$$I_\ell(x) := \int_{B(x,\ell)^c} d(x, y)^{(\gamma-1)r}\, d\mu(y) = \int_0^\infty \mu\big(E_{\ell,\lambda,x}\big)\, d\lambda,$$

where for every $\lambda > 0$,

$$E_{\ell,\lambda,x} := \big\{y \in B(x,\ell)^c;\ d(x, y) < \lambda^{1/\eta}\big\}.$$

Observe that the set $E_{\ell,\lambda,x}$ is an empty set whenever $\lambda \geqslant \ell^\eta$. This implies that

$$I_\ell(x) = \int_0^{\ell^\eta} \mu(E_{\ell,\lambda,x})\, d\lambda.$$

Clearly, $E_{\ell,\lambda,x} \subset B(x, \lambda^{1/\eta})$ and so by our hypothesis on μ, there exists a constant $C_0 > 0$ such that

$$\mu(E_{\ell,\lambda,x}) \leqslant C_0 \lambda^{\beta/\eta}, \quad \lambda > 0.$$

Combining the above facts with $\beta/\eta > -1$ yields:

$$I_\ell(x) \leqslant C_0 \int_0^{\ell^\eta} \lambda^{\beta/\eta}\, d\lambda = \frac{C_0}{\beta/\eta + 1} \ell^{\beta+\eta},$$

and this completes the proof. □

Lemma 3.44 *Let μ be a upper Ahlfors β-regular measure on a quasi-metric space (X, d, μ) with $\beta < (1-\gamma)r$ for some $\gamma \in (0, 1)$ and $r > 0$. Then, there is a positive constant $C = C(r, \gamma)$ such that, for all ℓ and s satisfying the condition $0 < s < \ell < d_X$, and all $x \in X$:*

$$I_{\ell,s}(x) = \int_{B(x,\ell)\setminus B(x,s)} d(x, y)^{(\gamma-1)r}\, d\mu(y) \leqslant C\, \Omega_{\beta,\ell}(x)\big(s^{(\gamma-1)r+\beta} + \ell^{(\gamma-1)r+\beta}\big).$$

Proof Following the proof of Lemma 3.43, we let $\eta := (\gamma-1)r$. Then, for any fixed $0 < s < \ell < d_X$ and $x \in X$, we have:

$$\begin{aligned}
I_{\ell,s}(x) &= \int_0^{s^\eta} \mu\big(\big(B(x,\ell) \setminus B(x,s)\big) \cap B(x, \lambda^{1/\eta})\big)\, d\lambda \\
&= \int_{\ell^\eta}^{s^\eta} \mu\big(\big(B(x,\ell) \setminus B(x,s)\big) \cap B(x, \lambda^{1/\eta})\big)\, d\lambda \\
&\quad + \int_0^{\ell^\eta} \mu\big(\big(B(x,\ell) \setminus B(x,s)\big) \cap B(x, \lambda^{1/\eta})\big)\, d\lambda \\
&\leqslant \int_{\ell^\eta}^{s^\eta} \mu\big(B\big(x, \lambda^{1/\eta}\big)\big)\, d\lambda + \int_0^{\ell^\eta} \mu\big(B(x,\ell)\big) d\lambda \\
&=: f_{\ell,s}(x) + g_{\ell,s}(x).
\end{aligned}$$

Now we observe that $\lambda > \ell^\eta$ is equivalent to $\lambda^{1/\eta} \leqslant \ell$ (by $\eta < 0$) and so

$$\mu\big(B(x, \lambda^{1/\eta})\big) \leqslant \Omega_{\beta,\ell}(x)\lambda^{1/\eta}.$$

This estimate combined with $\beta/\eta > -1$ yields with $C = C(r, \gamma) > 0$:

$$f_{\ell,s}(x) \leqslant C\, \Omega_{\beta,\ell}(x) s^{\beta+\eta}.$$

To finish it is enough to observe that

$$g_{\ell,s}(x) \leqslant \ell^{\eta} \mu(B(x, \ell)) \leqslant \ell^{\eta+\beta} \Omega_{\beta,\ell}(x),$$

which ends the proof. □

Now we are ready to prove our main result.

Proof of Theorem 3.42 For simplicity of notation, we put $T := J_{\gamma,\mu}$ and $L^r := L^r(X, \mu)$ for any $1 \leqslant r < \infty$. We have explained that the boundedness of T from L^p to L^q is equivalent to the Ahlfors β-regularity of μ. We prove sufficiency of the condition:

$$\lim_{\ell \to 0} \sup_{x \in X} \Omega_{\beta,\ell}(x) = 0.$$

Fix $\ell \in (0, d_X)$. Following Hedberg [135], for every $f \in L^p$, we decompose Tf as follows:

$$\begin{aligned} Tf(x) &= \int\limits_{B(x,\ell)} f(y) d(x, y)^{\gamma-1} d\mu(y) + \int\limits_{B(x,\ell)^c} f(y) d(x, y)^{\gamma-1}\, d\mu(y) \\ &=: U_\ell f(x) + V_\ell f(x), \quad x \in X. \end{aligned}$$

Observe that operator V_ℓ is an integral operator generated by kernel K_ℓ given by

$$K_\ell(x, y) := \chi_{B(x,\ell)^c}(y) d(x, y)^{\gamma-1}, \quad (x, y) \in X \times X.$$

We claim that $V_\ell \colon L^p \to L^q$ is compact for sufficiently small ℓ. To prove this, we show that $K_\ell \in L^{p'}[L^q]$ and then we apply Corollary 3.41. First we note that our hypothesis on μ implies that $\beta < (1 - \gamma)p'$. Applying Lemma 3.43 with $r = p'$, we conclude that there exists a positive constant $C = C(p, q, \gamma)$ such that, for all $x \in X$ and $\ell \in (0, d_X)$,

$$\int\limits_{B(x,\ell)^c} d(x, y)^{(\gamma-1)p'}\, d\mu(y) \leqslant C \ell^{(\gamma-1)p'+\beta}.$$

This proves that $K_\ell \in L^\infty[L^q] \subset L^{p'}[L^q]$ and so the claim is proved.

We will now estimate the norm of the operator U_ℓ. For any $0 \leqslant f \in L^p$ with $\|f\|_{L^p} \leqslant 1$ and all $x \in X$, we have:

$$
\begin{aligned}
U_\ell f(x) &= \sum_{k=0}^{\infty} \int\limits_{B(x,2^{-k}\ell)\setminus B(x,2^{-(k+1)}\ell)} f(y)d(x,y)^{\gamma-1}d\mu(y) \\
&\leqslant C_\gamma \sum_{k=0}^{\infty} (2^{-k}\ell)^{\gamma-1} \frac{\mu(B(x,N_0 2^{-k}\ell))}{\mu(B(x,N_0 2^{-k}\ell))} \int\limits_{B(x,2^{-k}\ell)} f(y)d\mu(y) \\
&\leqslant C_\gamma \ell^{\gamma-1+\beta} \widetilde{M}_\ell f(x) \Omega_{\beta,\ell}(x),
\end{aligned}
$$

where $C_\gamma = 2^{1-\gamma}$ and

$$
\widetilde{M}_\ell f(x) = \sup_{0<r<\ell} \frac{1}{\mu(B(x,N_0 r))} \int\limits_{B(x,r)} |f|\, d\mu, \qquad x \in X.
$$

For a given $x \in X$, we put:

$$
r_x := \big(\widetilde{M}_\ell f(x)\big)^{-p/\beta} \Omega_{\beta,\ell}(x)^{-1/\beta}.
$$

We consider two cases:
Case (i): $\ell \leqslant r_x$. Then we have that for some constant $C > 0$ which depends on p, q and γ:

$$
\begin{aligned}
U_\ell f(x) &\leqslant C\, \Omega_{\beta,\ell}(x)^{(1-\gamma)/\beta} \big(\widetilde{M}_\ell f(x)\big)^{1-\frac{p(\gamma-1+\beta)}{\beta}} \\
&= C\, \Omega_{\beta,\ell}(x)^{(1-\gamma)/\beta} \big(\widetilde{M}_\ell f(x)\big)^{p/q}.
\end{aligned}
$$

Case (ii): $\ell > r_x$. We represent $U_\ell f(x)$ as follows:

$$
\begin{aligned}
U_\ell f(x) &= \int\limits_{B(x,r_x)} f(y)d(x,y)^{\gamma-1}\, d\mu(y) + \int\limits_{B(x,\ell)\setminus B(x,r_x)} f(y)d(x,y)^{\gamma-1}d\mu(y) \\
&=: R_\ell f(x) + S_\ell f(x).
\end{aligned}
$$

Further, repeating the arguments used above, we deduce that there exists a constant $C = C_{p,q,\gamma} > 0$ such that

$$
\begin{aligned}
R_\ell f(x) &\leqslant C\, \Omega_{\beta,\ell}(x)^{(1-\gamma)/\beta} \big(\widetilde{M}_\ell f(x)\big)^{1-\frac{p(\gamma-1+\beta)}{\beta}} \\
&= C\, \Omega_{\beta,\ell}(x)^{(1-\gamma)/\beta} \big(\widetilde{M}_\ell f(x)\big)^{p/q}.
\end{aligned}
$$

Now we estimate $S_\ell f(x)$. By Lemma 3.44 with $s = r_x$, Hölder's inequality, the fact that $\beta + (\gamma - 1)p' < 0$ and the condition $\|f\|_{L^p} \leqslant 1$, for some $C > 0$, we have:

$$S_\ell f(x) \leqslant \Big(\int_{B(x,\ell)\setminus B(x,r_x)} d(x,y)^{(\gamma-1)p'} d\mu(y) \Big)^{1/p'}$$

$$\leqslant C\Omega_{\beta,\ell}(x)^{1/p'} \Big(r_x^{(\gamma-1)p'+\beta} + \ell^{(\gamma-1)p'+\beta} \Big)^{1/p'}$$

$$\leqslant 2C\Omega_{\beta,\ell}(x)^{1/p'} r_x^{\gamma-1+\beta/p'} = 2C\Omega_{\beta,\ell}(x)^{(1-\gamma)/\beta} \big(\widetilde{M}_\ell f(x)\big)^{p/q}.$$

Hence,

$$S_\ell f(x) \leqslant 2C\big(\sup_{y\in X} \Omega_{\beta,\ell}(y)\big)^{(1-\gamma)/\beta} \big(\widetilde{M}_\ell f(x)\big)^{p/q}.$$

Consequently, there exists $C' > 0$ such that for all $x \in X$, we have:

$$U_\ell f(x) \leqslant C'\big(\sup_{y\in X} \Omega_{\beta,\ell}(y)\big)^{(1-\gamma)/\beta} \big(\widetilde{M}_\ell f(x)\big)^{p/q}.$$

Combining the above estimates, we conclude that there exists a constant $\widetilde{C} > 0$ such that

$$\big\| U_\ell \big\|_{L^p \to L^q} \leqslant \widetilde{C}\big(\sup_{x\in X} \Omega_{\beta,\ell}(x)\big)^{(1-\gamma)/\beta}.$$

Thus we have proved that

$$\big\| T - V_\ell \big\|_{L^p \to L^q} \leqslant \big\| U_\ell \big\|_{L^p \to L^q} \leqslant \widetilde{C}\big(\sup_{x\in X} \Omega_{\beta,\ell}(x)\big)^{(1-\gamma)/\beta}.$$

Therefore, our hypothesis yields that

$$\big\| T - V_\ell \big\|_{L^p \to L^q} \to 0 \quad \text{as } \ell \to 0.$$

Since $\{V_\ell\}_{\ell\in(0,d_X)}$ is a net of compact operators, T is necessarily compact. This completes the proof of sufficiency.

Now we prove necessity. Assume that $T\colon L_p \to L_q$ is compact operator. We claim that $\lim_{\ell\to 0} \sup_{x\in X} \Omega_{\beta,\ell}(x) = 0$. Suppose the contrary. Then, there exists a sequence (ℓ_n) with $\ell_n \to 0$ as $n \to \infty$ such that

$$\sup_{x\in X} \Omega_{\beta,\ell_n}(x) > 2\varepsilon, \qquad n \in \mathbb{N}$$

for some $\varepsilon > 0$. This implies that there exist sequence (x_n) in X and real sequence (r_n) with $r_n \to 0$ as $n \to \infty$ (by $r_n \leqslant \ell_n$) such that, for each $n \in \mathbb{N}$, we have:

$$\frac{\mu(B(x_n, r_n))}{r_n^\beta} > \varepsilon. \tag{3.39}$$

Consider a sequence (f_n) given for each $n \in \mathbb{N}$ by

$$f_n := \frac{\chi_{B(x_n,r_n)}}{\mu(B(x_n, r_n))^{1/p}}.$$

Clearly, $\|f_n\|_{L^p} = 1$ for each n. We claim that $f_n \to 0$ weakly in L^p as $n \to \infty$. In fact, for any functional x^* in $(L^p)^*$, there exists a function $g \in L^{p'}$ such that

$$x^*(f) = \int_X fg \, d\mu, \qquad f \in L^p.$$

Combining Hölder's inequality with $\mu(B(x,r)) \leqslant Cr^\beta$ for all $x \in X$ and $r < d_X$ yields:

$$|x^*(f_n)| \leqslant \|f_n\|_{L^p} \|\chi_{B(x_n,r_n)}\|_{L^{p'}} = \mu(B(x_n, r_n))^{1/p'} \leqslant Cr_n^{\beta/p'}.$$

Hence, $f_n \to 0$ weakly in L^p as required. Since T is a compact operator, it follows that $Tf_n \to 0$ in L^q. To conclude observe that by (3.39), we get that

$$\begin{aligned}
\|Tf_n\|_{L^q} &\geqslant \Bigg(\int_{B(x_n,r_n)} \Bigg(\int_{B(x_n,r_n)} \frac{f_n(y)}{d(x,y)^{1-\gamma}} \, d\mu(y) \Bigg)^q d\mu(x) \Bigg)^{1/q} \\
&\geqslant \frac{1}{(2\kappa r_n)^{1-\gamma}} \mu(B(x_n, r_n))^{1-\frac{1}{p}+\frac{1}{q}} \\
&= (2\kappa)^{\gamma-1} \Bigg(\frac{\mu(B(x_n, r_n))}{r_n^\beta} \Bigg)^{\frac{1-\gamma}{\beta}} \geqslant (2\kappa)^{\gamma-1} \varepsilon^{\frac{1-\gamma}{\beta}},
\end{aligned}$$

which is a contradiction. This completes the proof. □

To show some applications of Theorem 3.42, we need some further notation. Let G (resp., Γ) be a subset in $\mathbb{R}^n$ (resp., a rectifiable curve in the complex plain $\mathbb{C}$), then we let

$$G(x,r) := B(x,r) \cap G, \qquad (\text{resp., } \Gamma(z,r) := D(z,r) \cap \Gamma),$$

where $B(x,r)$ (resp., $D(z,r)$) is a ball in $\mathbb{R}^n$ with radius $r > 0$ and center x (resp., disc in the complex plain $\mathbb{C}$ with radius r and center z).

If A is a measurable subset of a quasi-metric space (X, d, μ), then we consider A with the induced quasi-metric and with a measure μ restricted to the induced σ-algebra on A. In particular, balls in A have a form $A \cap B(x,r)$, where $B(x,r)$ is any ball in X. Consequently, we have that $G(x,r)$ and $\Gamma(z,r)$ are balls in $X = G$ and $X = \Gamma$, respectively.

We emphasize that basic mapping properties of fractional integral operators on Euclidean spaces defined with respect to the Lebesgue measure can be found, e.g., in the monograph Samko et al. [296].

Theorem 3.45 *Let $1 < p < q < \infty$ and let $n(1/p - 1/q) < \alpha < n$. Suppose that G is a bounded domain in $\mathbb{R}^n$ and v is a weight function on G. Then the operator $I_{\alpha,G}$ is compact from $L^p_{v^{1-p}}(G)$ to $L^q_v(G)$ if and only if μ given by $d\mu = v d\lambda_n$ is upper Ahlfors σ-regular with σ defined by $(n - \gamma)/\sigma = 1/p' + 1/q$, and*

$$\lim_{\ell \to 0} \sup_{x \in G} \Omega_{\sigma,\ell}(x) = 0.$$

Proof It is easy to see that the operator $I_{\alpha,G}$ is compact from $L^p_{v^{1-p}}(G)$ to $L^q_v(G)$ if and only if the operator

$$J_{\gamma,G,\mu} f(x) = \int_G \frac{f(y)}{d(x,y)^{1-\gamma}} d\mu(x)$$

is compact from $L^p(G,\mu)$ to $L^q(G,\mu)$ with $d\mu = v d\lambda_n$, $d(x,y) = |x - y|^n$ and $\gamma = \alpha/n$. By Theorem 3.42 we have that the latter compactness holds if and only if

$$\sup_{x \in G,\, r < d_G} \frac{\mu(\overline{G}(x,r))}{r^\eta} < \infty \quad \text{and} \quad \lim_{\ell \to 0} \sup_{x \in G} \overline{\Omega}_{\eta,\ell}(x) = 0,$$

where

$$\overline{\Omega}_{\eta,\ell}(x) := \sup_{r \leqslant \ell < d_G} \frac{\mu(\overline{G}(x,r))}{r^\eta}, \qquad x \in G$$

with

$$\frac{1-\gamma}{\eta} = \frac{1}{p'} + \frac{1}{q}, \qquad \overline{G}(x,r) := \{y \in G;\ |x - y| < r^{1/n}\}.$$

To conclude it is enough to observe that for $R = r^n$, we have:

$$\frac{\mu(\overline{G}(x,r))}{r^\eta} = \frac{\mu(G(x,R))}{R^{\frac{pq(n-\alpha)}{p+q-pq}}} = \frac{\mu(G(x,R))}{R^\sigma}.$$

□

Similarly we obtain the following result.

Theorem 3.46 *Let $1 < p < q < \infty$ and let $1/p - 1/q < \alpha < 1$. Suppose that Γ is a rectifiable curve on $\mathbb{C}$ and v is a weight function on Γ. Then the operator $K_{\alpha,\Gamma}$*

is compact from $L^p_{v^{1-p}}(\Gamma)$ *to* $L^q_v(\Gamma)$ *if and only if* μ *defined by* $d\mu = v d\nu$ *is upper Ahlfors* β*-regular with* β *given by* $(1-\alpha)/\beta = 1/p' + 1/q$, *and*

$$\lim_{\ell \to 0} \sup_{z \in \Gamma} \Omega_{\beta,\ell}(z) = 0.$$

We conclude with the following remark that our results could be applied to get sufficient conditions for compactness of positive operators from $L^p(X,\mu)$ into $L^q(X,\mu)$, which are dominated by discussed potential operators. For example, we shall associate with a given metric measure space (X, d, μ) a fractional kernel $K_\gamma \in L^0(\mu \times \mu)$ of type $\gamma \in (0,1)$, i.e., such that K_γ satisfies the following condition:

$$|K_\gamma(x,y)| \leqslant \frac{C}{d(x,y)^{1-\gamma}}$$

for all $x, y \in X$ with $x \neq y$ and some $C > 0$. The corresponding kernel K_γ generates a fractional integral operator given by

$$K_\gamma f(x) = \int_X k_\gamma(x,y) f(y)\, d\mu, \qquad f \in L^p(\mu), \quad x \in X.$$

We conclude with the following remark that the sufficient condition for compactness of $J_{\gamma,\mu}$ from $L^p(X,\mu)$ into $L^q(X,\mu)$ for $1 < p < q < \infty$ assures the compactness of the fractional integral operator K_γ. To get this statement, it is enough to apply a deep result due to Dodds and Fremlin [78] true even in the setting of abstract Banach lattices, which states that if E and F are Banach lattices with E^* and F having order continuous norms and $S\colon E \to F$ is a positive operator dominated by a compact operator T (i.e., $0 \leqslant S \leqslant T$), then S is necessarily a compact operator.

3.5 Compactness of Fractional-Type Integral Operators Defined on an SHT

We deal with the following fractional type integral operator:

$$K_\alpha f(x) := \int_X k(x,y) f(y)\, d\mu(y), \quad k(x,y) \geqslant 0, \tag{3.40}$$

where the kernel $k(x,y)$ satisfies the condition: there are positive constant c_1 and c_2 such that for all $x, y \in X$, $x \neq y$,

$$c_1 d(x,y)^{\alpha - N} \leqslant k(x,y) \leqslant c_2 d(x,y)^{\alpha - N}, \quad 0 < \alpha < N,$$

and (X, d, μ) is an SHT with finite measure μ of dimension N, i.e., the measure μ satisfies the condition: there are positive constants b_1 and b_2 such that

$$b_1 r^N \leqslant \mu(B(x,r) \leqslant b_2 r^N, \quad 0 < r < d_X. \tag{3.41}$$

Here d_X denotes diameter of X.

For instance, the following fractional integrals are examples of operators of type (3.40)

(i)

$$J_{\alpha,\mu,N} f(x) := \int\limits_X \frac{f(y)}{d(x,y)^{N-\alpha}} d\mu(y), \quad 0 < \alpha < N;$$

(ii)

$$I_{\gamma,X} f(x) := \int\limits_X \frac{f(y)}{\mu B(x, d(x,y))^{1-\gamma}} d\mu(y), \ 0 < \gamma < 1.$$

Let Ω be a domain in $\mathbb{R}^n$, and let I_γ^Ω be a fractional integral operator defined by

$$I_{\gamma,\Omega} f(x) = \frac{1}{c_\gamma} \int\limits_\Omega \frac{f(y)}{|x-y|^{n-\gamma}}, \quad 0 < \gamma < n,$$

where $c_\gamma = \frac{\pi^{n/2} 2^\gamma \Gamma(\gamma/2)}{\Gamma\left(\frac{n-\gamma}{2}\right)}$.

If $\Omega := \mathbb{R}^n$, then we have the Riesz potential operator $I_\gamma := I_{\gamma,\mathbb{R}^n}$.

The classical Sobolev inequality says that if $1 < s < \infty$, $0 < \gamma < n/s$, $s^* := \frac{ns}{n-\gamma s}$, then

$$\|I_\gamma f\|_{L^{s^*}(\mathbb{R}^n)} \leqslant C \|f\|_{L^s(\mathbb{R}^n)}, \quad f \in L^s(\mathbb{R}^n).$$

It is also known that $I_{\gamma,\Omega}$ does not act compactly from $L^s(\Omega)$ to $L^{s^*}(\Omega)$, but it is compact (completely continuous) from $L^s(\Omega)$ to $L^r(\Omega)$ for $r < s^*$ provided that Ω is a bounded domain (see, e.g., Krasnosel'skii et al. [200, Section 8]).

Let $1 \leqslant p, q < \infty$. As before, we denote the Lebesgue space defined with respect to a measure μ by $L^p(\mu)$. Our aim is to establish necessary and sufficient condition on exponent q governing the compactness of the fractional integral operator $J_{\alpha,\mu,N}$ from $L^p(\mu)$ into $L^q(\mu)$. In particular, this condition on q reads as follows: $q < p^*$, where p^* is the Sobolev exponent for p given by the condition:

$$p^* := \frac{pN}{N - \alpha p}. \tag{3.42}$$

We follow definitions and results from Sect. 3.4.1, restating some for the convenience of the reader.

Let (X, d, μ) be a quasi-metric measure space.

Recall that a measure μ is said to satisfy a doubling condition ($\mu \in DC(X)$) if there is a constant $C_{DC} > 0$ such that

$$\mu B(x, 2r) \leqslant C_{DC} \mu B(x, r) \tag{3.43}$$

for every $x \in X$ and $r > 0$.

The best possible constant in (3.43) is called doubling constant and is denoted again by C_{DC}.

Any quasi-metric d defines a topology, for which the balls $B(x, r)$ form a basis. Unlike the metric case, i.e., $\kappa = 1$, the open balls defined by a quasi-metric d may not be open sets. Further, there exist positive constants c_1 and c_2 and a quasi-distance d' such that d is equivalent to d' in the sense that $c_1 d' \leqslant d \leqslant c_2 d'$ and such that the d'-balls are open sets in the topology induced by both d and d'. Furthermore, $d' = \rho^\alpha$, where ρ is a metric on X and α is a positive constant.

Recall that a space of homogeneous type (SHT) is a triple (X, d, μ), where X is a set, d is a quasi-metric on X and μ is a doubling measure.

Further, recall that a quasi-metric measure spaces (X, d, μ) is called normal SHT if condition (3.41) holds for μ. The constant N is called dimension of normal SHT.

Throughout the section we will assume that condition (3.41) is satisfied. In this case the doubling condition is automatically satisfied for μ. Also, it is clear that all balls in such an SHT have finite measure and that $\mu\{x\} = 0$ for all $x \in X$.

We list some examples of normal SHT:

(a) the domains in $\mathbb{R}^N \subset \mathbb{R}^n$ satisfying so-called $\mathcal{A}$ condition: there is a positive constant $C > 0$ such that $|\Omega \cap B(x, r)| \geqslant C r^N$, where $|E|$ is the Lebesgue measure induced on Ω;

(b) regular curves, i.e., rectifiable curves Γ satisfying the condition:

$$\nu\big(\Gamma \cap D(x, r)\big) \leqslant Cr, \tag{3.44}$$

where $D(x, r)$ is the disc with center x and radius $r > 0$ and ν is the arc-length measure on Γ;

(c) nilpotent Lie group G with Haar measure is a normal SHT with $N = Q$, where Q is a homogeneous dimension of G.

To prove the main result, we will need the following lemma (cf. Lemma 3.44).

Lemma 3.47 *Let $0 < \gamma < \frac{N}{r_0}$, where r_0 be a constant such that $r_0 > 1$. Suppose that $0 < \ell_2 < \ell_1 < d_X$ and then there is a positive constant $C_{\gamma,N,r}$ such that the following estimate holds:*

$$\int\limits_{B(x,\ell_1)\backslash B(x,\ell_2)} d(x, y)^{(\gamma-N)r_0'} \leqslant C_{\gamma,N,r_0}\left(\ell_1^{(\gamma-N)r_0'+N} + \ell_2^{(\gamma-N)r_0'+N}\right).$$

Proof Let $\eta := (\gamma - N)r_0'$. Let us fix $0 < \ell_1 < \ell_2 < d_X$ and $x \in X$. Then

$$I_{\ell,s}(x) = \int_0^{\ell_1^\eta} \mu\big(\big(B(x,\ell_1) \setminus B(x,\ell_2)\big) \cap B(x,\lambda^{1/\eta})\big)\, d\lambda$$

$$= \int_{\ell_1^\eta}^{\ell_2^\eta} \mu\big(\big(B(x,\ell_1) \setminus B(x,\ell_2)\big) \cap B(x,\lambda^{1/\eta})\big)\, d\lambda$$

$$+ \int_0^{\ell_1^\eta} \mu\big(\big(B(x,\ell_1) \setminus B(x,\ell_2)\big) \cap B(x,\lambda^{1/\eta})\big)\, d\lambda$$

$$\leqslant \int_{\ell_1^\eta}^{\ell_2^\eta} \mu\big(B\big(x,\lambda^{1/\eta}\big)\big)\, d\lambda + \int_0^{\ell_1^\eta} \mu\big(B(x,\ell_1)\big) d\lambda$$

$$=: f_{\ell_1,\ell_2}(x) + g_{\ell_1}(x).$$

Now we observe that $\lambda > \ell_1^\eta$ is equivalent to $\lambda^{1/\eta} \leqslant \ell_1$ (by $\eta < 0$) and so

$$\mu\big(B(x,\lambda^{1/\eta})\big) \leqslant C\lambda^{N/\eta}.$$

This estimate combined with $N/\eta > -1$ yields:

$$f_{\ell_1,\ell_2}(x) \leqslant C\ell_2^{N+\eta}.$$

with $C = C(r_0, \gamma) > 0$.

To finish the proof, it is enough to

$$g_{\ell_1}(x) \leqslant \ell_1^\eta \mu(B(x,\ell_1)) \leqslant C\ell_1^{\eta+N}.$$

□

The following statement is known in the literature (see, e.g., Andô [19], Krasnosel'skii et al. [200], Sections 5.3 and 5.4) for Euclidean spaces, but analyzing the proof, it can be also formulated in the case of a SHT.

Theorem 3.48 *Let p and q be constants such that $1 \leqslant q < p < \infty$. If the operator $Kf(x) = \int_X k(x,y)f(y)d\mu(y)$ is bounded from $L^q(X)$ to $L^p(X)$, then is also compact.*

Recall that the centered Hardy–Littlewood maximal operator on X is given by

$$M^{(c)}f(x) := \sup_{r>0} M_r^{(c)}f(x) = \sup_{r>0} \frac{1}{\mu B(x,r)} \int\limits_{B(x,r)} |f(y)|d\mu(y).$$

It is known that (see, e.g., Strömberg and Torchinsky [316]) $M^{(c)}$ is bounded in $L^{s_0}(\mu)$ for $1 < s_0 < \infty$.

3.5.1 Compactness of K_α

Our main statement in this section reads as follows.

Theorem 3.49 *Let $1 \leqslant p, q < \infty$. Suppose that $0 < \alpha < \frac{N}{p}$. Let p^* be defined by (3.42). Then K_α is compact from $L^p(\mu)$ to $L^q(\mu)$ if and only if $q < p^*$.*

Proof *Necessity.* Suppose the contrary: $q \geqslant p^*$. Fix $a \in X$. Take the sequence of balls $B_n := B(a, r_n)$, where $a \in X$:

$$f_n(x) = \frac{\chi_{B_n}(x)}{(\mu(B_n))^{1/p}}, \quad r_n \to 0, \quad n \to \infty.$$

Then applying Hölder's inequality and the fact that $\mu\{a\} = 0$, for $g \in L^{p'}(\mu)$, we have:

$$\left| \int\limits_X f_n(x)g(x)d\mu(x) \right| \leqslant \|g\|_{L^{p'}(B_n)} \to 0$$

as $n \to \infty$. Hence, f_n is weakly convergent to 0. Consequently, since K_α is compact, we have that $\|K_\alpha f\|_{L^q(\mu)} \to 0$ at $n \to \infty$. On the other hand,

$$\|K_\alpha f\|_{L^q(\mu)} \geqslant b_1 \left\| \chi_{B_n}(\cdot) \int\limits_{B_n} \frac{f_n(y)}{d(\cdot, y)^{N-\alpha}} d\mu(y) \right\|_{L^q(\mu)} \geqslant C_{q,N,\alpha} r_n^{\alpha+\frac{N}{q}-\frac{N}{p}}.$$

Since $q \geqslant p^*$, we have that $r_n^{\alpha-\frac{N}{q}-\frac{N}{p}}$ does not converge to 0 which contradicts the fact that $\|K_\alpha f\|_{L^q(\mu)} \to 0$. This completes the proof of necessity.

Sufficiency. Let $q < p^*$. The boundedness of K_α from $L^p(\mu)$ to $L^q(\mu)$ holds since Hölder's and Sobolev's inequalities yield that

$$\|K_\alpha f\|_{L^q(\mu)} \leqslant C\|K_\alpha f\|_{L^{p^*}(\mu)} \leqslant C\|f\|_{L^p(\mu)}.$$

Suppose that $p \leqslant q < p^*$. We use arguments of the proof of Theorem 3.42. We split $K_\alpha f(x)$ as follows:

$$
\begin{aligned}
K_\alpha f(x) &= \int\limits_{B(x,\ell)} f(y)k(x,y)\,d\mu(y) + \int\limits_{X\setminus B(x,\ell)} f(y)k(x,y)\,d\mu(y) \\
&:= K^{(1)}_{\alpha,\ell} f(x) + K^{(2)}_{\alpha,\ell} f(x).
\end{aligned}
$$

Observe that (3.41) and condition $\alpha < N/p$ yield:

$$
\begin{aligned}
\|\chi_{X\setminus B(x,\ell)} k(x,\cdot)\|_{L^{p'}(\mu)} &\leqslant b_2 \|\chi_{X\setminus B(x,\ell)} d(x,\cdot)^{\alpha-N}\|_{L^{p'}(\mu)} \\
&= b_2 \Bigg(\int\limits_{X\setminus B(x,\ell)} d(x,y)^{(\alpha-N)p'} d\mu(y) \Bigg)^{1/p'} \\
&= b_2 \Bigg(\sum_{k=0}^{m_0} \int\limits_{B(x,2^{k+1}\ell)\setminus B(x,2^k\ell)} d(x,\cdot)^{(\alpha-N)p'} d\mu(y) \Bigg)^{1/p'} \\
&\leqslant C_p \ell^{\frac{(\alpha-N)p'+N}{p'}} = C_p \ell^{\alpha-\frac{N}{p}}.
\end{aligned}
$$

where m_0 is an integer depending on ℓ. Hence, $K^{(2)}_{\alpha,\ell}$ is compact for all $\ell \in (0, d_X)$.

To estimate $K^{(1)}_{\alpha,\ell} f(x)$, without loss of generality, we can assume that $f \geqslant 0$ and $\|f\|_{L^p(\mu)} \leqslant 1$. Applying condition (3.41), we have:

$$
\begin{aligned}
K^{(1)}_{\alpha,\ell} f(x) &\leqslant c_1 \sum_{k=0}^{\infty} \int\limits_{B(x,\ell/2^k)\setminus B(x,\ell/2^{k+1})} \frac{f(y)}{d(x,y)^{N-\alpha}}\, d\mu(y) \\
&\leqslant C \sum_{k=0}^{\infty} 2^{-\alpha k} \ell^\alpha M^{(c)} f(x) = C_{\alpha,N} \ell^\alpha M^{(c)} f(x) = C_{\alpha,N} \ell^{\alpha-\beta} \ell^\beta M^{(c)} f(x),
\end{aligned}
$$

where $M^{(c)}$ is the centered Hardy–Littlewood maximal operator on X, and

$$
\beta := \frac{N}{p} - \frac{N}{q}.
$$

It is clear that $0 < \beta < \alpha$ because $q < p^*$.

Let us set:

$$
r_x := M^{(c)} f(x)^{-p/N}.
$$

Now we consider two cases:

Case 1: $\ell^\beta \leqslant r_x$. Then

$$K^{(1)}_{\alpha,\ell} f(x) \leqslant M^{(c)} f(x)^{-\frac{\beta p}{N}+1} = M^{(c)} f(x)^{p/q}.$$

Case 2: $\ell^\beta > r_x$. Then

$$\begin{aligned} K^{(1)}_{\alpha,\ell} f(x) &\leqslant \ell^{\alpha-\beta} \int\limits_{B(x,\ell)} f(y) d(x,y)^{\beta-N} d\mu(y) \\ &= \ell^{\alpha-\beta}\Bigg(\int\limits_{B(x,r_x)} f(y) d(x,y)^{\beta-N} d\mu(y) \\ &\quad + \int\limits_{B(x,\ell)\setminus B(x,r_x)} f(y) d(x,y)^{\alpha-N} d\mu(y)\Bigg) \\ &:= \ell^{\alpha-\beta}(R_1 f(x) + R_2 f(x)). \end{aligned}$$

We now estimate $R_1 f(x)$ and $R_2 f(x)$ separately.
Repeating the arguments used in the estimates of $K^{(1)}_{\alpha,\ell} f(x)$ above, we see that

$$R_1 f(x) \leqslant C_{\alpha,N} r_x^\beta M^{(c)} f(x) \leqslant C_{\alpha,N} M^{(c)} f(x)^{1-\beta p/N} = C_{\alpha,N} M^{(c)} f(x)^{p/q}.$$

Now we estimate $R_2 f(x)$. By Hölder's inequality and Lemma 3.47, we have:

$$\begin{aligned} R_2 f(x) &\leqslant \Bigg(\int\limits_{B(x,\ell)\setminus B(x,r_x)} d(x,y)^{(\beta-N)p'} d\mu(y)\Bigg)^{1/p'} \\ &\leqslant C\Big(\ell^{(\beta-N)p'+N} + r_x^{(\beta-N)p'+N}\Big)^{1/p'} \\ &\leqslant C r_x^{\beta-N/p} = C M^{(c)} f(x)^{p/q}. \end{aligned}$$

Summarizing the estimates for $R_1 f(x)$ and $R_2 f(x)$, we conclude that

$$\int\limits_{B(x,\ell)} f(y) d(x,y)^{\alpha-N} d\mu(y) \leqslant C\ell^{\alpha-\beta} M^{(c)} f(x)^{p/q}.$$

Consequently, by the boundedness of $M^{(c)}$ in $L^p(X)$, we find that

$$\begin{aligned} \|K_\alpha - K^{(2)}_{\alpha,\ell}\|_{L^p(\mu)\to L^q(\mu)} &\leqslant C\|K^{(1)}_{\alpha,\ell}\|_{L^p(\mu)\to L^q(\mu)} \\ &\leqslant C\|M^{(c)}\|^{p/q}_{L^p(\mu)\to L^p(\mu)} \ell^{\alpha-\beta} \to 0 \end{aligned}$$

as $\ell \to 0$. The remaining part (i.e., the case $q < p$) of the theorem follows from Theorem 3.48. □

As a special case of Theorem 3.49, we have the following statement, for example, for the fractional integral operator

$$I_{\gamma,\Gamma} f(t) = \int_{\Gamma} \frac{f(\tau)}{|t-\tau|^{1-\gamma}} d\nu(\tau), \quad 0 < \gamma < 1,$$

defined on a bounded rectifiable regular curve Γ (see (3.44)) with the arc-length measure ν on Γ.

Corollary 3.50 *Let $1 \leqslant p, q < \infty$ and let $0 < \gamma < 1/p$. Suppose that Γ is a regular curve. We set $p^* := \frac{p}{1-\gamma p}$. Then $I_{\gamma,\Gamma}$ is compact from $L^p(\Gamma)$ to $L^q(\Gamma)$ if and only if $q < p^*$.*

3.6 Measure of Noncompactness of Multilinear Operators

The aim of this section is to define variants of measure of noncompactntess in the setting of multilinear operators on the product of Banach spaces and show some applications to the multilinear variants of Hilbert and Riesz transforms.

3.6.1 *Preliminaries*

In Banach space theory of operators, the quantities which measure the noncompactness of operators play an important role. Throughout the section we use standard notation from Banach space theory and operator theory. We recall that for a bounded subset A of a Banach space E, the measure of noncompactness of A, denoted by $\gamma(A)$, is the infimum over all r such that A can be covered with a finite number of balls $B(x,r) := \{y \in E;\ \|x-y\|_E \leqslant r\}$ with radius $r > 0$. We note that $\gamma(A)$ can be given by the formula:

$$\gamma(A) = \inf\{r > 0;\ A \subset K + rB_E,\ \ K \text{ compact}\},$$

where $B_E := B(0,1)$ denotes the closed unit ball of E.

Let E and F be Banach spaces and let $\mathcal{L}(E,F)$ be the space of bounded linear operators between Banach spaces E and F. The measure of noncompactness of an operator $T \in \mathcal{L}(E,F)$ is given by $\gamma(T) = \gamma(T(B_E))$. Clearly, $\gamma(T) = 0$ if T is a compact linear operator and that

$$\gamma(S+T) \leqslant \gamma(S) + \gamma(T), \qquad \gamma(\lambda T) = |\lambda|\gamma(T)$$

for every $S, T \in \mathcal{L}(E, F)$ and all scalars λ. These properties imply that $\gamma(T)$ induces the norm of the coset $[T]$ in the quotient Banach space $\mathcal{L}(E, F)/\mathcal{K}(E, F)$, where $\mathcal{K}(E, F)$ is the space of all compact linear operators from E into F.

The next important characteristic of the space $\mathcal{L}(E, F)$ is the so-called the essential norm defined for $T \in \mathcal{L}(E, F)$ by

$$\|T\|_e := \|[T]\|_{\mathcal{L}(E,F)/\mathcal{K}(E,F)} = \inf\{\|T - S\|_{E\to F};\ S \in \mathcal{K}(E, F)\}.$$

Observe that $\gamma(T) = \gamma(T - S)$ for arbitrary $S \in \mathcal{K}(E, F)$ and thus

$$\gamma(T) \leqslant \|T\|_e.$$

We note that Lebow and Schechter [206] proved that if the Banach space F has the λ-CAP, then for every Banach space E, we have:

$$\|T\|_e \leqslant \lambda\gamma(T), \qquad T \in \mathcal{L}(E, F).$$

We recall that a Banach space E is said to have the λ-*compact approximation property* (λ-CAP for short) if for every compact subset $A \subset E$ and for every $\varepsilon > 0$, there exists a compact operator $S \in \mathcal{K}(E)$ such that

$$\sup_{x\in A} \|Sx - x\|_E \leqslant \varepsilon \quad \text{and} \quad \|S - I\| \leqslant \lambda.$$

A Banach space E is said to have the bounded compact approximation property ($BCAP$) if it has the λ-CAP for some $\lambda \geqslant 1$. For an example of a Banach space without the $BCAP$, we refer to Lindenstrauss and Tzafriri [211, p. 111].

We point out that in the theory of Banach spaces, an important role plays the bounded approximation property. Recall that the Banach space E is said to have the *bounded approximation property* (BAP for short) if there exists $\lambda > 0$ such that for every compact subset A and for every $\varepsilon > 0$ one can find a finite rank operator $V : E \to E$ with $\|V\| \leqslant \lambda$ and

$$\|Vx - x\|_E \leqslant \varepsilon, \qquad x \in A.$$

Clearly, the essential norm of $T \in \mathcal{L}(E, F)$ is dominated by the measure of *non-approximability* $a(T)$ defined by

$$a(T) := \inf\{\|T - S\|_{E\to F};\ S \in \mathcal{L}(E, F),\ \text{rank}(S) < \infty\}.$$

3.6.2 Results

We first fix some notation and collect a few definitions and results that will be needed in the sequel. Let $m \geqslant 2$ be an integer and let $X_1,...,X_m$ and Y be Banach spaces. We equipped the product $X_1 \times \cdots \times X_m$ with the norm $\|(x_1, ..., x_m)\| = \max_{1 \leqslant j \leqslant m} \|x_j\|_{X_j}$. We denote by $\mathcal{L}_m(X_1 \times \cdots \times X_m, Y)$ the Banach space of multilinear and continuous operators defined on $X_1 \times \cdots \times X_m$ with values in Y equipped with the norm:

$$\|T\|_{X_1 \times \cdots \times X_m \to Y} := \sup\big\{\|T(x_1, ..., x_m)\|_Y;\ (x_1, ..., x_m) \in B_{X_1} \times \cdots \times B_{X_m}\big\}.$$

In the case Y is the scalar field ($\mathbb{R}$ or $\mathbb{C}$), we denote the space of multilinear forms by $\mathcal{L}_m(X_1 \times \cdots \times X_m)$.

The measure of noncompactness of $T \in \mathcal{L}_m(X_1 \times \cdots \times X_m)$ is given by

$$\gamma_m(T) := \gamma\big(T(B_{X_1} \times \cdots \times B_{X_m})\big).$$

An m-linear operator $T\colon X_1 \times \cdots \times X_m \to Y$ is said to be compact if $T(B_{X_1} \times \cdots \times B_{X_m})$ is relatively compact in Y; the space of all such operators is denoted by $\mathcal{K}_m(X_1 \times \cdots \times X_m, Y)$. Clearly that $\gamma_m(T) = 0$ if T is compact.

For any $T \in \mathcal{L}_m(X_1, \ldots, X_m, Y)$, we denote the image of T by $\mathrm{Im}(T)$. Since in general $\mathrm{Im}(T)$ is not a linear subspace of Y, we define $\mathrm{rank}(T)$ as the dimension of $[\mathrm{Im}(T)]$, where $[S]$ denotes the linear span of a subset S in a linear vector space W.

The space of all $T \in \mathcal{L}_m(X_1 \times \ldots \times X_m, Y)$ such that $\mathrm{rank}(T) < \infty$ is denoted by $\mathcal{F}_m(X_1 \times \ldots \times X_m, Y)$. We let:

$$\|T\|_{e,m} := \inf\big\{\|T - S\|;\ S \in \mathcal{K}_m(X_1 \times \cdots \times X_m, Y)\big\}$$

and

$$a_{(m)}(T) := \inf\big\{\|T - S\|;\ S \in \mathcal{F}_m(X_1 \times \cdots \times X_m, Y)\big\}.$$

We have the following multilinear variant of the mentioned Lebow–Schechter result in the linear setting.

Lemma 3.51 *Let $X_1, \ldots, X_m$ and Y be Banach spaces. Assume that Y has the λ-compact approximation property. Then for every $T \in \mathcal{L}_m(X_1 \times \cdots \times X_m, Y)$:*

$$\|T\|_{e,m} \leqslant \lambda\, \gamma_m(T).$$

Proof Fix $\varepsilon > 0$. There exists a finite set $K = \{y_1, \cdots, y_k\}$ in Y such that

$$T(B_{X_1} \times \cdots \times B_{X_m}) \subset K + (\gamma_m(T) + \varepsilon)B_Y.$$

By hypothesis there exists a compact operator $P\colon Y \to Y$ such that $\|P - I\| \leqslant \lambda$ and

$$\|Py_j - y_j\|_Y < \varepsilon, \quad 1 \leqslant j \leqslant k.$$

Let $x = (x_1, \dots, x_m) \in B_{X_1} \times \cdots \times B_{X_m}$ be given. Then there is a $y_j \in K$ such that

$$\|Tx - y_j\|_Y \leqslant \gamma_m(T) + \varepsilon.$$

Combining the above estimates, we get:

$$\begin{aligned}\|(T - PT)x\|_Y &\leqslant \|(I - P)(Tx - y_j)\|_Y + \|(I - P)y_j\|_Y\\ &\leqslant \|I - P\|\, \|\|Tx - y_j\|_Y + \varepsilon\\ &\leqslant \lambda(\gamma_m(T) + \varepsilon) + \varepsilon.\end{aligned}$$

Clearly that $PT \in \mathcal{K}_m(X_1 \times \cdots \times X_m, Y)$ and so

$$\|T\|_{e,m} \leqslant \lambda\gamma_m(T) + \varepsilon(\lambda + 1).$$

Since ε is arbitrary, the required estimate follows. □

We will need the following technical result.

Proposition 3.52 *Let $X_1, \dots, X_m$ and Y be Banach spaces. Assume that Y has the bounded approximation property. Then for every $T \in \mathcal{L}_m(X_1 \times \cdots \times X_m, Y)$:*

$$\|T\|_{e,m} = a_{(m)}(T).$$

Proof For given $\varepsilon > 0$, we can find $S \in \mathcal{K}_m(X_1 \times \cdots \times X_m, Y)$ such that

$$\|T - S\| < \|T\|_{e,m} + \varepsilon/2.$$

Our hypothesis $Y \in$ (BAP) easily yields that there exists $P \in \mathcal{F}_m(X_1 \times \cdots \times X_m, Y)$ such that $\|S - P\| \leqslant \varepsilon/2$. Hence

$$\|T - P\| \leqslant \|T - S\| + \|S - P\| \leqslant \|T\|_{e,m} + \varepsilon.$$

Since $\varepsilon > 0$ is arbitrary, we get:

$$a_{(m)}(T) \leqslant \|T\|_{e,m}.$$

Clearly that $\|T\|_{e,m} \leqslant a_{(m)}(T)$ and so the proof is complete. □

We will consider BFLs. Recall that if $(\Omega, \mathcal{S}, \mu)$ is a complete and σ-finite measure space, then $L^0(\mu) := L^0(\Omega, \mathcal{S}, \mu)$ denotes the space of all equivalence classes of real-valued measurable functions on Ω. A BFL $(E, \|\cdot\|_E)$ on $(\Omega, \mathcal{S}, \mu)$ is defined to be a subspace of $L^0(\mu)$, endowed with a complete norm $\|\cdot\|_E$ such that there exists $u \in E$ with $u > 0$ a.e. It is also assumed that E is an ideal in $L^0(\mu)$, i.e., if $f \in L^0(\mu)$ satisfies $|f| \leqslant |g|$ a.e., for some $g \in E$, then $f \in E$ and $\|f\|_E \leqslant \|g\|_E$. A BFL E is said to be order continuous whenever for every non-negative sequence (f_n) in E satisfying $f_n \downarrow 0$, we have $\|f_n\|_E \to 0$. We will use the well-known and easily verifiable fact that E is order continuous whenever for any sequence (f_n) satisfying $|f_n| \leqslant g \in E$ for each $n \in \mathbb{N}$ and $f_n \to g$ μ-a.e., then $\|f_n - g\|_E \to 0$.

An important class of BFLs are *rearrangement invariant* (r.i. for short) spaces. We recall its definition below. Given $f \in L^0(\mu)$, its distribution function is defined by $\mu_f(\lambda) = \mu\{t \in \Omega; |f(t)| > \lambda\}$, $\lambda \geqslant 0$. Then a Banach lattice E on $(\Omega, \mathcal{S}, \mu)$ is a r.i. space if $g \in E$ and $\|f\|_E = \|g\|_E$ whenever $\mu_f = \mu_g$ and $f \in E$. It is well-known (see for instance Bennett and Sharpley [25], Krein et al. [201]) that for any r.i. space E, we have:

$$L_1 \cap L_\infty \subset E \subset L_1 + L_\infty$$

and so in particular this implies that any function $f \in E$ is locally integrable. Clearly, for a r.i. space E, $\|\chi_A\|_E$ depends only on $\mu(A)$ for any measurable set A of finite measure. If $(\Omega, \mathcal{S}, \mu)$ is an atomless measure space, then for every $t \in \mathbb{R}_+$ with $t \leqslant \mu(\Omega)$, we may define the function ϕ_E by $\phi_E(t) = \|\chi_A\|_E$, where A is any measurable set with $\mu(A) = t$. This function is called the *fundamental function* of E.

Let X be a Hausdorff topological space and $\mathcal{B}(X)$ the Borel subsets of X. Let Σ be a σ-algebra containing $\mathcal{B}(X)$ and $\mu\colon \Sigma \to [0, \infty]$ be finitely additive. We recall that $A \in \Sigma$ is inner regular if

$$\mu(A) = \sup\left\{\mu(K);\ K \subset A, K \text{ compact}\right\};$$

$E \in \Sigma$ is outer regular if $\mu(A) = \inf\{\mu(G);\ G \supset A,\ G \text{ open}\}$. A set A is said to be regular if it is both inner and outer regular. A Borel measure is a measure defined on the Borel sets, which is finite on compact sets. A Borel measure is regular if every set is outer regular and every open set is inner regular.

We also will work on metric measure spaces. We refer to Björn and Björn [28], Heinonen [136] for the theory and application of these spaces in potential theory. We recall that if (X, d) is a metric space and μ is a measure on a σ algebra $\mathcal{B}(X)$ of Borel sets, then the measure space $(X, \mathcal{B}(X), \mu)$ denoted by (X, d, μ) is called a metric measure space. As usual a ball with a center at $x \in X$ and radius $r > 0$ in a metric space (X, d) is denoted by

$$B := B(x, r) = \{y \in X; d(x, y) < r\}.$$

To avoid trivial measures, we will always assume throughout the section that $0 < \mu(B) < \infty$ for every ball B. Consequently, μ is a σ-finite measure.

Lemma 3.53 *Let $X_1, \dots, X_m$ be Banach spaces and let E be order continuous BFL on $\mathbb{R}^n$. If $T \in \mathcal{L}_m(X_1 \times \cdots \times X_m, E)$ with $N = \operatorname{rank}(T) < \infty$, then for every $a \in \mathbb{R}^n$ and $\varepsilon > 0$, there exist $S \in \mathcal{L}_m(X_1 \times \cdots \times X_m, E)$ and $0 < r < R < \infty$ such that* $\operatorname{rank}(S) \leqslant N$,

$$\|T - S\|_{X_1 \times \cdots \times X_m \to E} \leqslant \varepsilon$$

and $\operatorname{supp}(S(\vec{g})) \subset B(a, R) \setminus B(a, r)$ *for all* $\vec{g} \in X_1 \times \cdots \times X_m$, $\vec{g} := (g_1, \dots, g_m)$.

Proof Let $V = [\operatorname{Im}(T)] \subset E$ with $\dim(V) = N$. We use Auerbach's lemma which states that there are unit basis vectors $f_1, \dots, f_N \in V$ and unit vectors $f_1^*, \dots, f_N^* \in V^*$ such that

$$\langle f_i, f_j^* \rangle = \delta_{ij}, \qquad 1 \leqslant i, j \leqslant N.$$

By the Hahn–Banach theorem, it follows that there exists $\varphi_j^* \in E^*$ such that $\varphi_j^*|_V = f_j^*$. Since $Tx \in V$ for all $x \in X_1 \times \cdots \times X_m$

$$Tx = \sum_{j=1}^{N} \varphi_j^*(Tx) f_j = \sum_{j=1}^{N} \varphi_j^* \circ T(x) f_j.$$

Hence

$$T = \sum_{j=1}^{N} B_j \otimes f_j,$$

where $B_j := \varphi_j \circ T \in \mathcal{L}_m(X_1 \times \cdots \times X_m)$ with $\|B_j\| \leqslant \|T\|$ for each $1 \leqslant j \leqslant m$. This implies that, for every $x = (x_1, \dots, x_m) \in X_1 \times \cdots \times X_m$, we have:

$$\sum_{j=1}^{N} |B_j(x)| \leqslant N \|T\| \, \|x_1\|_{X_1} \cdots \|x_m\|_{X_m}.$$

Now we fix two positive sequences $\{r_n\}$ and $\{R_n\}$ such that $r_n < R_n$ for each $n \in \mathbb{N}$ and $r_n \downarrow 0$ and $R_n \uparrow \infty$. For a given $a \in X$ and for each $n \in \mathbb{N}$, we let $A_n := B(a, R_n) \setminus B(a, r_n)$. Combining $f\chi_{A_n} \to f$ μ-a.e. with order continuity of E, it follows that for all $f \in E$, we have:

$$\|f - f\chi_{A_n}\|_E \to 0 \quad \text{as} \quad n \to \infty.$$

This implies that there exist positive r and R with $r < R$ such that

$$\|f_j - f_j\chi_{B(a,R)\setminus B(a,r)}\|_E \leqslant \frac{\varepsilon}{N\|T\|}, \qquad 1 \leqslant j \leqslant N.$$

We let $g_j := f_j\chi_{B(a,R)\setminus B(a,r)}$ for each $1 \leqslant j \leqslant N$ and define:

$$Sx := \sum_{j=1}^{N} B_j(x)g_j, \qquad x \in X_1 \times \cdots \times X_m.$$

Clearly that $S \in \mathcal{L}_m(X_1 \times \cdots \times X_m, E)$, $\operatorname{rank}(S) \leqslant N$ and $\operatorname{supp}(S(x)) \subset B(a, R) \setminus B(a, r)$ for all $x \in X_1 \times \cdots \times X_m$. Combining the above estimates yields:

$$\begin{aligned}\|Tx - Sx\|_E &\leqslant \sum_{j=1}^{N} |B_j(x)|\|f_j - g_j\|_E \leqslant \frac{\varepsilon}{N\|T\|}\sum_{j=1}^{N} |B_j(x)| \\ &\leqslant \varepsilon\|x_1\|_{X_1}\cdots\|x_m\|_{X_m}\end{aligned}$$

for all $x = (x_1, \ldots, x_m) \in B_{X_1} \times \cdots \times B_{X_m}$. Thus $\|T - S\| \leqslant \varepsilon$ as required. □

We show that a large class of Banach lattices on some metric measure spaces have compact approximation property. Before we state the result, we need to introduce some more notion and definitions. Let (X, d, μ) be a metric measure space. A μ-measurable function $f : X \to \mathbb{R}$ is said to be locally integrable ($f \in L^1_{\text{loc}}(\mu)$ for short) if f is integrable over any balls of X. If (X, d, μ) is a measure metric space such that $0 < \mu(B) < \infty$, we define the Hardy–Littlewood operator M by

$$Mf(x) = \sup_{B \ni x} \frac{1}{\mu(B)} \int_B |f|\, d\mu, \qquad f \in L^1_{\text{loc}}(\mu),$$

where the supremum is taken over all balls B containing x.

It is known (see, e.g., Bennett and Sharpley [25], p. 154) that, for example, if $X = \mathbb{R}^n$ and μ is a Lebesgue measure, then the maximal operator M is bounded in r.i. space E if and only if upper index of E is less than 1.

A metric measure space (X, d, μ) is said to be a *Vitali space* if all balls in X have finite measure and if for every set A in X and every family $\mathcal{B}$ of balls in X satisfying

$$\inf\{r > 0;\ B(x, r) \in \mathcal{F}\} = 0 \quad \text{for} \quad x \in A,$$

there is a countable disjoint subfamily $\{B_n\}$ of $\mathcal{B}$ such that

$$\mu\Big(A \setminus \bigcup_n B_n\Big) = 0.$$

It is well-known that if (X, d, μ) with μ a doubling measure, then it follows by the well-known Vitali's theorem that X is a Vitali space (see Heinonen [136, Theorem 1.6]). We note that if μ is a Radon measure on $\mathbb{R}^n$, then $(\mathbb{R}^n, \mu)$ is also a Vitali space (see Mattila [232, Theorem 2.8]).

We recall that a measure μ is doubling if there is a constant $C_{DC} > 0$ such that for all balls $\mu(2B) \leqslant C_{DC}\mu(B)$, where $2B(x, r) = B(x, 2r)$.

Let X be locally compact Hausdorff space. We denote by $C_c(X)$ the space of all real-valued continuous functions on X which have compact support. In what follows we will need the following proposition.

Proposition 3.54 *Let X be a locally compact Hausdorff space and let μ be a Radon measure on $\mathcal{B}(X)$ and let $E \subset L^0(\mu)$ be order continuous BFL such that $\chi_A \in E$ for every μ-finite Borel set. Then $C_c(X)$ is dense in E.*

Proof Clearly, the order continuity of E implies that the set of simple functions is dense in E. If $\{\varepsilon_n\}$ is a positive sequence with $\varepsilon_n \downarrow 0$ and s is a simple function in E, then it follows from Luzin's theorem that there exist a sequence $\{f_n\}$ of functions in $C_c(X)$ such that except on a measurable set A_n with $\mu(A_n) < \varepsilon_n$, $|f_n| \leqslant \|s\|_\infty$. This implies that, for each n, we have:

$$\|f_n - s\|_E \leqslant \|s\|_\infty \|\chi_{A_n}\|_E$$

and so $f_n \to s$ in E by order continuity. □

We are ready to prove the following result, which shows that a large class of Banach lattices on metric measure spaces have the compact approximation property.

Theorem 3.55 *Let (X, d, μ) be a locally compact Vitali measure metric space such that μ is a regular measure with $\mu(\partial B) = 0$ for every ball in X. Assume that E is an order continuous Banach function lattice in $L^0(X, \mu)$ with $\chi_A \in E$ for any μ-measurable set A and that the Hardy–Littlewood maximal operator M is bounded in E. Then E has λ-compact approximation property with $\lambda = \|M\| + 1$.*

Proof Fix $\varepsilon > 0$ and let $f_1, \ldots, f_N$ be given functions in E. By Proposition 3.54, it follows that $C_c(X)$ is dense in E; we may assume that f_j are continuous functions with compact supports for each $1 \leqslant j \leqslant N$. Let K be a compact set outside, which all f_j vanish. The uniform continuity on K of each f_j yields that there exists $\delta > 0$ such that, for every $x, y \in K$ with $d(x, y) < \delta$, we have:

$$|f_j(x) - f_j(y)| < \frac{\varepsilon}{2\|\chi_K\|_E}, \qquad 1 \leqslant j \leqslant N.$$

Since (X, d, μ) is a Vitali space, it follows that the family $\mathcal{B}$ of closed balls

$$\mathcal{B} := \left\{\bar{B}(x, r) : x \in K,\ B(x, r) \subset \operatorname{int} K,\ r \in (0, \delta/2)\right\} \\ \cup \left\{\bar{B}(x, r) : x \in \partial K,\ r \in (0, \delta/2)\right\}$$

contains a countable disjoint subfamily $\{\bar{B}_n\}_{n=1}^{\infty}$ of closed balls such that

$$\mu\Big(K \setminus \bigcup_{n=1}^{\infty} \bar{B}_n\Big) = 0.$$

We note that $d(x, y) < \delta$ for all $x, y \in B_n$ and each $n \in \mathbb{N}$. Since all balls in X have positive measure, $f_j(z) = 0$ for all $z \in \partial K$ by continuity of f_j for each $1 \leqslant j \leqslant N$. Combining we obtain that, for each $n \in \mathbb{N}$ and each $1 \leqslant j \leqslant N$

$$|f_j(x) - f_j(y)| \leqslant \frac{\varepsilon}{2\|\chi_K\|_E} \chi_K(x), \qquad x \in B_n \cap K, \ y \in B_n \setminus K. \tag{3.45}$$

Now we define a linear mapping S on E by

$$Sf := \sum_{n=1}^{\infty} \Big(\frac{1}{\mu(B_n)} \int_{B_n} f \, d\mu\Big)\chi_{K \cap B_n}, \qquad f \in E.$$

Observe that for every $x \in X$, we have:

$$|Sf(x)| \leqslant \sum_{n=1}^{\infty} \Big(\frac{1}{\mu(B_n)} \int_{B_n} |f| \, d\mu\Big)\chi_{B_n}(x) \leqslant Mf(x)$$

and so $S\colon E \to E$ is bounded with $\|S\|_{E\to E} \leqslant \|M\|_{E\to E}$. Thus, $\|S - I\|_{E\to E} \leqslant \|M\|_{E\to E} + 1$.

We claim that S is a compact operator. To see this observe that the sequence $\{S_n\}$ of finite-dimensional operators given by

$$S_n f := \sum_{j=1}^{n} \Big(\frac{1}{\mu(B_j)} \int_{B_j} f \, d\mu\Big)\chi_{K \cap B_j}, \qquad f \in E$$

satisfies $S_n f \to Sf$ and $|S_n f| \leqslant S|f| \in E$. This implies by order continuity of E that

$$\|S_n f - Sf\|_E \to 0 \quad \text{as } n \to \infty.$$

To finish the proof, it is enough to show that

$$\|Sf_j - f_j\|_E \leqslant \varepsilon, \qquad 1 \leqslant j \leqslant N.$$

To see this we recall that $d(x, y) < \delta$ for all $x, y \in B_n$ and each $n \in \mathbb{N}$. Since $(K \cap B_i) \cap (K \cap B_k) = \emptyset$ for each $i \neq k$ and $\mu(K \setminus \bigcup_{n=1}^{\infty} B_n) = 0$, $f_j =$

$\sum_{n=1}^{\infty} f_j \chi_{K \cap K_n}$ a.e. for each $1 \leqslant j \leqslant N$. Combining these with the estimates (3.45), we obtain that, for almost all $x \in X$ and for each $1 \leqslant j \leqslant N$

$$
\begin{aligned}
|Sf_j(x) - f_j(x)| &= \left| \sum_{n=1}^{\infty} \left(\frac{1}{\mu(B_n)} \int\limits_{K \cap B_n} f_j \, d\mu \right) \chi_{K \cap B_n}(x) - f_j(x) \right| \\
&\leqslant \sum_{n=1}^{\infty} \left(\frac{1}{\mu(B_n)} \int\limits_{K \cap B_n} \left| f_j(y) - f_j(x) \right| d\mu(y) \right) \chi_{K \cap B_n}(x) \\
&= \sum_{n=1}^{\infty} \left(\frac{1}{\mu(B_n)} \int\limits_{K \cap B_n} \left| f_j(x) - f_j(y) \right| d\mu(y) \right) \chi_{K \cap B_n}(x) \\
&+ \sum_{n=1}^{\infty} \left(\frac{1}{\mu(B_n)} \int\limits_{B_n \setminus K} \left| f_j(x) - f_j(y) \right| d\mu(y) \right) \chi_{K \cap B_n}(x) \\
&\leqslant \frac{\varepsilon}{\|\chi_K\|_E} \chi_K(x)
\end{aligned}
$$

and so the required estimate follows. □

We finish with the following and obvious remark that the condition $\mu(\partial B) = 0$ for any ball B holds in a wide class of metric measure spaces (X, d, μ). For example, this is true whenever the function $(0, d_X) \ni r \mapsto \mu(B(x, r))$ is continuous for every x, then $\mu(\partial B(x, r)) = 0$, where $d_X := \sup_{x, y \in X} d(x, y)$. In fact we have:

$$\partial B(x, r) \subset B(x, r + \varepsilon_n) \setminus B(x, r - \varepsilon_n)$$

for any sequence $\{\varepsilon_n\}$ in $(0, r - d_X)$ with $\varepsilon_n \to 0$ as $n \to \infty$ and whence,

$$\mu(\partial B(x, r)) \leqslant \mu(B(x, r + \varepsilon_n)) - \mu(B(x, r - \varepsilon_n)), \qquad n \in \mathbb{N}.$$

Clearly this yields $\mu(\partial B(x, r)) = 0$.

3.6.3 Application to Multilinear Variant of Hilbert and Riesz Transforms

In this section we apply the derived results to obtain the lower estimate of the essential norm for multilinear variants of Hilbert and Riesz transforms bounded in r.i. Banach spaces.

For each positive integer m, we define a variant of multilinear Hilbert transform H_m on $L^1_{\text{loc}}(\mathbb{R}^m, \lambda_m)$ given by the formula:

$$H_m(f_1, \ldots, f_m)(x) := \text{p.v.} \int\limits_{\mathbb{R}^m} \frac{f_1(t_1) \cdots f_m(t_m)}{\big((x - t_1) + \ldots + (x - t_m)\big)^m} \, d\lambda_m, \qquad x \in \mathbb{R},$$

where λ_m is the Lebesgue measure in $\mathbb{R}^m$.

Theorem 3.56 *Let $X_1, \ldots, X_m$ and E be r.i. Banach spaces on $\mathbb{R}$ such that $H_m \in \mathcal{L}_m(X_1 \times \cdots \times X_m, E)$. If E has bounded approximation property and $\varphi_{X_1}(t) \cdots \varphi_{X_m}(t) \leqslant C\varphi_E(t)$ for all $t > 0$ and some $C > 0$, then*

$$\|H_m\|_{e,m} \geqslant \frac{1}{C(2m)^m}.$$

Proof Let $\delta > \|H_m\|_{e,m}$. From Lemma 3.52, it follows that $\delta > a_{(m)}(T)$. Thus, by the definition, there exists $T \in \mathcal{F}_m(X_1 \times \cdots \times X_m, E)$ such that

$$\|H_m - T\| < \delta.$$

Applying Lemma 3.53 with $a = 0$, we conclude that there are positive number $\beta > 0$ and $S \in \mathcal{F}_m(X_1 \times \cdots \times X_m, E)$ such that

$$\|T - S\| < \frac{1}{2}\big(\delta - \|H_m - S\|\big)$$

and for all $f \in X_1 \times \cdots \times X_m$,

$$\operatorname{supp}(Sf) \subset \mathbb{R} \setminus (-\beta, \beta).$$

Hence

$$\|(H_m - S)f\|_E \leqslant \delta\|f_1\|_{X_1} \cdots \|f_m\|_{X_m} \tag{3.46}$$

and so for every $\tau \in (0, \beta)$,

$$\|\chi_{(-\tau,\tau)} H_m f\|_E \leqslant \delta\|f_1\|_{X_1} \cdots \|f_m\|_{X_m}.$$

Now observe that for each $1 \leqslant j \leqslant m$ and for all $t_j \in (0, \tau)$, $x \in (-\tau, 0)$, we have that $0 < (t_j - x) < 2\tau$. Then from (3.46), taking $f_j = \chi_{(0,\tau)}$ for all $j \in \{1, \ldots, m\}$, it follows that

$$\begin{aligned}
\delta \,\|\chi_{(0,\tau)}\|_{X_1} \cdots \|\chi_{(0,\tau)}\|_{X_m} &\geqslant \|\chi_{(-\tau,0)} H_m \chi_{(0,\tau)}\|_E \\
&\geqslant \Big\|\chi_{(-\tau,0)} \int\limits_{(0,\tau)^m} \big((x - t_1) + \ldots + (x - t_m)\big)^{-m} \, d\lambda_m\Big\|_E \\
&\geqslant \frac{1}{(2m\tau)^m}\Big(\int\limits_{(0,\tau)^m} d\lambda_m\Big)\|\chi_{(-\tau,0)}\|_E.
\end{aligned}$$

By hypothesis on fundamental functions, we get:

$$\delta \geqslant \frac{1}{C(2m)^m}.$$

Since $\delta > \|H_m\|_{e,m}$ was arbitrary, the required estimate follows. □

Further, we are also interested in the estimate of the essential norm for the multilinear Riesz transform. Let us recall (see, e.g., Grafakos and Torres [122]) that for each positive integer m and for each $1 \leqslant i \leqslant m$, an m-linear i-th Riesz transforms $R_i^{(m)}$ on $L^1_{\text{loc}}(\mathbb{R}^{mn}, \lambda_{mn})$ is defined by the formula:

$$R_i^{(m)}(f_1, \dots, f_m)(x)$$
$$:= \text{p.v.} \int_{(\mathbb{R}^n)^m} \frac{\sum_{j=1}^m (x_i - (y_j)_i)}{\left(\sum_{j=1}^m |x - y_j|^2\right)^{\frac{mn+1}{2}}} f_1(y_1) \cdots f_m(y_m)\, d\lambda_{mn}, \quad x \in \mathbb{R}^n,$$

where λ_{mn} is be the Lebesgue measure in $\mathbb{R}^{mn}$. Here x_i and $(y_j)_i$ are i-th coordinates of x and y_j, respectively.

We note that if $n = m = 1$, then $R_i^{(1)}$ is the linear Hilbert transform H. Now we are ready to proof the following statement.

Theorem 3.57 *Let $X_1, \dots, X_m$ and E be rearrangement invariant Banach spaces defined on $\mathbb{R}^n$ such that $R_i^{(m)} \in \mathcal{L}_m(X_1 \times \cdots \times X_m, E)$ for each $i \in \{1, \dots, n\}$. If E has bounded approximation property and $\varphi_{X_1}(t) \cdots \varphi_{X_m}(t) \leqslant C\varphi_E(t)$ for all $t > 0$ and some $C > 0$, then the following estimate holds for the sum of the m-linear Riesz transforms: $R^{(m)} := \sum_{i=1}^n R_i^{(m)}$,*

$$\|R^{(m)}\|_{e,m} \geqslant \frac{1}{C(2m)^{mn} n^{(mn+1)/2}}.$$

Proof Following to the proof of Theorem 3.56, we take $\delta > \|H_m\|_{e,m}$. Arguing as in the proof of Theorem 3.56, we find that the inequality

$$\|(R^{(m)} - S)(\vec{f})\|_E \leqslant \delta \|f_1\|_{X_1} \cdots \|f_m\|_{X_m}$$

holds for all $f = (f_1, \dots, f_m) \in X_1 \times \cdots \times X_m$, with $\text{supp}(Sf) \subset \mathbb{R}^n \setminus Q(0, \beta)$, where $Q(0, \beta)$ is the cube with center 0 and side length 2β.

Let $0 < \tau < \beta$ and let $Q := Q(0, \tau)$. Following García-Cuerva and Rubio de Francia [112, p. 417], denote by Q' the cubes touching Q and having the same side length as Q, and such that $x_i \geqslant (y_j)_i$ for any $x \in Q$ and $y_j \in Q'$. Then, for every $x \in Q$ and $f_k = \chi_{Q'}$ for each $1 \leqslant k \leqslant m$, we have:

$$R^{(m)}(f_1,\ldots,f_m)(x) \geqslant \sum_{i=1}^{n} \int\limits_{(Q')^m} \frac{\sum_{j=1}^{m}(x_i-(y_j)_i)}{\left(\sum_{j=1}^{m}|x-y_j|\right)^{mn+1}} f_1(y_1)\cdots f_m(y_m)\,d\lambda_{mn}$$

$$= \int\limits_{(Q')^m} \frac{\sum_{j=1}^{m}\sum_{i=1}^{n}(x_i-(y_j)_i)}{\left(\sum_{j=1}^{m}|x-y_j|\right)^{mn+1}}\,d\lambda_{mn}$$

$$\geqslant n^{-1/2} \int\limits_{(Q')^m} \frac{\sum_{j=1}^{m}|x-y_j|}{\left(\sum_{j=1}^{m}|x-y_j|\right)^{mn+1}}\,d\lambda_{mn}$$

$$\geqslant n^{-1/2} \int\limits_{(Q')^m} \frac{1}{\left(\sum_{j=1}^{m}|x-y_j|\right)^{mn}}\,d\lambda_{mn}$$

$$\geqslant \frac{(2\tau)^{mn}n^{-1/2}}{(4m\sqrt{n}\tau)^{mn}} = \frac{1}{(2m)^{mn}n^{(mn+1)/2}}.$$

Hence, taking $f_j = \chi_{Q'}$, $j \in \{1,\ldots,m\}$, we find that

$$\delta\,\|\chi_{Q'}\|_{X_1}\cdots\|\chi_{Q'}\|_{X_m} \geqslant \left\|\chi_Q R^{(m)}\chi_{Q'}\right\|_E \geqslant \frac{1}{(2m\sqrt{n})^{mn}}\|\chi_Q\|_E.$$

Thus, we have:

$$\delta \geqslant \frac{1}{C(2m)^{mn}n^{(mn+1)/2}}$$

and the proof is complete. □

We conclude with the following remark that order continuous r.i. spaces have bounded approximation property (see Mastyło [230]).

3.7 Essential Norm of Multilinear Operators: Two-Weight Estimates

Our aim is to derive lower estimates of the essential norm (measure of noncompactness) for the multilinear Hilbert transform H_m, the sum of multilinear Riesz transforms $R^{(m)} = \sum_{i=1}^{n} R_i^{(m)}$ (see Sect. 3.6 for the definition) and multilinear fractional integral operator:

$$\mathcal{I}_\alpha^{(m)}(f_1,\ldots,f_m)(x) = \int\limits_{(\mathbb{R}^n)^m} \frac{f_1(y_1)\cdots f_m(y_m)}{(|x-y_1|+\cdots+|x-y_m|)^{mn-\alpha}}d\vec{y},\ 0<\alpha<nm.$$

In this section we use the notation of Sect. 3.6.

Our objective, stated more precisely, is to obtain two-weight lower estimates for the essential norm of H_m and $R^{(m)}$ acting boundedly from the product of weighted BFLs $X_1(w_1) \times \cdots \times X_m(w_m)$ to another weighted BFL $E(v)$. Similar problem is investigated for $\mathcal{I}_\alpha^{(m)}$. As a consequence we have appropriate estimates in weighted Lebesgue spaces, i.e., for these operators acting between $L^{p_1}(w_1) \times \cdots \times L^{p_m}(w_m)$ and $L^q(v)$. From the results derived in the section, we conclude that there are no a.e. positive functions v, w_j, $j = 1, \ldots, m$, such that these operators are compact from $L^{p_1}(w_1) \times \cdots \times L^{p_m}(w_m)$ to $L^q(v)$.

To prove the main statements, we are based on the statement claimed that if a BFL E has bounded approximation property, then for every m- linear bounded operator T (acting from the product of BFLs $X_1 \times \cdots \times X_m$ to E), the essential norm $\|T\|_{e,m}$ is equal to the measure of non-approximability of T (see Sect. 3.6).

In the sequel, sometimes, we will use the symbol $\|T\|_{e,m,X_1,\ldots,X_m,Y}$ for $\|T\|_{e,m}$ to underline that T acts from $X_1 \times \cdots \times X_m$ to Y.

3.7.1 Two-Weight Estimates

Let E be a BFL defined on a domain $\Omega \subseteq \mathbb{R}^n$, and let w be a weight function on Ω (i.e., w be a.e. positive function on Ω). We denote by $E(w)$ weighted BFL which is a class of all functions $f : \Omega \to \mathbb{R}$ for which

$$\|f\|_{E(w)} := \|fw\|_E < \infty.$$

For a domain $\Omega \subset \mathbb{R}^n$ and weight $w \in L^p_{loc}(\Omega)$, we denote by $L^p(w, \Omega)$ (or by $L^p(w)$, $1 < p < \infty$, the weighted Lebesgue space equipped with the standard norm

$$\|f\|_{L^p(w,\Omega)} = \left(\int_\Omega (|f(x)|w(x))^p dx \right)^{1/p}.$$

In the sequel, we will denote by $Q(a, r)$ a cube with center a and side length $2r$ (all cubes are assumed to have sides parallel to the coordinate axes). Following the argument similar to that used for the linear case in García-Cuerva and Rubio de Francia [112] (see p. 417), for a cube Q, we denote by Q' the cube touching Q having the same side length r as Q such that $y_j \geqslant z_j$ for any $(y_1, \ldots, y_n) \in Q'$ and $(z_1, \ldots, z_n) \in Q$ and $j = 1, \ldots, n$. It is clear that Q and Q' touch each other at a single point. If we denote that point by a, then the couple (Q, Q') will be denoted by (Q_a, Q'_a). If side length of Q_a is equal to r, then we denote the couple (Q_a, Q'_a) by $(Q_a(r), Q'_a(r))$.

To formulate the main statements of this section, we introduce the following (one-sided) characteristics of weights defined on $\mathbb{R}^n$:

$$[v, \vec{w}]^+_{X_1,\dots,X_m,E}(\mathbb{R}^n) := \sup_{a\in\mathbb{R}^n} \overline{\lim_{r\to 0}}\, r^{-mn} \|\chi_{Q_a(r)} v\|_E \prod_{j=1}^m \|w_j^{-1}(\cdot)\chi_{Q'_a(r)}\|_{X'_j};$$

and

$$[v, \vec{w}]^-_{X_1,\dots,X_m,E}(\mathbb{R}^n) := \sup_{a\in\mathbb{R}^n} \overline{\lim_{r\to 0}}\, r^{-mn} \|\chi_{Q'_a(r)} v\|_E \prod_{j=1}^m \|w_j^{-1}(\cdot)\chi_{Q_a(r)}\|_{X'_j},$$

where $\vec{w} = (w_1, \dots, w_m)$.

If $n = 1$, then these characteristics have the following form:

$$[v, \vec{w}]^+_{X_1,\dots,X_m,E}(\mathbb{R}) := \sup_{a\in\mathbb{R}} \overline{\lim_{r\to 0}}\, r^{-m} \|\chi_{(a-r,a)} v\|_E \prod_{j=1}^m \|w_j^{-1}(\cdot)\chi_{(a,a+r)}\|_{X'_j};$$

and

$$[v, \vec{w}]^-_{X_1,\dots,X_m,E}(\mathbb{R}) := \sup_{a\in\mathbb{R}} \overline{\lim_{r\to 0}}\, r^{-m} \|\chi_{(a,a+r)} v\|_E \prod_{j=1}^m \|w_j^{-1}(\cdot)\chi_{(a-r,a)}\|_{X'_j}.$$

For $X_k = L^{p_k}$, $k = 1, \cdots, m$, and $E = L^q$, we will use the notation:

$$[v, \vec{w}]^+_{p_1,\dots,p_m,q}(\mathbb{R}^n) := [v, \vec{w}]^+_{X_1,\dots,X_m,E}(\mathbb{R}^n);$$

and

$$[v, \vec{w}]^-_{p_1,\dots,p_m,q}(\mathbb{R}^n) := [v, \vec{w}]^-_{X_1,\dots,X_m,E}(\mathbb{R}^n).$$

Theorem 3.58 *Let $X_1, \dots, X_m$ and E be $BFLs$, and let $w_1, \cdots, w_m$ and v be weight functions on $\mathbb{R}$ such that $H_m \in \mathcal{L}_m(X_1(w_1) \times \dots \times X_m(w_m), E(v))$. If E has bounded approximation property, then the following estimate holds:*

$$\|H_m\|_{e,m} \geqslant \frac{1}{(2m)^m} \max\left\{[v, \vec{w}]^+_{X_1,\dots,X_m,E}(\mathbb{R}); [v, \vec{w}]^-_{X_1,\dots,X_m,E}(\mathbb{R})\right\}. \tag{3.47}$$

Theorem 3.59 *Let $X_1, \dots, X_m$ and E be $BFLs$ on $\mathbb{R}^n$. Let $w_1, \cdots, w_m$ and v be weight functions on $\mathbb{R}^n$ such that $R^{(m)} \in \mathcal{L}_m(X_1(w_1) \times \cdots \times X_m(w_m), E(v))$. If E has bounded approximation property, then the following estimate holds:*

$$\|R^{(m)}\|_{e,m} \geqslant (2\sqrt{m})^{-mn} \max\left\{[v, \vec{w}]^+_{X_1,\dots,X_m,E}(\mathbb{R}^n); [v, \vec{w}]^-_{X_1,\dots,X_m,E}(\mathbb{R}^n)\right\}.$$

Corollary 3.60 *Let $1 < p_j < \infty$, $j = 1, \dots, m$. Let p be defined by*

$$\frac{1}{p} = \sum_{j=1}^{m} \frac{1}{p_j} \tag{3.48}$$

Suppose that v and w_j, $j = 1, \dots, m$, are weights such that $v \in L^p_{\mathrm{loc}}(\mathbb{R})$, $w_j \in L^{p_j}_{\mathrm{loc}}(\mathbb{R})$, $j = 1, \dots, m$. Assume that H_m is bounded from $L^{p_1}(w_1, \mathbb{R}) \times \cdots \times L^{p_m}(w_m, \mathbb{R})$ to $L^p(v, \mathbb{R})$. Then the following estimate holds:

$$\|H_m\|_{e,m} \geqslant \frac{1}{(2m)^m} \max\left\{[v, \vec{w}]^+_{p_1,\dots,p_m,p}(\mathbb{R}); [v, \vec{w}]^-_{p_1,\dots,p_m,p}(\mathbb{R})\right\}. \tag{3.49}$$

Moreover, there are no weights v, w_j, $j = 1, \dots, m$, such that H_m is compact from $L^{p_1}(w_1, \mathbb{R}) \times \cdots \times L^{p_m}(w_m, \mathbb{R})$ to $L^p(v, \mathbb{R})$.

Corollary 3.61 *Let $1 < p_j < \infty$, $j = 1, \cdots, m$. Suppose that p is defined by (3.48). Let v and w_j, $j = 1, \dots, m$, be weights on $\mathbb{R}^n$ such that $v \in L^p_{\mathrm{loc}}(\mathbb{R}^n)$, $w_j \in L^{p_j}_{\mathrm{loc}}(\mathbb{R}^n)$, $j = 1, \dots, m$. Assume that $R^{(m)}$ is bounded from $L^{p_1}(w_1, \mathbb{R}^n) \times \cdots \times L^{p_m}(w_m, \mathbb{R}^n)$ to $L^p(v, \mathbb{R}^n)$. Then the following estimate holds:*

$$\|R^{(m)}\|_{e,m} \geqslant \frac{1}{(2\sqrt{m})^{mn}} \max\left\{[v, \vec{w}]^+_{p_1,\dots,p_m,p}(\mathbb{R}^n); [v, \vec{w}]^-_{p_1,\dots,p_m,p}(\mathbb{R}^n)\right\}.$$

Moreover, there are no weights v, w_j, $j = 1, \cdots, m$, such that $R^{(m)}$ is compact from $L^{p_1}(w_1, \mathbb{R}^n) \times \cdots \times L^{p_m}(w_m, \mathbb{R}^n)$ to $L^p(v, \mathbb{R}^n)$.

Regarding multilinear Riesz potentials, we have the following statement.

Theorem 3.62 *Let $X_1, \dots, X_m$ and E be $BFLs$ defined on $\mathbb{R}^n$, and let $w_1, \dots, w_m$ and v be a.e. positive functions on $\mathbb{R}^n$ such that $\mathcal{I}^{(m)}_\alpha \in \mathcal{L}_m(X_1(w_1) \times \cdots \times X_m(w_m), E(v))$. If E has bounded approximation property, then the following estimate holds:*

$$\|\mathcal{I}^{(m)}_\alpha\|_{e,m} \geqslant \frac{1}{(m\sqrt{n})^{mn-\alpha}} \sup_{a\in\mathbb{R}^n} \overline{\lim_{r\to 0}} \frac{1}{(2r)^{mn-\alpha}} \|v\chi_{Q(a,r)}\|_E \prod_{j=1}^{m} \|w_j^{-1}\chi_{Q(a,r)}\|_{X'_j}.$$

Corollary 3.63 *Suppose that $0 < \alpha < mn$ and that $1 < p_j < \infty$, $j = 1, \dots, m$ with $\frac{1}{m} < p < \frac{n}{\alpha}$, where p is defined by (3.48). Let q be an exponent defined by $q = \frac{np}{\alpha - np}$. Suppose that v and w_j, $j = 1, \dots, m$, are weights on $\mathbb{R}^n$ such that $v \in L^q_{\mathrm{loc}}(\mathbb{R}^n)$, $w_j \in L^{p_j}_{\mathrm{loc}}(\mathbb{R}^n)$, $j = 1, \dots, m$. If $\mathcal{I}^{(m)}_\alpha \in \mathcal{L}_m(L^{p_1}(w_1, \mathbb{R}^n) \times \cdots \times L^{p_m}(w_m\mathbb{R}^n), L^q(v, \mathbb{R}^n))$, then the following estimate hold:*

$$\|\mathcal{I}_\alpha^{(m)}\|_{e,m} \geqslant \frac{1}{(m\sqrt{n})^{mn-\alpha}} \sup_{a\in\mathbb{R}^n} \overline{\lim_{r\to 0}} \left(\frac{1}{|Q(a,r)|}\int_{Q(a,r)} v^q(y)dy\right)^{\frac{1}{q}}$$

$$\times \prod_{j=1}^{m}\left(\frac{1}{|Q(a,r)|}\int_{Q(a,r)} w_j^{-p'}(y_j)dy_j\right)^{\frac{1}{p'_j}}.$$

Moreover, there are no weights v, w_j, $j = 1, \ldots, m$, such that $\mathcal{I}_\alpha$ is compact from $L^{p_1}(w_1, \mathbb{R}^n) \times \cdots \times L^{p_m}(w_m, \mathbb{R}^n)$ to $L^q(v, \mathbb{R}^n)$.

3.7.2 Proofs

Proof of Theorem 3.58 We use the arguments from Chapter 6 of Meskhi [240], and also Sect. 3.6. First we prove that

$$\|H_m\|_{e,m} \geqslant \frac{1}{(2m)^m}[v, \vec{w}]^-_{X_1,\ldots,X_m,E}(\mathbb{R}). \tag{3.50}$$

Observe that

$$\|H_m\|_{e,m,X_1(w_1),\ldots,X_m(w_m),E(v)} = \|H_{m,w_1,\ldots,w_m,v}\|_{e,m,X_1,\ldots,X_m,E},$$

where

$$H_{m,w_1,\ldots,w_m,v}\vec{f} = vH_m(f_1/w_1, \ldots, f_m/w_m), \quad \vec{f} := (f_1, \ldots, f_m).$$

That is why we will estimate $\|H_{m,w_1,\ldots,w_m,v}\|_{e,m,X_1,\ldots,X_m,E}$ from below by the desired expression.

Taking

$$\delta > \|H_{m,w_1,\ldots,w_m,v}\|_{e,m,X_1,\ldots,X_m,E},$$

by Proposition 3.52 we find that $\delta > a_{(m)}(H_{m,w_1,\ldots,w_m,v})$. Then, there exists $T \in \mathcal{F}_m(X_1 \times \cdots \times X_m, E)$, such that

$$\|H_{m,w_1,\ldots,w_m,v} - T\| < \delta.$$

Let $a \in \mathbb{R}$. Applying Lemma 3.53, we have that there are positive number $\beta > 0$ and $S \in \mathcal{F}_m(X_1 \times \cdots \times X_m, E)$ with supp $S\vec{f} \subset \mathbb{R} \setminus (a-\beta, a+\beta)$ for all $\vec{f} \in X_1 \times \cdots \times X_m$ such that

$$\|T - S\| < \frac{1}{2}\big(\delta - \|H_{m,w_1,\ldots,w_m,v} - T\|\big).$$

Hence, by simple estimates we get:

$$\|(H_{m,w_1,\dots,w_m,v} - S)\vec{f}\|_E \leqslant \delta$$

for all $\vec{f} = (f_1, \dots, f_m)$ with $f_j \in B_{X_j}$, $j = 1, \dots, m$.

Thus, for all r and $\vec{f}$ with $0 < r < \beta$ and $\|f_j\|_{X_j} \leqslant 1$ $(j = 1, \dots, m)$ respectively, we find that

$$\|\chi_{(a,a+r)}\big(H_{m,w_1,\dots,w_m,v}\vec{f}\big)\|_E \leqslant \delta.$$

Further, observe that for all non-negative $f_j \in X_j$ with supp $f_j \subset (a-r, a)$ and $\|f_j\|_{X_j} \leqslant 1$, $j = 1, \dots, m$, the following estimates hold:

$$\begin{aligned}
\delta &\geqslant \|\chi_{(a,a+r)}\big(H_{m,w_1,\dots w_m,v}\vec{f}\big)\|_E \\
&\geqslant \left\|\chi_{(a,a+r)}(x)v(x)\int\limits_{a-r}^{r}\cdots\int\limits_{a-r}^{r}\frac{f_1(t_1)\cdots f_m(t_m)(w_1(t_1)\cdots w_m(t_m))^{-1}}{[(x-t_1)+\cdots+(x-t_m)]^m}dt_1\cdots dt_m\right\|_E \\
&\geqslant \frac{1}{(2m)^m}r^{-m}\|\chi_{(a,a+r)}(\cdot)v(\cdot)\|_E \\
&\qquad\times\int\limits_{(a-r,r)^m} f_1(t_1)\cdots f_m(t_m)(w_1(t_1)\cdots w_m(t_m))^{-1}dt_1\cdots dt_m.
\end{aligned}$$

Taking the supremum over all such functions f_j, $j = 1, \dots, m$, and passing δ to $\|H_{m,w_1,\dots,w_m,v}\|_{e,m,X_1,\dots,X_m,E}$, we derive (3.50).

In a similar way, it follows that

$$\|H_m\|_{e,m} \geqslant \frac{1}{(2m)^m}[v, \vec{w}]^+_{X_1,\dots,X_m,E}(\mathbb{R}). \tag{3.51}$$

Combining (3.50) and (3.51), we get the desired estimate. □

Proof of Theorem 3.59 We use the arguments of the proof of Theorem 3.58 (see also Theorem 3.2 of Kokilashvili et al. [187] for unweighted case). First we prove that

$$\|R^{(m)}\|_{e,m} \geqslant \frac{1}{(2\sqrt{m})^{mn}}[v, \vec{w}]^+_{X_1,\dots,X_m,E}(\mathbb{R}^n). \tag{3.52}$$

We have:

$$\|R^{(m)}\|_{e,m,X_1(w_1),\dots,X_m(w_m),E(v)} = \|R^{(m)}_{w_1,\dots,w_m,v}\|_{e,m,X_1,\dots,X_m,E},$$

where

$$R^{(m)}_{w_1,\ldots,w_m,v}(\vec{f}) = vR^{(m)}(f_1/w_1,\ldots,f_m/w_m).$$

For $\delta > \|R^{(m)}_{w_1,\ldots,w_m,v}\|_{e,m,X_1,\ldots,X_m,E}$, applying Proposition 3.52 we find that there exists $T \in \mathcal{F}_m(X_1 \times \cdots \times X_m, E)$ such that

$$\|R^{(m)}_{w_1,\ldots,w_m,v} - T\| < \delta.$$

Take $a \in \mathbb{R}^n$. By virtue of Lemma 3.53 and the arguments of the proof of Theorem 3.58, we conclude that there are positive number $\beta > 0$ and $S \in \mathcal{F}_m(X_1 \times \cdots \times X_m, E)$ with supp $S\vec{g} \subset \mathbb{R}^n \setminus Q(a,\beta)$ for all $\vec{g} \in X_1 \times \cdots \times X_m$, such that

$$\|R^{(m)}_{w_1,\ldots,w_m,v} - S\|_E < \delta.$$

Take now $0 < r < \beta$ and a couple of cubes $(Q_a(r), Q'_a(r))$. Observe that $Q_a(r) \subset Q(a,\beta)$ and $Q'_a(r) \subset Q(a,\beta)$. Consequently, for such an r, we have:

$$\left\|\chi_{Q_a(r)}\left(R^{(m)}_{w_1,\ldots,w_m,v}\vec{f}\right)\right\|_E \leqslant \delta, \quad f_j \in B_{X_j}, \quad j = 1,\ldots,m.$$

Hence, for all $f_k \geqslant 0$ with $f_k \in B_{X_k}$ and supp $f_k \subset Q'_a(r)$, $1 \leqslant k \leqslant m$, we have

$$\delta > \left\|\chi_{Q_a(r)}(x)v(x)\sum_{i=1}^{n}\int_{(Q'_a(r))^m}\frac{\sum_{j=1}^m(x_i-(y_j)_i)}{\left(\sum_{j=1}^m|x-y_j|^2\right)^{\frac{mn+1}{2}}}\frac{f_1(y_1)\cdots f_m(y_m)}{w_1(y_1)\cdots w_m(y_m)}\,d\vec{y}\right\|_E$$

$$= \left\|\chi_{Q_a(r)}(x)v(x)\int_{(Q'_a(r))^m}\frac{\sum_{j=1}^m\sum_{i=1}^n((y_j)_i-x_i)}{\left(\sum_{j=1}^m|x-y_j|^2\right)^{\frac{mn+1}{2}}}\frac{f_1(y_1)\cdots f_m(y_m)}{w_1(y_1)\cdots w_m(y_m)}\,d\vec{y}\right\|_E$$

$$\geqslant \left\|\chi_{Q_a(r)}(x)v(x)\int_{(Q'_a(r))^m}\frac{f_1(y_1)\cdots f_m(y_m)}{\left(\sum_{j=1}^m|x-y_j|^2\right)^{\frac{mn}{2}}}\left(w_1(y_1)\cdots w_m(y_m)\right)^{-1}d\vec{y}\right\|_E$$

$$\geqslant (2r\sqrt{m})^{-mn}\|\chi_{Q_a(r)}\|_{E(v)}\int_{(Q'_a(r))^m}\frac{f_1(y_1)\cdots f_m(y_m)}{w_1(y_1)\cdots w_m(y_m)}\,d\vec{y}.$$

Here we used the easily verifiable estimates:

$$\sum_{j=1}^m\sum_{i=1}^n((y_j)_i-x_i) \geqslant \sum_{j=1}^m\left(\sum_{i=1}^n[(y_j)_i-x_i]^2\right)^{1/2} = \sum_{j=1}^m|y_j-x| \geqslant \left(\sum_{j=1}^m|x-y_i|^2\right)^{1/2}$$

and

$$\Big(\sum_{j=1}^{m} |x-y_j|^2\Big)^{\frac{1}{2}} \leqslant 2r\sqrt{m},\ \ y_j \in Q'_a(r),\ x \in Q_a(r),\ j=1,\dots,n.$$

Taking now supremum in the latter inequality with respect to all $f_1, \dots, f_m$ satisfying the abovementioned conditions, we conclude that (3.52) holds.

In a similar way, we can obtain that

$$\|R^{(m)}\|_{e,m} \geqslant \frac{1}{(2\sqrt{m})^{mn}} [v, \vec{w}]^{-}_{X_1,\dots,X_m,E}(\mathbb{R}^n).$$

Finally the result follows. □

Proof of Corollary 3.60 The main estimate is a direct consequence of Theorem 3.58, in particular, (3.49) follows from (3.47). We only need to check the noncompactness argument. By using the Lebesgue differentiation theorem, we find that (3.49) implies

$$\|H_m\|_{e,m,L^{p_1}(w_1),\dots,L^{p_m}(w_m),L^p(v)} \geqslant \frac{1}{(2m)^m} \operatorname*{ess\,sup}_{a\in\mathbb{R}^n} \frac{v(a)}{\prod_{j=1}^m w_j(a)}.$$

Since $v, w_1, \dots, w_m$ are a.e. positive functions, we conclude that there are no weights v, w_j, $j = 1, \dots, n$, such that H_m is compact from $L^{p_1}(w_1, \mathbb{R}) \times \cdots \times L^{p_m}(w_m, \mathbb{R})$ to $L^p(v, \mathbb{R})$. □

Proof of Corollary 3.61 It is similar to that of Corollary 3.60. We only check the differentiate argument. Observe that in this case, we have:

$$\frac{1}{r^n}\int_{Q_a(r)} |v(y)-v(a)|dy \leqslant \frac{2^n}{|Q(a,r)|}\int_{Q(a,r)} |v(y)-v(a)|dy \to 0,\ \ r\to 0$$

for every Lebesgue points a of v. The same conclusion holds for w_j, $j = 1, \dots, m$. □

Theorem 3.62 and Corollary 3.63 follow in the same manner as the previous statements; that is why we omit all details. We only observe that to prove Theorem 3.62 we estimate the kernel of $\mathcal{I}_\alpha^{(m)}$ as follows:

$$\Big(\sum_{j=1}^{m} |x-y_j|\Big)^{mn-\alpha} \leqslant (2m\sqrt{n}r)^{mn-\alpha},\ \ x, y_j \in Q(a,r),\ \ j=1,\dots,m,$$

which implies that

$$\begin{aligned}
&\left\| \chi_{Q(a,r)} \left(\mathcal{I}^{(m)}_{\alpha,w_1,\ldots,w_m,v} \vec{f} \right) \right\|_E \\
&= \left\| \chi_{Q(a,r)}(x) v(x) \int\limits_{(Q(a,r))^m} \frac{f_1(y_1)/w_1(y_1) \cdots f_m(y_m)/w_1(y_1)}{\left(\sum_{j=1}^m |x - y_j| \right)^{mn-\alpha}} \, d\vec{y} \right\|_E \\
&\geqslant \frac{1}{(2m\sqrt{n}r)^{mn-\alpha}} \| \chi_{Q(a,r)} \|_{E(v)} \int\limits_{(Q(a,r))^m} \frac{f_1(y_1) \cdots f_m(y_m)}{w_1(y_1) \cdots w_m(y_m)} \, d\vec{y}
\end{aligned}$$

for weight functions $w_1, \ldots, w_n$ and all non-negative $f_k \in B_{X_k}$ with supp $f_k \subset Q(a, r)$, $j = 1, \ldots, m$, where

$$\mathcal{I}^{(m)}_{\alpha,w_1,\ldots,w_m,v}(\vec{f}) = v \, \mathcal{I}^{(m)}_{\alpha}(f_1/w_1, \ldots, f_m/w_m).$$

Hence, repeating the arguments of Theorem 3.58 (or Theorem 3.59), we have the desired estimate.

The first part of Corollary 3.63 is a direct consequence of Theorem 3.62. The noncompactness of $\mathcal{I}^{(m)}_{\alpha}$ follows from the Lebesgue differentiation theorem (see the proof of Corollary 3.61 for details).

3.8 Comments to Chap. 3

The integral operators play an important role in many areas, including the theory of Fourier series, the theory of eigenvalue distributions of Riesz operators, etc. (see, e.g., Johnson [156], König [193, 193], Kühn and Mastyło [203]). The fractional integral operators have applications in harmonic analysis, PDEs, and the theory of Sobolev embeddings (see, e.g., Adams and Hedberg [7], Maz'ya [235]).

Complete characterization of those measures μ for which the fractional integral operator $J_{\gamma,\mu}$ defined with respect to μ is bounded from $L^p(X, \mu)$ into $L^q(X, \mu)$, $1 < p < q < \infty$, was established in Kokilashvili and Meskhi [171] (we refer also to García-Cuerva and Gatto [110] for the case $q = \frac{p}{1-\gamma p}$).

For compactness criteria of fractional integral operators on Lebesgue spaces defined on the Euclidean spaces in terms of capacities, we refer to Adams and Hedberg [7].

We note that the results for linear operators in the classical and some nonstandard Lebesgue spaces, analogous to those presented in Sect. 3.6, can be found in the monographs Edmunds and Evans [85, Ch. V], Meskhi [240], and references listed therein. In the monograph Meskhi [240], the author delivered estimates of the measure of noncompactness (essential norm) for a wide range of operators in Lebesgue spaces and their generalizations in terms of weights (see also Oniani [267] for related topics in symmetric spaces).

The one-weight problem (L^p_w boundedness) for the linear Hilbert transform $Hf(x) = \text{p.v.} \int\limits_{\mathbb{R}} \frac{f(t)}{x-t} dt$ was characterized by Hunt et al. [140] under the well-known Muckenhoupt [253] A_p condition on w.

To solve the two-weight problem even for linear singular integrals, in particular, for the linear Hilbert transform H turned out more complicated than the one-weight one. Recently, Hytönen [146] characterized the L^2 Two-weight inequality for H for an arbitrary pair of positive Radon measures. In particular, that characterization involves Sawyer-type testing conditions and a variant of the Muckenhoupt–Poisson two-weight A_2 condition (we refer also to Lacey et al. [205] for related topics).

Two-weight characterizations for the linear Riesz potential operators I_α are known in terms of different conditions. We refer to Sawyer [309] for the condition involving the operator itself, and Gabidzashvili and Kokilashvili [108] (see also Kokilashvili and Krbec [170]) for integral-type conditions.

Finally we mention that the results presented in Sect. 3.1 were obtained in Kokilashvili and Meskhi [174], the content of Sect. 3.2 is based on Meskhi and Natelashvili [245] (see also Meskhi and Natelashvili [246]), while Sect. 3.3 follows Kokilashvili et al. [180]. Section 3.4 is based on the work presented in Kokilashvili et al. [188], and Sect. 3.5 relies on Kokilashvili and Meskhi [175]. Additionally, Sect. 3.6 is derived from Kokilashvili et al. [187], and finally, the results of Sect. 3.7 were established in Meskhi [243].

Chapter 4
Sharp Bounds in Weighted Inequalities

In this chapter, we establish sharp weighted norm estimates (Buckley-type estimates), generally speaking, for positive kernel operators on spaces of homogeneous type. We discuss the operators involving one-sided maximal and fractional integral operators, Hilbert transforms, multiple integral operators, potential operators, etc. Some of the results are derived via two-weight estimates for integral operators.

4.1 A Brief Overview on Sharp Weighted Inequalities

Let w be an almost everywhere positive locally integrable function on an open subset Ω of $\mathbb{R}^n$. We denote by $L^p_w(\Omega)$, $1 < p < \infty$, the set of all measurable functions $f : \Omega \to \mathbb{R}$ by the standard norm $\|f\|_{L^p_w(\Omega)} = (\int\limits_\Omega |f(x)|^p w(x)dx)^{\frac{1}{p}}$. If $w \equiv$ is constant function, then we denote $L^p_w(\Omega)$ by $L^p(\Omega)$. Suppose that $L^{p,\infty}_w(\Omega)$ is the weighted weak Lebesgue space defined by the quasi-norm:

$$\|f\|_{L^{p,\infty}_w(\Omega)} = \sup_{\lambda>0} \lambda[w(\{x \in \Omega : |f(x)| > \lambda\})]^{1/p}.$$

Let X and Y be two Banach spaces. Given a bounded operator $T : X \to Y$, we denote the operator norm by $\|T\|_{X\to Y}$ which is defined in the standard way, i.e., $\sup_{\|f\|_X\leqslant 1} \|Tf\|_Y$. If $X = Y$, then we use the symbol $\|T\|_X$.

Recall that a non-negative, locally integrable function w defined on $\mathbb{R}^n$ is said to satisfy $A_p(\mathbb{R}^n)$ condition ($w \in A_p(\mathbb{R}^n)$) for $1 < p < \infty$, if

$$\|w\|_{A_p(\mathbb{R}^n)} := \sup_Q \left(\frac{1}{|Q|}\int\limits_Q w(x)dx\right)\left(\frac{1}{|Q|}\int\limits_Q w(x)^{1-p'}dx\right)^{p-1} < \infty,$$

A. Meskhi et al., *Integral Operators in Non-Standard Function Spaces*,
Operator Theory: Advances and Applications 310,
https://doi.org/10.1007/978-3-032-03691-9_4

where $p' = \frac{p}{p-1}$, and supremum is taken over all cubes Q in $\mathbb{R}^n$ with sides parallel to the coordinate axes. We call that by the symbol $\|w\|_{A_p(\mathbb{R}^n)}$ is denoted the A_p characteristic of w.

In 1972, Muckenhoupt [254] showed that if $w \in A_p(\mathbb{R}^n)$, where $1 < p < \infty$, then the Hardy–Littlewood maximal operator

$$Mf(x) = \sup_{x\in Q} \frac{1}{|Q|} \int_Q |f(y)|dy$$

is bounded in $L^p_w(\mathbb{R}^n)$.

Later, Buckley [37] investigated the sharp A_p bound for the operator M and established the inequality:

$$\|M\|_{L^p_w(\mathbb{R}^n)} \leqslant C\|w\|^{\frac{1}{p-1}}_{A_p(\mathbb{R}^n)}, \quad 1 < p < \infty. \tag{4.1}$$

Moreover, he showed that the exponent $\frac{1}{p-1}$ is best possible in the sense that we cannot replace $\|w\|^{\frac{1}{p-1}}_{A_p}$ by $\psi(\|w\|_{A_p})$ for any positive non-decreasing function ψ growing slowly than $x^{\frac{1}{p-1}}$. From here it follows that for any $\lambda > 0$,

$$\sup_{w\in A_p} \frac{\|M\|_{L^p_w \to L^p_w}}{\|w\|^{\frac{1}{p-1}-\lambda}_{A_p}} = \infty.$$

It is worth mentioning the work of Lerner [208], who provided a simple approach to obtain a Buckley-type estimate for the maximal operator.

We say that $w \in A_\infty(\mathbb{R}^n)$ if $w \in A_p(\mathbb{R})$ for some $p > 1$. In what follows, we will use the symbol $\|\rho\|_{A_\infty}$ for the A_∞ characteristic of a weight function ρ:

$$\|\rho\|_{A_\infty} = \sup_I \frac{1}{\rho(I)} \int_I M(\rho\chi_I)(x)dx.$$

This characteristic appeared first in the papers by Wilson [335, 336] and is lower than that the one introduced by Khrushchev [165]:

$$[\rho]_{A_\infty} = \sup_I \left(\frac{1}{|I|}\int_I \rho(x)dx\right) \exp\left(\frac{1}{|I|}\int_I \log \rho^{-1}(x)dx\right).$$

In 2012, Hytönen et al. [144] improved Buckley's result and obtained a sharp weighted bound involving A_∞ constant:

$$\|M\|_{L^p_w} \leqslant c_n\left(\frac{1}{p-1}\|w\|_{A_p}\|\sigma\|_{A_\infty}\right)^{1/p}, \quad 1 < p < \infty, \quad \sigma = w^{1-p'}.$$

In 1973 Hunt et al. [140] proved that the Hilbert transform

$$Hf(x) = \text{p.v.} \int_{\mathbb{R}} \frac{f(y)}{x-y} dy, \quad x \in \mathbb{R},$$

is also bounded in $L_w^p(\mathbb{R})$ if and only if $w \in A_p(\mathbb{R})$. In particular, they established the inequality

$$\|Hf\|_{L_w^p(\mathbb{R})} \leqslant c_p \|w\|_{A_p(\mathbb{R})}^{\beta} \|f\|_{L_w^p(\mathbb{R})} \tag{4.2}$$

for some positive constant β and some constant c_p depending on p. S. Petermichl showed that the value of the exponent $\beta = \max\{1, p'/p\}$ in (4.2) is sharp. In particular, the following statement holds (see Petermichl [272] and Dragičević et al. [79] for $p \neq 2$):

Theorem 4.1 *Let $1 < p < \infty$ and let w be a weight function on $\mathbb{R}$. Then there is a positive constant c_p depending only on p such that*

$$\|H\|_{L_w^p(\mathbb{R})} \leqslant c_p \|w\|_{A_p(\mathbb{R})}^{\beta}, \tag{4.3}$$

where $\beta = \max\{1, p'/p\}$. Moreover, the exponent in (4.3) is sharp.

Later, in Hytönen and Pérez [143], it was proved that the sharp weighted bound involving the A_∞ characteristic for the Calderón–Zygmund operator provides an improved estimate than the one obtained in the celebrated paper Hytönen [145] about the A_2 conjecture. We recall the result of Hytönen et al. [144] for the Hilbert transform H in the following theorem.

Theorem 4.2 *Let H be the Hilbert transform and let $p \in (1, \infty)$. Then if $w \in A_p(\mathbb{R})$, we have*

$$\|H\|_{L_w^p(\mathbb{R})} \leqslant \begin{cases} \|w\|_{A_p(\mathbb{R})}^{2/p} \|\sigma\|_{A_\infty(\mathbb{R})}^{2/p-1}, & \text{if } p \in (1,2], \\ \|w\|_{A_p(\mathbb{R})}^{2/p} \|w\|_{A_\infty(\mathbb{R})}^{1-2/p}, & \text{if } p \in [2,\infty), \end{cases} \tag{4.4}$$

where $\sigma := w^{1-p'}$.

It is known (see Hytönen and Pérez [143]) that

$$c_n \|\rho\|_{A_\infty(\mathbb{R}^n)} \leqslant [\rho]_{A_\infty(\mathbb{R}^n)} \leqslant \|\rho\|_{A_p(\mathbb{R}^n)}. \tag{4.5}$$

It can be checked that

$$[\sigma]_{A_\infty(\mathbb{R}^n)}^{p-1} \leqslant \|\sigma\|_{A'_p(\mathbb{R}^n)}^{p-1} = \|w\|_{A_p(\mathbb{R}^n)}.$$

It was also shown by S. Buckley that for $1 < p < \infty$, the convolution Calderón–Zygmund singular operator satisfies

$$\|T\|_{L^p_w(\mathbb{R}^n)} \leqslant c\|w\|^{\beta}_{A_p(\mathbb{R}^n)}$$

and the best possible exponent β is at least $\max\{1, 1/(p-1)\}$.

For the Riesz transforms

$$R_j f(x) = c_n \,\text{p.v.} \int_{\mathbb{R}^n} \frac{x_j - y_j}{|x-y|^{n+1}} f(y)dy,$$

the following sharp estimate holds:

$$\|R_j\|_{L^p_w(\mathbb{R}^n)} \leqslant c\|w\|^{\max\{1,\frac{1}{p-1}\}}_{A_p(\mathbb{R}^n)}.$$

S. Petermichl obtained the results for $p = 2$. The general case $p \neq 2$ then follows by the sharp version of the Rubio de Francia extrapolation theorem given by Dragičević et al. [79] (see also Hytönen [145] regarding the A_2 conjecture for Calderón–Zygmund operators which, in fact, implies an appropriate estimate for all exponents $1 < p < \infty$ by applying a sharp version of the Rubio de Francia extrapolation theorem).

In 1974, Muckenhoupt and Wheeden [256] found necessary and sufficient condition for the one-weight inequality; namely, they proved that the Riesz potential operator $I_\alpha f(x) = \int_{\mathbb{R}^n} \frac{f(y)}{|x-y|^{n-\alpha}} dy$, or fractional maximal operator $M_\alpha f(x) = \sup_{Q \ni x} \frac{1}{|Q|^{1-\frac{\alpha}{n}}} \int_Q |f(y)|dy$, is bounded from $L^p_{w^p}(\mathbb{R}^n)$ to $L^q_{w^q}(\mathbb{R}^n)$, where $1 < p < \infty$, $0 < \alpha < n/p$, $q = \frac{np}{n-\alpha p}$ if and only if w satisfies the so-called $A_{p,q}(\mathbb{R}^n)$ condition (see the definition below). Moreover, from their result, it follows that there is a positive constant c depending only on p and α such that

$$\|K_\alpha\|_{L^p_{w^p} \to L^q_{w^q}} \leqslant c\|w\|^{\beta}_{A_{p,q}}, \tag{4.6}$$

for some positive exponent β, where K_α is either I_α or M_α and $\|w\|_{A_{p,q}}$ is the $A_{p,q}$ characteristic of w. Lacey et al. [204] proved that the best possible value of β in (4.6) is $(1-\alpha/n)p'/q$ (resp. $(1-\alpha/n)\max\{1, p'/q\}$) for M_α (resp. for I_α) (see also Cruz-Uribe and Moen [60] for this and other sharp results).

Sawyer proved the following inequality for the one-sided (right) maximal operator M^+:

$$\|M^+ f\|_{L^p_w(\mathbb{R})} \leqslant C_p \|w\|^{\beta}_{A^+_p(\mathbb{R})} \|f\|_{L^p_w(\mathbb{R})}, \quad f \in L^p_w(\mathbb{R}), \tag{4.7}$$

with some positive exponent β, where $\|w\|_{A^+_p(\mathbb{R})}$ is A^+_p characteristic of a weight w (see the appropriate definitions in Sect. 4.2).

Later, in their celebrated work, Andersen and Sawyer [18] completely characterized the one-weight boundedness for one-sided fractional operators. In particular, they proved that if $1 < p < \infty$, $0 < \alpha < 1/p$, $q = \frac{p}{1-\alpha p}$, then

$$\|w\mathcal{N}_\alpha^+ f\|_{L^q(\mathbb{R})} \leqslant C_{p,\alpha}\|w\|^\beta_{A^+_{p,q}(\mathbb{R})}\|wf\|_{L^p(\mathbb{R})}, \quad f \in L^p_{w^p}(\mathbb{R}), \tag{4.8}$$

for some positive β, where $\mathcal{N}_\alpha^+$ denotes either the Weyl transform $\mathcal{W}_\alpha$ or the right fractional maximal operator M_α^+ and $\|w\|_{A^+_{p,q}}$ is the right $A^+_{p,q}$ characteristic of a weight w (see the appropriate definitions in Sect. 4.2).

Two-weight characterization for fractional integrals was given by Sawyer [307, 309] under the conditions involving the operator itself. Two-weight criteria for the Riesz potential I_α under more transparent conditions were derived by M. Gabidzashvili and V. Kokilashvili (see, e.g., Kokilashvili and Krbec [170]). Using the latter result, Wheeden [333] characterized the two-weight norm inequality for fractional maximal operator M_α. Similar problem for integral operators with positive kernels on space of homogeneous type (SHT) was studied in Sawyer and Wheeden [310] and Genebashvili et al. [114] under different type conditions (see also Kairema [159] for the Sawyer-type result without any additional geometric condition on SHT).

Cruz-Uribe and Moen [60] obtained the following results for the Riesz potentials and fractional maximal operator defined on $\mathbb{R}^n$.

Theorem 4.3 *Given $0 < \alpha < n$ and $1 < p \leqslant q < \infty$, suppose $[v, u] \in A^\alpha_{p,q}(\mathbb{R}^n)$ and $u^{1-p'} \in A_\infty(\mathbb{R}^n)$. Then*

$$\|M_\alpha\|_{L^p_u \to L^q_v} \leqslant c[v, u]_{A^\alpha_{p,q}(\mathbb{R}^n) A^{\exp}_\infty(u^{1-p'})^{1/q}} \tag{4.9}$$

and

$$\|M_\alpha\|_{L^p_u \to L^q_v} \leqslant c[v, u]_{A_{p,q}(\mathbb{R}^n)}\|u^{1-p'}\|^{1/q}_{A^M_\infty(\mathbb{R}^n)}, \tag{4.10}$$

where

$$[v, u]_{A^\alpha_{p,q}(\mathbb{R}^n) A^{\exp}_\infty(u)} := \sup_Q A^\alpha_{p,q}(v, u, Q) A^{\exp}_\infty(u, Q),$$

$$[v, u]_{A^\alpha_{p,q}(\mathbb{R}^n)} := \sup_Q A^\alpha_{p,q}(v, u, Q)$$

$$:= \sup_Q |Q|^{\frac{\alpha}{n}+\frac{1}{q}-\frac{1}{p}} \left(\frac{1}{|Q|}\int_Q v\right)^{\frac{1}{q}} \left(\frac{1}{|Q|}\int_Q u^{1-p'}\right)^{\frac{1}{p'}},$$

$$\|\omega\|_{A^M_\infty(\mathbb{R}^n)} := \sup_Q A^M_\infty(\omega, Q) := \sup_Q \frac{1}{\omega(Q)}\int_Q M(\omega\chi_Q)(x)dx,$$

and

$$\|\omega\|_{A_\infty^{\exp}(\mathbb{R}^n)} := \sup_Q A_\infty^{\exp}(\omega, Q)$$
$$:= \sup_Q \left(\frac{1}{|Q|}\int_Q \omega(x)dx\right)\exp\left(-\frac{1}{|Q|}\int_Q \log\omega(x)dx\right).$$

Theorem 4.4 *Given $0 < \alpha < n$ and $1 < p < q < \infty$, suppose that $[v, u] \in A^\alpha_{p,q}$ and $u^{1-p'}, v \in A_\infty$. Then*

$$\|I_\alpha\|_{L^p_u \to L^q_v} \leqslant c[v, u]_{A^\alpha_{p,q}(\mathbb{R}^n)} A_\infty^{\exp}(u^{1-p'})^{1/q} + [v, u]_{A^\alpha_{q',p'}(\mathbb{R}^n)} A_\infty^{\exp}(v)^{1/p'} \tag{4.11}$$

and

$$\|I_\alpha\|_{L^p_u \to L^q_v} \leqslant c[v, u^{1-p'}]_{A_{p,q}(\mathbb{R}^n)}(\|v\|^{1/p'}_{A^M_\infty(\mathbb{R}^n)} + \|u^{1-p'}\|^{1/q}_{A^M_\infty(\mathbb{R}^n)}), \tag{4.12}$$

where the characteristics $A^\alpha_{p,q} A^{\exp}_\infty$ and A^M_∞ are defined in the previous theorem.

These results, in particular, imply the Buckley-type theorems (i.e., sharp weighted bounds in terms of $A_{p,q}$ characteristics) for fractional integrals established in Lacey et al. [204].

Let (X, d, μ) be an SHT (see Sect. 2.1.1 for the definition of an SHT) with a quasi-metric d and doubling measure μ. The doubling constant for μ will be denoted by C_{DC} (see 2.10). For the definition of the reverse doubling condition (RD space), we refer to Definition 2.7. The reverse doubling constants will be denoted by a and b.

Remark 4.5 It is known that if $\mu \in DC(X)$, then $\mu \in RD(X)$ (see, e.g., Strömberg and Torchinsky [316], p.11, Lemma 20).

Regarding sharp one-weighted estimate for the fractional integral operator

$$I_{\alpha,X} f(x) := \int_X \frac{f(y)}{\mu(B(x, d(x, y)))^{1-\alpha}} d\mu(y), \qquad 0 < \alpha < 1, \tag{4.13}$$

defined an SHT, it is known the following theorem due to Kairema [159].

Theorem 4.6 *Suppose that (X, d, μ) is an SHT. Let $0 < \alpha < 1$ and let $1 < p < 1/\alpha$. We set $1/p - 1/q = \alpha$. Let $w \in A_{p,q}(X)$. Then*

(a)

$$\|I_{\alpha,X}\|_{L^p_{w^p}(X) \to L^{q,\infty}_{w^q}(X)} \leqslant c\|w\|^{1-\alpha}_{A_{p,q}(X)};$$

(b)

$$\|I_{\alpha,X}\|_{L^p_{w^p}(X)\to L^q_{w^q}(X)} \leqslant c\|w\|^{(1-\alpha)\max\{1,\frac{p'}{q}\}}_{A_{p,q}(X)}.$$

Furthermore, these estimate are sharp in the sense that $1-\alpha$ *and* $(1-\alpha)\max\{1, \frac{p'}{q}\}$ *are best possible in* (*a*) *and* (*b*)*, respectively.*

In this theorem $\|w\|_{A_{p,q}(X)}$ is the $A_{p,q}$ characteristic of the weight w defined on X (see the definition below).

Mixed A_p-A_∞ type estimates for Calderón–Zygmund operators were established by Lerner and Moen [209]. Hytönen et al. [144] derived mixed A_p-A_∞ type estimates for the Hardy–Littlewood maximal operator

$$Mf(x) := \sup_{x\in B}\frac{1}{\mu(B)}\int_B |f(t)|d\mu(t),$$

defined on an SHT. In particular, they showed that

$$\|M\|_{L^p_w(X)} \leqslant C\|w\|_{A_p(X)}\|w^{1-p'}\|_{A^M_\infty(X)}, \quad 1<p<\infty, \tag{4.14}$$

holds, where $\|\omega\|_{A^M_\infty(X)}$ is one of the A_∞ characteristics of a weight ω (see the definition below).

4.2 Sharp Estimates for One-Sided Integral Operators

Despite the fact that the two-weight problem for one-sided fractional integrals are studies under different type conditions (see Martín-Reyes and de la Torre [228], Lorente [213], Edmunds et al. [92, §2.2]), it is useful to have the sharp estimate norms of these operators under the so-called Muckenhoupt characteristics.

4.2.1 One-Sided Hardy–Littlewood Maximal Operator

Let f be a locally integrable function on $\mathbb{R}$. Then we define one-sided Hardy–Littlewood maximal functions as

$$M^+f(x) = \sup_{h>0}\frac{1}{h}\int_x^{x+h}|f(t)|dt, \qquad M^-f(x) = \sup_{h>0}\frac{1}{h}\int_{x-h}^{h}|f(t)|dt, \quad x\in\mathbb{R}.$$

Sawyer [308] characterized the one-weighted inequality for M^+ and M^- under the so-called one-sided Muckenhoupt condition; namely, he proved that if $1 < p < \infty$, then the following holds:

(i) The inequality

$$\|M^+ f\|_{L^p_w(\mathbb{R})} \leqslant C\|f\|_{L^p_w(\mathbb{R})}, \quad f \in L^p_w(\mathbb{R}),$$

holds if and only if $w \in A^+_p(\mathbb{R})$, i.e.,

$$\|w\|_{A^+_p(\mathbb{R})} := \sup_{x\in\mathbb{R}, h>0} \left(\frac{1}{h}\int_{x-h}^{x} w(t)dt\right)\left(\frac{1}{h}\int_{x}^{x+h} w^{1-p'}(t)dt\right)^{p-1} < \infty.$$

(ii) The inequality

$$\|M^- f\|_{L^p_w(\mathbb{R})} \leqslant C\|f\|_{L^p_w(\mathbb{R})}, \quad f \in L^p_w(\mathbb{R}),$$

holds if and only if $w \in A^-_p(\mathbb{R})$, i.e.,

$$\|w\|_{A^-_p(\mathbb{R})} := \sup_{x\in\mathbb{R}, h>0} \left(\frac{1}{h}\int_{x}^{x+h} w(t)dt\right)\left(\frac{1}{h}\int_{x-h}^{x} w^{1-p'}(t)dt\right)^{p-1} < \infty.$$

Our main result of this section reads as follows.

Theorem 4.7 *Let* $1 < p < \infty$.

(i) The inequality

$$\|M^+\|_{L^p_w(\mathbb{R})} \leqslant c\|w\|^{1/(p-1)}_{A^+_p(\mathbb{R})}$$

holds and the exponent $\frac{1}{p-1}$ *is the best possible.*

(ii) The inequality

$$\|M^-\|_{L^p_w(\mathbb{R})} \leqslant c\|w\|^{1/(p-1)}_{A^-_p(\mathbb{R})}$$

holds and the exponent $\frac{1}{p-1}$ *is the best possible.*

To prove Theorem 4.7 we need some auxiliary statements. The fact that the A^+_p (resp. A^-_p) condition implies $A^+_{p-\varepsilon}$ (resp. $A^-_{p-\varepsilon}$) is well-known (see, e.g., Martín-Reyes [227] and Sawyer [308]), but we formulate and prove the next lemma because of sharp estimates.

Lemma 4.8 *If $w \in A_p^+(\mathbb{R})$ (resp. $w \in A_p^-(\mathbb{R})$), then $w \in A_{p-\varepsilon}^+(\mathbb{R})$ (resp. $w \in A_{p-\varepsilon}^-(\mathbb{R})$), where $\varepsilon \approx \|w\|_{A_p^+(\mathbb{R})}^{1-p'}$ (resp. $\varepsilon \approx \|w\|_{A_p^-(\mathbb{R})}^{1-p'}$) and $\|w\|_{A_{p-\varepsilon}^+(\mathbb{R})} \leqslant c\|w\|_{A_p^+(\mathbb{R})}$ (resp. $\|w\|_{A_{p-\varepsilon}^-(\mathbb{R})} \leqslant c\|w\|_{A_p^-(\mathbb{R})}$).*

We only give the proof of this lemma for $w \in A_p^+(\mathbb{R})$, the case $w \in A_p^-(\mathbb{R})$ follows analogously.

It should be noted that if $w \in A_p^+(\mathbb{R})$, then w, in general, does not satisfy the reverse Hölder inequality (see Sawyer [308]), but nevertheless, it satisfies the so-called weak reverse Hölder inequality which is stated as follows.

Lemma 4.9 (Martín-Reyes [227]) *Let $1 < p < \infty$. If w satisfies $A_p^+(\mathbb{R})$, then there exist positive constants C and c independent of w, a, and b and such that*

$$\int_a^b w^{1+\delta}(x)dx \leqslant C(M^-(w\chi_{(a,b)}(b)))^\delta \int_a^b w(x)dx,$$

where $\delta = \frac{1}{c\|w\|_{A_p^+(\mathbb{R})}}$, and therefore

$$M^-(w^{1+\delta}\chi_{(a,b)})(b) \leqslant C(M^-(w\chi_{(a,b)})(b))^{1+\delta}$$

with the same constants C and c.

Proof The proof of this lemma is given in Martín-Reyes [227], while here we present the proof only for the estimate of δ. Following the proof of Lemma 5 of Martín-Reyes [227], we have the estimate

$$\left(1 - \frac{\delta}{(1+\delta)\alpha\beta^{1+\delta}}\right) \int_a^b w^{1+\delta}(x)dx \leqslant (M^-(w\chi_{(a,b)}(b)))^\delta \int_a^b w(x)dx,$$

where $\alpha = 1 - 2\big(4\|w\|_{A_p^+(\mathbb{R})}\beta\big)^{1/(p-1)}$ and β can be chosen appropriately later. Ultimately, assuming that $\beta = 1/(2^{2p}\|w\|_{A_p^+(\mathbb{R})})$ (in this case $\alpha = 1/2$), we have that

$$\frac{\delta}{(1+\delta)\alpha\beta^{1+\delta}} \leqslant 1$$

taking $\delta = 1/(c\|w\|_{A_p^+(\mathbb{R})})$ with a constant c large enough and independent of w, a, and b. □

The following theorem is in Sawyer [308].

Theorem 4.10 *Suppose that* $1 < p < \infty$. *Then there is a positive constant* c_p *depending only on* p *such that the weak type inequalities*

$$w(\{M^+ f > \alpha\}) \leqslant c_p \|w\|_{A_p^+(\mathbb{R})} \left(\frac{\|f\|_{L_w^p(\mathbb{R})}}{\alpha} \right)^p$$

and

$$w(\{M^- f > \alpha\}) \leqslant c_p \|w\|_{A_p^-(\mathbb{R})} \left(\frac{\|f\|_{L_w^p(\mathbb{R})}}{\alpha} \right)^p$$

hold.

Now we recall the Marcinkiewicz interpolation theorem.

Theorem 4.11 *Suppose that* $1 \leqslant p_0 < p_1 < \infty$ *and* T *is a sublinear operator of weak-type* p_0 *and* p_1, *with respect to the measure* $d\mu = w dx$, *with norms* R_0 *and* R_1, *respectively. Then* T *is bounded on* $L_w^p(\mathbb{R})$ *for all* $p_0 < p < p_1$. *More precisely, for any* $0 < t < 1$,

$$\|Tf\|_{L_w^{p_t}(\mathbb{R})} \leqslant C_t R_0^{1-t} R_1^t \|f\|_{L_w^{p_t}(\mathbb{R})},$$

where

$$\frac{1}{p_t} = \frac{1-t}{p_0} + \frac{t}{p_1} \quad \text{and} \quad C_t^p = \frac{2^{p_t}}{p_t}\left(\frac{p_1}{p_1 - p_t} + \frac{p_0}{p_t - p_0} \right).$$

Proof of Lemma 4.8 We follow the proof of Proposition 3 in Martín-Reyes [227]. For simplicity, let us denote $s = p - \varepsilon$. To show $w \in A_s^+(\mathbb{R})$, $1 < s < p$, it is enough to show that

$$\begin{aligned} \mathcal{M}_s^+(w) &:= \sup_{a,b} \sup_{a<x<b} \frac{1}{(b-a)^s} \left(\int_a^x w(t)dt \right) \left(\int_x^b w^{-1/(s-1)}(t)dt \right)^{s-1} \\ &\leqslant C \|w\|_{A_+^s(\mathbb{R})} \end{aligned}$$

because (see Martín-Reyes [227])

$$\mathcal{M}_r^+(w) \approx \|w\|_{A_r^+(\mathbb{R})}, \quad 1 < r < \infty.$$

Observe that $w \in A_p^+(\mathbb{R})$ if and only if $\sigma \in A_{p'}^-(\mathbb{R})$. Then, by analogy with Lemma 4.9 for $A_{p'}^-(\mathbb{R})$ classes, we have that there is a constant C independent of w, x and b and such that

$$M^+(\sigma^{1+\delta} \chi_{(x,b)})(x) \leqslant C (M^+(\sigma_{\chi_{(x,b)}})(x))^{1+\delta}, \tag{4.15}$$

where

$$\delta = \frac{1}{c\|\sigma\|_{A^-_{p'}(\mathbb{R})}} = \frac{1}{c\|w\|^{p'-1}_{A^+_p(\mathbb{R})}}.$$

Now we claim that $w \in A^+_s(\mathbb{R})$, where $s = \frac{p+\delta}{1+\delta}$. Fix $a < x < b$. Since σ is locally integrable, it follows from (4.15) that the same holds for $\sigma^{1+\delta}$. Hence, there exists a finite decreasing sequence $x_0 = x > x_1 > \cdots > x_N \geqslant a = x_{N+1}$ such that

$$\int_a^b \sigma^{1+\delta} = 2^k \int_x^b \sigma^{1+\delta} \quad \text{if} \quad k = 0, \dots, N \quad \text{and} \quad \int_a^{x_N} \sigma^{1+\delta} < 2^N \int_x^b \sigma^{1+\delta}. \tag{4.16}$$

From (4.16) it follows that for every $k = 0, \dots, N$,

$$\int_{x_{k+1}}^b \sigma^{1+\delta} \leqslant 2^{k+1} \int_x^b \sigma^{1+\delta}. \tag{4.17}$$

On the other hand, (4.15) and (4.16) yield

$$\begin{aligned}
\int_a^x w\left(\frac{1}{b-a}\int_x^b \sigma^{1+\delta}\right)^s &= \sum_{k=0}^N \frac{1}{2^{ks}} \int_{x_{k+1}}^{x_k} w\left(\frac{1}{b-a}\int_{x_k}^b \sigma^{1+\delta}\right)^s \\
&\leqslant \sum_{k=0}^N \frac{1}{2^{ks}} \int_{x_{k+1}}^{x_k} w(y)(M^+(\sigma^{1+\delta}\chi_{(y,b)})(y))^s dy \\
&\leqslant \sum_{k=0}^N \frac{2}{2^{ks}} \int_{x_{k+1}}^{x_k} w(y)(M^+(\sigma\chi_{(x_{k+1},b)})(y))^{p+\delta} dy.
\end{aligned}$$

Since $w \in A^+_p(\mathbb{R})$, we know by Theorem 4.10 that M^+ maps $L^p_w(\mathbb{R})$ into weak $L^p_w(\mathbb{R})$ with the norm less than or equal to $C_p\|w\|^{1/p}_{A^+_p(\mathbb{R})}$. Then, by the Marcinkiewicz interpolation theorem, the operator M^+ maps $L^{p+\delta}_w(\mathbb{R})$ into $L^{p+\delta}_w(\mathbb{R})$ with the norm $\leqslant C\|w\|^{1/(p+\delta)}_{A^+_p(\mathbb{R})}$. This, together with (4.17) and the fact that $s > 1$, gives

$$\int_a^x w\left(\frac{1}{b-a}\int_x^b \sigma^{1+\delta}\right)^s \leqslant C\|w\|_{A_p^+(\mathbb{R})}\sum_{k=0}^N \frac{1}{2^{ks}}\int_{x_{k+1}}^b \sigma^{1+\delta}$$

$$\leqslant C\|w\|_{A_p^+(\mathbb{R})}\sum_{k=0}^N \frac{2^{k+1}}{2^{ks}}\int_x^b \sigma^{1+\delta}$$

$$\leqslant c_p\|w\|_{A_p^+(\mathbb{R})}\int_x^b \sigma^{1+\delta} < \infty,$$

where the constant c_p does not depend on w and depends only on p. Now the lemma follows. Furthermore, note that

$$\varepsilon = \frac{p-1}{1+c\|w\|^{p'-1}_{A_p^+(\mathbb{R})}} \approx c_p\|w\|^{1-p'}_{A_p^+(\mathbb{R})}.$$

□

Proof of Theorem 4.7 (i) We follow Buckley [37]. Let $w \in A_p^+(\mathbb{R})$. Then by Lemma 4.8, $w \in A^+_{p-\varepsilon}(\mathbb{R})$, and trivially, w is also an $A^+_{p+\varepsilon}(\mathbb{R})$ weight. Moreover, the inequalities

$$\|w\|_{A^+_{p-\varepsilon}(\mathbb{R})} \leqslant c\|w\|_{A_p^+(\mathbb{R})}, \quad \|w\|_{A^+_{p+\varepsilon}(\mathbb{R})} \leqslant \|w\|_{A_p^+(\mathbb{R})}$$

hold. Applying the Marcinkiewicz interpolation theorem corresponding to weak type results at $p-\varepsilon$ and $p+\varepsilon$, we get the estimate

$$\|M^+ f\|_{L^p_w(\mathbb{R})} \leqslant c\|w\|^{1/(p-1)}_{A_p^+(\mathbb{R})}\|f\|_{L^p_w(\mathbb{R})}.$$

Sharpness. Let us take $0<\varepsilon<1$. Let $w(x) = |1-x|^{(1-\varepsilon)(p-1)}$. Then it is easy to check that

$$\|w\|^{1/(p-1)}_{A_p^+(\mathbb{R})} \approx \frac{1}{\varepsilon}.$$

Also observe that for

$$f(t) = (1-t)^{\varepsilon(p-1)-1}\chi_{(0,1)}(t),$$

we have $f \in L^p_w$. Now let $0<x<1$. Then we find that the estimate

$$M^+ f(x) \geqslant \frac{1}{1-x}\int_x^1 f(t)dt = c\frac{1}{\varepsilon}f(x)$$

holds. Finally, we conclude

$$\|M^+ f\|_{L^p_w(\mathbb{R})} \geqslant c\frac{1}{\varepsilon}\|f\|_{L^p_w(\mathbb{R})}.$$

Thus the exponent $1/(p-1)$ is sharp.

The proof of (ii) is similar. We show only *sharpness*. For that let us take $0 < \varepsilon < 1$. Suppose now that $w(x) = |x|^{(1-\varepsilon)(p-1)}$. Then it is easy to check that $\|w\|^{1/(p-1)}_{A^-_p(\mathbb{R})} \approx \frac{1}{\varepsilon}$. Also observe that for $f(t) = t^{\varepsilon(p-1)-1}\chi_{(0,1)}(t)$, there holds $f \in L^p_w(\mathbb{R})$.

Now let $0 < x < 1$. Then we have the estimate

$$M^- f(x) \geqslant \frac{1}{x}\int_0^x f(t)dt = c\frac{1}{\varepsilon}f(x).$$

Finally,

$$\|M^- f\|_{L^p_w(\mathbb{R})} \geqslant c\frac{1}{\varepsilon}\|f\|_{L^p_w(\mathbb{R})}.$$

Thus the exponent $1/(p-1)$ is sharp. □

4.2.2 One-Sided Fractional Integrals

Let f be a locally integrable function on $\mathbb{R}$ and let $0 < \alpha < 1$. Then we define one-sided fractional integrals

$$\mathcal{W}_\alpha f(x) = \int_x^\infty \frac{f(t)}{(t-x)^{1-\alpha}}dt, \qquad \mathcal{R}_\alpha f(x) = \int_{-\infty}^x \frac{f(t)}{(x-t)^{1-\alpha}}dt, \qquad x \in \mathbb{R},$$

and the corresponding fractional maximal functions

$$M^+_\alpha f(x) = \sup_{h>0}\frac{1}{h^{1-\alpha}}\int_x^{x+h}|f(t)|dt,$$

$$M^-_\alpha f(x) = \sup_{h>0}\frac{1}{h^{1-\alpha}}\int_{x-h}^{h}|f(t)|dt, \qquad x \in \mathbb{R}.$$

Taking formally $\alpha = 0$, we have one-sided Hardy–Littlewood maximal operators.

As mentioned above (see the introduction), necessary and sufficient conditions governing the one-weight inequality for one-sided fractional integrals under the $A^{\pm}_{p,q}$ conditions were established in Andersen and Sawyer [18]. Now we give the definition of $A^{\pm}_{p,q}$ classes.

Definition 4.12 Let $1 < p, q < \infty$. We say that a weight function w defined on $\mathbb{R}$ satisfies the $A^{+}_{p,q}(\mathbb{R})$ condition ($w \in A^{+}_{p,q}(\mathbb{R})$) if

$$\|w\|_{A^{+}_{p,q}(\mathbb{R})} := \sup_{\substack{x\in\mathbb{R}\\ h>0}} \left(\frac{1}{h}\int_{x-h}^{x} w^{q}(t)dt\right)\left(\frac{1}{h}\int_{x}^{x+h} w^{-p'}(t)dt\right)^{q/p'} < \infty;$$

the weight function w satisfies the $A^{-}_{p,q}(\mathbb{R})$ condition ($w \in A^{-}_{p,q}(\mathbb{R})$) if

$$\|w\|_{A^{-}_{p,q}(\mathbb{R})} := \sup_{\substack{x\in\mathbb{R}\\ h>0}} \left(\frac{1}{h}\int_{x}^{x+h} w^{q}(t)dt\right)\left(\frac{1}{h}\int_{x-h}^{x} w^{-p'}(t)dt\right)^{q/p'} < \infty.$$

The Two-Weight Problem

Two-weight criteria of various type for one-sided fractional integrals are well-known. The Sawyer type two-weight characterization for the operators M^{+}_{α}, M^{-}_{α} reads as follows (see Martín-Reyes and de la Torre [228]):

Theorem 4.13 *Let $1 < p \leqslant q < \infty$ and let $0 < \alpha < 1$. Suppose that v and w are weights functions on $\mathbb{R}$. Then*

(i) M^{+}_{α} is bounded from $L^{p}_{w}(\mathbb{R})$ to $L^{q}_{v}(\mathbb{R})$ if and only if $(v, w) \in S^{+}_{p,q}$, i.e.,

$$A^{+}_{MT}(v, w, p, q) := \sup_{I} \frac{\|M^{+}_{\alpha}(\sigma\chi_I)\|_{L^{q}_{v}(\mathbb{R})}}{\big(\sigma(I)\big)^{1/p}} < \infty,$$

where the supremum is taken over all intervals I with $\sigma(I) := w^{1-p'}(I) < \infty$. Moreover, $\|M^{+}_{\alpha}\|_{L^{p}_{w}\to L^{q}_{v}} \approx A^{+}_{MT}(v, w, p, q)$.

(ii) M^{-}_{α} is bounded from $L^{p}_{w}(\mathbb{R})$ to $L^{q}_{v}(\mathbb{R})$ if and only if $(v, w) \in S^{-}_{p,q}$, i.e.,

$$A^{-}_{MT}(v, w, p, q) := \sup_{I} \frac{\|M^{-}_{\alpha}(\sigma\chi_I)\|_{L^{q}_{v}(\mathbb{R})}}{\big(\sigma(I)\big)^{1/p}} < \infty,$$

where the supremum is taken over all intervals I with $\sigma(I) := w^{1-p'}(I) < \infty$. Moreover, $\|M_\alpha^-\|_{L_w^p \to L_v^q} \approx A_{MT}^-(v, w, p, q)$.

Lorente and de la Torre [214] gave necessary and sufficient conditions for the boundedness of one-sided potentials from $L_w^p(\mathbb{R})$ to $L_v^q(\mathbb{R})$ (resp. to $L^{q,\infty}(\mathbb{R})$) under the Sawyer type conditions.

In what follows, $\mathcal{N}_\alpha$ is one of the operators $\mathcal{W}_\alpha$ or $\mathcal{R}_\alpha$. Observe that if $\mathcal{N}_\alpha^*$ denotes the adjoint of $\mathcal{N}_\alpha$, then it is easy to see that $\mathcal{N}_\alpha^* = \mathcal{R}_\alpha$ if $\mathcal{N}_\alpha = \mathcal{W}_\alpha$ and vice versa.

Theorem 4.14 *Let $1 < p \leqslant q < \infty$ and let $0 < \alpha < 1$. Suppose that v and w are weight functions on $\mathbb{R}$. Then*

(i) $\mathcal{N}_\alpha$ is bounded from $L_w^p(\mathbb{R})$ to $L_v^{q,\infty}(\mathbb{R})$ if and only if there is a positive constant C such that

$$A_{LT}^*(v, w, p, q) := \sup_I \frac{\|\mathcal{N}_\alpha^*(v\chi_I)\|_{L^{p'}_{w^{1-p'}}(\mathbb{R})}}{\big(v(I)\big)^{1/q'}} < \infty,$$

where the supremum is taken over all intervals I with $v(I) < \infty$. Moreover, $\|\mathcal{N}_\alpha\|_{L_w^p \to L_v^{q,\infty}} \approx A_{LT}^*(v, w, p, q)$.

(ii) $\mathcal{N}_\alpha$ is bounded from $L_w^p(\mathbb{R})$ to $L_v^q(\mathbb{R})$ if and only if

$$\sup_I \frac{\|\mathcal{N}_\alpha(\sigma\chi_I)\|_{L_v^q(\mathbb{R})}}{\big(\sigma(I)\big)^{1/p}} < \infty,$$

where the supremum is taken over all intervals I with $\sigma(I) < \infty$, and

$$\sup_I \frac{\|\mathcal{N}_\alpha^*(\sigma\chi_I)\|_{L_\sigma^{p'}(\mathbb{R})}}{\big(v(I)\big)^{1/p}} < \infty,$$

where $\sigma := w^{1-p'}$ and $v(I) < \infty$.

The following statements are from Edmunds et al. [93] (see also Edmunds et al. [92, §2.2]) and give Gabidzashvili–Kokilashvili type two-weight criteria for one-sided potentials (we refer to Kokilashvili and Krbec [170, pp. 189-190] for the latter conditions in the case of Riesz potentials).

Theorem 4.15 *Let $1 < p < q < \infty$ and let $0 < \alpha < 1$. Suppose that v and w are weight functions on $\mathbb{R}$. Then $\mathcal{W}_\alpha$ is bounded from $L_w^p(\mathbb{R})$ to $L_v^{q,\infty}(\mathbb{R})$ if and only if*

$$[v, w]_{\mathrm{Glo}}^+(p, q) := \sup_{\substack{a\in\mathbb{R}\\ h>0}} \Bigg(\int_{a-h}^{a+h} v(t)dt\Bigg)^{1/q} \Bigg(\int_{a+h}^{\infty} (t-a)^{(\alpha-1)p'} w^{1-p'}(t)dt\Bigg)^{1/p'} < \infty.$$

Moreover, $\|\mathcal{W}_\alpha\|_{L^p_w \to L^{q,\infty}_v} \approx [v, w]^+_{\mathrm{Glo}}(p, q)$.

Theorem 4.16 *Let* $1 < p < q < \infty$ *and let* $0 < \alpha < 1$. *Suppose that* v *and* w *are weight functions on* $\mathbb{R}$. *Then* $\mathcal{R}_\alpha$ *is bounded from* $L^p_w(\mathbb{R})$ *to* $L^{q,\infty}_v(\mathbb{R})$ *if and only if*

$$[v, w]^-_{\mathrm{Glo}}(p, q) := \sup_{\substack{a\in\mathbb{R}\\ h>0}} \left(\int_{a-h}^{a+h} v(t)dt\right)^{1/q} \left(\int_{-\infty}^{a-h} (a-t)^{(\alpha-1)p'} w^{1-p'}(t)dt\right)^{1/p'} < \infty.$$

Moreover, $\|\mathcal{R}_\alpha\|_{L^p_w \to L^{q,\infty}_v} \approx [v, w]^-_{\mathrm{Glo}}(p, q)$.

Theorem 4.17 *Let* $1 < p < q < \infty$ *and let* $0 < \alpha < 1$. *Suppose that* v *and* w *are weight functions on* $\mathbb{R}$. *Then* $\mathcal{W}_\alpha$ *is bounded from* $L^p_w(\mathbb{R})$ *to* $L^q_v(\mathbb{R})$ *if and only if*

(i) $[v, w]^+_{\mathrm{Glo}}(p, q) < \infty$.
(ii)

$$A^+_{GK}(v, w, p, q) := \sup_{\substack{a\in\mathbb{R}\\ h>0}} \left(\int_{a-h}^{a+h} w^{1-p'}(t)dt\right)^{1/p'} \left(\int_{-\infty}^{a-h} \frac{v(y)}{(a-y)^{(1-\alpha)q}}dy\right)^{1/q} < \infty. \tag{4.18}$$

Moreover, $\|\mathcal{W}_\alpha\|_{L^p_w \to L^q_v} \approx [v, w]^+_{\mathrm{Glo}}(p, q) + A^+_{GK}(v, w, p, q)$.

Theorem 4.18 *Let* $1 < p < q < \infty$ *and let* $0 < \alpha < 1$. *Suppose that* v *and* w *are weight functions on* $\mathbb{R}$. *Then* $\mathcal{R}_\alpha$ *is bounded from* $L^p_w(\mathbb{R})$ *to* $L^q_v(\mathbb{R})$ *if and only if*

(i) $[v, w]^-_{\mathrm{Glo}}(p, q) < \infty$.
(ii)

$$A^-_{GK}(v, w, p, q) := \sup_{\substack{a\in\mathbb{R}\\ h>0}} \left(\int_{a-h}^{a+h} w^{1-p'}(t)dt\right)^{1/p'} \left(\int_{a+h}^{\infty} \frac{v(y)}{(y-a)^{(1-\alpha)q}}dy\right)^{1/q} < \infty. \tag{4.19}$$

Moreover, $\|\mathcal{R}_\alpha\|_{L^p_w \to L^q_v} \approx [v, w]^-_{\mathrm{Glo}}(p, q) + A^-_{GK}(v, w, p, q)$.

Remark 4.19 The strong type statements (see Theorems 4.17 and 4.18) follow from appropriate weak type results (see Theorems 4.15 and 4.16) and the Sawyer type two-weight criteria (see Theorem 4.14). We refer to Edmunds et al. [92, §2.2] for details.

Remark 4.20 It is easy to see that

$$\text{(i)} \qquad \|\mathcal{R}_\alpha\|_{L^p_w \to L^q_v} \approx \|\mathcal{R}_\alpha\|_{L^p_w \to L^{q,\infty}_v} + \|\mathcal{W}_\alpha\|_{L^{q'}_{v^{1-q'}} \to L^{p',\infty}_{w^{1-p'}}}; \tag{4.20}$$

(ii) $$\|\mathcal{W}_\alpha\|_{L^p_w \to L^q_v} \approx \|\mathcal{W}_\alpha\|_{L^p_w \to L^{q,\infty}_v} + \|\mathcal{R}_\alpha\|_{L^{q'}_{v^{1-q'}} \to L^{p',\infty}_{w^{1-p'}}}. \tag{4.21}$$

We need similar criteria for one-sided fractional maximal operators defined on $\mathbb{R}$ (cf. Wheeden [333]).

Proposition 4.21 *Suppose that* $1 < p < q < \infty$ *and that* $0 < \alpha < 1$. *Then*

(i) M^+_α *is bounded from* $L^p_w(\mathbb{R})$ *to* $L^q_v(\mathbb{R})$ *if and only if* (4.18) *holds. Moreover,* $\|M^+_\alpha\|_{L^p_w \to L^q_v} \approx A^+_{GK}(v, w, p, q)$.

(ii) M^-_α *is bounded from* $L^p_w(\mathbb{R})$ *to* $L^q_v(\mathbb{R})$ *if and only if* (4.19) *holds. Moreover,* $\|M^-_\alpha\|_{L^p_w \to L^q_v} \approx A^-_{GK}(v, w, p, q)$.

Proof We prove (i). The proof for (ii) is similar. Let $A^+_{GK}(v, w, p, q) < \infty$. Observe that

$$A^+_{GK}(v, w, p, q) = [w^{1-p'}, v^{1-q'}]^-_{\mathrm{Glo}}(q', p').$$

Hence, by Theorem 4.16 we have that

$$\|\mathcal{R}_\alpha\|_{L^{q'}_{v^{1-q'}} \to L^{p',\infty}_{w^{1-p'}}} \approx A^+_{GK}(v, w, p, q).$$

Applying now Theorem 4.14 (i) for $\mathcal{N}_\alpha = \mathcal{R}_\alpha$ (in that case $\mathcal{N}^*_\alpha = \mathcal{W}_\alpha$), we conclude that

$$A^*_{LT}(w^{1-p'}, v^{1-q'}, q', p') \leqslant C\|\mathcal{R}_\alpha\|_{L^{q'}_{v^{1-q'}} \to L^{p',\infty}_{w^{1-p'}}} \leqslant CA^+_{GK}(v, w, p, q) < \infty.$$

Consequently, by Theorem 4.13, we have that

$$\begin{aligned} \|M^+_\alpha\|_{L^p_w \to L^q_v} &\leqslant CA^+_{MT}(v, w, p, q) \\ &\leqslant CA^*_{LT}(w^{1-p'}, v^{1-q'}, q', p') \leqslant CA^+_{GK}(v, w, p, q) < \infty. \end{aligned}$$

The necessity and, consequently, the inequality

$$A^+_{GK}(v, w, p, q) \leqslant C\|\mathcal{R}_\alpha\|_{L^p_w \to L^q_v}$$

follow in the standard way by choosing appropriate test functions. □

The Main Results

Now we are in the position to formulate our main results of this section.

Theorem 4.22 *Suppose that* $0 < \alpha < 1$, $1 < p < 1/\alpha$ *and* q *is such that* $1/p - 1/q - \alpha = 0$.

(i) *There exists a positive constant c depending only on p and α such that*

$$\|M_\alpha^+\|_{L^p_{w^p}\to L^q_{w^q}} \leqslant c\|w\|_{A^+_{p,q}(\mathbb{R})}^{\frac{p'}{q}(1-\alpha)}. \tag{4.22}$$

Moreover, the exponent $\frac{p'}{q}(1-\alpha)$ *is the best possible.*

(ii) *There exists a positive constant c depending only on p and α such that*

$$\|M_\alpha^-\|_{L^p_{w^p}\to L^q_{w^q}} \leqslant c\|w\|_{A^-_{p,q}(\mathbb{R})}^{\frac{p'}{q}(1-\alpha)}. \tag{4.23}$$

Moreover, the exponent $\frac{p'}{q}(1-\alpha)$ *is the best possible.*

Theorem 4.23 *Let* $1 < p < \frac{1}{\alpha}$, *where* $0 < \alpha < 1$. *We set* $q = \frac{p}{1-\alpha p}$.

(i) *The inequality*

$$\|\mathcal{R}_\alpha\|_{L^p_{w^p}\to L^{q,\infty}_{w^q}} \leqslant c\|w\|_{A^-_{p,q}(\mathbb{R})}^{1-\alpha}, \tag{4.24}$$

holds, where the positive constant c depends only on p and α.

(ii) *The inequality*

$$\|\mathcal{W}_\alpha\|_{L^p_{w^p}\to L^{q,\infty}_{w^q}} \leqslant c\|w\|_{A^+_{p,q}(\mathbb{R})}^{1-\alpha}, \tag{4.25}$$

holds, where the positive constant c depends only on p and α.

Theorem 4.24 *Let* $0 < \alpha < 1$, $1 < p < 1/\alpha$ *and let q satisfy* $q = \frac{p}{1-\alpha p}$.

(i) *There is a positive constant c depending only on p and α such that*

$$\|\mathcal{R}_\alpha\|_{L^p_{w^p}\to L^q_{w^q}} \leqslant c\|w\|_{A^-_{p,q}(\mathbb{R})}^{(1-\alpha)\max\{1,p'/q\}}. \tag{4.26}$$

Moreover, this estimate is sharp.

(ii) *There is a positive constant c depending only on p and α such that*

$$\|\mathcal{W}_\alpha\|_{L^p_{w^p}\to L^q_{w^q}} \leqslant c\|w\|_{A^+_{p,q}(\mathbb{R})}^{(1-\alpha)\max\{1,p'/q\}}. \tag{4.27}$$

Moreover, this estimate is sharp.

Main Lemmas

In this subsection, we prove some lemmas regarding the weights satisfying one-sided conditions.

Lemma 4.25 *Let $u \in A^-_{p,q}(\mathbb{R})$. Then for every $a \in \mathbb{R}$ and $h > 0$,*

$$\int_a^{a+h} u^q(x)dx \leqslant \|u\|_{A^-_{p,q}(\mathbb{R})} \int_{a-h}^{a} u^q(x)dx.$$

Further, let $u \in A^+_{p,q}(\mathbb{R})$. Then for every $a \in \mathbb{R}$ and $h > 0$,

$$\int_{a-h}^{a} u^q(x)dx \leqslant \|u\|_{A^+_{p,q}(\mathbb{R})} \int_a^{a+h} u^q(x)dx.$$

Proof Let us prove the first part. The second one follows analogously. Let $u \in A^-_{p,q}(\mathbb{R})$. Using Hölder's inequality, we have

$$\begin{aligned}\int_a^{a+h} u^q(x)dx &\leqslant \Big(\int_a^{a+h} u^q(x)dx\Big)h^{-q}\Big(\int_{a-h}^{a} u^q(x)dx\Big)\Big(\int_{a-h}^{a} u^{-q'}(x)dx\Big)^{q/q'}\\ &\leqslant h^{-1}\Big(\int_a^{a+h} u^q(x)dx\Big)\Big(\int_{a-h}^{a} u^q(x)dx\Big)\Big(\frac{1}{h}\int_{a-h}^{a} u^{-p'}(x)dx\Big)^{q/p'}\\ &\leqslant \|u\|_{A^-_{p,q}(\mathbb{R})} \int_{a-h}^{a} u^q(x)dx.\end{aligned}$$

□

Lemma 4.26 *Let $u \in A^-_{p,q}(\mathbb{R})$. Then for all $a \in \mathbb{R}$ and $r > 0$, we have*

$$\frac{\int_{a-r}^{a} u^q(x)dx}{\int_{a-2r}^{a} u^q(x)dx} \leqslant \frac{\|u\|_{A^-_{p,q}(\mathbb{R})}}{\|u\|_{A^-_{p,q}(\mathbb{R})}+1}.$$

Further, let $u \in A^+_{p,q}(\mathbb{R})$. Then for all $a \in \mathbb{R}$ and $r > 0$, we have

$$\frac{\int_{a}^{a+r} u^q(x)dx}{\int_{a}^{a+2r} u^q(x)dx} \leqslant \frac{\|u\|_{A^+_{p,q}(\mathbb{R})}}{\|u\|_{A^+_{p,q}(\mathbb{R})}+1}.$$

Proof Let $u \in A^-_{p,q}(\mathbb{R})$. By Lemma 4.25, we find that

$$\int\limits_{a-2r}^{a} u^q(x)dx = \int\limits_{a-2r}^{a-r} u^q(x)dx + \int\limits_{a-r}^{a} u^q(x)dx \geqslant \left(\frac{1}{\|u\|_{A^-_{p,q}(\mathbb{R})}} + 1\right) \int\limits_{a-r}^{a} u^q(x)dx.$$

The remaining part of the lemma follows analogously. □

To prove Theorems 4.22 and 4.23, we need to prove also the following lemma.

Lemma 4.27 *Let* $0 < \alpha < 1$, $1 < p < 1/\alpha$ *and let* q *be such that* $\frac{1}{q} = \frac{1}{p} - \alpha$. *Then*

$$[w^q, w^p]^+_{\text{Glo}}(p, q) \leqslant C\|w\|^{1-\alpha}_{A^+_{p,q}(\mathbb{R})}, \tag{4.28}$$

$$[w^q, w^p]^-_{\text{Glo}}(p, q) \leqslant C\|w\|^{1-\alpha}_{A^-_{p,q}(\mathbb{R})}. \tag{4.29}$$

Proof We prove (4.29). The proof for (4.28) follows in the same manner. First observe that $p'(1-\alpha) = 1 + \frac{p'}{q}$. Further, let us take $a \in \mathbb{R}$ and $h > 0$. Then, by Lemma 4.26,

$$\left(\int\limits_{a-2h}^{a} w^q(x)dx\right)^{1/q} \left(\int\limits_{-\infty}^{a-2h} (a-x-h)^{(\alpha-1)p'} w^{-p'}(x)dx\right)^{1/p'}$$

$$\leqslant c\left(\int\limits_{a-2h}^{a} w^q(x)dx\right)^{1/q} \left(\sum_{j=1}^{\infty}(2^j h)^{(\alpha-1)p'} \int\limits_{a-2^{j+1}h}^{a-2^j h} w^{-p'}(x)dx\right)^{1/p'}$$

$$= c\left[\sum_{j=1}^{\infty}(2^j h)^{(\alpha-1)p'}\left(\int\limits_{a-2h}^{a} w^q(x)dx\right)^{p'/q}\left(\int\limits_{a-2^{j+1}h}^{a-2^j h} w^{-p'}(x)dx\right)\right]^{1/p'}$$

$$= c\left[\sum_{j=1}^{\infty}\left(\frac{\int_{a-2h}^{a} w^q(x)dx}{\int_{a-2^j h}^{a} w^q(x)dx}\right)^{p'/q}\left(\frac{1}{2^j h}\int\limits_{a-2^j h}^{a} w^q(x)dx\right)^{p'/q}\right.$$

$$\left.\times\left(\frac{1}{2^j h}\int\limits_{a-2^{j+1}h}^{a-2^j h} w^{-p'}(x)dx\right)\right]^{1/p'}$$

$$\leqslant c\|w\|^{1/q}_{A^-_{p,q}(\mathbb{R})}\left(\sum_{j=1}^{\infty}\left(\frac{\int\limits_{a-2h}^{a} w^q(x)dx}{\int\limits_{a-2^jh}^{a} w^q(x)dx}\right)^{p'/q}\right)^{1/p'}$$

$$\leqslant c\|w\|^{1/q}_{A^-_{p,q}(\mathbb{R})}\left(\sum_{j=0}^{\infty}\left(\frac{\|w\|_{A^-_{p,q}(\mathbb{R})}}{1+\|w\|_{A^-_{p,q}(\mathbb{R})}}\right)^{p'j/q}\right)^{1/p'}$$

$$= c\|w\|^{1/q}_{A^-_{p,q}(\mathbb{R})}\left(\frac{1}{1-\left(\frac{\|w\|_{A^-_{p,q}(\mathbb{R})}}{1+\|w\|_{A^-_{p,q}(\mathbb{R})}}\right)^{p'/q}}\right)^{1/p'}$$

$$\leqslant c\|w\|^{1/q}_{A^-_{p,q}(\mathbb{R})}\|w\|^{1/p'}_{A^-_{p,q}(\mathbb{R})} = c\|w\|^{1-\alpha}_{A^-_{p,q}(\mathbb{R})}.$$

In the latter inequality, we have used the fact that $\|w\|_{A^-_{p,q}(\mathbb{R})} \geqslant 1$ and the relation $\left(1-\left(\frac{s}{1+s}\right)^{p'/q}\right)^{-1} = O(s)$ as $s \to \infty$. □

Proofs of the Main Results

Proof of Theorem 4.22 (i) *Sharpness.* Let us take $0 < \varepsilon < 1$. Suppose that $w(x) = |1-x|^{(1-\varepsilon)/p'}$. Then it is easy to check that

$$\|w\|_{A^+_{p,q}(\mathbb{R})} = \|w^q\|_{A^+_{1+q/p'}(\mathbb{R})} \approx \frac{1}{\varepsilon^{q/p'}}.$$

Also observe that for $f(t) = (1-t)^{\varepsilon-1}\chi_{(0,1)}(t)$, there holds $\|wf\|_{L^p} \approx c\frac{1}{\varepsilon^{1/p}}$. Now let $0 < x < 1$. Then we find that

$$M^+_\alpha f(x) \geqslant \frac{1}{(1-x)^{1-\alpha}}\int\limits_x^1 f(t)dt = c\frac{1}{\varepsilon}(1-x)^{\varepsilon-1+\alpha}.$$

Finally,

$$\|wM^+_\alpha f\|_{L^q(\mathbb{R})} \geqslant \varepsilon^{-1-1/q}. \tag{4.30}$$

Thus, letting $\varepsilon \to 0$, we conclude that the exponent $p'/q(1-\alpha)$ is sharp.

Inequality (4.22) follows from Proposition 4.21, Lemma 4.27 (i), and the following observations:

$$A^+_{GK}(w^q, w^p, p, q) = \left[w^{-p'}, w^{-q'}\right]^-_{\mathrm{Glo}}(q', p') \leqslant C\|w^{-1}\|^{1-\alpha}_{A^-_{q',p'}} = \|w\|^{(1-\alpha)p'/q}_{A^+_{p,q}}.$$

Also notice that $\frac{1}{q'} - \frac{1}{p'} = \alpha$.

For part (ii), we show only the sharpness. Inequality (4.23) follows in the same way as (4.22) was proved. Let $0 < \varepsilon < 1$. Let $w(x) = |x|^{(1-\varepsilon)/p'}$. Then

$$\|w\|_{A^-_{p,q}(\mathbb{R})} = \|w^q\|_{A^-_{1+q/p'}(\mathbb{R})} \approx \frac{1}{\varepsilon^{q/p'}}.$$

Suppose that $f(t) = t^{\varepsilon-1}\chi_{(0,1)}(t)$. Suppose that $\|wf\|_{L^p} \approx c\frac{1}{\varepsilon^{1/p}}$. If $0 < x < 1$, then we find that

$$M^-_\alpha f(x) \geqslant \frac{1}{x^{1-\alpha}} \int_0^x f(t)dt = c\frac{1}{\varepsilon}x^{\varepsilon-1+\alpha}.$$

Finally,

$$\|wM^-_\alpha f\|_{L^q(\mathbb{R})} \geqslant \varepsilon^{-1-1/q}. \tag{4.31}$$

Letting $\varepsilon \to 0$, we conclude that the exponent $p'/q(1-\alpha)$ is sharp. □

Proof of Theorem 4.23 This follows immediately from Lemma 4.27 and Theorems 4.15 and 4.16. □

Proof of Theorem 4.24 We prove only (4.26); the proof of estimate (4.27) follows analogously.

For estimate (4.26), observe that Theorem 4.18 and Theorem 4.23 yield

$$\begin{aligned}\|\mathcal{R}_\alpha\|_{L^p_{w^p}\to L^q_{v^q}} &\leqslant C\Big(\|w\|^{1-\alpha}_{A^-_{p,q}(\mathbb{R})} + A^-_{GK}(w^q, w^p, p, q)\Big)\\ &\leqslant C\Big(\|w\|^{1-\alpha}_{A^-_{p,q}(\mathbb{R})} + \|w\|^{p'/q(1-\alpha)}_{A^-_{p,q}(\mathbb{R})}\Big) \leqslant C\|w\|^{(1-\alpha)\max\{1,p'/q\}}_{A^-_{p,q}(\mathbb{R})}.\end{aligned}$$

Here we have used the relations

$$A^-_{GK}(w^q, w^p, p, q) \leqslant C\|w^{-1}\|^{1-\alpha}_{A^+_{q',p'}(\mathbb{R})} = c\|w\|^{p'/q(1-\alpha)}_{A^-_{p,q}(\mathbb{R})}.$$

The first inequality is due to Lemma 4.27, while the latter equality is easy to check.

For sharpness first we assume that $p'/q \geqslant 1$. Observe that the pointwise estimate

$$M^-_\alpha f \leqslant \mathcal{R}_\alpha f, \quad f \geqslant 0,$$

holds.

Then, using the same w and f as in the proof of Theorem 4.22, we have estimate (4.31) with M_α^- replaced by $\mathcal{R}_\alpha$ showing sharpness. Similarly, for $\mathcal{W}_\alpha$ observe that the pointwise estimate

$$M_\alpha^+ f \leqslant \mathcal{W}_\alpha f, \quad f \geqslant 0,$$

holds.

Then, applying the same w and f as in the proof of (ii) of Theorem 4.22, we have estimate (4.30) with M_α^+ replaced by $\mathcal{W}_\alpha$, which shows sharpness. The case when $p'/q < 1$ follows from duality arguments.

Indeed,

$$\|\mathcal{R}_\alpha\|_{L^p_{w^p} \to L^q_{w^q}} = \|\mathcal{W}_\alpha\|_{L^{q'}_{w^{-q'}} \to L^{p'}_{w^{-p'}}} \leqslant c\|w^{-1}\|^{(1-\alpha)q/p'}_{A^+_{q',p'}(\mathbb{R})} = c\|w\|^{1-\alpha}_{A^-_{p,q}(\mathbb{R})}.$$

□

4.3 Sharp Weighted Bounds for Fractional Integrals

In this section, we study mixed-type weighted bounds for fractional integrals on spaces of homogeneous type (SHT). The main results are obtained by means of the two-weight theory of integral operators with positive kernels. Similar problem is investigated for one-sided fractional integrals. In the latter case, sharp bounds involve one-sided A_∞ characteristics of weights.

4.3.1 Preliminaries

Let (X, d, μ) be an SHT, with the quasi-metric constant κ and doubling constant C_{DC} (see Sect. 2.1.1).

Let $d_X := \operatorname{diam}(X) = \sup\{d(x, y) : x, y \in X\}$ and $B(x,r) := \{y \in X : d(x, y) < r\}$ be diameter of X and the ball with center x and radius r, respectively. The dilation of a ball $B(x, \lambda r)$ with $\lambda > 0$ will be denoted by λB. We will assume that $\mu\{x\} = 0$ for all $x \in X$.

Throughout this section, we assume that $B(x, r_2)\backslash B(x, r_1) \neq \phi$ for any $x \in X$ and for all r_1, r_2 with $0 < r_1 < r_2 < d_X$.

Although previously mentioned in Sect. 4.1 for the Euclidean space, we now recall similar concepts in a more abstract setting for the convenience of the reader.

Let w be a weight on X, i.e., it is μ a.e. positive and locally integrable on X. We denote by $L^p_w(X)$, $1 < p < \infty$, the set of all measurable functions $f : X \to \mathbb{R}$ with the norm $\|f\|_{L^p_\omega(X)} = (\int\limits_X |f(x)|^p \omega(x) d\mu(x))^{\frac{1}{p}}$. Suppose that $L^{p,\infty}_w(X)$ be the weighted weak Lebesgue space with respect to the (quasi-)norm:

$\|f\|_{L_w^{p,\infty}(X)} = \sup_{\lambda>0} \lambda[w(\{x \in X : |f(x)| > \lambda\})]^{1/p}$, where $w(E) := \int_E w d\mu$ for all E measurable subsets of X.

Now we recall the definitions of A_p and $A_{p,q}$ classes of weights defined on an SHT.

Definition 4.28 Let $1 < p < \infty$. We say that $w \in A_p(X)$ if

$$\|w\|_{A_p(X)} := \sup_B A_p(w, B) = \sup_B \left(\frac{1}{\mu(B)} \int_B w d\mu\right)\left(\frac{1}{\mu(B)} \int_B w^{1-p'} d\mu\right)^{p-1} < \infty.$$

Definition 4.29 Let $1 < p, q < \infty$. We say that $w \in A_{p,q}(X)$ if

$$\begin{aligned}\|w\|_{A_{p,q}(X)} &:= \sup_B A_{p,q}(w, B)\\ &= \sup_B \left(\frac{1}{\mu(B)} \int_B w^q d\mu\right)\left(\frac{1}{\mu(B)} \int_B w^{-p'} d\mu\right)^{q/p'} < \infty.\end{aligned}$$

By definition, the class of weights $A_\infty(X)$ is defined as $A_\infty(X) = \bigcup_{p\geqslant 1} A_p(X)$. For an $A_\infty(X)$ weight, it is defined the following characteristic due to Khrushchev [165].

Definition 4.30 For a weight $w \in A_\infty(X)$, we define

$$\begin{aligned}\|w\|_{A_\infty^{\exp}(X)} &:= \sup_B A_\infty^{\exp}(w, B)\\ &= \sup_B \left(\frac{1}{\mu(B)} \int_B w d\mu\right) \exp\left(\frac{1}{\mu(B)} \int_B \log w^{-1} d\mu\right) < \infty.\end{aligned}$$

Another type of A_∞ characteristic in terms of the maximal function was originally introduced by Fujii [107] and later by Wilson [335]:

$$\|w\|_{A_\infty^M(X)} := \sup_B A_\infty^M(w, B) = \sup_B \left(\frac{1}{w(B)} \int_B M(w\chi_B)(x) d\mu\right) < \infty.$$

Further, it is also known that if $X = \mathbb{R}^n$, then $\|w\|_{A_\infty^M(X)} \leqslant C\|w\|_{A_\infty^{\exp}(X)}$ (see, e.g., Hytönen et al. [144]).

For $0 < \alpha < 1$, the fractional maximal operator is defined as

$$M_{\alpha,X} f(x) := \sup_{B\ni x} \frac{1}{\mu(B)^{1-\alpha}} \int_B |f(t)| d\mu(t).$$

The two-weight characterization for $M_{\alpha,X}$ was proved in Wheeden [333] for Euclidean spaces ($X = \mathbb{R}^n$) and in Genebashvili et al. [114] (see p. 158) for an SHT.

Theorem 4.31 *Let $1 < p < q < \infty$ and let $0 < \alpha < 1$. Suppose that v and w be weights on X. Then*

$$\|M_{\alpha,X} f\|_{L^q_v(X)} \leqslant c\|f\|_{L^p_u(X)} \tag{4.32}$$

if and only if

$$A_{GK}(v,u,\alpha,p,q) := \sup_{x\in X, r>0} \left(u^{1-p'}(B(x,2N_0 r))\right)^{1/p'} \times \left(\int_{X\setminus B(x,r)} \mu(B(x,d(x,y)))^{(\alpha-1)q} v(y)d\mu(y)\right)^{1/q} < \infty. \tag{4.33}$$

Moreover,

$$\|M_{\alpha,X}\|_{L^p_u \to L^q_v} \approx A_{GK}(v,u,\alpha,p,q)$$

with constants depending only on p, q, and α.

Now we define a class of kernels on X^2 (see Sawyer and Wheeden [310], Genebashvili et al. [114, Chap. 3]).

Definition 4.32 We say a positive measurable kernel $k : X^2 \to \mathbb{R}$ belongs to V ($k \in V$) if there exists $c > 0$ such that

$$k(x,y) \leqslant ck(x',y)$$

for all x, y and x' in X such that $d(x,x') \leqslant 2N_0 d(x,y)$.

Consider the positive kernel operator defined on X

$$Kf(x) = \int_X k(x,y)f(y)d\mu(y), \qquad x \in X.$$

Let

$$K^* f(x) = \int_X k^*(x,y)f(y)d\mu(y), \qquad x \in X,$$

where $k^*(x,y) = k(y,x)$.

The following two-weight characterization for the positive kernel operator on an SHT was given in Genebashvili et al. [114, Chap. 3].

Theorem 4.33 *Let* $1 < p < q < \infty$ *and let both* $k, k^* \in V$. *Then there exists* $c > 0$ *such that*

$$\|Kf\|_{L^q_v(X)} \leqslant c\|f\|_{L^p_u(X)}$$

holds if and only if

(i)

$$B(v,u,k,p,q) := \sup_{x\in X, r>0} \big(v(B(x,2N_0 r))\big)^{1/q} \times \left(\int_{X\setminus B(x,r)} k^{p'}(x,y)u^{1-p'}(y)d\mu(y)\right)^{1/p'} < \infty; \tag{4.34}$$

(ii)

$$D(v,u,k,p,q) := \sup_{x\in X, r>0} \big(u^{1-p'}(B(x,2N_0 r))\big)^{1/p'} \times \left(\int_{X\setminus B(x,r)} k^{q}(y,x)v(y)d\mu(y)\right)^{1/q} < \infty,$$

where N_0 *is the constant depending only on quasi-metric constants* κ.

Moreover,

$$\|K\|_{L^p_u(X)\to L^q_v(X)} \approx B(v,u,k,p,q) + D(v,u,k,p,q). \tag{4.35}$$

The following two-weight characterization for the positive kernel operator on an SHT was given in Genebashvili et al. [114, Chap. 3].

Theorem 4.34 *Let* $1 < p < q < \infty$ *and let* $k \in V$. *Then there exists* $c > 0$ *such that*

$$\|Kf\|_{L^{q,\infty}_v(X)} \leqslant c\|f\|_{L^p_u(X)}$$

holds if and only if (4.34) *holds. Moreover,*

$$\|K\|_{L^p_u(X)\to L^{q,\infty}_v(X)} \approx B(v,u,k,p,q). \tag{4.36}$$

It follows from Theorems 4.33, 4.34 the next remark.

Remark 4.35 Let conditions of Theorem 4.33 be satisfied. Then

$$\|K\|_{L^p_u(X)\to L^q_v(X)} \approx \|K\|_{L^p_u(X)\to L^{q,\infty}_v(X)} + \|K^*\|_{L^{q'}_{v^{1-q'}}(X)\to L^{p',\infty}_{u^{1-p'}}(X)}.$$

Let k be a positive kernel. Then by the symbols $\phi(B)$ and $\phi^*(B)$, we denote (see Sawyer and Wheeden [310], Genebashvili et al. [114, Chap. 3])

$$\phi(B) := \sup\{k(x, y) : x, y \in B, d(x, y) \geqslant cr(B)\},$$

$$\phi^*(B) := \sup\{k^*(x, y) : x, y \in B, d(x, y) \geqslant cr(B)\},$$

respectively, where $r(B)$ is the radius of the ball B and c is sufficiently small positive constant depending on κ.

Let us denote:

$$\begin{aligned}[v, u]_{A^\phi_{p,q}(X)} &:= \sup_B A^\phi_{p,q}(v, u, B)\\ &= \sup_B \phi(B)\Big(\int_B v(x)d\mu(x)\Big)^{1/q}\Big(\int_B u^{1-p'}(x)d\mu(x)\Big)^{1/p'}.\end{aligned}$$

It is easy to see that $[v, u]_{A^\phi_{p,q}(X)} = [u^{1-p'}, v^{1-q'}]_{A^\phi_{q',p'}(X)}$.

Further, if $\phi(B) = \mu(B)^{\alpha-1}$, then we denote

$$[v, u]_{A^\phi_{p,q}(X)} := [v, u]_{A^\alpha_{p,q}(X)}. \tag{4.37}$$

4.3.2 Integral Operators on an SHT

Initially we prove some lemmas similar to those in Lacey et al. [204] (see also Kokilashvili et al. [186], Lemma 3.1).

Lemma 4.36 *Let $1 < r < \infty$. Suppose that $\omega \in A_r(X)$. Let a and b be as in Definition 2.7. Then for any balls $B \subset X$, the estimate*

$$\frac{\omega(B)}{\omega(aB)} \leqslant 1 - c_{b,r}\|w\|^{-1}_{A_r(X)} \tag{4.38}$$

holds for a constant $c_{b,r}$ depending only on b and r.

Proof Let $E \subset aB$. We show that

$$\Big(\frac{\mu(E)}{\mu(aB)}\Big)^r (A_r(\omega, aB))^{-1} \leqslant \frac{\omega(E)}{\omega(aB)} \tag{4.39}$$

holds. Indeed, denoting $\sigma = \omega^{1-r'}$ and using Hölder's inequality, we see that

$$\frac{\mu(E)}{\mu(aB)} = \frac{\int_E \omega\omega^{-1}d\mu}{\mu(aB)}$$
$$\leqslant \left[\frac{\omega(E)}{\mu(aB)}\right]^{1/r}\left[\frac{\sigma(E)}{\mu(aB)}\right]^{1/r'}$$
$$\leqslant \left[\frac{\omega(E)}{\omega(aB)}\right]^{1/r}\left[\frac{\omega(aB)}{\mu(aB)}\right]^{1/r}\left[\frac{\sigma(aB)}{\mu(aB)}\right]^{1/r'}$$
$$\leqslant \left[\frac{\omega(E)}{\omega(aB)}\right]^{1/r}(A_r(\omega, aB))^{1/r}$$
$$\leqslant \left[\frac{\omega(E)}{\omega(aB)}\right]^{1/r}\|w\|^{1/r}_{A_r(X)}.$$

Since μ satisfies the doubling condition, then by Remark 4.5, we have $\mu(B) \leqslant \frac{1}{b}\mu(aB)$. Taking $E = aB - B$ in (4.39), we get the required result with $c_{b,r} = (1 - 1/b)^r$. □

Lemma 4.37 *Let* $1 < p \leqslant q < \infty$ *and let* $s > 1$. *Further, let* $v \in A_s(X)$ *and* $k \in V$. *Then the following estimate*

$$B(v, u, k, p, q) \leqslant C[v, u]_{A^{\phi}_{p,q}(X)}\|v\|^{1/p'}_{A_s(X)}$$

holds.

Proof Let $s > 0$ and let a be the same as in Definition 2.7. By Lemma 4.36 we have

$$\Big(v(B(x, N_0r))\Big)^{1/q}\Bigg(\int\limits_{X\setminus B(x,r)} k^{p'}(x, y)u^{1-p'}(y)d\mu(y)\Bigg)^{1/p'}$$
$$= c\Big(v(B(x, N_0r))\Big)^{1/q}\Bigg(\sum_{k=0}^{\infty}\int\limits_{B(x,a^kN_0r)\setminus B(x,a^{k-1}N_0r)} k^{p'}(x, y)u^{1-p'}(y)d\mu(y)\Bigg)^{1/p'}$$
$$\leqslant c\Big(v(B(x, N_0r))\Big)^{1/q}\Bigg(\sum_{k=0}^{\infty}\phi(B(x, a^kN_0r))^{p'}\int\limits_{B(x,a^kN_0r)} u^{1-p'}(y)d\mu(y)\Bigg)^{1/p'}$$
$$= c\Bigg[\sum_{k=0}^{\infty}\left(\frac{v(B(x, N_0r))}{v(B(x, a^kN_0r)}\right)^{p'/q}\phi(B(x, a^kN_0r))^{p'}$$
$$\times\Bigg(\int\limits_{B(x,a^kN_0r)} vd\mu\Bigg)^{p'/q}\Bigg(\int\limits_{B(x,a^kN_0r)} u^{1-p'}d\mu\Bigg)\Bigg]^{1/p'}$$

$$\leqslant c[v, u]_{A^{\phi}_{p,q}(X)} \left[\sum_{k=0}^{\infty} (1 - c\|v\|^{-1}_{A_s(X)})^{p'k/q} \right]^{1/p'}$$

$$\leqslant c[v, u]_{A^{\phi}_{p,q}(X)} \|v\|^{1/p'}_{A_s(X)}.$$

Here we used the relation:

$$\frac{1}{1 - \left(1 - \lambda^{-1}\right)^{p'/q}} = O(\lambda), \quad \lambda \to \infty.$$

□

Theorem 4.38 *Let $r, s > 1$. Let p, q, k, k^* be as in Theorem 4.33. Further suppose that $v \in A_r(X)$ and $u^{1-p'} \in A_s(X)$. Then there exists $c > 0$ depending only on r, s, μ and quasi-metric d such that the following inequalities hold:*

$$\|K\|_{L^p_u(X) \to L^{q,\infty}_v(X)} \leqslant c[v, u]_{A^{\phi}_{p,q}} \|v\|^{1/p'}_{A_r(X)}; \tag{4.40}$$

$$\|K\|_{L^p_u(X) \to L^q_v(X)} \leqslant c\left([v, u]_{A^{\phi}_{p,q}} \|v\|^{1/p'}_{A_r(X)} + [v, u]_{A^{\phi *}_{p,q}} \|u^{1-p'}\|^{1/q}_{A_s(X)} \right) \tag{4.41}$$

hold.

Proof By Theorem 4.34 we have that

$$\|Kf\|_{L^p_u(X) \to L^{q,\infty}_v(X)} \leqslant CB(v, u, k, p, q).$$

Now using Lemma 4.37, we find that estimate (4.40) holds.

Estimate (4.35), Remark 4.35, Lemma 4.37, and the relation $[v, u]_{A^{\phi *}_{p,q}(X)} = [u^{1-p'}, v^{1-q'}]_{A^{\phi *}_{q',p'}(X)}$ imply (4.41). □

From the two-weight estimate in Theorem 4.38, we can derive the sharp weighted bounds for the potential operator $I_{\alpha,X}$ (see (4.13) for the definition).

Corollary 4.39 *Let $r, s > 1$. Let $0 < \alpha < 1$ and $1 < p < q < \infty$. Suppose that $[v, u] \in A^{\alpha}_{p,q}(X)$. Further suppose that $v \in A_r(X)$ and $u^{1-p'} \in A_s(X)$. Then*

(a)

$$\|I_{\alpha,X}\|_{L^p_u(X) \to L^{q,\infty}_v(X)} \leqslant C[v, u]_{A^{\alpha}_{p,q}(X)} \|v\|^{1/p'}_{A_r(X)};$$

(b)

$$\|I_{\alpha,X}\|_{L^p_u(X) \to L^q_v(X)} \leqslant C[v, u]_{A^{\alpha}_{p,q}(X)} \left(\|v\|^{1/p'}_{A_r(X)} + \|u^{1-p'}\|^{1/q}_{A_s(X)} \right),$$

where $[v, u]_{A^{\alpha}_{p,q}(X)}$ is defined by (4.37).

Proof Firstly note that for $k(x, y) = \mu(B(x, d(x, y)))^{\alpha-1}$, the doubling and reverse doubling conditions imply that $\phi(B) \approx \mu(B)^{\alpha-1}$ and $k^*(x, y) \approx k(y, x)$ for $x, y \in X$. Taking these estimate into account and using Theorem 4.38, we have the required inequalities. □

Corollary 4.40 *Let* $r, s > 1$. *Let* $0 < \alpha < 1$ *and* $1 < p < q < \infty$. *Suppose that* $w \in A_{p,q}(X)$. *Further suppose that* $w^q \in A_r(X)$ *and* $w^{-p'} \in A_s(X)$. *Then:*

(a)

$$\|I_{\alpha,X}\|_{L^p_{w^p}(X)\to L^{q,\infty}_{w^q}(X)} \leqslant c\|w\|^{1/q}_{A_{p,q}(X)}\|w^q\|^{1/p'}_{A_r(X)};$$

(b)

$$\|I_{\alpha,X}\|_{L^p_{w^p}(X)\to L^q_{w^q}(X)} \leqslant c\|w\|^{1/q}_{A_{p,q}(X)}\left(\|w^q\|^{1/p'}_{A_r(X)} + \|w^{-p'}\|^{1/q}_{A_s(X)}\right).$$

Proof Taking $v = w^q$ and $u = w^p$ and taking into account that $[w^q, w^p]_{A^\alpha_{p,q}(X)} = \|w\|^{1/q}_{A_{p,q}(X)}$ in Theorem 4.38, we have the required estimate. □

Observe that by taking $r = 1 + \frac{q}{p'}$ and $s = 1 + \frac{p'}{q}$ in Corollary 4.40, we have the sharp estimate given in Theorem 4.6. In this case $\|w^q\|_{A_r(X)} = \|w\|_{A_{p,q}(X)}$, $\|w^{-p'}\|_{A_s(X)} = \|w\|^{p'/q}_{A_{p,q}(X)}$.

Now we investigate sharp bounds for the fractional maximal operator $M_{\alpha,X}$ using the two-weight theory.

First observe now that

$$A_{GK}(v, u, \alpha, p, q) = D(v, u, k, p, q),$$

where $k(x, y) = \Big(\mu B(x, d(x, y))\Big)^{\alpha-1}$.

Theorem 4.41 *Let* $s > 1$. *Let* $0 < \alpha < 1$ *and* $1 < p < q < \infty$. *Suppose that* $[v, u] \in A^\alpha_{p,q}(X)$. *Further suppose that* $u^{1-p'} \in A_s(X)$. *Then*

$$\|M_{\alpha,X}\|_{L^p_u(X)\to L^q_v(X)} \leqslant C[v, u]_{A^\alpha_{p,q}(X)}\|u^{1-p'}\|^{1/q}_{A_s(X)}.$$

Proof By Corollary 4.39 and simple observations, we have that

$$\begin{aligned}\|M_{\alpha,X}\|_{L^p_u(X)\to L^q_v(X)} &\leqslant C[u^{1-p'}, v^{1-q'}]_{A^\alpha_{q',p'}(X)}\|u^{1-p'}\|^{1/q}_{A_s(X)}\\ &\leqslant C[v, w]_{A^\alpha_{p,q}(X)}\|u^{1-p'}\|^{1/q}_{A_s(X)}.\end{aligned}$$

□

Corollary 4.42 *Let $s > 1$. Let $0 < \alpha < 1$ and $1 < p < q < \infty$. Suppose that $[v, u] \in A^{\alpha}_{p,q}(X)$. Further suppose that $u^{1-p'} \in A_s(X)$. Then*

$$\|M_{\alpha,X}\|_{L^p_u(X)\to L^q_v(X)} \leqslant C[v,u]_{A^{\alpha}_{p,q}(X)}\|u^{1-p'}\|^{1/q}_{A_s(X)}.$$

Corollary 4.43 *Let $s > 1$. Let $0 < \alpha < 1$ and $1 < p < q < \infty$. Suppose that $w \in A_{p,q}(X)$. Further suppose that $w^{-p'} \in A_s(X)$. Then*

$$\|M_{\alpha,X}\|_{L^p_{w^p}(X)\to L^q_{w^q}(X)} \leqslant C\|w\|^{1/q}_{A_{p,q}(X)}\|w^{-p'}\|^{1/q}_{A_s(X)}.$$

Finally we have

Theorem 4.44 *Let $0 < \alpha < 1$ and let $1 < p < \frac{1}{\alpha}$. Suppose that $w \in A_{p,q}(X)$. We set $q = \frac{p}{1-\alpha p}$.*

Then the following estimate holds:

$$\|M_{\alpha,X}\|_{L^p_{w^p}(X)\to L^q_{w^q}(X)} \leqslant C\|w\|^{(1-\alpha)p'/q}_{A_{p,q}(X)}.$$

Moreover, the exponent $(1-\alpha)p'/q$ is sharp.

Proof Taking $s = 1 + p'/q$ in Corollary 4.43 and observing that $\|w^{-p'}\|_{A_s(X)} = \|w\|^{p'/q}_{A_{p,q}(X)}$, we have the desired result.

The sharpness follows from Theorem 4.6 (part (a)) and the observation

$$\|M_{\alpha,X}\|_{L^p_{w^p}(X)\to L^q_{w^q}(X)} \approx \|T_\alpha\|_{L^{q'}_{w^{-q'}}(X)\to L^{p'}_{w^{-p'}}(X)},$$

which ends the proof. □

4.3.3 One-Sided Fractional Integrals

This section can be considered as a continuation of Sect. 4.2.1.

We are interested also in so-called two-weighted $A^{\pm}_{p,q,\alpha}$ characteristics of Muchenhoupt type (cf. Cruz-Uribe and Moen [60]).

Definition 4.45 Let $1 < p, q < \infty$. We say that a weight pair (v, w) defined on $\mathbb{R}$ satisfies the $A^{+}_{p,q,\alpha}$ condition ($w \in A^{+}_{p,q,\alpha}$), if

$$A^{+}_{p,q,\alpha}(v,u) := \sup_{\substack{x\in\mathbb{R}\\ h>0}} h^{\alpha+1/q-1/p}\left(\frac{1}{h}\int\limits_{x-h}^{x} v(t)dt\right)^{1/q}\left(\frac{1}{h}\int\limits_{x}^{x+h} u^{1-p'}(t)dt\right)^{1/p'} < \infty;$$

a weight function w satisfies the $A^-_{p,q,\alpha}$ condition ($w \in A^-_{p,q,\alpha}$), if

$$A^-_{p,q,\alpha}(v,u) := \sup_{\substack{x\in\mathbb{R}\\ h>0}} h^{\alpha+1/q-1/p}\left(\frac{1}{h}\int_x^{x+h} v(t)dt\right)^{1/q}\left(\frac{1}{h}\int_{x-h}^{x} u^{1-p'}(t)dt\right)^{1/p'} < \infty.$$

If $0 < \alpha < 1/p$, $q = \frac{p}{1-\alpha p}$, $v = w^q$, $u = w^p$, then $A^{\pm}_{p,q,\alpha}$ coincide with $A^{\pm}_{p,q}$ characteristics.

Let us recall $A^{\pm}_p$ characteristics for a locally integrable weight ω:

$$\begin{aligned} \|\omega\|_{A^+_p(\mathbb{R})} &:= \sup_{x\in\mathbb{R},h>0} A^+_p(\omega,x,h) \\ &:= \sup_{x\in\mathbb{R},h>0}\left(\frac{1}{h}\int_{x-h}^{x}\omega(t)dt\right)\left(\frac{1}{h}\int_{x}^{x+h}\omega^{1-p'}(t)dt\right)^{p-1}, \end{aligned} \tag{4.42}$$

and

$$\begin{aligned} \|\omega\|_{A^-_p(\mathbb{R})} &:= \sup_{x\in\mathbb{R},h>0} A^-_p(\omega,x,h) \\ &:= \sup_{x\in\mathbb{R},h>0}\left(\frac{1}{h}\int_{x}^{x+h}\omega(t)dt\right)\left(\frac{1}{h}\int_{x-h}^{x}\omega^{1-p'}(t)dt\right)^{p-1}. \end{aligned} \tag{4.43}$$

It is clear that for fixed $x \in \mathbb{R}$ and $h > 0$,

$$\lim_{p\to\infty} A^+_p(\omega,x,h) = \left(\frac{1}{h}\int_{x-h}^{x}\omega(t)dt\right)\exp\left(\frac{1}{h}\int_{x}^{x+h}\log\frac{1}{w(t)}dt\right)$$

and

$$\lim_{p\to\infty} A^-_p(\omega,x,h) = \left(\frac{1}{h}\int_{x}^{x+h}\omega(t)dt\right)\exp\left(\frac{1}{h}\int_{x-h}^{x}\log\frac{1}{w(t)}dt\right).$$

Therefore there naturally arises question regarding the estimates of the one-sided operator norm by the mixed type ($A^{\pm}_{p,q,\alpha}$, $A^{\pm}_{\infty,\exp}$) characteristics (see Hytönen and Pérez [143], Cruz-Uribe and Moen [60]), where $A^{\pm}_{\infty,\exp}$ is one-sided Hruščev-type characteristic defined as follows:

$$\|\sigma\|_{A^+_{\infty,\exp}} := \sup_{a\in\mathbb{R},\, r>0} A^+_{\infty,\exp}(\sigma, a, r)$$
$$= \sup_{a,r}\left(\frac{1}{r}\int_{a-r}^{a}\sigma(t)dt\right)\exp\left(\frac{1}{r}\int_{a}^{a+r}\log\frac{1}{\sigma(t)}dt\right);$$

and

$$\|\sigma\|_{A^-_{\infty,\exp}} := \sup_{a\in\mathbb{R},\, r>0} A^-_{\infty,\exp}(\sigma, a, r)$$
$$= \sup_{a,r}\left(\frac{1}{r}\int_{a}^{a+r}\sigma(t)dt\right)\exp\left(\frac{1}{r}\int_{a-r}^{a}\log\frac{1}{\sigma(t)}dt\right).$$

We say that $w \in A^{\pm}_{\infty}$ if $w \in A^{\pm}_{r}$, for some r. For the definition and properties of one-sided A_∞ weights, we refer to Martín-Reyes et al. [229].

By Jensen's inequality, we have

$$\|w\|_{A^{\pm}_{\infty,\exp}} \leqslant \|w\|_{A^{\pm}_{r}}, \quad r > 1. \tag{4.44}$$

The main statements of this section read as follows.

Theorem 4.46 *Suppose that* $0 < \alpha < 1$ *and let* $1 < p < q < \infty$. *Then the following statements hold:*

(i) If $u^{1-p'} \in A^-_\infty$, *then*

$$\|M^+_\alpha\|_{L^p_u \to L^q_v} \leqslant c_{p,q,\alpha} A^+_{p,q,\alpha}(v, u)\|u^{1-p'}\|^{1/q}_{A^-_{\infty,\exp}};$$

(ii) If $u^{1-p'} \in A^+_\infty$, *then*

$$\|M^-_\alpha\|_{L^p_u \to L^q_v} \leqslant c_{p,q,\alpha} A^-_{p,q,\alpha}(v, u)\|u^{1-p'}\|^{1/q}_{A^+_{\infty,\exp}}.$$

Theorem 4.47 *Let* $0 < \alpha < 1$ *and let* $1 < p < q < \infty$.

(a) If $v \in A^-_\infty$, *then*

$$\|\mathcal{R}_\alpha\|_{L^p_u \to L^{q,\infty}_v} \leqslant c A^-_{p,q,\alpha}(v, u)\|v\|^{1/p'}_{A^-_{\infty,\exp}},$$

where the positive constant c *depends only on* p, q, *and* α.

(b) If $v \in A^+_\infty$, *then*

$$\|\mathcal{W}_\alpha\|_{L^p_u \to L^{q,\infty}_v} \leqslant c A^+_{p,q,\alpha}(v, u)\|v\|^{1/p'}_{A^+_{\infty,\exp}},$$

where the positive constant c depends only on p, q and α.

Theorem 4.48 *Let* $0 < \alpha < 1$ *and let* $1 < p < q < \infty$.

(a) *If* $v \in A_\infty^-$ *and* $u^{1-p'} \in A_\infty^+$, *then there is a positive constant c depending only on p, q, and* α *such that*

$$\|\mathcal{R}_\alpha\|_{L_u^p \to L_v^q} \leqslant c A_{p,q,\alpha}^-(v,u)\Big(\|v\|_{A_{\infty,\exp}^-}^{1/p'} + \|u^{1-p'}\|_{A_{\infty,\exp}^+}^{1/q}\Big). \tag{4.45}$$

(b) *If* $v \in A_\infty^+$ *and* $u^{1-p'} \in A_\infty^-$, *then there is a positive constant c depending only on p, q, and* α *such that*

$$\|\mathcal{W}_\alpha\|_{L_u^p \to L_v^q} \leqslant c A_{p,q,\alpha}^+(v,u)\Big(\|v\|_{A_{\infty,\exp}^+}^{1/p'} + \|u^{1-p'}\|_{A_{\infty,\exp}^-}^{1/q}\Big). \tag{4.46}$$

To show that these statements are true we need to prove some lemmas:

Lemma 4.49 *Let* ω *be a locally integrable function.*

(i) *If* $\omega \in A_\infty^-$, *then for every* $a \in \mathbb{R}$, $h > 0$ *and* $r > 1$,

$$\omega\big((a,a+h)\big) \leqslant A_r^-(\omega,a,h)\omega\big((a-h,a)\big).$$

Hence,

$$\omega\big((a,a+h)\big) \leqslant A_{\infty,\exp}^-(\omega,a,h)\omega\big((a-h,a)\big).$$

(ii) *Let* $\omega \in A_\infty^+$. *Then for every* $a \in \mathbb{R}$, $h > 0$ *and* $r > 1$,

$$\omega\big((a-h,a)\big) \leqslant A_r^+(\omega,a,h)\omega\big((a,a+h)\big).$$

Hence,

$$\omega\big((a-h,a)\big) \leqslant A_{\infty,\exp}^+(\omega,a,h)\omega\big((a,a+h)\big).$$

Proof (i) Let $\omega \in A_\infty^-$ and let $r > 1$. Following the proof of Lemma 3.8 of Kokilashvili et al. [185], we see that for $r > 0$,

$$\begin{aligned}
\omega\big((a,a+h)\big) &\leqslant \Big(\int_a^{a+h}\omega(x)dx\Big)h^{-r}\Big(\int_{a-h}^{a}\omega(x)dx\Big)\Big(\int_{a-h}^{a}\omega^{1-r'}(x)dx\Big)^{r-1}\\
&\leqslant h^{-1}\Big(\int_a^{a+h}\omega(x)dx\Big)\Big(\int_{a-h}^{a}\omega(x)dx\Big)\Big(\frac{1}{h}\int_{a-h}^{a}\omega^{1-r'}(x)dx\Big)^{r-1}\\
&\leqslant A_r^-(\omega,x,h)\omega\big((a-h,a)\big).
\end{aligned}$$

Passing now to the limit when $r \to \infty$ we have the desired result.

The remaining part of the lemma is proved analogously. □

This lemma immediately implies (see also the proof of Lemma 3.9 in Kokilashvili et al. [185]) the next statement.

Lemma 4.50 *Let $\omega \in A_\infty^-$. Then for all $a \in \mathbb{R}$, $h > 0$ and $r > 1$, we have that*

$$\frac{\omega\big((a-h,a)\big)}{\omega\big((a-2h,a)\big)} \leqslant \frac{A_r^-(\omega, a-h, h)}{A_r^-(\omega, a-h, h)+1}.$$

Consequently,

$$\frac{\omega\big((a-h,a)\big)}{\omega\big((a-2h,a)\big)} \leqslant \frac{A_{\infty,\exp}^-(\omega, a-h, h)}{A_{\infty,\exp}^-(\omega, a-h, h)+1}.$$

Further, let $\omega \in A_\infty^+$. Then for all $a \in \mathbb{R}$, $h > 0$ and $r > 1$, we have

$$\frac{\omega\big((a,a+h)\big)}{\omega\big((a,a+2h)\big)} \leqslant \frac{A_r^+(\omega, a+h, h)}{A_r^+(\omega, a+h, h)+1}.$$

Consequently,

$$\frac{\omega\big((a,a+h)\big)}{\omega\big((a,a+2h)\big)} \leqslant \frac{A_{\infty,\exp}^+(\omega, a+h, h)}{A_{\infty,\exp}^+(\omega, a+h, h)+1}.$$

We need also the following Lemma:

Lemma 4.51 *Let $0 < \alpha < 1$ and let $1 < p < \infty$. Then*

$$[v,u]_{\mathrm{Glo}}^-(p,q) \leqslant C A_{p,q,\alpha}^-(v,u)\|v\|_{A_{\infty,\exp}^-}^{1/p'}; \tag{4.47}$$

$$[v,u]_{\mathrm{Glo}}^+(p,q) \leqslant C A_{p,q,\alpha}^+(v,u)\|v\|_{A_{\infty,\exp}^+}^{1/p'}. \tag{4.48}$$

Proof Let us show (4.47). The proof for (4.48) is similar. Denote $\sigma = u^{1-p'}$. Taking Lemma 4.50 into account, we have for $a \in \mathbb{R}$ and $h > 0$

$$\big(v((a-2h,a)\big)^{1/q}\bigg(\int_{-\infty}^{a-2h}(a-x-h)^{(\alpha-1)p'}\sigma(x)dx\bigg)^{1/p'}$$

$$\leqslant c\big(v((a-2h,a)\big)^{1/q}\bigg(\sum_{j=1}^{\infty}(2^j h)^{(\alpha-1)p'}\sigma\big(a-2^{j+1}h, a-2^j h)\big)\bigg)^{1/p'}$$

$$= c\left[\sum_{j=1}^{\infty}(2^j h)^{(\alpha-1)p'}\Big(v\big((a-2h,a)\big)^{p'/q}\big(\sigma\big(a-2^{j+1}h,a-2^j h\big)\big)\Big)^{1/p'}\right.$$

$$= c\left[\sum_{j=1}^{\infty}\left(\frac{v\big((a-2h,)\big)}{v\big((a-2^j h,a)\big)}\right)^{p'/q}\left(\frac{1}{2^j h}\int_{a-2^j h}^{a} v(x)dx\right)^{p'/q}\right.$$

$$\left.\left(\frac{1}{2^j h}\int_{a-2^{j+1}h}^{a-2^j h}\sigma(x)dx\right)\right]^{1/p'}$$

$$\leqslant cA^{-}_{p,q,\alpha}(v,u)\left(\sum_{j=1}^{\infty}\left(\frac{v\big((a-2h,a)\big)}{v\big((a-2^j h,a)\big)}\right)^{p'/q}\right)^{1/p'}$$

$$\leqslant cA^{-}_{p,q,\alpha}(v,u)\left(\sum_{j=0}^{\infty}\left(\frac{\|v\|_{A^{-}_{\infty,\exp}}}{1+\|v\|_{A^{-}_{\infty,\exp}}}\right)^{p'j/q}\right)^{1/p'}$$

$$= cA^{-}_{p,q,\alpha}(v,u)\left(\frac{1}{1-\left(\frac{\|v\|_{A^{-}_{\infty,\exp}}}{1+\|v\|_{A^{-}_{\infty,\exp}}}\right)^{p'/q}}\right)^{1/p'} = cA_{p,q,\alpha}(v,u)\|v\|^{1/p'}_{A^{-}_{\infty,\exp}},$$

and the proof is complete. □

Now using this lemma, proofs of Theorems 4.46–4.48 follow easily. For example, the proof of (i) of Theorem 4.46 can be derived from Lemma 4.51 and Proposition 4.21 by observing that

$$\begin{aligned}A^{+}_{GK}(v,u,p,q) &= \big[u^{1-p'},v^{1-q'}\big]^{-}_{\mathrm{Glo}}(q',p')\\ &\leqslant cA^{-}_{q',p',\alpha}(u^{1-p'},v^{1-q'})\|u^{1-p'}\|^{1/q}_{A^{-}_{\infty,\exp}}\\ &= cA^{+}_{p,q,\alpha}(v,u)\|u^{1-p'}\|^{1/q}_{A^{-}_{\infty,\exp}}.\end{aligned}$$

By similar manner, we conclude that Theorem 4.48 (part (a)) is a consequence of Lemma 4.51 and Theorem 4.18.

Taking $0<\alpha<1/p$ and $q=\frac{p}{1-\alpha p}$, $v=w^q$ and $u=w^p$, Theorems 4.46–4.48 yield the following corollaries.

Corollary 4.52 *Suppose that* $1<p<\infty$, $0<\alpha<1/p$ *and* $q=\frac{p}{1-\alpha p}$. *Then the following statements hold:*

(i) there is a positive constant c depending only on p and α such that

$$\|M^{+}_{\alpha}\|_{L^p_u\to L^q_v}\leqslant c\|w\|^{1/q}_{A^{+}_{p,q}}\|w^{-p'}\|^{1/q}_{A^{-}_{\infty,\exp}};$$

(*ii*) *there is a positive constant c depending only on p and* α *such that*

$$\|M_\alpha^-\|_{L_u^p \to L_v^q} \leqslant c\|w\|_{A_{p,q}^-}^{1/q} \|w^{-p'}\|_{A_{\infty,\exp}^+}^{1/q}.$$

Corollary 4.53 *Let* $1 < p < \infty$ *and* $0 < \alpha < 1/p$. *We set* $q = \frac{p}{1-\alpha p}$. *Then:*

(*a*)

$$\|\mathcal{R}_\alpha\|_{L_u^p \to L_v^{q,\infty}} \leqslant c\|w\|_{A_{p,q}^-}^{1/q} \|w^q\|_{A_{\infty,\exp}^-}^{1/p'},$$

where the positive constant c depends only on p and α.

(*b*) $v \in A_\infty^+$, *then*

$$\|\mathcal{W}_\alpha\|_{L_u^p \to L_v^{q,\infty}} \leqslant c\|w\|_{A_{p,q}^+}^{1/q} \|w^q\|_{A_{\infty,\exp}^+}^{1/p'},$$

where the positive constant c depends only on p and α.

Corollary 4.54 *Let* $1 < p < \infty$ *and* $0 < \alpha < 1/p$. *We set* $q = \frac{p}{1-\alpha p}$. *Then:*

(*a*) *then there is a positive constant c depending only on p and* α *such that*

$$\|\mathcal{R}_\alpha\|_{L_u^p \to L_v^q} \leqslant c\|w\|_{A_{p,q}^-}^{1/q} \left(\|w^q\|_{A_{\infty,\exp}^-}^{1/p'} + \|w^{-p'}\|_{A_{\infty,\exp}^+}^{1/q}\right); \tag{4.49}$$

(*b*) *if* $v \in A_\infty^+$ *and* $u^{1-p'} \in A_\infty^-$, *then there is a positive constant c depending only on p and* α *such that*

$$\|\mathcal{W}_\alpha\|_{L_u^p \to L_v^q} \leqslant c\|w\|_{A_{p,q}^+}^{1/q} \left(\|w^q\|_{A_{\infty,\exp}^+}^{1/p'} + \|w^{-p'}\|_{A_{\infty,\exp}^-}^{1/q}\right). \tag{4.50}$$

Remark 4.55 Observe that one-weighted sharp estimates (see Theorems 4.22–4.24) can be also obtained from Corollaries 4.52–4.54- For example, Theorem 4.24 (Part (a)) can be derived from Corollary 4.54 (Part (a)). Indeed, by Jensen's inequality we have that $\|v\|_{A_{\infty,\exp}}^- \leqslant \|v\|_{A_s}^-$ for any $s > 1$, and $\|u^{1-p'}\|_{A_{\infty,\exp}}^+ \leqslant \|u^{1-p'}\|_{A_r}^+$ for any $r > 1$. Taking $s = 1 + q/p'$ and $r = 1 + p'/q$, we have that

$$\begin{aligned}
\|\mathcal{R}_\alpha\|_{L^p_{w^p} \to L^q_{w^q}} &\leqslant c\|w\|_{A_{p,q}^-}^{1/q} \left(\|w^q\|_{A_s^-}^{1/p'} + \|w^{p(1-p')}\|_{A_r^+}^{\frac{p'}{q^2}}\right) \\
&= c\|w\|_{A_{p,q}}^{1/q} \left(\|w\|_{A_{p,q}^-}^{1/p'} + \|w\|_{A_{p,q}^-}^{p'/q^2}\right) \\
&= c\|w\|^{(1-\alpha)\max\{1,p'/q\}}.
\end{aligned}$$

4.4 Sharp Weighted Bounds for Hilbert Transform of Odd and Even Functions

In this section, we investigate sharp weighted bounds, involving A_p and A_∞ characteristics of weights, for the Hilbert transform of odd and even functions. Following general results, we derive these sharp weighted A_p bounds in terms of so-called Andersen–Muckenhoupt characteristics.

4.4.1 *Preliminaries*

Let $f : \mathbb{R} \to \mathbb{R}_+$ be odd. Then it is easy to check that Hf is even and given by $(Hf)(x) = (H_0 f)(x)$ for $x > 0$, where

$$(H_0 f)(x) = \frac{2}{\pi} \int_0^\infty \frac{tf(t)}{t^2 - x^2} dt, \qquad x > 0.$$

If f is even, then Hf is odd and is given by $(Hf)(x) = (H_e f)(x)$ for $x > 0$, where

$$(H_e f)(x) = \frac{2}{\pi} \int_0^\infty \frac{xf(t)}{t^2 - x^2} dt$$

Our aim is to investigate the sharp weighted bound of the type (4.4) for operators H_0 and H_e and to derive sharp estimates of the type

$$\|H_0 f\|_{L^p_W(\mathbb{R}_+)} \leqslant c_p \|W\|^{\beta}_{A^0_p(\mathbb{R}_+)} \|f\|_{L^p_W(\mathbb{R}_+)}, \tag{4.51}$$

$$\|H_e f\|_{L^p_W(\mathbb{R}_+)} \leqslant c_p \|W\|^{\gamma}_{A^e_p(\mathbb{R}_+)} \|f\|_{L^p_W(\mathbb{R}_+)} \tag{4.52}$$

where $1 < p < \infty$ with

$$\|W\|_{A^0_p(\mathbb{R}_+)} := \sup_{[a,b]\subset(0,\infty)} \left(\frac{1}{b^2 - a^2} \int_a^b W(x)dx\right) \left(\frac{1}{b^2 - a^2} \int_a^b x^{p'} W^{1-p'}(x)dx\right)^{p-1}$$

and

$$\|W\|_{A^e_p(\mathbb{R}_+)} := \sup_{[a,b]\subset(0,\infty)} \left(\frac{1}{b^2 - a^2} \int_a^b x^p W(x)dx\right) \left(\frac{1}{b^2 - a^2} \int_a^b W^{1-p'}(x)dx\right)^{p-1}.$$

Andersen [17] showed that if $1 < p < \infty$, then:

(i) H_0 is bounded in $L^p_W(\mathbb{R}_+)$ if and only if $\|W\|_{A^0_p(\mathbb{R}_+)} < \infty$;
(ii) H_e is bounded in $L^p_W(\mathbb{R}_+)$ if and only if $\|W\|_{A^e_p(\mathbb{R}_+)} < \infty$.

In the sequel, we will use the following relation between weights $w : \mathbb{R} \to \mathbb{R}_+$ and $W : \mathbb{R}_+ \to \mathbb{R}_+$ (resp. between $\sigma : \mathbb{R} \to \mathbb{R}_+$ and $\Sigma : \mathbb{R}_+ \to \mathbb{R}_+$)

$$w(x) := \frac{W(\sqrt{|x|})}{2\sqrt{|x|}} \quad \left(\text{resp. } \sigma(x) := \frac{\Sigma(\sqrt{|x|})}{2\sqrt{|x|}}\right),$$

where $x \neq 0$.

The following lemma was proved in Andersen [17], but we give the proof because of the exponents of characteristics of weights.

Lemma 4.56 *Let $1 < r < \infty$ and let w be a non-negative measurable function on $(0, \infty)$. Then*

$$\|W\|_{A^0_r(\mathbb{R}_+)} \approx \|w\|_{A_r(\mathbb{R})}$$

with constants depending only on r.

Proof First we show that

$$\|w\|_{A_r(\mathbb{R})} \leqslant c_r \|W\|_{A^0_r(\mathbb{R}_+)}.$$

Let $[a, b] \subset (0, \infty)$. Then

$$\Big(\int_a^b w(x)dx\Big)\Big(\int_a^b w^{1-r'}(x)dx\Big)^{r-1}$$
$$= \Big(\int_a^b W(\sqrt{x})\frac{dx}{2\sqrt{x}}\Big)\Big(\int_a^b W^{1-r'}(\sqrt{x})\frac{dx}{(2\sqrt{x})^{1-r'}}\Big)^{r-1}$$
$$= 2^r\Big(\int_{\sqrt{a}}^{\sqrt{b}} W(x)dx\Big)\Big(\int_{\sqrt{a}}^{\sqrt{b}} x^{r'} W^{1-r'}(x)dx\Big)^{r-1}$$

If $\|W\|_{A^0_r(\mathbb{R}_+)} < \infty$, then the latter expression is bounded by

$$2^r \|W\|_{A^0_r(\mathbb{R}_+)}((\sqrt{b})^2 - (\sqrt{a})^2)^r = 2^r \|W\|_{A^0_r(\mathbb{R}_+)}(b - a)^r.$$

This follows from the definition of $\|W\|_{A^0_r(\mathbb{R}_+)}$.

Suppose now that $[a,b] \subset (-\infty, 0)$. Arguing as before, we see that

$$\Big(\int_a^b w(x)dx\Big)\Big(\int_a^b w^{1-r'}(x)dx\Big)^{r-1}$$

$$= 2^r\Big(\int_{\sqrt{-b}}^{\sqrt{-a}} W(x)dx\Big)\Big(\int_{\sqrt{-b}}^{\sqrt{-a}} x^{r'}W^{1-r'}(x)dx\Big)^{r-1}$$

$$\leqslant 2^r \|W\|_{A_r^0(\mathbb{R}_+)}(b-a)^r.$$

Now let $a < 0 < b$. Suppose that $c > 0$ is a number such that $[a,b] \subset [-c,c]$ and $[a,b]$ and $[-c,c]$ have at least one common endpoint. Then by using the above arguments, we see that

$$\Big(\int_a^b w(x)dx\Big)\Big(\int_a^b w^{1-r'}(x)dx\Big)^{r-1} \leqslant 2^r\Big(\int_0^c w(x)dx\Big)\Big(\int_0^c w^{1-r'}(x)dx\Big)^{r-1}$$

$$\leqslant c_r \|W\|_{A_r^0(\mathbb{R}_+)}(b-a)^r.$$

where c_r is a positive constant depending only on r. Finally,

$$\|w\|_{A_r(\mathbb{R})} \leqslant c_r \|W\|_{A_r^0(\mathbb{R}_+)}.$$

Inequality $\|W\|_{A_r^0(\mathbb{R}_+)} \leqslant c_r \|w\|_{A_r(\mathbb{R})}$ follows from the arguments similar to those used above. □

Now we introduce Wilson type A_∞ characteristic for weights defined on $\mathbb{R}_+$. The classes A_∞^0 and A_∞^e are defined as follows:

$$A_\infty^0 = \cup_{p>1} A_p^0; \quad A_\infty^e = \cup_{p>1} A_p^e.$$

Let $\|W\|_{A_\infty^0}$ be the A_∞^0 characteristic of a W on $\mathbb{R}_+$ defined as follows:

$$\|W\|_{A_\infty^0} = \sup_{(a,b)\subset\mathbb{R}_+} \frac{1}{W([a,b])}\int_a^b x\big(\overline{M}(W\chi_{(a,b)})(x)dx,$$

where

$$\overline{M}f(x) = \sup_{(c,d)\ni x} \frac{1}{d^2-c^2}\int_c^d W(t)dt. \tag{4.53}$$

Here the supremum is taken over all interval $(c, d) \subset \mathbb{R}_+$ containing x.

The next statement will be useful to prove the main theorem.

Lemma 4.57 *Let w be a weight on $\mathbb{R}$. Then the following relation holds:*

$$\|w\|_{A_\infty(\mathbb{R})} \approx \|W\|_{A^0_\infty(\mathbb{R}_+)}, \tag{4.54}$$

with constants independent of w.

Proof At first suppose that $I := (a, b) \subset \mathbb{R}_+$. Then it is easy to see that

$$\frac{1}{w(I)} \int\limits_I M(w\chi_I)(x)dx \approx \frac{1}{W([\sqrt{a}, \sqrt{b}])} \int\limits_{\sqrt{a}}^{\sqrt{b}} x\overline{M}\big(W\chi_{[\sqrt{a},\sqrt{b}]}\big)(x)dx, \tag{4.55}$$

with constants independent of I and w, where $\overline{M}$ is defined by formula (4.53).

Next, we use the following observation: let $x \in (a, b)$,

$$M\big(w\chi_{(a,b)}\big)(x) \approx \overline{M}\big(W\chi_{(\sqrt{|a|},\sqrt{|b|})}\big)(\sqrt{x})$$

which can be obtained from the relation between w and W. In a similar manner, if $I := (a, b) \subset \mathbb{R}_-$, we have

$$\frac{1}{w(I)} \int\limits_a^b M(w\chi_I)(x)dx \approx \frac{1}{W([\sqrt{-a}, \sqrt{-b}])} \int\limits_{\sqrt{-b}}^{\sqrt{-a}} x\overline{M}\big(W\chi_{(\sqrt{a},\sqrt{b})}\big)(x)dx. \tag{4.56}$$

Let now $0 \in I$. Then we represent $I = (a, 0] \cup (0, b)$ to get

$$\begin{aligned}\frac{1}{w(I)} \int\limits_I M(w\chi_I)(x)dx &\leqslant \frac{1}{w(I)} \int\limits_{(a,0)} M(w\chi_{(a,0)})(x)dx \\ &+ \frac{1}{w(I)} \int\limits_{(a,0)} M(w\chi_{(0,b)})(x)dx + \frac{1}{w(I)} \int\limits_{(0,b)} M(w\chi_{(a,0)})(x)dx \\ &+ \frac{1}{w(I)} \int\limits_{(0,b)} M(w\chi_{(0,b)})(x)dx := S_1 + S_2 + S_3 + S_4.\end{aligned}$$

We have to estimate S_2 and S_3. Estimates for S_1 and S_4 can be derived in a similar manner by using the estimates

$$\frac{1}{w(I)} \int\limits_a^0 M(w\chi_{[a,0]})(x)dx \leqslant \frac{1}{w([a, 0])} \int\limits_a^0 M(w\chi_{[a,0]})(x)dx$$

and

$$\frac{1}{w(I)}\int_0^b M(w\chi_{[0,b]})(x)dx \leqslant \frac{1}{w([0,b])}\int_0^b M(w\chi_{[0,b]})(x)dx.$$

Simple observations lead us to the estimates:

$$S_i \leqslant C\frac{1}{W([0,\sqrt{A}])}\int_0^{\sqrt{A}} x\overline{M}\big(W\chi_{[0,\sqrt{A}]}\big)(x)dx \leqslant C\|W\|_{A_\infty^0}, \quad i=2,3,$$

where $A := \max\{|a|,|b|\}$. Finally we have that

$$\|w\|_{A_\infty(\mathbb{R})} \leqslant C\|W\|_{A_\infty^0}$$

with a constant C independent of w. The reverse estimate can be obtained in a similar manner. □

The next lemma is a consequence of (4.5), Lemma 4.57, and Lemma 4.56.

Lemma 4.58 *Let* $1 < p < \infty$. *Then*

$$\|W\|_{A_\infty^0} \leqslant C\|W\|_{A_p^0}.$$

In the sequel we assume that $\sigma = w^{1-p'}$. Taking into account the definition of Σ, we have that

$$\Sigma(u) = W^{1-p'}(u)(2u)^{p'}. \tag{4.57}$$

Theorem 4.59 *Let* $1 < p < \infty$. *Then*

(i)

$$\|H_0\|_{L^p_W} \leqslant \begin{cases} \|W\|_{A_p^0}^{2/p}(\|\Sigma\|_{A_\infty^0})^{2/p-1}, & \text{if } p\in(1,2], \\ \|W\|_{A_p^0}^{2/p}(\|W\|_{A_\infty^0})^{1-2/p}, & \text{if } p\in[2,\infty), \end{cases} \tag{4.58}$$

(ii)

$$\|H_e\|_{L^p_W} \leqslant \begin{cases} \|W\|_{A_p^e}^{2/p}(\|W^{1-p'}\|_{A_\infty^0})^{1-2/p'}, & \text{if } p\in(1,2], \\ \|W\|_{A_p^e}^{2/p}(\|W_p\|_{A_\infty^0})^{2/p'-1}, & \text{if } p\in[2,\infty), \end{cases} \tag{4.59}$$

where W *and* Σ *are related by* (4.57) *and* $W_p(x) = W(x)(2x)^p$.

Proof Let us prove (i). The proof for (ii) is a consequence of the dual arguments and will be discussed afterward.

Let us denote $g(x) := f(\sqrt{x})$, $x > 0$, $g(x) = 0$ otherwise. Suppose that w and W are related as in Lemma 4.56, we have

$$\int_{-\infty}^{+\infty} |g(x)|^p w(x)dx = \int_0^\infty |f(\sqrt{x})|^p w(x)dx$$

$$= \int_0^\infty |f(\sqrt{x})|^p \frac{W(\sqrt{x})}{2\sqrt{x}} dx = \int_0^\infty |f(u)|^p W(u)du.$$

Furthermore, for $x > 0$,

$$(Hg)(x) = \frac{1}{\pi} \int_0^\infty \frac{f(\sqrt{t})}{t - x} dt = \frac{1}{\pi} \int_0^\infty \frac{2tf(t)}{t^2 - x} dt = (H_0 f)(\sqrt{x}).$$

By definition, we have

$$\|H_0 f\|^p_{L^p_W(\mathbb{R}_+)} = \int_0^\infty |(H_0 f)(x)|^p W(x)dx = \int_0^\infty |(H_0 f)(\sqrt{u})|^p W(\sqrt{u}) \frac{du}{2\sqrt{u}}$$

$$= \int_0^\infty |(H_0 f)(\sqrt{u})|^p w(u)du$$

$$= \int_0^\infty |(Hg)(u)|^p w(u)du \leqslant \|Hg\|^p_{L^p_w(\mathbb{R})}.$$

Let $1 < p \leqslant 2$. Then by Theorem 4.2 and Lemmas 4.56 and 4.57, we have that

$$\|H\|_{L^p_w(\mathbb{R})} \leqslant \|w\|^{2/p}_{A_p(\mathbb{R})} \|\sigma\|^{2/p-1}_{A_\infty(\mathbb{R})} \approx \|W\|^{2/p}_{A^0_p(\mathbb{R}_+)} \|\Sigma\|^{2/p-1}_{A^0_\infty(\mathbb{R}_+)}$$

where $\sigma = w^{1-p'}$, $\sigma(x) = \Sigma(\sqrt{|x|})/(2\sqrt{|x|})$. Observe that W and Σ are related also by (4.57). The case $p \geqslant 2$ follows analogously. Thus we have (4.58).

To prove (4.59) we use the duality arguments. First observe that the Riesz identity for the classical Hilbert transform H and the appropriate substitution of the variable yields that

$$\int_{\mathbb{R}_+} (H_0 f)(x) g(x)dx = -\int_{\mathbb{R}_+} (H_e g)(x) f(x)dx.$$

Hence, it follows that the adjoint of H_o is H_e with the equation

$$\|H_e\|_{L^p_w(\mathbb{R}_+)} = \|H_o\|_{L^{p'}_\sigma(\mathbb{R}_+)}.$$

By applying case (i) and Lemmas 4.56 and 4.57, we have the desired result also for (ii). □

The next statement gives sharp weighted bound in terms of A_p characteristics.

Theorem 4.60 *Let* $1 < p < \infty$ *and let* W *be a weight function on* $\mathbb{R}_+$*. Then the following estimates hold:*

(a)

$$\|H_0\|_{L^p_W(\mathbb{R}_+)} \leqslant c_p \|W\|^{\beta}_{A^0_p(\mathbb{R})}; \tag{4.60}$$

(b)

$$\|H_e\|_{L^p_W(\mathbb{R}_+)} \leqslant C_p \|W\|^{\beta}_{A^e_p(\mathbb{R}_+)} \tag{4.61}$$

with some positive constants c_p *and* C_p*, respectively, depending only on* p*, where* $\beta = \max\{1, \frac{p'}{p}\}$*. Moreover the exponent* β *in* (4.60) *and* (4.61) *is best possible.*

Proof We prove (a). The estimate (b) follows from the duality arguments. Let $1 < p \leqslant 2$. To show the validity of (a), we use (4.58), Lemma 4.56, and relations

$$\|\Sigma\|_{A^0_\infty(\mathbb{R}_+)} \approx \|\sigma\|_{A_\infty(\mathbb{R})} \leqslant \|\sigma\|_{A_{p'}(\mathbb{R})} = \|w\|^{p'-1}_{A_p(\mathbb{R})} \approx \|W\|^{p'-1}_{A^0_p(\mathbb{R}_+)}.$$

The case $p > 2$ follows from the estimates:

$$\|W\|_{A^0_\infty(\mathbb{R}_+)} \approx \|w\|_{A_\infty(\mathbb{R})} \leqslant \|w\|_{A_p(\mathbb{R})} \approx \|W\|_{A^0_p(\mathbb{R})}.$$

Sharpness: First we will show the sharpness for $p = 2$. Let

$$g(x) = x^{\varepsilon-1}\chi_{(0,1)}, \quad w(x) = |x|^{1-\varepsilon}.$$

Then (see Petermichl [272]) the following estimate hold:

$$\|g\|_{L^2(\mathbb{R})} \approx \frac{1}{\varepsilon}; \quad \|w\|_{A_2(\mathbb{R})} \approx \frac{1}{\varepsilon}; \quad \|Hg\|_{L^2_w(\mathbb{R})} \geqslant 4\varepsilon^{-3}.$$

Let now

$$f(x) = x^{2(\varepsilon-1)}\chi_{(0,1)}, \quad W(x) = |x|^{3-\varepsilon}.$$

Hence, by using the same changing of variable, we find that

$$\|f\|^2_{L^2_W(\mathbb{R})} \approx \frac{1}{\varepsilon}; \quad \|H_0 f\|^2_{L^2_W(\mathbb{R}_+)} \geqslant \varepsilon^{-3}.$$

Consequently, if the exponent $1-\varepsilon$ is the best possible for the A_2^0 characteristic in the one-weight inequality for some $\lambda > 0$, we have

$$4\varepsilon^{-3} \leqslant \|H_0 f\|_{L^2_W(\mathbb{R}_+)} \leqslant C\|W\|^{1-\lambda}_{A_2^0} \|f\|_{L^2_W(\mathbb{R})} \leqslant C\|W\|^{1-\varepsilon}_{A_2^0} \leqslant C\varepsilon^{\lambda-3}.$$

Let $1 < p < 2$. Suppose that $0 < \varepsilon < 1$ and that $w(x) = |x|^{(1-\varepsilon)(p-1)}$. Then it is easy to check that (see also Petermichl [272])

$$\|w\|^{1/(p-1)}_{A_p} \approx \frac{1}{\varepsilon}.$$

Observe also that for the function defined by

$$f(x) = x^{\varepsilon-1}\chi_{(0,1)}, \tag{4.62}$$

the relation $\|f\|_{L^p_w} \approx \frac{1}{\varepsilon^{\frac{1}{p}}}$ holds. Let

$$g(x) = x^{2(\varepsilon-1)}, \quad W(x) = |x|^{2(1-\varepsilon)(p-1)}.$$

Then the following estimates can be checked easily by using the appropriate change of variables:

$$\begin{aligned}\|H_0 g\|_{L^p_w(\mathbb{R}_+)} &= 2^{-1/p}\|Hf\|_{L^p_w(\mathbb{R})} \geqslant 2^{-1/p}\frac{1}{\varepsilon}\|f\|_{L^p_w(\mathbb{R})} \\ &\approx \|w\|^{p'/p}_{A_p}\|f\|_{L^p_w(\mathbb{R})} \approx \|W\|^{p'/p}_{A_p^0}\|g\|_{L^p_W(\mathbb{R}_+)}\end{aligned}$$

are fulfilled. Thus we have sharpness in (4.60) for $1 < p < 2$.

It remains to consider the case when $p > 2$. In the same manner as above, we can argue for the operator H_e and obtain the sharpness in (4.61) for $1 < p < \infty$. The duality arguments now imply the sharpness in (4.60) for $2 < p < \infty$. □

4.5 Sharp Weighted Bounds for Multiple Integral Operators

In this section, we establish sharp weighted bounds for multiple operators (strong maximal functions, potentials and singular integrals with product kernels) in terms of A_p-type characteristics of weights.

To explain better the point of sharp estimates for multiple operators in this section, let us discuss, for example, the strong Hardy–Littlewood maximal operator $M^{(s)}$ defined on $\mathbb{R}^2$. Denote by $A_p^{(s)}(\mathbb{R}^2)$ the Muckenhoupt class taken with respect to the rectangles with sides parallel to the coordinate axes (see Sect. 4.5.1 for the definitions). Let $\|w\|_{A_p^{(s)}(\mathbb{R}^2)}$ be $A_p^{(s)}$ characteristic of w. There arises a natural questions regarding the sharp bound in the inequality

$$\|M^{(s)}\|_{L^p_w(\mathbb{R}^2)} \leqslant c\|w\|^{\beta}_{A_p^{(s)}(\mathbb{R}^2)}. \tag{4.63}$$

We show that the following estimate is sharp

$$\|M^{(s)}\|_{L^p_w(\mathbb{R}^2)} \leqslant c\Big(\|w\|_{A_p(x_1)}\|w\|_{A_p(x_2)}\Big)^{1/(p-1)}, \tag{4.64}$$

where $\|w\|_{A_p(x_i)}$ be the characteristics of the weight w defined with respect to the i-th variable uniformly to another one $i = 1, 2$ (see, e.g., Fefferman and Stein [101], Kokilashvili [167], García-Cuerva and Rubio de Francia [112], Ch. IV for the one-weight theory for multiple integral operators). Inequality (4.64) together with the Lebesgue differentiation theorem implies that (4.63) holds for $\beta = \frac{2}{p-1}$.

4.5.1 Strong Maximal and Multiple Integral Operators

Let T be an operator acting on a class of functions defined on $\mathbb{R}^n$. We denote by T^k, $k = 1 \cdots n$, the operator defined on functions in $\mathbb{R}$ by letting T act on the k-th variable while keeping the remaining variable fixed. Formally,

$$(T^k f)(x) = (Tf)(x_1, x_2 \cdots, x_{k-1}, \cdot, x_{k+1}, \cdots, x_n)(x_k).$$

Definition 4.61 A weight function w satisfies $A_p^{(s)}(\mathbb{R}^n)$ condition ($w \in A_p^{(s)}(\mathbb{R}^n)$), $1 < p < \infty$, if

$$\|w\|_{A_p^{(s)}} := \sup_P \left(\frac{1}{|P|}\int_P w(x)dx\right)\left(\frac{1}{|P|}\int_P w(x)^{-1/(p-1)}dx\right)^{p-1} < \infty,$$

where the supremum is taken over all parallelepipeds P in $\mathbb{R}^n$ with sides parallel to the coordinate axes.

Definition 4.62 Let $1 < p < \infty$. A weight function $w = w(x_1, \cdots, x_n)$ defined on $\mathbb{R}^n$ is said to satisfy A_p condition in x_i uniformly with respect to other variables ($w \in A_p(x_i)$) if

$$\|w\|_{A_p(x_i)} := \operatorname*{ess\,sup}_{\substack{(x_1,\cdots,x_{i-1},\\ x_{i+1}\cdots,x_n)\in\mathbb{R}^{n-1}}} \sup_I \left(\frac{1}{|I|}\int\limits_Q w(x_1,\cdots,x_n)dx_i\right) \times \left(\frac{1}{|I|}\int\limits_I w^{1-p'}(x_1,\cdots,x_n)dx_i\right)^{p-1} < \infty,$$

where by I we denote a bounded interval in $\mathbb{R}$.

Remark 4.63 $w(x_1,\cdots,x_n) \in A_p^{(s)}(\mathbb{R}^n) \Leftrightarrow w \in \bigcap_{i=1}^n A_p(x_i)$ (see, e.g., pp. 453-454 of García-Cuerva and Rubio de Francia [112], Kokilashvili [167]).

Proposition 4.64 *Let T be an operator defined on $\mathbb{R}^n$ such that*

$$\|T^k\|_{L^p_w(\mathbb{R})} \leqslant c\|w\|^{\gamma(p)}_{A_p(x_k)} \qquad k=1,\cdots,n, \tag{4.65}$$

holds, where $\gamma(p)$ is a constant depending only on p. Then the following estimate:

$$\|T\|_{L^p_w(\mathbb{R}^n)} \leqslant c(\|w\|_{A_p(x_1)}\cdots\|w\|_{A_p(x_n)})^{\gamma(p)}.$$

holds.

Proof For simplicity we give proof for $n=2$ the proof general case is the same. Suppose that $f \geqslant 0$. Using (4.65) two times (i.e., for $k=1,2$) and Fubini's theorem, we have

$$\begin{aligned}
\|Tf\|^p_{L^p_w(\mathbb{R}^2)} &= \iint\limits_{\mathbb{R}^2} (Tf(x_1,x_2))^p w(x_1,x_2)dx_1dx_2 \\
&= \int\limits_{\mathbb{R}}\left(\int\limits_{\mathbb{R}} (T^1(T^2 f(\cdot,x_2)))(x_1)^p w(x_1,x_2)dx_1\right)dx_2 \\
&\leqslant c\|w\|^{p\gamma(p)}_{A_p(x_1)} \int\limits_{\mathbb{R}}\left(\int\limits_{\mathbb{R}} (T^2 f(x_1,x_2))^p w(x_1,x_2)dx_1\right)dx_2 \\
&= c\|w\|^{p\gamma(p)}_{A_p(x_1)} \int\limits_{\mathbb{R}}\left(\int\limits_{\mathbb{R}} (T^2 f(x_1,x_2))^p w(x_1,x_2)dx_2\right)dx_1 \\
&= c(\|w\|_{A_p(x_1)}\|w\|_{A_p(x_2)})^{p\gamma(p)}\|f\|^p_{L^p_w(\mathbb{R}^2)},
\end{aligned}$$

which ends the proof. □

Strong Hardy–Littlewood Maximal Functions and Multiple Singular Integrals

Let f be a locally integrable function on $\mathbb{R}^n$. Then we define strong Hardy–Littlewood maximal operator as

$$\big(M^{(s)}f\big)(x) = \sup_{P \ni x} \frac{1}{|P|} \int_P |f(y)|dy, \qquad x \in \mathbb{R}^n,$$

where the supremum is taken over all parallelepipeds $P \ni x$ in $\mathbb{R}^n$ with sides parallel to the coordinate axes.

Theorem 4.65 *Let $1 < p < \infty$ and w be a weight function on $\mathbb{R}^n$ such that $w \in A_p^{(s)}(\mathbb{R}^n)$. Then there exists a constant c depending only on n and p such that the following inequality*

$$\|M^{(s)}f\|_{L_w^p(\mathbb{R}^n)} \leqslant c\bigg(\prod_{i=1}^{n} \|w\|_{A_p(x_i)}\bigg)^{1/(p-1)} \|f\|_{L_w^p(\mathbb{R}^n)} \tag{4.66}$$

holds, for all $f \in L_w^p(\mathbb{R}^n)$. Furthermore, the exponent $1/(p-1)$ in the estimate (4.66) is sharp.

Proof We can estimate $M^{(s)}$ as follows:

$$\big(M^{(s)}f\big)(x) \leqslant \big(M^1 \circ M^2 \circ \cdots \circ M^n\big)f(x),$$

where

$$\big(M^k f\big)(x_1, \cdots x_n) = \sup_{I_k \ni x_k} \frac{1}{|I_k|} \int_{I_k} |f(x_1, \cdots, x_{k-1}, t, x_{k+1}, \cdots, x_n)|dt.$$

Now by Buckley estimate (4.1) and Proposition 4.64 (for $\gamma(p) = \frac{1}{p-1}$), we find that

$$\|M^{(s)}f\|_{L_w^p(\mathbb{R}^n)} \leqslant c\bigg(\prod_{i=1}^{n} \|w\|_{A_p(x_i)}\bigg)^{1/(p-1)} \|f\|_{L_w^p(\mathbb{R}^n)}.$$

For sharpness, we consider the case for $n = 2$. Observe that when w is of product type, i.e., $w(x_1, x_2) = w_1(x_1)w_2(x_2)$, then

$$\|w\|_{A_p(x_1)} = \|w_1\|_{A_p(\mathbb{R})}, \qquad \|w\|_{A_p(x_2)} = \|w_2\|_{A_p(\mathbb{R})}. \tag{4.67}$$

Let us take $0 < \varepsilon < 1$. Suppose that $w(x_1, x_2) = |x_1|^{(1-\varepsilon)(p-1)}|x_2|^{(1-\varepsilon)(p-1)}$. Then it is easy to check that

$$(\|w\|_{A_{p(x_1)}}\|w\|_{A_{p(x_2)}})^{1/(p-1)} \approx \frac{1}{\varepsilon^2}.$$

Observe also that for

$$f(x_1, x_2) = {x_1}^{\varepsilon-1}\chi_{(0,1)}(x_1){x_2}^{\varepsilon-1}\chi_{(0,1)}(x_2),$$

we have $\|f\|^p_{L^p_w} \approx \frac{1}{\varepsilon^2}$. Now let $0 < x_1, x_2 < 1$. Then we find that the following estimate

$$\big(M^{(s)} f\big)(x_1, x_2) \geqslant \frac{1}{x_1 x_2}\int_0^{x_1}\int_0^{x_2} f(t,\tau)dt d\tau = \frac{1}{\varepsilon^2} f(x, y)$$

holds. Finally

$$\|M^{(s)} f\|_{L^p_w} \geqslant \frac{1}{\varepsilon^2}\|f\|_{L^p_w},$$

thus we have the sharpness in (4.66). □

Let us denote by $\mathcal{H}^{(n)}$ the Hilbert transform with product kernels (or n-dimensional Hilbert transform) defined by

$$\big(\mathcal{H}^{(n)} f\big)(x) = \lim_{\substack{\varepsilon_1\to 0\\ \cdots \\ \varepsilon_n\to 0}} \int\limits_{|x_1-t_1|>\varepsilon_1} \cdots \int\limits_{|x_n-t_n|>\varepsilon_n} \frac{f(t_1,\cdots,t_n)}{(x_1-t_1)\cdots(x_n-t_n)} dt_1\cdots dt_n.$$

We denote $\mathcal{H}^{(1)} =: \mathcal{H}$. Notice that for each $x \in \mathbb{R}^n$, we can write

$$\big(\mathcal{H}^{(n)} f\big)(x) = \big(\mathcal{H}^1 \circ \cdots \circ \mathcal{H}^n\big) f(x) \tag{4.68}$$

where

$$\big(\mathcal{H}^k f\big)(x) = \lim_{\varepsilon_k\to 0} \int\limits_{|x_k-y_k|>\varepsilon_k} \frac{f(x_1,\cdots,y_k,\cdots,x_n)}{x_k-y_k} dy.$$

The following theorem is due to Petermichl [272].

Theorem 4.66 *Let $1 < p < \infty$ and w be a weight function on $\mathbb{R}^n$ such that $w \in A_p^{(s)}(\mathbb{R}^n)$. Then there exists a constant c depending only on n and p such that the following inequality*

$$\|\mathcal{H}^{(n)} f\|_{L^p_w(\mathbb{R}^n)} \leqslant c(\|w\|_{A_p(x_1)}\cdots\|w\|_{A_p(x_n)})^{\max\{1,p'/p\}}\|f\|_{L^p_w(\mathbb{R}^n)} \tag{4.69}$$

holds for all $f \in L^p_w$. *Furthermore, the exponent* $\max\{1, p'/p\}$ *in estimate* (4.69) *is sharp.*

Proof Using representation (4.68), Proposition 4.64, and Theorem 4.1, we have that

$$\|\mathcal{H}^{(n)} f\|_{L^p_w(\mathbb{R}^n)} \leqslant C\big(\|w\|_{A_p(x_1)} \cdots \|w\|_{A_p(x_n)}\big)^{\max\{1,p'/p\}} \|f\|_{L^p_w(\mathbb{R}^n)}.$$

Let $n = 2$. For sharpness, we observe that when w is of product type, i.e., $w(x_1, x_2) = w_1(x_1)w_2(x_2)$, then inequality (4.67) holds. Let us first derive sharpness for $p = 2$. Let us take $0 < \varepsilon < 1$ and let $w(x_1, x_2) = w_1(x_1)w_2(x_2)$, where $w_1(x_1) = |x_1|^{1-\varepsilon}$ and $w_2(x_2) = |x_2|^{1-\varepsilon}$. Then it is easy to check that (4.67) holds. Observe also that for

$$f(x_1, x_2) = x_1{}^{\varepsilon-1}\chi_{(0,1)}(x_1)x_2{}^{\varepsilon-1}\chi_{(0,1)}(x_2), \tag{4.70}$$

$\|f\|^2_{L^2_w} \approx \frac{1}{\varepsilon}$. Now let $0 < x_1, x_2 < 1$. Then we find that

$$\|\mathcal{H}^{(2)} f\|_{L^2_w(\mathbb{R}^2)} \geqslant 4\varepsilon^{-3}.$$

Letting $\varepsilon \to 0$ we have sharpness in (4.69) for $p = 2$, i.e., the estimate

$$\|\mathcal{H}^{(2)}\|_{L^2_w(\mathbb{R}^2)} \leqslant C\|w\|_{A_2(x_1)}\|w\|_{A_2(x_2)}$$

is sharp.

Let $1 < p < 2$. Suppose that $w(x_1, x_2) = |x_1|^{(1-\varepsilon)(p-1)}|x_2|^{(1-\varepsilon)(p-1)}$ and $0 < \varepsilon < 1$. Then it is easy to check that

$$(\|w\|_{A_{p(x_1)}}\|w\|_{A_{p(x_2)}})^{1/(p-1)} \approx \frac{1}{\varepsilon^2}.$$

Observe also that for the function defined by (4.70) the relation $\|f\|_{L^p_w} \approx (\frac{1}{\varepsilon^2})^{\frac{1}{p}}$ holds. Now let $0 < x_1, x_2 < 1$. Then we find that the following estimates

$$\|\mathcal{H}^{(2)} f\|_{L^p_w(\mathbb{R}^2)} \geqslant \frac{1}{\varepsilon^2}\|f\|_{L^p_w(\mathbb{R}^2)} \approx \big(\|w\|_{A_p(x_1)}\|w\|_{A_p(x_2)}\big)^{p'/p}\|f\|_{L^p_w(\mathbb{R}^2)}$$

are fulfilled. Thus we have sharpness in (4.69) for $1 < p < 2$. Using the fact that n-dimensional Hilbert transform is essentially self-adjoint and applying duality argument together with the obvious equality $\|u^{1-p'}\|_{A_{p'}(\mathbb{R}^n} = \|u\|^{1/(p-1)}_{A_p(\mathbb{R}^n)}$, $u \in A_p$, we have sharpness for $p > 2$. This completes the proof. □

Let $x = (x^{(1)}, \cdots, x^{(n)}) \in \mathbb{R}^{d_1} \times \cdots \times \mathbb{R}^{d_n}$, where $d_1, d_2, \cdots, d_n \in \mathbb{N}$. Suppose that $x^{(k)}_{j_k}$ are components of $x^{(k)}$, $k = 1, \cdots, n$, $1 \leqslant j_k \leqslant d_k$. Then we define the n-fold Riesz transform

$$\big(R^{(n)}_{(j_1,\cdots,j_n)} f\big)(x) = \text{p.v.} \int\limits_{\mathbb{R}^{d_1}} \cdots \int\limits_{\mathbb{R}^{d_n}} \prod_{k=1}^{n} \frac{(x^{(k)}_{j_k} - y^{(k)}_{j_k})}{|x^{(k)} - y^{(k)}|^{d_k+1}}$$

$$\times f(y^{(1)}, \cdots, y^{(n)}) dy^{(1)} \cdots dy^{(n)},$$

where $1 \leqslant j_k \leqslant d_k$, $k = 1, \cdots, n$. It can be noticed that

$$\big(R^{(n)}_{(j_1,\cdots,j_n)} f\big)(x) = \big(R^1_{j_1} \circ \cdots \circ R^n_{j_n} f\big)(x)$$

where

$$\big(R^{k}_{(j_1,\cdots,j_n)} f\big)(x) = \text{p.v.} \int\limits_{\mathbb{R}^{d_k}} \frac{x^{(k)}_{j_k} - y^{(k)}_{j_k}}{|x^{(k)} - y^{(k)}|^{d_k+1}}$$

$$\times f(x^{(1)}, \cdots, x^{(k-1)}, y^{(k)}, x^{(k+1)} \cdots, x^{(n)}) dy^{(k)}.$$

Theorem 4.67 *Let $1 < p < \infty$ and w be a weight function on $\mathbb{R}^{d_1} \times \cdots \times \mathbb{R}^{d_n}$ satisfy the condition $w \in A^{(s)}_p(\mathbb{R}^{d_1} \times \cdots \times \mathbb{R}^{d_n})$. Then there exist a constant c depending only on n and p such that the following inequalities*

$$\|R^{(n)}_{(j_1,\cdots,j_n)} f\|_{L^p_w(\mathbb{R}^d)} \leqslant c(\|w\|_{A_p(x^{(1)})} \cdots \|w\|_{A_p(x^{(n)})})^{\max\{1,p'/p\}} \|f\|_{L^p_w(\mathbb{R}^d)} \tag{4.71}$$

hold for all $f \in L^p_w(\mathbb{R}^d)$ and $1 \leqslant j_k \leqslant d_k$, $k = 1 \cdots n$, where $d = d_1 + \cdots, +d_n$. Further, the exponent $\max\{1, p'/p\}$ *in estimate* (4.71) *is sharp.*

Proof The proof of is similar to that of the previous one; we need to apply Proposition 4.64 and the results of Petermichl [273]. □

Example 4.68 Let $-1 < \gamma < p - 1$. It is known that $w(x) = |x|^\gamma$ belongs to $A^{(s)}_p(\mathbb{R}^n)$. Let $\overline{w}(t) = |t|^\gamma$, $t \in \mathbb{R}$. We set $b_\gamma := \max\{2^{\frac{\gamma}{2}}, 1\}$ and $d_\gamma := \max\{2^{\frac{-\gamma}{2}-p+1}, 1\}$.

(i) Then it follows from Theorem 4.65 that

$$\|M^{(s)}\|_{L^p_w(\mathbb{R}^n)} \leqslant C^n C_\gamma^{\frac{n}{p-1}} \Big(1 + \|\overline{w}\|_{A_p(\mathbb{R})}\Big)^{\frac{n}{p-1}} \tag{4.72}$$

where C is the constant from the Buckley's estimate (see (4.1)) and

$$C_\gamma = \begin{cases} b_\gamma, & 0 \leqslant \gamma < p - 1, \\ d_\gamma, & -1 < \gamma < 0; \end{cases} \tag{4.73}$$

It is known (see Grafakos [119, pp. 287–289]) that C in (4.72) can be taken as $C = 3^{p+p'}2^{p'-p}p'24^{\frac{2}{p}}p^{\frac{1}{p-1}}$.

(ii) From Theorem 4.69 it follows that

$$\|\mathcal{H}^{(n)}\|_{L^p_w(\mathbb{R}^n)} \leqslant c^n C_\gamma^{n\max\{1,p'/p\}}\left(1 + \|\overline{w}\|_{A_p(\mathbb{R})}\right)^{n\max\{1,\frac{p'}{p}\}}$$

holds, where c is the constant from (4.1) and C_γ is defined in (4.73).

Following Grafakos [119, pp. 285–286], it can be verified that

$$\|\overline{w}\|_{A_p(\mathbb{R})} \leqslant \max\left\{2^{|\gamma|}, \frac{4^p}{(\gamma+1)(\gamma(1-p')+1)^{p-1}}\right\}.$$

We can get also another type of estimate of the norms in L^p_w. By using the same arguments as in Grafakos [119, pp. 285–286], we find that

$$\|w\|_{A_p(x_i)} \leqslant \begin{cases} \Gamma_\gamma, & 0 \leqslant \gamma < p-1, \\ G_\gamma, & -1 < \gamma < 0; \end{cases} \qquad i = 1, 2,$$

where

$$\Gamma_\gamma = \max\left\{((4/3)^2+1)^{\gamma/2}(2/3)^\gamma, b_\gamma 4^p\left((\gamma+1)^{-1}(\gamma(1-p')+1)^{1-p}+1\right)\right\},$$

$$G_\gamma = \max\left\{((4/3)^2+1)^{-\gamma/2}(2/3)^\gamma, d_0 d_\gamma 4^p\left((\gamma+1)^{-1}(\gamma(1-p')+1)^{1-p}+1\right)\right\}.$$

Consequently, using directly Theorems 4.65 and 4.69, we have the following estimate:

$$\|M^{(s)}\|_{L^p_w(\mathbb{R}^n)} \leqslant C^n \begin{cases} \Gamma_\gamma^{n/(p-1)}, & 0 \leqslant \gamma < p-1, \\ G_\gamma^{n/(p-1)}, & -1 < \gamma < 0; \end{cases}$$

$$\|\mathcal{H}^{(n)}\|_{L^p_w(\mathbb{R}^n)} \leqslant c^n \begin{cases} \Gamma_\gamma^{n\max\{1,p'/p\}}, & 0 \leqslant \gamma < p-1, \\ G_\gamma^{n\max\{1,p'/p\}}, & -1 < \gamma < 0; \end{cases}$$

where C and c are constants in (4.1) and (4.1), respectively.

Strong Fractional Maximal Functions and Riesz Potentials with Product Kernels

In this subsection, we state and prove sharp weighted norm estimates for strong fractional maximal and Riesz potential with product kernels. To get the main results, we use the ideas of the previous subsection.

Definition 4.69 A weight function w satisfies $A^{(s)}_{p,q}$ condition ($w \in A^{(s)}_{p,q}$), $1 < p < \infty$, if

$$\|w\|_{A^{(s)}_{p,q}} := \sup_{P \ni x} \left(\frac{1}{|P|} \int_P w^q(x)dx \right)^{1/q} \left(\frac{1}{|P|} \int_P w(x)^{-p'} dx \right)^{1/p'} < \infty,$$

where the supremum is taken over all parallelepipeds P in $\mathbb{R}^n$ with sides parallel to the coordinate axes.

Definition 4.70 Let $1 < p \leqslant q < \infty$. A weight function $w = w(x_1, \cdots, x_n)$ defined on $\mathbb{R}^n$ is said to satisfy $A_{p,q}$ condition in x_i uniformly with respect to other variables ($w \in A_{p,q}(x_i)$) if

$$\|w\|_{A_{p,q}(x_i)} := \operatorname*{ess\,sup}_{\substack{(x_1,\cdots,x_{i-1},\\ x_{i+1}\cdots,x_n)\in\mathbb{R}^{n-1}}} \sup_{I \ni x_i} \left(\frac{1}{|I|} \int_I w^q(x_1, \cdots, x_n)dx_i \right)^{1/q}$$
$$\times \left(\frac{1}{|I|} \int_I w^{-p'}(x_1, \cdots, x_n)dx_i \right)^{1/p'} < \infty,$$

where I is a bounded interval.

Remark 4.71 Like $A^{(s)}_p(\mathbb{R}^n)$ weights for given $w(x_1, \cdots, x_n) \in A^{(s)}_{p,q} \Leftrightarrow w \in \bigcap_{i=1}^n A_{p,q}(x_i)$.

Proposition 4.72 *Let $1 < p \leqslant q < \infty$ and let T be an operator defined on $\mathbb{R}^n$. Suppose that w is weight, $w \in A^{(s)}_{p,q}$. Let*

$$\|T^k\|_{L^p_{w^p}(\mathbb{R}) \to L^q_{w^q}(\mathbb{R})} \leqslant c\|w\|^{\gamma(p,q)}_{A_{p,q}(x_k)} \qquad k = 1, \cdots, n, \tag{4.74}$$

hold, where $\gamma(p,q)$ is a constant depending only on p and q. Then

$$\|T\|_{L^p_{w^p}(\mathbb{R}^n) \to L^q_{w^q}(\mathbb{R}^n)} \leqslant c(\|w\|_{A_p(x_1)} \cdots \|w\|_{A_p(x_n)})^{\gamma(p,q)}.$$

The proof of Proposition 4.72 is similar to that of Proposition 4.64; therefore it is omitted.

The following Theorem is from Lacey et al. [204].

Theorem 4.73 *Suppose that* $0 < \alpha < n$, $1 < p < n/\alpha$ *and* q *is defined by the relationship* $1/q = 1/p - \alpha/n$. *If* $w \in A_{p,q}(\mathbb{R}^n)$, *then*

$$\|wM_\alpha f\|_{L^q(\mathbb{R}^n)} \leqslant c\|w\|_{A_{p,q}(\mathbb{R}^n)}^{\frac{p'}{q}(1-\alpha/n)} \|wf\|_{L^p(\mathbb{R}^n)}.$$

Furthermore, the exponent $\frac{p'}{q}(1-\alpha/n)$ *is sharp.*

Let f be a locally integrable function and let $0 < \alpha < 1$. The strong fractional maximal operator is defined by

$$\big(M_\alpha^{(s)} f\big)(x) = \sup_{P \ni x} \frac{1}{|P|^{1-\alpha}} \int_P |f(y)|dy,$$

where the supremum is taken over all parallelepipeds P in $\mathbb{R}^n$ with sides parallel to the coordinate axes. It is easy to see that

$$\big(M_\alpha^{(s)} f\big)(x) \leqslant \big(M_\alpha^1 \circ M_\alpha^2 \circ \cdots \circ M_\alpha^n f\big)(x), \tag{4.75}$$

where

$$\big(M_\alpha^k f\big)(x_1, \cdots x_n) = \sup_{I_k \ni x_k} \frac{1}{|I_k|^{1-\alpha}} \int_{I_k} |f(x_1, \cdots, x_{k-1}, t, x_{k+1}, \cdots, x_n)|dt,$$

where I_k are intervals in $\mathbb{R}$ such that $P = I_1 \times \cdots \times I_k$.

Theorem 4.74 *Let* $0 < \alpha < 1$, $1 < p < \frac{1}{\alpha}$, $q = \frac{p}{1-\alpha p}$ *and* w *be a weight function on* $\mathbb{R}^n$ *such that* $w \in A_{p,q}^{(s)}(\mathbb{R}^n)$. *Then there exists a constant* c *depending only on* n, p *and* α *such that the following inequality*

$$\|wM_\alpha^{(s)} f\|_{L^q(\mathbb{R}^n)} \leqslant c\Big(\prod_{i=1}^n \|w\|_{A_{p,q}(x_i)}\Big)^{\frac{p'}{q}(1-\alpha)} \|wf\|_{L^p(\mathbb{R}^n)} \tag{4.76}$$

holds, for all $f \in L^p_{w^p}(\mathbb{R}^n)$. *Furthermore, the exponent* $\frac{p'}{q}(1-\alpha)$ *in estimate* (4.76) *is sharp.*

Proof Using estimate (4.75), Theorem 4.73, and Proposition 4.72, we get easily (4.76). The main "difficulty" here is to derive sharpness. Let, for simplicity, $n = 2$. Let us take $0 < \varepsilon < 1$. suppose that w is of product type $w(x_1, x_2) = w_1(x_1)w_2(x_2)$, where $w_1(x_1) = |x_1|^{(1-\varepsilon)/p'}$ and $w_2(x_2) = |x_2|^{(1-\varepsilon)/p'}$. Then it is easy to see that

$$\|w\|_{A_{p,q}(x_1)} = \|w_1\|_{A_{1+q/p'}(\mathbb{R})} \approx \varepsilon^{-q/p'}; \quad \|w\|_{A_{p,q}(x_2)} = \|w_2\|_{A_{1+q/p'}(\mathbb{R})} \approx \varepsilon^{-q/p'}.$$

Further, if

$$f(t_1, t_2) = |t_1|^{\varepsilon-1}\chi_{(0,1)}(t_1)|t_2|^{\varepsilon-1}\chi_{(0,1)}(t_2),$$

then $\|wf\|_{L^p(\mathbb{R}^2)} \approx \frac{1}{\varepsilon^{2/p}}$. Let $0 < x_1, x_2 < 1$. Then we find that

$$M_\alpha^{(s)} f(x_1, x_2) \geqslant \frac{1}{|x_1|^{1-\alpha}|x_2|^{1-\alpha}} \int_0^{x_1}\int_0^{x_2} f(t_1, t_2)dt_1dt_2 \approx c\frac{|x_1|^{\varepsilon-1+\alpha}|x_2|^{\varepsilon-1+\alpha}}{\varepsilon^2}.$$

Finally, we conclude that

$$\|wM_\alpha^{(s)} f\|_{L^q(\mathbb{R}^2)} \geqslant \varepsilon^{-2-2/q}. \tag{4.77}$$

Thus letting $\varepsilon \to 0$ we have sharpness. □

Let $0 < \alpha < 1$. We define Riesz potential with product kernels on $\mathbb{R}^n$ as follows:

$$(I_\alpha^{(n)} f)(x) = \int_{\mathbb{R}^n} \frac{f(t_1, \cdots, t_n)}{\prod_{i=1}^n |x_i - t_i|^{1-\alpha}} dt_1 \cdots dt_n, \quad x = (x_1, \cdots, x_n) \in \mathbb{R}^n.$$

When $n = 1$ we use the symbol I_α for $I_\alpha^{(1)}$. The following theorem is from Lacey et al. [204].

Theorem 4.75 *Let $0 < \alpha < n$, $1 < p < n/\alpha$. We put $q = \frac{np}{n-\alpha p}$. Suppose that $w \in A_{p,q}(\mathbb{R}^n)$. Then*

$$\|wI_\alpha f\|_{L^q(\mathbb{R}^n)} \leqslant c\|w\|_{A_{p,q}(\mathbb{R}^n)}^{(1-\alpha/n)\max\{1,p'/q\}} \|wf\|_{L^p(\mathbb{R}^n)}.$$

Furthermore, this estimate is sharp.

Our result regarding $I_\alpha^{(n)}$ reads as follows.

Theorem 4.76 *Let $0 < \alpha < 1$, $1 < p < 1/\alpha$. We put $q = \frac{p}{1-\alpha p}$. Let w be a weight function on $\mathbb{R}^n$ such that $w \in A_{p,q}^{(s)}(\mathbb{R}^n)$. Then there exists a constant c depending only on n, p and α such that the following inequality*

$$\|wI_\alpha^{(n)} f\|_{L^q(\mathbb{R}^n)} \leqslant c\left(\prod_{i=1}^n \|w\|_{A_{p,q}(x_i)}\right)^{\max\{1,\frac{p'}{q}\}(1-\alpha)} \|wf\|_{L^p(\mathbb{R}^n)} \tag{4.78}$$

holds for all $f \in L^p_{w^p}(\mathbb{R}^n)$. Furthermore, the exponent $\max\{1, \frac{p'}{q}\}(1-\alpha)$ in estimate (4.78) *is sharp.*

The proof of Theorem 4.78) follows using the same arguments as in the proof of Theorem 4.74 together with Theorem 4.75.

Strong One-Sided Maximal Operators

Based on the results derived in this section, we give sharp bounds for one-sided strong maximal operators.

Let f be a locally integrable function on $\mathbb{R}^n$. We define one-sided strong fractional maximal operators as

$$M^{+(s)} f(x_1, \cdots, x_n) = \frac{1}{\prod_{i=1}^n h_i} \int_{x_1}^{x_1+h_1} \cdots \int_{x_n}^{x_n+h_n} |f(y_1, \cdots y_n)| dy_1 \cdots dy_n, \tag{4.79}$$

$$M^{-(s)} f(x_1, \cdots, x_n) = \frac{1}{\prod_{i=1}^n h_i} \int_{x_1-h_1}^{x_1} \cdots \int_{x_n-h_n}^{x_n} |f(y_1, \cdots y_n)| dy_1 \cdots dy_n. \tag{4.80}$$

Let $1 < p < \infty$. We say that a weight function w belongs to the class $A_p^{-(s)}(\mathbb{R}^n)$ if

$$\|w\|_{A_p^{-(s)}(\mathbb{R}^n)} := \sup_{\substack{h_1,\cdots,h_n>0 \\ x_1,\cdots x_n \in \mathbb{R}}} \left(\frac{1}{h_1 \cdots h_n} \int_{x_1}^{x_1+h_1} \cdots \int_{x_n}^{x_n+h_n} w(t_1, \cdots, t_n) dt_1 \cdots dt_n \right)$$
$$\times \left(\frac{1}{h_1 \cdots h_n} \int_{x_1-h_1}^{x_1} \cdots \int_{x_n-h_n}^{x_n} w^{1-p'}(t_1, \cdots, t_n) dt_1 \cdots dt_n \right)^{p-1} < \infty;$$

further, $w \in A_p^{-(s)}(\mathbb{R}^n)$ if

$$\|w\|_{A_p^{+(s)}(\mathbb{R}^n)} := o \sup_{\substack{h_1,\cdots,h_n>0 \\ x_1,\cdots x_n \in \mathbb{R}}} \left(\frac{1}{h_1 \cdots h_n} \int_{x_1-h_1}^{x_1} \cdots \int_{x_n-h_n}^{x_n} w(t_1, \cdots, t_n) dt_1 \cdots dt_n \right)$$
$$\times \left(\frac{1}{h_1 \cdots h_n} \int_{x_1}^{x_1+h_1} \cdots \int_{x_n}^{x_n+h_n} w^{1-p'}(t_1, \cdots, t_n) dt_1 \cdots dt_n \right)^{p-1} < \infty.$$

Definition 4.77 Let $1 < p < \infty$. A weight function $w = w(x_1, \cdots, x_n)$ defined on $\mathbb{R}^n$ is said to satisfy A_p^- condition in x_i uniformly with respect to other variables ($w \in A_p^-(x_i)$) if

$$\|w\|_{A_p^-(x_i)} := \operatorname*{ess\,sup}_{\substack{(x_1,\cdots x_{i-1},\\ x_{i+1}\cdots,x_n)\in\mathbb{R}^{n-1}}} \sup_{h_i>0} \left(\frac{1}{h_i}\int_{x_i}^{x_i+h_i} w(x_1,\cdots,x_{i-1},t,x_{i-1},\cdots,x_n)dt\right)$$

$$\times \left(\frac{1}{h_i}\int_{x_i-h_i}^{x_i} w(x_1,\cdots,x_{i-1},t,x_{i-1},\cdots,x_n)^{-1/(p-1)}dt\right)^{p-1} < \infty.$$

further, $w \in A_p^+(x_i)$ if

$$\|w\|_{A_p^+(x_i)} := \operatorname*{ess\,sup}_{\substack{(x_1,\cdots x_{i-1},\\ x_{i+1}\cdots,x_n)\in\mathbb{R}^{n-1}}} \sup_{h_i>0} \left(\frac{1}{h_i}\int_{x_i-h_i}^{x_i} w(x_1,\cdots,x_{i-1},t,x_{i-1},\cdots,x_n)dt\right)$$

$$\times \left(\frac{1}{h_i}\int_{x_i}^{x_i+h_i} w(x_1,\cdots,x_{i-1},t,x_{i-1},\cdots,x_n)^{-1/(p-1)}dt\right)^{p-1} < \infty.$$

Remark 4.78 It is known that (see Kokilashvili et al. [177, Chap. 5]) that $w(x_1, \cdots, x_n) \in A_p^{\pm(s)} \Leftrightarrow w \in \bigcap_{i=1}^n A_p^{\pm}(x_i)$.

Theorem 4.79 *Let* $1 < p < \infty$.

(i) Suppose that a weight function w *on* $\mathbb{R}^n$ *belongs to the class* $A_p^{+(s)}(\mathbb{R}^n)$. *Then there exists a constant* c *depending only on* n *and* p *such that the following inequality*

$$\|M^{+(s)}f\|_{L_w^p(\mathbb{R}^n)} \leqslant c\left(\prod_{i=1}^n \|w\|_{A_p^+(x_i)}\right)^{1/(p-1)} \|f\|_{L_w^p(\mathbb{R}^n)} \tag{4.81}$$

holds for all $f \in L_w^p(\mathbb{R}^n)$. *Furthermore, the exponent* $1/(p-1)$ *in estimate* (4.81) *is sharp.*

(ii) Let $w \in A_p^{-(s)}(\mathbb{R}^n)$. *Then there exists a constant* c *depending only on* n *and* p *such that the following inequality*

$$\|M^{-(s)}f\|_{L_w^p(\mathbb{R}^n)} \leqslant c\left(\prod_{i=1}^n \|w\|_{A_p^-(x_i)}\right)^{1/(p-1)} \|f\|_{L_w^p(\mathbb{R}^n)} \tag{4.82}$$

holds for all $f \in L_w^p(\mathbb{R}^n)$. *Furthermore, the exponent* $1/(p-1)$ *in estimate* (4.82) *is sharp.*

Proof We show (i). The proof of (ii) is similar. Since the proof of inequality (4.81) follows in the same way as in the case of $M^{(s)}$ (see Theorem 4.65), we show only sharpness. Let $n = 2$. We take $0 < \varepsilon < 1$. Let $w(x_1, x_2) = |1 - x_1|^{(1-\varepsilon)(p-1)}|1 - x_2|^{(1-\varepsilon)(p-1)}$. Then it is easy to check that

$$(\|w\|_{A_p^+(x_1)}\|w\|_{A_p^+(x_2)})^{1/(p-1)} \approx \frac{1}{\varepsilon^2}.$$

Observe also that for

$$f(x_1, x_2) = (1 - x_1)^{\varepsilon(p-1)-1}\chi_{(0,1)}(x_1)(1 - x_2)^{\varepsilon(p-1)-1}\chi_{(0,1)}(x_2),$$

we have $\|f\|_{L^p_w} \approx \frac{1}{\varepsilon^2}$. Now let $0 < x_1, x_2 < 1$. Then

$$M^{+(s)}f(x_1, x_2) \geqslant \frac{1}{(1 - x_1)(1 - x_2)} \int_{x_1}^{1}\int_{x_2}^{1} f(t, \tau)dtd\tau = c\frac{1}{\varepsilon^2}f(x_1, x_2).$$

Finally

$$\|M^{+(s)}f\|_{L^p_w(\mathbb{R}^2)} \geqslant c\frac{1}{\varepsilon^2}\|f\|_{L^p_w}.$$

Thus we have the sharpness in (4.81). □

One-Sided Multiple Fractional Integrals

Now we discuss sharp bounds for one-sided strong maximal potential operators with product kernels.

Let f be a locally integrable function on $\mathbb{R}^n$ and let $0 < \alpha < 1$. We define one-sided strong fractional maximal operators as

$$M_\alpha^{+(s)}f(x_1, \cdots, x_n)$$
$$= \sup_{h_1,\cdots,h_n>0} \frac{1}{\prod_{i=1}^n h_i^{1-\alpha}} \int_{x_1}^{x_1+h_1} \cdots \int_{x_n}^{x_n+h_n} |f(y_1, \cdots y_n)|dy_1 \cdots dy_n,$$

and

$$M_\alpha^{-(s)}f(x_1, \cdots, x_n)$$
$$= \sup_{h_1,\cdots,h_n>0} \frac{1}{\prod_{i=1}^n h_i^{1-\alpha}} \int_{x_1-h_1}^{x_1} \cdots \int_{x_n-h_n}^{x_n} |f(y_1, \cdots y_n)|dy_1 \cdots dy_n. \qquad (4.83)$$

Let $1 < p \leqslant q < \infty$. We say that a weight function w belongs to the class $A_{p,q}^{-(s)}(\mathbb{R}^n)$ if

$$\|w\|_{A_{p,q}^{-(s)}(\mathbb{R}^n)} := \sup_{\substack{h_1,\cdots,h_n>0\\ x_1,\cdots x_n\in\mathbb{R}}} \left(\frac{1}{h_1\cdots h_n}\int_{x_1}^{x_1+h_1}\cdots\int_{x_n}^{x_n+h_n} w^q(t_1,\cdots,t_n)dt_1\cdots dt_n\right)^{\frac{1}{q}}$$
$$\times\left(\frac{1}{h_1\cdots h_n}\int_{x_1-h_1}^{x_1}\cdots\int_{x_n-h_n}^{x_n} w^{-p'}(t_1,\cdots,t_n)dt_1\cdots dt_n\right)^{1/p'} < \infty;$$

further, $w \in A_{p,q}^{+(s)}(\mathbb{R}^n)$ if

$$\|w\|_{A_{p,q}^{+(s)}(\mathbb{R}^n)} := \sup_{\substack{h_1,\cdots,h_n>0\\ x_1,\cdots x_n\in\mathbb{R}}} \left(\frac{1}{h_1\cdots h_n}\int_{x_1-h_1}^{x_1}\cdots\int_{x_n-h_n}^{x_n} w^q(t_1,\cdots,t_n)dt_1\cdots dt_n\right)^{\frac{1}{q}}$$
$$\times\left(\frac{1}{h_1\cdots h_n}\int_{x_1}^{x_1+h_1}\cdots\int_{x_n}^{x_n+h_n} w^{-p'}(t_1,\cdots,t_n)dt_1\cdots dt_n\right)^{1/p'} < \infty.$$

Definition 4.80 Let $1 < p \leqslant q < \infty$. A weight function $w = w(x_1,\cdots,x_n)$ defined on $\mathbb{R}^n$ is said to satisfy $A_{p,q}^{-}$ condition in x_i uniformly with respect to other variables ($w \in A_{p,q}^{+}(x_i)$) if

$$\|w\|_{A_{p,q}^{+}(x_i)}$$
$$:= \operatorname*{ess\,sup}_{\substack{(x_1,\cdots x_{i-1},\\ x_{i+1}\cdots,x_n)\in\mathbb{R}^{n-1}}} \sup_{h_i>0}\left(\frac{1}{h_i}\int_{x_i}^{x_i+h_i} w^q(x_1,\cdots,x_{i-1},t,x_{i+1}\cdots,x_n)dt\right)^{\frac{1}{q}}$$
$$\times\left(\frac{1}{h_i}\int_{x_i-h_i}^{x_i} w^{-p'}(x_1,\cdots,x_{i-1},t,x_{i+1}\cdots,x_n)dt\right)^{1/p'} < \infty,$$

further, $w \in A_{p,q}^{-}(x_i)$ if

$$\|w\|_{A_{p,q}^{-}(x_i)}$$
$$\equiv \sup_{\substack{(x_1,\cdots x_{i-1},\\ x_{i+1}\cdots,x_n)\in\mathbb{R}^{n-1}}} \sup_{h_i>0}\left(\frac{1}{h_i}\int_{x_i-h_i}^{x_i} w^q(x_1,\cdots,x_{i-1},t,x_{i+1}\cdots,x_n)dt\right)^{\frac{1}{q}}$$
$$\times\left(\frac{1}{h_i}\int_{x_i}^{x_i+h_i} w^{-p'}(x_1,\cdots,x_{i-1},t,x_{i+1}\cdots,x_n)dt\right)^{1/p'} < \infty.$$

Remark 4.81 It is easy to check that $w(x_1, \cdots, x_n) \in A^{\pm(s)}_{p,q} \Leftrightarrow w \in \bigcap_{i=1}^n A^{\pm}_{p,q}(x_i)$.

Theorem 4.82 *Let $0 < \alpha < 1$, $1 < p < 1/\alpha$. We put $q = \frac{p}{1-\alpha p}$. Suppose that w is a weight function defined on $\mathbb{R}^n$ such that $w \in A^{+(s)}_{p,q}(\mathbb{R}^n)$. Then there exists a constant c depending only on n, p, and α such that the following inequality*

$$\|wM^{+(s)}_{\alpha} f\|_{L^q(\mathbb{R}^n)} \leqslant c\big(\|w\|_{A^+_{p,q}(x_1)} \cdots \|w\|_{A^+_{p,q}(x_n)}\big)^{\frac{p'}{q}(1-\alpha)} \|wf\|_{L^p(\mathbb{R}^n)} \tag{4.84}$$

holds for all $f \in L^p_{w^p}(\mathbb{R}^n)$. Furthermore, the exponent $\frac{p'}{q}(1-\alpha)$ in estimate (4.84) *is sharp.*

Proof Estimate (4.84) follows in the same way as in the previous cases. For sharpness we take $n = 2$ and $w(x_1, x_2) = w_1(x_1)w_2(x_2)$, where $w_1(x_1) = |1 - x_1|^{(1-\varepsilon)p'}$; $w_2(x_2) = |1 - x_2|^{(1-\varepsilon)p'}$, $0 < \varepsilon < 1$. Then

$$\begin{aligned}\|w\|_{A^+_{p,q}(x_1)}\|w\|_{A^+_{p,q}(x_2)} &= \|w_1\|_{A^+_{p,q}(\mathbb{R})}\|w\|_{A^+_{p,q}(\mathbb{R})} = \|w_1^q\|_{A^+_{1+q/p'}(\mathbb{R})}\|w_2\|_{A^+_{1+q/p'}(\mathbb{R})} \\ &\approx \varepsilon^{2q/p'}.\end{aligned}$$

If

$$f(t_1, t_2) = (1-t_1)^{\varepsilon-1}\chi_{(0,1)}(t_1)(1-t_2)^{\varepsilon-1}\chi_{(0,1)}(t_2),$$

then $\|wf\|_{L^p(\mathbb{R})^2} \approx \frac{1}{\varepsilon^{2/p}}$. Now let $0 < x < 1$. Then we find that the following estimate:

$$\begin{aligned}M^{+(s)}_{\alpha} f(x_1, x_2) &\geqslant \frac{1}{|1-x_1|^{1-\alpha}|1-x_2|^{1-\alpha}} \int_{x_1}^{1}\int_{x_2}^{1} f(t_1, t_2)dt_1 dt_2 \\ &\approx \frac{|1-x_1|^{\varepsilon-1+\alpha}|1-x_2|^{\varepsilon-1+\alpha}}{\varepsilon^2}\end{aligned}$$

holds. Finally

$$\|wM^{+(s)}_{\alpha} f\|_{L^q(\mathbb{R}^2)} \geqslant \varepsilon^{-2-2/q}. \tag{4.85}$$

Thus, letting $\varepsilon \to 0$ we are done. □

The next statement can be proved analogously. Details are omitted.

Theorem 4.83 *Let α, p, and q satisfy the conditions of Theorem 4.82. Let w be a weight function on $\mathbb{R}^n$ such that $w \in A_{p,q}^{-(s)}(\mathbb{R}^n)$. Then there exist a constant c depending only on n, p, and α such that the following inequality*

$$\|wM_\alpha^{-(s)} f\|_{L^q(\mathbb{R}^n)} \leqslant c(\|w\|_{A_{p,q}^-(x_1)} \cdots \|w\|_{A_{p,q}^+(x_n)})^{\frac{p'}{q}(1-\alpha)} \|wf\|_{L^p(\mathbb{R}^n)} \tag{4.86}$$

holds for all $f \in L^p_{w^p}(\mathbb{R}^n)$. Further, the exponent $\frac{p'}{q}(1-\alpha)$ in estimate (4.86) is sharp.

Let f be a measurable function on $\mathbb{R}^n$ and let $0 < \alpha < 1$. We define one-sided potentials $\mathcal{R}_\alpha^{(n)}$ and $\mathcal{W}_\alpha^{(n)}$ with product kernels

$$\mathcal{W}_\alpha^{(n)} f(x_1, \cdots, x_n) = \int_{x_1}^{\infty} \cdots \int_{x_n}^{\infty} \frac{f(t_1, \cdots, t_n)}{(t_1 - x_1)^{1-\alpha} \cdots (t_n - x_n)^{1-\alpha}} dt_1 \cdots dt_n,$$

$$\mathcal{R}_\alpha^{(n)} f(x_1, \cdots, x_n) = \int_{-\infty}^{x_1} \cdots \int_{-\infty}^{x_n} \frac{f(t_1, \cdots, t_n)}{(x_1 - t_1)^{1-\alpha} \cdots (x_n - t_n)^{1-\alpha}} dt_1 \cdots dt_n,$$

where $x_i \in \mathbb{R}$, $i = 1, \cdots, n$.

Finally we formulate the "sharp result" for one-sided potentials with product kernels. We do not repeat the arguments using above and therefore omit the proof of the next statement.

Theorem 4.84 *Let α, p, and q satisfy the conditions of Theorem 4.82. Suppose that w be a weight function on $\mathbb{R}^n$ such that $w \in A_{p,q}^{-(s)}(\mathbb{R}^n)$. Then*

(i) there exists a constant c depending only on n, p, and α such that the following inequality

$$\|w\mathcal{R}_\alpha^{(n)} f\|_{L^q(\mathbb{R}^n)} \leqslant c\left(\prod_{i=1}^n \|w\|_{A_{p,q}^-(x_i)}\right)^{\max\{1, \frac{p'}{q}\}(1-\alpha)} \|wf\|_{L^p(\mathbb{R}^n)} \tag{4.87}$$

holds for all $f \in L^p_{w^p}(\mathbb{R}^n)$. Further, the exponent $\max\{1, \frac{p'}{q}\}(1-\alpha)$ in estimate (4.87) is sharp,

(ii) there is a constant c depending only on n, p and α such that

$$\|w\mathcal{W}_\alpha^{(n)} f\|_{L^q(\mathbb{R}^n)} \leqslant c\left(\prod_{i=1}^n \|w\|_{A_{p,q}^+(x_i)}\right)^{\max\{1, \frac{p'}{q}\}(1-\alpha)} \|wf\|_{L^p(\mathbb{R}^n)} \tag{4.88}$$

for all $f \in L^p_{w^p}(\mathbb{R}^n)$*. Further, the exponent* $\max\{1, \frac{p'}{q}\}(1-\alpha)$ *in estimate* (4.88) *is sharp.*

4.6 Comments to Chap. 4

The history of sharp weighted estimates for operators of harmonic analysis, in terms of weight characteristics, was described in Sect. 4.1. Additionally, we note that sharp estimates for maximal operators defined with respect to general Muckenhoupt bases, in terms of appropriate weight characteristics, were discussed in Luque et al. [217].

The results presented in Sect. 4.2 were obtained in Kokilashvili et al. [185], and Sect. 4.3 follows from Kokilashvili et al. [186]. Section 4.4 relies on Gilles and Meskhi [116], while Sect. 4.5 is taken from Kokilashvili et al. [184].

Chapter 5
Olsen Inequality

In this chapter, we establish the multilinear inequality

$$\|g\,\mathcal{I}_\alpha(\vec{f}\,)\|_{L^q_r} \leqslant C\|g\|_{L^q_\ell}\prod_{j=1}^{m}\|f_j\|_{L^{p_j}_{s_j}}, \tag{5.1}$$

where

$$\mathcal{I}_\alpha(\vec{f}\,)(x) = \int\limits_{(\mathbb{R}^n)^m} \frac{f_1(y_1)\cdots f_m(y_m)}{(|x-y_1|+\cdots+|x-y_m|)^{mn-\alpha}}d\vec{y}, \quad x\in\mathbb{R}^n,$$

with $0<\alpha<mn$, where L^q_r, L^q_ℓ, $L^{p_j}_{s_j}$, $j=1,\ldots,m$, are Morrey space with indices satisfying certain homogeneity conditions. This inequality is sharp in the sense that the Morrey norm in $\|g\|_{L^q_\ell}$ cannot be replaced by the smaller one. This estimate, in particular, gives necessary and sufficient condition on a weight function V for which the inequality

$$\|\mathcal{I}_\alpha(\vec{f}\,)\|_{L^q_r(V)} \leqslant C\prod_{j=1}^{m}\|f_j\|_{L^{p_j}_{s_j}}. \tag{5.2}$$

holds.

We also derive a characterization of the trace inequality

$$\|B_\alpha(f_1,f_2)\|_{L^q_r(d\mu)} \leqslant C\prod_{j=1}^{2}\|f_j\|_{L^{p_j}_{s_j}},$$

A. Meskhi et al., *Integral Operators in Non-Standard Function Spaces*,
Operator Theory: Advances and Applications 310,
https://doi.org/10.1007/978-3-032-03691-9_5

in terms of a Borel measure μ, where B_α is the bilinear fractional integral operator given by the formula

$$B_\alpha(f_1, f_2)(x) = \int\limits_{\mathbb{R}^n} \frac{f_1(x+t)f_2(x-t)}{|t|^{n-\alpha}} dt, \quad 0 < \alpha < n.$$

5.1 Olsen Inequality in Multilinear Setting

Our aim is to study the behavior of the operator $\mathcal{I}_\alpha$ in Morrey spaces. As a consequence, we establish a sharp Olsen-type inequality (5.1), where parameters of spaces satisfy the conditions: $1 < q \leqslant r < \infty$, $1 < p_j \leqslant s_j < \infty$, $j = 1, \dots, m$, $p < q < \infty$, $0 < \alpha < \frac{n}{s}$, $\frac{1}{p} - \frac{1}{q} = \frac{1}{s} - \frac{1}{r} = \frac{\alpha}{n} - \frac{1}{\ell}$. Throughout this chapter, we use the following notation:

$$\frac{1}{p} := \sum_{i=1}^{m} \frac{1}{p_i}, \quad \frac{1}{s} := \sum_{i=1}^{m} \frac{1}{s_i}, \quad m \geqslant 2. \tag{5.3}$$

Estimate (5.1) in the sharp form is used to obtain a complete characterization of weight functions V such that the estimate (5.2) holds. In particular, we show that if $1 < p < q < \infty$, $\frac{1}{p} - \frac{1}{q} = \frac{1}{s} - \frac{1}{r}$, $0 < \alpha < \frac{n}{s}$, then (5.2) holds for all $f_j \in L^{p_j}_{s_j}$, $j = 1, \dots, m$, if and only if (Adams-type) condition is satisfied:

$$[V]_{p,q,\alpha} := \sup_{Q \in \mathcal{Q}} \left(\int\limits_Q V(x)dx \right)^{\frac{1}{q}} |Q|^{\frac{\alpha}{n} - \frac{1}{p}} < \infty \tag{5.4}$$

Here $\mathcal{Q}$ denotes the class of all cubes in $\mathbb{R}^n$ with sides parallel to the coordinate axis.

For the linear case ($m = 1$ and $\mathcal{I}_\alpha = I_\alpha$), and in the setting of Lebesgue spaces ($q = r$, $p_i = s_i = p$, $i = 1, \dots, m$), the aforementioned characterization goes back to Adams [2]. This result in the multilinear setting was derived in Kokilashvili et al. [179] (see also Kokilashvili et al. [191], Ch. 24) for Lebesgue spaces. We refer to Eridani et al. [98] for necessary and sufficient conditions governing the trace inequality for linear fractional integral operators in Morrey spaces, in terms of the $[V]_{p,q,\alpha}$ characteristics.

5.1.1 Preliminary Background

Let $1 \leqslant q \leqslant r < \infty$ and let $d\mu$ be a Borel measure on $\mathbb{R}^n$. We denote by $L_r^q(d\mu)$ the Morrey space of all measurable functions f on $\mathbb{R}^n$ such that

$$\|f\|_{L_r^q(d\mu)} := \sup_{Q\in\mathcal{Q}} \frac{1}{|Q|^{\frac{1}{q}-\frac{1}{r}}} \left(\int_Q |f(x)|^q d\mu(x) \right)^{1/q} < \infty. \tag{5.5}$$

In this definition, cubes can be replaced by balls, and the supremum will then be over all balls B in $\mathbb{R}^n$. This yields a norm equivalent to $\|\cdot\|_{L_r^q(d\mu)}$. If V is a locally integrable a.e. positive function on $\mathbb{R}^n$, i.e., a weight on $\mathbb{R}^n$, then we denote $L_r^q(d\mu)$ by $L_r^q(V)$. In the case $q = r$ (i.e., for Lebesgue spaces), the symbols $L^q(d\mu)$ and $L^q(V)$ will be used for $L_r^q(d\mu)$ by $L_r^q(V)$, respectively.

The following equivalent form of Morrey space norm appears in the literature:

$$\|f\|_{L^{q,\lambda}(V)} := \sup_{Q\in\mathcal{Q}} \left(\frac{1}{|Q|^\lambda} \int_Q |f(x)|^q V(x)dx \right)^{1/q}. \tag{5.6}$$

Note that $\|\cdot\|_{L_r^q(V)}$ coincides with $\|\cdot\|_{L^{q,\lambda/q}(V)}$ given in (5.6) when $\lambda = 1 - \frac{q}{r}$.

The weak weighted Morrey space $WL_r^q(V)$ is defined as the space of all measurable functions f such that

$$\|f\|_{WL_r^q(V)} := \sup_{Q\in\mathcal{Q}} \frac{1}{|Q|^{\frac{1}{q}-\frac{1}{r}}} \sup_{\lambda>0} \lambda \left(\int_{\{x\in Q:|f(x|>\lambda\}} V(x)dx \right)^{1/q} < \infty.$$

Obviously, one has $WL_r^q(V) \hookrightarrow L_r^q(V)$.

If V is a constant function, then we denote $L_r^q(V)$ and $WL_r^q(V)$ by L_r^q and WL_r^q, respectively. In the case $q = r$, we have weighted Lebesgue spaces $L^q(V)$ and $WL^q(V)$, respectively.

The following boundedness results for fractional integral operator on Morrey spaces are known.

Theorem 5.1 (Spanne, unpublished) *Let $0 < \alpha < n$, $1 < p_0 \leqslant s_0 < \infty$, $1 < q_0 \leqslant r_0 < \infty$. Suppose that $\frac{1}{s_0} - \frac{1}{r_0} = \frac{1}{p_0} - \frac{1}{q_0} = \frac{\alpha}{n}$. Then I_α is bounded from $L_{s_0}^{p_0}$ to $L_{r_0}^{q_0}$.*

Theorem 5.2 (Adams [4]) *Let $0 < \alpha < n$, $1 < p_0 \leqslant s_0 < \infty$, $1 < q_0 \leqslant r_0 < \infty$. Suppose that $\frac{1}{r_0} = \frac{1}{s_0} - \frac{\alpha}{n}$, $\frac{q_0}{r_0} = \frac{p_0}{s_0}$. Then I_α is bounded from $L_{s_0}^{p_0}$ to $L_{r_0}^{q_0}$.*

In the unweighted case, the following multilinear result is also known.

Theorem 5.3 (Tang [317]) *Let* $0 < \alpha < mn$, $1 < q \leqslant r < \infty$, $1 < p_i \leqslant s_i < \infty$, $i = 1, \ldots, m$ *be such that*

$$\frac{1}{s} - \frac{1}{r} = \frac{1}{p} - \frac{1}{q} = \frac{\alpha}{n},$$

where p *and* s *are defined by* (5.3)*. Then there exists a positive constant* C *such that for all* $f_j \in L^{p_j}_{s_j}$, $j = 1, \ldots, m$, *we have*

$$\|\mathcal{I}_\alpha(\vec{f})\|_{L^q_r} \leqslant C \prod_{j=1}^{m} \|f_j\|_{L^{p_j}_{s_j}}.$$

Adams [2] proved the trace inequality for the Riesz potential I_α

$$I_\alpha f(x) := \int\limits_{\mathbb{R}^n} \frac{f(y)}{|x-y|^{n-\alpha}}\, dy, \quad x \in \mathbb{R}^n, \quad 0 < \alpha < n. \tag{5.7}$$

Theorem 5.4 (Adams [2]) *Let* $1 < p < q < \infty$ *and let* $0 < \alpha < n/p$*. Suppose that* μ *is a Borel measure on* $\mathbb{R}^n$*. Then the inequality*

$$\|I_\alpha f\|_{L^q(d\mu)} \leqslant C\|f\|_{L^p}$$

holds if and only if

$$[\mu] := \sup_Q (\mu(Q))^{\frac{1}{q}} |Q|^{\frac{\alpha}{n} - \frac{1}{p}} < \infty. \tag{5.8}$$

Moreover, $\|I_\alpha\|_{L^p \mapsto L^q(d\mu)} \approx [\mu]$.

We need the following sharpening of the statement given in Kokilashvili et al. [179] (see also Kokilashvili et al. [191, Chapter 24]).

Proposition 5.5 *Let* $1 < p_i < \infty$, $i = 1, \ldots, m$*. Assume that* $\alpha < n/p$ *and* $p < q < \infty$*. Then the following estimate holds:*

$$\left\|\mathcal{I}_\alpha(\vec{f})\right\|_{L^q(V)} \leqslant C\,[V]_{p,q,\alpha} \prod_{j=1}^{m} \left\|f_j\right\|_{L^{p_j}}. \tag{5.9}$$

5.1.2 Boundedness Results

Theorem 5.6 *Let* $1 < q \leqslant r < \infty$, $1 < p_i \leqslant s_i < \infty$, $i = 1, \ldots, m$, $1 < p < q$, $0 < \alpha < \frac{n}{s}$. *Let* $\frac{1}{p} - \frac{1}{q} = \frac{1}{s} - \frac{1}{r} = \frac{\alpha}{n} - \frac{1}{\ell}$, *where* $\frac{1}{s} = \sum_{j=1}^{m} \frac{1}{s_j}$, $\frac{1}{p} = \sum_{j=1}^{m} \frac{1}{p_j}$. *Then there exists a positive constant* C *depending only on* $n, \alpha, q, r, p_i, s_i$, $i = 1, \ldots, m$, *such that for all* $f_j \in L_{s_j}^{p_j}$, $j = 1, \ldots, m$, *inequality* (5.1) *holds.*

Proof First observe that $p < q < \ell$ and $\frac{n}{\ell} < \alpha < \frac{n}{s} < \frac{n}{p}$. Without loss of generality we assume that $g \geqslant 0$, $f_j \geqslant 0$, $j = 1, \ldots, m$. For any ball $B := B(a, r)$, let $2B := B(a, 2r)$ be the ball with center a and radius $2r$. We write $f_j = f_j^0 + f_j^\infty$, where

$$f_j^0 = f_j \chi_{2B}, \quad f_j^\infty = f_j \chi_{(2B)^c}, \quad j = 1, \ldots, m.$$

Let $f_j \geqslant 0$, $j = 1, \ldots, m$. In view of this representation we write

$$\mathcal{I}_\alpha \vec{f}(x) \leqslant \mathcal{I}_\alpha(f_1^0, \ldots, f_m^0)(x) + \mathcal{I}_\alpha(f_1^\infty, \ldots, f_m^\infty)(x) + \sum_{j=1}^{m} \mathcal{I}_\alpha(f_1^{\beta_1}, \ldots, f_m^{\beta_m})(x),$$

where $\beta_1, \ldots, \beta_m \in \{0, \infty\}$ and the sum contains at least one $\beta_j = 0$ and $\beta_j = \infty$. Consequently,

$$\begin{aligned}\|g\mathcal{I}_\alpha(\vec{f})\|_{L^q(B)} &\leqslant \|g\mathcal{I}_\alpha(f_1^0, \ldots, f_m^0)\|_{L^q(B)} + \|gI_\alpha(f_1^\infty, \ldots, f_m^\infty)\|_{L^q(B)} \\ &\quad + \sum_{\beta_1, \ldots, \beta_m} \|\mathcal{I}_\alpha(f_1^{\beta_1}, \ldots, f_m^{\beta_m})\|_{L^q(B)} := N_1 + N_2 + \sum.\end{aligned}$$

Using Proposition 5.5 for $V = |g|^q$, we write

$$\begin{aligned}N_1 &\leqslant C\|g\|_{L_\ell^q} \prod_{j=1}^{m} \|\chi_{2B} f_j\|_{L^{p_j}} \leqslant C\|g\|_{L_\ell^q} \prod_{j=1}^{m} \|\chi_{2B} f_j\|_{L_{s_j}^{p_j}} r^{n\sum_{j=1}^{m}\left(\frac{1}{p_j} - \frac{1}{s_j}\right)} \\ &= C\|g\|_{L_\ell^q} \prod_{j=1}^{m} \|\chi_{2B} f_j\|_{M_{s_j}^{p_j}} r^{n\left(\frac{1}{p} - \frac{1}{s}\right)} = C\|g\|_{L_\ell^q} \prod_{j=1}^{m} \|\chi_{2B} f_j\|_{M_{s_j}^{p_j}} r^{n\left(\frac{1}{q} - \frac{1}{r}\right)}.\end{aligned}$$

Let us estimate N_2. First observe that if $x \in B$ and $y_j \in (2B)^c$, then by simple geometric observations, we find that $\frac{1}{2}|a - y_j| \leqslant |x - y_j| \leqslant \frac{3}{2}|a - y_j|$. Thus, we get

$$\mathcal{I}_\alpha(f_1^\infty, \ldots, f_m^\infty)(x) \leqslant C \int_{2r}^{\infty} s^{\alpha - mn - 1} \left(\prod_{j=1}^{m} \int_{\{y_j : |x - y_j| < s\}} f_j^\infty(y_j) dy_j \right) ds$$

$$\leqslant C\int_{2r}^{\infty} s^{\alpha-mn-1}\left(\prod_{j=1}^{m}\int_{\{y_j:|a-y_j|<2s\}} f_j^{\infty}(y_j)dy_j\right)ds$$

$$\leqslant C\int_{2r}^{\infty} s^{\alpha-mn-1}\left(\prod_{j=1}^{m}\|f_j\|_{L^{p_j}(B(a,2s))}\right)\prod_{j=1}^{m}(s^{n/p_j'})ds$$

$$\leqslant C\prod_{j=1}^{m}\|f_j\|_{L^{p_j}_{s_j}}\int_{2r}^{\infty} s^{\alpha-1-\sum_{j=1}^{m}\frac{n}{p_j}+n\sum_{j=1}^{m}\left(\frac{1}{p_j}-\frac{1}{s_j}\right)}ds$$

$$= C\prod_{j=1}^{m}\|f_j\|_{L^{p_j}_{s_j}}\int_{2r}^{\infty} s^{\alpha-1-\frac{n}{p}+n\left[\frac{1}{q}-\frac{1}{r}\right]}ds$$

$$= C\prod_{j=1}^{m}\|f_j\|_{L^{p_j}_{s_j}} r^{\alpha-\frac{n}{p}+n\left[\frac{1}{q}-\frac{1}{r}\right]}.$$

Here we used the fact that $\alpha < \frac{n}{s} = n\left[\frac{1}{p}-\frac{1}{q}+\frac{1}{r}\right] < \frac{n}{p}$. Hence,

$$\left(\int_B \mathcal{I}_\alpha(f_1^\infty,\ldots,f_m^\infty)^q(x)g^q(x)dx\right)^{1/q}$$

$$\leqslant Cr^{\alpha-\frac{n}{p}+n\left[\frac{1}{q}-\frac{1}{r}\right]}\left(\int_B g^q(x)dx\right)^{1/q}\prod_{j=1}^{m}\|f_j\|_{L^{p_j}_{s_j}}$$

$$\leqslant Cr^{n\left[\frac{1}{q}-\frac{1}{r}\right]}\|g\|_{L^q_\ell}\prod_{j=1}^{m}\|f_j\|_{L^{p_j}_{s_j}}.$$

In the last equality, we used the condition $\frac{1}{p}-\frac{1}{q}=\frac{\alpha}{n}-\frac{1}{\ell}$.

It remains to estimate $\sum$. For simplicity we take $m \geqslant 3$, $\beta_1=\beta_2=\infty$ and $\beta_3=\cdots=\beta_m=0$. Recall that $|x-y_j|\approx|a-y_j|$ for all $x\in B$ and $y_j\in(2B)^c$, $j=1,2$. Thus, without loss of generality, we have that one of the terms of Σ can be estimated as follows:

$$\mathcal{I}_\alpha(f_1^\infty,f_2^\infty,f_3^0,\ldots,f_m^0)(x)$$

$$=\int_{(2B)^c\times(2B)^c\times 2B\times\ldots\times 2B}\frac{f_1(y_1)f_2(y_2)f_3(y_3)\cdots f_m(y_m)}{(|x-y_1|+\cdots+|x-y_m|)^{mn-\alpha}}d\vec{y}$$

$$\leqslant C\Big(\int_{(2B)^c\times(2B)^c}\frac{f_1(y_1)f_2(y_2)dy_1dy_2}{(|a-y_1|+|a-y_2|)^{mn-\alpha}}\Big)$$

$$\Big(\int_{(2B)\times\cdots\times(2B)} f_3(y_3)\cdots f_m(y_m)dy_3\cdots dy_m\Big)$$

$$:= CI_1\cdot I_2.$$

Now we estimate I_1 and I_2 separately. By Hölder's inequality and simple observations, we obtain

$$I_1 = C\int_{(2B)^c\times(2B)^c}\Big(\int_{|a-y_1|+|a-y_2|}^{\infty} s^{-mn+\alpha-1}ds\Big) f_1(y_1)f_2(y_2)dy_1dy_2$$

$$\leqslant C\int_{2r}^{\infty}\Big(\int_{\{y_1,y_2:|a-y_1|+|a-y_2|<s\}} f_1(y_1)f_2(y_2)dy_1dy_2\Big)s^{-mn+\alpha-1}ds$$

$$\leqslant C\int_{2r}^{\infty}\prod_{i=1}^{2}\Big(\int_{B(a,s)} f_i^{p_i}(y_i)dy_i\Big)^{1/p_i} s^{\frac{n}{p_1'}+\frac{n}{p_2'}-mn+\alpha-1}ds$$

$$\leqslant C\int_{2r}^{\infty}\prod_{i=1}^{2}\Big(\frac{1}{|B(a,s)|^{1-\frac{p_i}{s_i}}}\int_{B(a,s)} f_i^{p_i}(y_i)dy_i\Big)^{1/p_i}$$

$$\times s^{\alpha-mn-1+n\left(\frac{1}{p_1'}+\frac{1}{p_2'}\right)+n\left[\frac{1}{p_1}+\frac{1}{p_1}-\left(\frac{1}{s_1}+\frac{1}{s_2}\right)\right]}ds$$

$$\leqslant C\prod_{i=1}^{2}\|f_i\|_{L^{p_i}_{s_i}} r^{\alpha-mn+2n-n\left[\frac{1}{p_1}+\frac{1}{p_2}\right]+n\left[\frac{1}{p_1}-\frac{1}{s_1}+\frac{1}{p_2}-\frac{1}{s_2}\right]}.$$

In the latter estimate, we used the fact that

$$\alpha-mn+2n-n\Big(\frac{1}{p_1}+\frac{1}{p_2}\Big)+n\Big[\frac{1}{p_1}+\frac{1}{p_2}-\Big(\frac{1}{s_1}+\frac{1}{s_2}\Big)\Big]$$

$$=\alpha-mn+2n-n\Big[\frac{1}{s_1}+\frac{1}{s_2}\Big]<0$$

which is a consequence of the condition $\alpha < \frac{n}{s}$. Further, by using Hölder's inequality again, we find that

$$I_2 \leqslant C\prod_{i=3}^{m}\|f_i\|_{L^{p_i}_{s_i}} r^{n\left(\sum_{k=3}^{m}\frac{1}{p'_k}\right)+n\left(\sum_{k=3}^{m}\left[\frac{1}{p_k}-\frac{1}{s_k}\right]\right)}$$
$$= C\prod_{i=3}^{m}\|f_i\|_{L^{p_i}_{s_i}} r^{n(m-2)-n\sum_{k=3}^{m}\frac{1}{p_k}} r^{n\left(\sum_{k=3}^{m}\left[\frac{1}{p_k}-\frac{1}{s_k}\right]\right)}.$$

Consequently, summarizing estimates for I_1 and I_2, we find that

$$\left(\int_B \mathcal{I}_\alpha(f_1^\infty, f_2^\infty f_3^0, \cdots, f_m^0)^q(x)g^q(x)dx\right)^{1/q}$$
$$\leqslant Cr^{n\left[\frac{\alpha}{n}-\frac{1}{p}\right]} r^{n\left[\frac{1}{p}-\frac{1}{s}\right]+n\left[\frac{1}{q}-\frac{1}{\ell}\right]}\|g\|_{L^q_\ell}\prod_{j=1}^{m}\|f_j\|_{L^{p_j}_{s_j}}$$
$$= Cr^{n\left[\frac{1}{q}-\frac{1}{r}\right]}\|g\|_{L^q_\ell}\prod_{j=1}^{m}\|f_j\|_{L^{p_j}_{s_j}}.$$

In the last equality, we again used the following condition: $\frac{1}{p}-\frac{1}{q}=\frac{1}{s}-\frac{1}{r}=\frac{\alpha}{n}-\frac{1}{\ell}$.
This completes the proof. □

Theorem 5.7 *Let $1 < q \leqslant r < \infty$, $1 < p_i \leqslant s_i < \infty$, $i = 1, \ldots, m$, $1 < p < q$, $0 < \alpha < \frac{n}{s}$. Let $\frac{1}{p}-\frac{1}{q}=\frac{1}{s}-\frac{1}{r}$, where $\frac{1}{s}=\sum_{j=1}^{m}\frac{1}{s_j}$, $\frac{1}{p}=\sum_{j=1}^{m}\frac{1}{p_j}$. Suppose that V is a weight function on $\mathbb{R}^n$. Then the following statements are equivalent:*

(i) there is a positive constant C such that for all measurable $\vec{f}$, we have

$$\|\mathcal{I}_\alpha(\vec{f}\,)\|_{L^q_r(V)} \leqslant C\prod_{j=1}^{m}\|f_j\|_{L^{p_j}_{s_j}}. \tag{5.10}$$

(ii) there is a positive constant C such that for all measurable $\vec{f}$, we have

$$\|\mathcal{I}_\alpha(\vec{f}\,)\|_{WL^q_r(V)} \leqslant C\prod_{j=1}^{m}\|f_j\|_{L^{p_j}_{s_j}}. \tag{5.11}$$

(iii) condition (5.4) is satisfied.

Moreover, under either assumption, we have the norm equivalence $\|\mathcal{I}_\alpha\| \approx [V]_{p,q,\alpha}$.

Proof The implication (iii) $\Rightarrow$ (i) follows from Theorem 5.6 taking $V = |g|^q$ there and observing that $\|g\|_{L^q_\ell} = [V]_{p,q,\alpha}$, where $\frac{1}{q} - \frac{1}{\ell} = \frac{1}{p} - \frac{\alpha}{n}$. The implication (ii) $\Rightarrow$ (iii) is a consequence of taking the test functions $f_j = \chi_B$, $j = 1, \ldots, m$ in (ii). Since (i) $\Rightarrow$ (ii), we are done. □

Taking $m = 1$ in Theorem 5.6, we deduce Corollary 5.8.

Corollary 5.8 *Let* $1 < q \leqslant r < \infty$, $1 < p \leqslant s < \infty$, $1 < p < q$ *and* $0 < \alpha < \frac{n}{s}$. *Let* $\frac{1}{p} - \frac{1}{q} = \frac{1}{s} - \frac{1}{r} = \frac{\alpha}{n} - \frac{1}{\ell}$. *Then there is a positive constant* C *depending only on* n, α, q, r, p, s *such that for all* $f \in L^p_s$ *and* $g \in L^q_\ell$, *we have*

$$\|g\, I_\alpha f\|_{L^q_r} \leqslant C\|g\|_{L^q_\ell}\|f\|_{L^p_s}.$$

We have the trace inequality for B_α analogous to that of Adams [2]; see also Eridani et al. [98] in the linear case.

Theorem 5.9 *Let* $1 < q \leqslant r$, $1 < p_i \leqslant s_i < \infty$, $i = 1, 2$, *and let* $1 < p < q < \infty$. *Let* $0 < \alpha < \min\{\frac{1}{s_1}, \frac{1}{s_2}\}$, $\frac{1}{p} - \frac{1}{q} = \frac{1}{s} - \frac{1}{r}$, *where* $\frac{1}{p} = \frac{1}{p_1} + \frac{1}{p_2}$, $\frac{1}{s} = \frac{1}{s_1} + \frac{1}{s_2}$. *Then there is a positive constant* C *depending on* $n, \alpha, q, r, p_1, p_2, s_1, s_2$ *such that for all* $f_1, f_2 \geqslant 0$,

$$\|B_\alpha(f_1, f_2)\|_{L^q_r(d\mu)} \leqslant C[\mu]\|f_1\|_{L^{p_1}_{s_1}}\|f_2\|_{L^{p_2}_{s_2}}, \tag{5.12}$$

holds, where $[\mu]$ *is defined in* (5.8).

Proof Let us fix a ball $B := B(a, r)$. Suppose that $f_1, f_2 \geqslant 0$. Using Hölder's inequality twice with exponents $\frac{p_1}{p}$ and $\frac{p_2}{p}$, we find that

$$\begin{aligned}\|\chi_B B_\alpha(f_1, f_2)\|_{L^q(d\mu)} &\leqslant \|\chi_B\big[I_\alpha(f_1^{p_1/p})\big]^{p/p_1}\big[I_\alpha(f_2^{p_2/p})\big]^{p/p_2}\|_{L^q(d\mu)}\\ &\leqslant \|\chi_B I_\alpha(f_1^{p_1/p})\|^{p/p_1}_{L^q(d\mu)}\|\chi_B I_\alpha(f_2^{p_2/p})\|^{p/p_2}_{L^q(d\mu)} := N_1 \cdot N_2,\end{aligned}$$

where I_α is the Riesz potential defined on $\mathbb{R}^n$; see (5.7).

Now we estimate N_1 and N_2 separately. Representing f_1 as $f_{1,1} + f_{1,2}$, where $f_{1,1} = f_1 \cdot \chi_{2B}$, $f_{1,2} = f_1 - f_{1,1}$, we find that

$$\begin{aligned}N_1 &\leqslant \Big(\int_B [I_\alpha(f_1^{p_1/p})(x)]^q d\mu(x)\Big)^{p/(qp_1)}\\ &\leqslant C\Bigg[\Big(\int_B [I_\alpha(f_{1,1}^{p_1/p})(x)]^q d\mu(x)\Big)^{p/(qp_1)} + \Big(\int_B [I_\alpha(f_{1,2}^{p_1/p})(x)]^q d\mu(x)\Big)^{p/(qp_1)}\Bigg]\\ &:= N_{1,1} + N_{1,2}.\end{aligned}$$

In view of Theorem 5.4, we have that

$$N_{1,1} \leqslant C[\mu]^{\frac{p}{p_1}} \|f_1\|_{L^{p_1}_{s_1}} |B|^{\frac{1}{p_1}-\frac{1}{s_1}}.$$

Now we estimate $N_{1,2}$. First observe that if $x \in B$ and $y \in (2B)^c$, then $|y-a| \leqslant 2|x-y|$. Consequently, by Hölder's inequality with respect to the exponents p and p', and the condition $0 < \alpha < \frac{1}{s_1}$, we get

$$\begin{aligned}
N_{2,1} &\leqslant C(\mu(B))^{p/(qp_1)} \Bigg(\int\limits_{(2B)^c} \frac{(f_{1,2})^{p_1/p}(y)}{|y-a|^{n-\alpha}} dy \Bigg)^{\frac{p}{p_1}} \\
&= C(\mu(B))^{p/(qp_1)} \Bigg[\sum_{k=1}^{\infty} \int\limits_{(2^{k+1}B)\setminus(2^k B)} \frac{(f_{1,2}(y))^{p_1/p}}{|y-a|^{n-\alpha}} dy \Bigg]^{\frac{p}{p_1}} \\
&\leqslant C(\mu(B))^{p/(qp_1)} \Bigg[\sum_{k=1}^{\infty} \Bigg(\int\limits_{(2^{k+1}B)\setminus(2^k B)} (f_{1,2}(y))^{p_1} dy \Bigg)^{\frac{1}{p}} \\
&\quad \times \Bigg(\int\limits_{(2^{k+1}B)\setminus(2^k B)} |y-a|^{(\alpha-n)p'} \Bigg)^{\frac{1}{p'}} \Bigg]^{\frac{p}{p_1}} \\
&\leqslant C(\mu(B))^{p/(qp_1)} \Bigg[\sum_{k=1}^{\infty} \Bigg(\int\limits_{(2^{k+1}B)\setminus(2^k B)} (f_{1,2}(y))^{p_1} dy \Bigg)^{\frac{1}{p}} |2^k B|^{\frac{\alpha}{n}-\frac{1}{p}} \Bigg]^{\frac{p}{p_1}} \\
&= C(\mu(B))^{p/(qp_1)} \Bigg[\sum_{k=1}^{\infty} \Bigg(\Bigg(\int\limits_{(2^{k+1}B)\setminus(2^k B)} (f_{1,2}(y))^{p_1} dy \Bigg)^{\frac{1}{p_1}} \\
&\quad \times |2^{k+1}B|^{\frac{1}{s_1}-\frac{1}{p_1}} \Bigg)^{\frac{p_1}{p}} |2^k B|^{\frac{\alpha}{n}-\frac{p_1}{s_1 p}} \Bigg]^{\frac{p}{p_1}} \\
&\leqslant C(\mu(B))^{p/(qp_1)} \|f_1\|_{L^{p_1}_{s_1}} |B|^{\frac{\alpha p}{np_1}-\frac{1}{s_1}} \leqslant C[\mu]^{p/p_1} \|f_1\|_{L^{p_1}_{s_1}} |B|^{\frac{1}{p_1}-\frac{1}{s_1}}.
\end{aligned}$$

Summarizing estimates for $N_{1,1}$ and $N_{1,2}$, we find that

$$N_1 \leqslant C[\mu]^{\frac{p}{p_1}} \|f_1\|_{L^{p_1}_{s_1}} |B|^{\frac{1}{p_1}-\frac{1}{s_1}}.$$

Analogously for N_2, we have that

$$N_2 \leqslant C[\mu]^{\frac{p}{p_2}} \|f_2\|_{L^{p_2}_{s_2}} |B|^{\frac{1}{p_2}-\frac{1}{s_2}}.$$

These estimates give

$$\|\chi_B B_\alpha(f_1, f_2)\|_{L^q(\mu)} \leqslant C[\mu]\|f_1\|_{L^{p_1}_{s_1}}\|f_2\|_{L^{p_2}_{s_2}}|B|^{\frac{1}{p}-\frac{1}{s}},$$

which implies the desired estimate. □

As a corollary, we have the trace inequality for the classical Lebesgue spaces.

Corollary 5.10 *Let $1 < p_i < \infty$, $1 < p < q < \infty$ and let $0 < \alpha < \min\{\frac{1}{p_1}, \frac{1}{p_2}\}$. Suppose that μ is a Borel measure on $\mathbb{R}^n$. Then there is a positive constant C such that for all $f_1, f_2 \geqslant 0$,*

$$\|B_\alpha(f_1, f_2)\|_{L^q(d\mu)} \leqslant C[\mu]\|f_1\|_{L^{p_1}}\|f_2\|_{L^{p_2}},$$

where $[\mu]$ is defined in (5.8).

Proof It follows from Theorem 5.9, taking $p_1 = s_1$, $p_2 = s_2$, $q = r$. □

We can further obtain a result for the bilinear fractional integral operator B_α based on Theorem 5.9.

Theorem 5.11 *Let $1 < q \leqslant r$, $1 < p_i \leqslant s_i < \infty$, $i = 1, 2$. Let $1 < p < q < \infty$ and $0 < \alpha < \min\{\frac{1}{s_1}, \frac{1}{s_2}\}$, $\frac{1}{p} - \frac{1}{q} = \frac{1}{s} - \frac{1}{r} = \frac{\alpha}{n} - \frac{1}{\ell}$, where $\frac{1}{p} = \frac{1}{p_1} + \frac{1}{p_2}$, $\frac{1}{s} = \frac{1}{s_1} + \frac{1}{s_2}$. Then there is a positive constant C depending only on n, α, q, r, r, p_1, p_2, s_1, s_2 such that for all $f_1, f_2, g \geqslant 0$, we have*

$$\|g\, B_\alpha(f_1, f_2)\|_{L^q_r} \leqslant C\|g\|_{L^q_\ell}\|f_1\|_{L^{p_1}_{s_1}}\|f_2\|_{L^{p_2}_{s_2}}. \tag{5.13}$$

Proof Taking $d\mu(x) = g^q(x)dx$ in Theorem 5.9, setting $\frac{1}{p} - \frac{1}{q} = \frac{\alpha}{n} - \frac{1}{\ell}$, and observing that $[\mu] = \|g\|_{L^q_\ell}$, we derive the claimed conclusion. □

5.2 Comments to Chap. 5

In the linear case ($m = 1$), inequalities of type (5.1) play an important role in the study of perturbed Schrödinger equation (see Olsen [266]). We refer to Sawano et al. [302, 303] for subsequent improvements of Olsen's original in equality and applications.

Unweighted and weighted problems for multilinear fractional integrals in Morrey spaces were studied in Iida et al. [147, 148, 149], Kokilashvili et al. [189], He and Yan [134]. In particular, in Iida et al. [147, 148], Olsen's type inequalities for multilinear fractional integrals have been derived without sharpness.

The results presented in this chapter follow Grafakos and Meskhi [121] (see also Grafakos and Meskhi [120]).

Chapter 6
More on Herz-Type Spaces

This chapter is a continuation of Kokilashvili et al. [183, Section 13.4]. We begin by establishing the boundedness of the Marcinkiewicz integral on continuous Herz spaces with variable exponent, where all defining parameters of the space are variable.

Next, utilizing the atomic decomposition of Herz-type Hardy spaces with variable smoothness and integrability, we derive boundedness results for central Calderón–Zygmund operators on Herz-type Hardy spaces with variable smoothness and integrability.

The chapter proceeds to extend these results to grand variable Herz spaces, where we obtain the boundedness of Marcinkiewicz integral once again, but now in the context of grand variable exponents.

We also introduce a novel class of grand variable Herz-type spaces, constructed via discrete grand spaces, and prove the boundedness of sublinear operators acting on these new function spaces.

Finally, we investigate the multilinear Calderón–Zygmund operators acting on the product of grand variable Herz spaces. These results generalize the boundedness of multilinear Calderón–Zygmund operators on product of variable exponent Lebesgue spaces and variable Herz spaces.

6.1 Preliminaries

6.1.1 Spaces of Variable Integrability

Let $E \subseteq \mathbb{R}^n$ be an open set and p be a real-valued measurable function on E with values in $[1, \infty)$. We will additionally suppose that

$$1 \leqslant p^- \leqslant p(x) \leqslant p^+ < \infty, \tag{6.1}$$

A. Meskhi et al., *Integral Operators in Non-Standard Function Spaces*,
Operator Theory: Advances and Applications 310,
https://doi.org/10.1007/978-3-032-03691-9_6

where $p^-(E) = p^- := \operatorname{ess\,inf}_{x\in E} p(x)$, $p^+(E) = p^+ := \operatorname{ess\,sup}_{x\in E} p(x)$. By $L^{p(\cdot)}(E)$, we denote the space of measurable functions f on E such that

$$\varrho_{p(\cdot),E}(f) = \varrho_{p(\cdot)}(f) = \int\limits_E |f(x)|^{p(x)} dx < \infty.$$

Equipped with the norm

$$\|f\|_{L^{p(\cdot)}(E)} = \inf\left\{\eta > 0 : \varrho_{p(\cdot)}\left(\frac{f}{\eta}\right) \leqslant 1\right\},$$

this is a Banach function space. By $p'(x) = p(x)/(p(x) - 1)$, we denote the conjugate exponent. For the following lemma, we refer to, e.g., Cruz-Uribe and Fiorenza [58].

Lemma 6.1 *Let E be a measurable subset of $\mathbb{R}^n$. Suppose that $1 \leqslant p^-(E) \leqslant p^+(E) < \infty$. Then*

(i)

$$\|f\|^{p^+(E)}_{L^{p(\cdot)}(E)} \leqslant \varrho_{p(\cdot),E}(f) \leqslant \|f\|^{p^-(E)}_{L^{p(\cdot)}(E)}, \quad \|f\|_{L^{p(\cdot)}(E)} \leqslant 1;$$

$$\|f\|^{p^-(E)}_{L^{p(\cdot)}(E)} \leqslant \varrho_{p(\cdot),E}(f) \leqslant \|f\|^{p^+(E)}_{L^{p(\cdot)}(E)}, \quad \|f\|_{L^{p(\cdot)}(E)} \geqslant 1.$$

(ii) Hölder's inequality

$$\left|\int\limits_E f(x)g(x)dx\right| \leqslant \left(\frac{1}{p^-(E)} + \frac{1}{(p^+(E))'}\right) \|f\|_{L^{p(\cdot)}(E)} \|g\|_{L^{p'(\cdot)}(E)}$$

holds, where $f \in L^{p(\cdot)}(E)$ and $g \in L^{p'(\cdot)}(E)$.

(iii) Generalized Hölder's inequality

$$\|fg\|_{L^{r(\cdot)}(E)} \lesssim \|f\|_{L^{p(\cdot)}(E)} \|g\|_{L^{q(\cdot)}(E)}$$

holds, where $f \in L^{p(\cdot)}(E)$, $g \in L^{q(\cdot)}(E)$ and $\frac{1}{r(x)} = \frac{1}{p(x)} + \frac{1}{q(x)}$ for every $x \in E$.

In the sequel, for $p : E \to [1, \infty)$, we use the well-known conditions.

Log-Condition: There exists $A = A(p) > 0$, independent of x, y, such that

$$|p(x) - p(y)| \leqslant \frac{A}{-\ln|x - y|}, \quad |x - y| \leqslant \frac{1}{2}, \quad x, y \in E. \tag{6.2}$$

Decay Condition (at Infinity): There exists a number $p_\infty \in (1, \infty)$ and a constant $A = A(p) > 0$, independent of x, such that

$$|p(x) - p_\infty| \leqslant \frac{A}{\ln(e + |x|)}, \quad x \in E. \tag{6.3}$$

Decay Condition (at the Origin): There is $A = A(p) > 0$, independent of x, such that

$$|p(x) - p(0)| \leqslant \frac{A}{\ln(e + 1/|x|)}, \quad x \in E. \tag{6.4}$$

With respect to classes of variable exponents used in this chapter, we adopt the following notation:

(a) $\mathcal{P}(E)$ is the class of all real-valued measurable functions q satisfying $q_- > 1$ and $q_+ < \infty$.
(b) $\mathcal{P}_0(E)$ is the class of all real-valued measurable functions q satisfying $q_- > 0$ and $q_+ < \infty$.
(c) $\mathbb{P}^{\log} = \mathbb{P}^{\log}(E)$ is the class of functions satisfying (6.1) and (6.2).
(d) When E is unbounded, $\mathbb{P}_\infty(E)$ and $\mathbb{P}_{0,\infty}(E)$ are the subsets of exponents in $L^\infty(E)$ with values in $[1, \infty)$ which satisfy condition (6.3) and both the conditions (6.3) and (6.4), respectively.
(e) $\mathbb{P}_\infty^{\log}(E)$ is the set of exponents $p \in \mathbb{P}_\infty(E)$ satisfying also the condition (6.2). $\mathbb{P}_0^{\log}(E)$ is similarly defined using condition (6.4).
(f) $\mathbb{P}_{0,\infty}^{\log}(\mathbb{R}^n) := \mathbb{P}_0^{\log}(\mathbb{R}^n) \cap \mathbb{P}_\infty^{\log}(\mathbb{R}^n)$.
(g) When $E = \mathbb{R}_{\nu+} = (\nu, +\infty)$ for $\nu \geqslant 0$, we also use the class $\mathscr{M}_\infty(\mathbb{R}_{\nu+})$ of functions g such that $g(t) = const + g_0(t)$, where $g_0 \in \mathbb{P}_\infty(\mathbb{R}_{\nu+})$. $\mathscr{M}_\infty(\mathbb{R}_+)$ is similarly defined.
(h) $\mathscr{M}_{0,\infty}(\mathbb{R}_+)$ is the class of functions $g \in \mathscr{M}_\infty(\mathbb{R}_+)$ which satisfy the decay condition also at the origin: that there exist a real numbers g_0 such that $|g(x) - g_0| \leqslant \frac{A}{|\ln x|}$, $0 < x \leqslant \frac{1}{2}$. We also write $g_0 = g(0)$, $g_\infty = g(\infty)$ in this case.
(i) $L^{q(\cdot)}_{\mathrm{loc}}(E) := \{f : f \in L^{q(\cdot)}(K) \text{ for all compact subsets } K \subset X\}$.
(j) $\mathcal{B}(E)$ is the set of $q \in \mathcal{P}(E)$ satisfying the condition that M is bounded on $L^{q(\cdot)}(E)$, where

$$Mf(x) := \sup_{r>0} r^{-n} \int_{B(x,r)} |f(y)|\, dy, \quad (x \in E).$$

Let $p, q \in \mathcal{P}_0(\mathbb{R}^n)$. The mixed variable exponent Lebesgue space $l^{q(\cdot)}(L^{p(\cdot)})$ is defined on sequence of $L^{p(\cdot)}$-function by the modular:

$$S_{p(\cdot),q(\cdot)}((f_\nu)_\nu) = \sum_\nu \inf\left\{\lambda_\nu > 0 : \varrho_{p(\cdot)}\left(\frac{f_\nu}{\lambda_\nu^{1/q(\cdot)}}\right) \leqslant 1\right\}.$$

The space is equipped with the quasi-norm:

$$\|(f_\nu)_\nu\|_{l^{q(\cdot)}(L^{p(\cdot)})} = \inf\left\{\mu > 0 : S_{p(\cdot),q(\cdot)}\left(\frac{1}{\mu}(f_\nu)_\nu\right) \leqslant 1\right\}. \tag{6.5}$$

We will also work with the variable exponent Lebesgue space with the Haar measure dt/t on $\mathbb{R}_{\nu+}$, $\nu \geqslant 0$, which is introduced in the usual way:

$$\|f\|_{L^{q(\cdot)}(\mathbb{R}_{\nu+};dt/t)} = \inf\left\{\eta > 0 : \int_\nu^\infty \left|\frac{f(t)}{\eta}\right|^{q(t)} \frac{dt}{t} \leqslant 1\right\}.$$

We will make frequent use of the following notation throughout this chapter:

- For any $k \in \mathbb{Z}$, we define the dyadic ball

$$B_k := \left\{x \in \mathbb{R}^n : |x| \leqslant 2^k\right\}.$$

- The spherical layer (or annulus) between two balls is given by

$$R_{t,\tau} := B_\tau \setminus B_t = \left\{x \in \mathbb{R}^n : t < |x| < \tau\right\}.$$

- In particular, we define the dyadic annulus $R_k := B_k \setminus B_{k-1}$.
- The characteristic functions of these regions are denoted by $\chi_k := \chi_{R_k}$ and $\chi_{t,\tau} := \chi_{R_{t,\tau}}$.

Technical Lemmata

In this subsection, we collect some technical lemmata that we will need.

Let

$$K\varphi(t) = \int_0^\infty \mathcal{K}\left(\frac{t}{\tau}\right)\varphi(\tau)\frac{d\tau}{\tau} \tag{6.6}$$

be an integral operator with the Haar measure $d\tau/\tau$ and the kernel homogeneous of order 0, known also as *Mellin convolution operator*. We refer to Samko [293] for Mellin convolution operators in variable exponent Lebesgue spaces.

Lemma 6.2 (Samko [294, Corollary 4.5] and Samko [293]) *Let* $q \in \mathbb{P}_{0,\infty}(\mathbb{R}_+)$ *and* $q(0) = q(\infty)$. *The operator* K *is bounded in the space* $L^{q(\cdot)}(\mathbb{R}_+; dt/t)$, *if*

$$\int_0^\infty |\mathcal{K}(t)|^s \frac{dt}{t} < \infty \quad \text{for} \quad s = 1 \text{ and } s = s_0, \tag{6.7}$$

where $\frac{1}{s_0} = 1 - \frac{1}{q_-} + \frac{1}{q_+}$.

Lemma 6.3 (Almeida and Drihem [12], Samko [294]) *Let* $D > 1$ *and* $p \in \mathbb{P}_{0,\infty}(\mathbb{R}^n)$. *Then*

$$\frac{1}{C_0} r^{\frac{n}{p(0)}} \leqslant \|\chi_{R_{r,Dr}}\|_{p(\cdot)} \leqslant C_0 r^{\frac{n}{p(0)}} \quad \text{for} \quad 0 < r \leqslant 1 \tag{6.8}$$

and

$$\frac{1}{C_\infty} r^{\frac{n}{p_\infty}} \leqslant \|\chi_{R_{r,Dr}}\|_{p(\cdot)} \leqslant C_\infty r^{\frac{n}{p_\infty}} \quad \text{for} \quad r \geqslant 1, \tag{6.9}$$

respectively, where $C_0 \geqslant 1$ *and* $C_\infty \geqslant 1$ *depend on* D, *but do not depend on* r.

Lemma 6.4 (Samko [294, Lemma 4.7]) *The following relations*

$$\int_{2a<|y|<t} |\Phi(y)|dy = \frac{1}{\ln 2} \int_a^t \frac{d\tau}{\tau} \int_{\max(2a,\tau)<|y|<\min(t,2\tau)} |\Phi(y)|dy, \; t > 2a > 0, \tag{6.10}$$

and

$$\int_{|y|>2t} |\Phi(y)|dy = \frac{1}{\ln 2} \int_t^\infty \frac{d\tau}{\tau} \int_{\max(\tau,2t)<|y|<2\tau} |\Phi(y)|dy, \; t > 0 \tag{6.11}$$

hold for every measurable function Φ *for which the integrals on the left-hand side exists.*

6.1.2 Continual Variable Exponent Herz Spaces

Let $\mathbb{R}_{+,\nu} = \{t \in \mathbb{R} : \nu < t < \infty\}$, where $\nu > 0$. By $\mathscr{M}_\infty^{\log}(\mathbb{R}_{+,\nu})$ and $\mathscr{M}_{0,\infty}^{\log}(\mathbb{R}_+)$, we denote the classes of functions g on $\mathbb{R}_{+,\nu}$ and $\mathbb{R}_+$, such that $g(t) = c + g_1(t)$, where c is a constant and $g_1 \in \mathbb{P}_\infty^{\log}(\mathbb{R}_{+,\nu})$ or $g_1 \in \mathbb{P}_{0,\infty}^{\log}(\mathbb{R}_+)$, respectively.

Definition 6.5 We define the *continual variable exponent Herz space*, denoted by $H_{\nu,\delta}^{p(\cdot),q(\cdot),\alpha(\cdot)}(\mathbb{R}^n)$, by the norm

$$\|f\|_{H_{\nu,\delta}^{p(\cdot),q(\cdot),\alpha(\cdot)}(\mathbb{R}^n)}$$

$$:= \|f\|_{L^{p(\cdot)}(B(0,\gamma\nu+\varepsilon))} + \left\| t^{\alpha(t)} \|f\chi_{R_{\gamma t,\delta t}}\|_{L^{p(\cdot)}} \right\|_{L^{q(\cdot)}((\gamma\nu,\infty);dt/t)} < \infty, \tag{6.12}$$

where $0 < \gamma < \delta < \infty$, $\varepsilon > 0$, $q : [\gamma\nu, \infty) \to [1, \infty)$, $p : \mathbb{R}^n \to [1, \infty)$ and $\alpha : [\gamma\nu, \infty) \to \mathbb{R}$ are variable exponents. The cases $\nu = 0$ and $\nu > 0$ correspond to *homogeneous* and *nonhomogeneous Herz spaces*, respectively. It is known (see Samko [294, Lemma 3.5]) that, when the exponent q is constant, this definition does not depend on the choice of γ, δ, or ε.

The following lemmas were proved in Samko [294]; see Lemmas 3.2 and 3.3 there.

Lemma 6.6 *Let* $0 < \varepsilon < 2$ *and* $4 \leqslant R_0 < \infty$. *Then*

$$\|f\|_{L^{p(\cdot)}(R_{2+\varepsilon,R_0})} \lesssim_{\varepsilon,R_0} \left\| t^{\alpha(t)} \|f\chi_{R_{t,2t}}\|_{L^{p(\cdot)}} \right\|_{L^q((2,\infty);dt/t)}.$$

Lemma 6.7 *Let* (6.1) *hold. Then the following norm equivalences hold:*

$$\|f\|_{H_{\nu,\delta}^{p(\cdot),q,\alpha(\cdot)}} \approx \|f\|_{L^{p(\cdot)}(B(0,\gamma\nu+\varepsilon))} + \left\| t^{\alpha_\infty} \|f\chi_{R_{\gamma t,\delta t}}\|_{L^{p(\cdot)}} \right\|_{L^q(\mathbb{R}_{+,\nu};dt/t)}, \quad \nu > 0, \tag{6.13}$$

and

$$\|f\|_{H_{0,\delta}^{p(\cdot),q,\alpha(\cdot)}} \approx \left\| t^{\alpha(0)}(1+t)^{\alpha_\infty-\alpha(0)} \|f\chi_{R_{\gamma t,\delta t}}\|_{L^{p(\cdot)}} \right\|_{L^{q(\cdot)}(\mathbb{R}_+;dt/t)}, \tag{6.14}$$

if $\alpha \in \mathscr{M}_\infty^{\log}(\mathbb{R}_{+,\nu})$ *in the case of* (6.13) *and* $\alpha \in \mathscr{M}_{0,\infty}^{\log}(\mathbb{R}_+)$ *in the case of* (6.14).

6.1.3 Discrete Variable Exponent Herz Spaces

In this subsection, we introduce variable exponent Herz spaces. In what follows, we denote $\chi_k = \chi_{R_k}$, $R_k = B_k \setminus B_{k-1}$ and $B_k = \{x \in \mathbb{R}^n : |x| \leqslant 2^k\}$ for all $k \in \mathbb{Z}$.

Definition 6.8 Let $1 < p < \infty$, $\alpha \in \mathbb{R}^n$ and $q \in \mathcal{P}(\mathbb{R}^n)$. The homogeneous Herz space $\dot{K}^{\alpha,p}_{q(\cdot)}(\mathbb{R}^n)$ is defined by

$$\dot{K}^{\alpha,p}_{q(\cdot)}(\mathbb{R}^n) = \left\{ f \in L^{q(\cdot)}_{\mathrm{loc}}(\mathbb{R}^n \setminus \{0\}) : \|f\|_{\dot{K}^{\alpha,p}_{q(\cdot)}(\mathbb{R}^n)} < \infty \right\},$$

where

$$\|f\|_{\dot{K}^{\alpha,p}_{q(\cdot)}(\mathbb{R}^n)} = \left(\sum_{k=-\infty}^{\infty} \|2^{k\alpha} f \chi_k\|^p_{L^{q(\cdot)}} \right)^{\frac{1}{p}}.$$

Definition 6.9 Let $1 < p < \infty$, $\alpha \in \mathbb{R}^n$ and $q \in \mathcal{P}(\mathbb{R}^n)$. The nonhomogeneous Herz space $K^{\alpha,p}_{q(\cdot)}(\mathbb{R}^n)$ is defined by

$$K^{\alpha,p}_{q(\cdot)}(\mathbb{R}^n) = \left\{ f \in L^{q(\cdot)}_{\mathrm{loc}}(\mathbb{R}^n) : \|f\|_{K^{\alpha,p}_{q(\cdot)}(\mathbb{R}^n)} < \infty \right\},$$

where

$$\|f\|_{K^{\alpha,p}_{q(\cdot)}(\mathbb{R}^n)} = \left(\sum_{k=1}^{\infty} \|2^{k\alpha} f \chi_k\|^p_{L^{q(\cdot)}} \right)^{\frac{1}{p}} + \|f\|_{L^{q(\cdot)}(B(0,1))}.$$

6.2 On the Boundedness of Marcinkiewicz Integral on Continual Variable Exponent Herz Spaces

6.2.1 *Marcinkiewicz Integral*

Let $\mathbb{S}^{n-1}$ denote the unit sphere in $\mathbb{R}^n$ equipped with a normalized Lebesgue measure. Let $\Omega \in L^r(\mathbb{S}^{n-1})$ be a homogeneous function of degree zero such that

$$\int_{\mathbb{S}^{n-1}} \Omega(y')d\sigma(y') = 0, \tag{6.15}$$

where $y' = y/|y|$ for any $y \neq 0$. The Marcinkiewicz integral was introduced by Stein [313] in connection with Littlewood–Paley g-function on $\mathbb{R}^n$ as

$$\mu_\Omega(f)(x) = \left(\int_0^\infty |F_{\Omega,s}(f)(x)|^2 \frac{ds}{s^3} \right)^{\frac{1}{2}},$$

where

$$F_{\Omega,s}(f)(x) = \int_{|x-y|\leqslant s} \frac{\Omega(x-y)}{|x-y|^{n-1}} f(y)dy.$$

Definition 6.10 Let w be a locally integrable function on $\mathbb{R}^n$. We say that a function $w \in A_p(\mathbb{R}^n)$, $1 < p < \infty$, if the following inequality

$$\sup_Q \left(\frac{1}{|Q|}\int_Q w(x)dx\right)\left(\frac{1}{|Q|}\int_Q w(x)^{1-p'}dx\right)^{p-1} < +\infty$$

holds, where $p' = \frac{p}{p-1}$ and the supremum is taken over all cubes Q in $\mathbb{R}^n$ with sides parallel to the coordinate axes.

The following theorem is due to Ding et al. [77].

Theorem 6.11 *Suppose that $\Omega \in L^r(\mathbb{S}^{n-1})$ $(r > 1)$ such that* (6.15) *holds. Furthermore, suppose that $w \in A_{p/r'}$, $r' < p < \infty$. Then there is a positive constant C_p such that the following inequality*

$$\int_{\mathbb{R}^n} |\mu_\Omega(f)(x)|^p w(x)dx \lesssim_p \int_{\mathbb{R}^n} |f(x)|^p w(x)dx,$$

holds.

For the following theorem, we refer to Cruz-Uribe et al. [63].

Theorem 6.12 *Suppose that we are given a family $\mathcal{F}$ and an open set $E \subset \mathbb{R}^n$ such that for some p_0, $0 < p_0 < \infty$ and every $w \in A_1$ and every $(f, g) \in \mathcal{F}$ the inequality*

$$\int_E f(x)^{p_0} w(x)dx \lesssim \int_E g(x)^{p_0} w(x)dx$$

holds, with the implicit constant depending only on p_0 and the A_1 constant of w. Let $p \in \mathbb{P}_\infty^{\log}(E)$ and $p_0 < p_-$. Then for all $(f, g) \in \mathcal{F}$ such that $f \in L^{p(\cdot)}(E)$

$$\|f\|_{L^{p(\cdot)}(E)} \lesssim \|g\|_{L^{p(\cdot)}(E)},$$

with the implicit constant independent of the pair (f, g).

Remark 6.13 Since $A_1 \subset A_{p/r'}$ by Theorems 6.11 and 6.12 for $r > (p^-)'$ and $\Omega \in L^r(\mathbb{S}^{n-1})$, one immediately obtains

$$\|\mu_\Omega f\|_{L^{p(\cdot)}(\mathbb{R}^n)} \lesssim \|f\|_{L^{p(\cdot)}(\mathbb{R}^n)}. \tag{6.16}$$

6.2.2 *Boundedness Results*

In Theorems 6.14 and 6.15, we obtain the conditions on the boundedness of the Marcinkiewicz integral in continual variable Herz space, where all the main parameters p, q, and α are variable, with different formulations for variable and constant q; see Sect. 6.1.2.

Recall that norms in Herz spaces with constant q are all equivalent to each other for different values of the auxiliary parameters γ and δ, $0 < \gamma < \delta < \infty$, which is not the case when q is variable. So the dependence on those auxiliary parameters come into play when q is variable. The main difference between the two cases is in the necessity to move the parameters γ and δ in the starting space when q is variable. By this reason in our main result, we slightly change the notation for Herz spaces, taking it in the form

$$H^{p(\cdot),q(\cdot),\alpha(\cdot)}_{\nu;(\gamma,\delta)}(\mathbb{R}^n).$$

Nonhomogeneous Herz Spaces (The Case $\nu = 0$)

Theorem 6.14 *Let $\nu > 0, 0 < \gamma < \delta < \infty$, $p \in \mathbb{P}_\infty(\mathbb{R}^n)$ with $1 < p^- \leqslant p^+ < \infty$, $q \in \mathbb{P}^{\log}_\infty(\mathbb{R}_{\nu+})$ with $1 \leqslant q_- \leqslant q_+ < \infty$, and $\alpha \in \mathscr{M}_\infty(\mathbb{R}_{\nu+})$. Let Ω be homogeneous of degree zero and $\Omega \in L^s(\mathbb{S}^{n-1})$ $(s > p'_-)$. If*

$$-\frac{n}{p(\infty)} < \alpha(\infty) < \frac{n}{p'(\infty)} - \frac{n}{s}, \tag{6.17}$$

then μ_Ω is bounded from $H^{p(\cdot),q(\cdot),\alpha(\cdot)}_{\nu;(\gamma',\delta')}(\mathbb{R}^n)$ to $H^{p(\cdot),q(\cdot),\alpha(\cdot)}_{\nu;(\gamma,\delta)}(\mathbb{R}^n)$ for any $0 < \gamma' < \gamma$ and $\delta < \delta' < \infty$.

When q is constant, then the same is true also with $\gamma' = \gamma$ and $\delta' = \delta$.

***Proof of Theorem 6.14—the Case of Constant** q* We choose $\nu = 2$ and $\gamma = 1, \delta = 2$ in (6.12) for simplicity, so that we will work with the norm

$$\|f\|_{H^{p(\cdot),q,\alpha(\cdot)}_{2,2}} = \|f\|_{L^{p(\cdot)}(B(0,2+\varepsilon))} + \mathcal{N}(f)_{p,q,\alpha} \tag{6.18}$$

where

$$\mathcal{N}(f)_{p,q,\alpha} = \left\| t^{\alpha_\infty} \|f\chi_{\mathsf{R}_{t,2t}}\|_{L^{p(\cdot)}} \right\|_{L^q((2,\infty);dt/t)}, \tag{6.19}$$

where $\mathsf{R}_{t,\tau}$ stands for the annulus $\mathsf{R}_{t,\tau} := B(0,\tau)\backslash B(0,t)$.

We first estimate $\|\mu_\Omega f\|_{L^{p(\cdot)}(B(0,2+\varepsilon))}$, where it suffices to consider $\varepsilon \in (0,1)$. This estimate follows from the $(L^{p(\cdot)} \to L^{p(\cdot)})$ boundedness of μ_Ω. Indeed, we have

$$\|\mu_\Omega f\|_{L^{p(\cdot)}(B(0,2+\varepsilon))} \lesssim \|f\|_{L^{p(\cdot)}(B(0,2+\varepsilon))} \lesssim \|f\|_{H_{2,2}^{p(\cdot),q,\alpha(\cdot)}}.$$

To estimate now the $\mathcal{N}(\mu_\Omega f)_{p,q,\alpha}$ term, we split the function $f(x)$ as

$$f(x) = f_0(x) + f_t(x) + g_t(x) + h_t(x),$$

where

$$f_0(x) = f(x)\chi_{B(0,1)}(x), \qquad f_t(x) = f(x)\chi_{B(0,t/2)\setminus B(0,1)}(x)$$
$$g_t(x) = f(x)\chi_{B(0,8t)\setminus B(0,t/2)}(x), \qquad h_t(x) = f(x)\chi_{\mathbb{R}^n\setminus B(0,8t)}(x)$$

so that we have the pointwise inequality

$$|\mu_\Omega f(x)| \leqslant |\mu_\Omega(f_0)(x)| + |\mu_\Omega(f_t)(x)| + |\mu_\Omega(g_t)(x)| + |\mu_\Omega(h_t)(x)|.$$

ESTIMATION OF $\mu_\Omega f_0$. Since $y \in B(0,1)$ and $x \in \mathsf{R}_{t,2t}$ by the integral Minkowski's inequality and using the estimate $|x-y| \geqslant |x| - |y| > t/2$, we have

$$\begin{aligned}|\mu_\Omega f_0(x)| &\lesssim t^{-n} \int_{B(0,1)} |\Omega(x-y)||f(y)|dy \\ &\lesssim t^{-n}\|f_0\|_{L^{p(\cdot)}(\mathbb{R}^n)}\|\Omega(x-\cdot)\chi_{B(0,1)}\|_{L^{p'(\cdot)}(\mathbb{R}^n)}\end{aligned}$$

with

$$\begin{aligned}\|\Omega(x-\cdot)\chi_{B(0,1)}\|_{L^{p'(\cdot)}} &\lesssim \|\Omega(x-\cdot)\chi_{B(0,1)}\|_{L^s(\mathbb{R}^n)}\|\chi_{B(0,1)}\|_{L^{P'(\cdot)}} \\ &\lesssim \left(\int_{|y|<1} |\Omega(x-y)|^s dy\right)^{\frac{1}{s}};\end{aligned}$$

where we used the generalized Hölder inequality with $\frac{1}{p'(x)} = \frac{1}{s} + \frac{1}{P'(x)}$. Now by change of variables $z = x - y$ and noting that $|z| \leqslant |x| + |y| \leqslant 2t + 1 \leqslant 3t$, we have

$$\|\Omega(x-\cdot)\chi_{B(0,1)}\|_{L^{p'(\cdot)}} \lesssim \left(\int_{|z|<3t} |\Omega(z)|^s dz\right)^{\frac{1}{s}}.$$

By change of variable in polar coordinates, we have

$$\|\Omega(x-\cdot)\chi_{B(0,1)}\|_{L^{p'(\cdot)}} \lesssim t^{\frac{n}{s}}\|\Omega\|_{L^s(\mathbb{S}^{n-1})},$$

and we obtain

$$\begin{aligned}
\mathcal{N}(\mu_\Omega f_0)_{p,q,\alpha} &= \left\| t^{\alpha_\infty}\|\chi_{\mathsf{R}_{t,2t}}\mu_\Omega f_0\|_{L^{p(\cdot)}} \right\|_{L^q((2,\infty);dt/t)} \\
&\lesssim \left\| t^{\alpha_\infty-n+\frac{n}{s}}\|\chi_{\mathsf{R}_{t,2t}}\|_{L^{p(\cdot)}} \right\|_{L^q((2,\infty);dt/t)} \|f_0\|_{L^{p(\cdot)}(\mathbb{R}^n)} \\
&\lesssim \left\| t^{\alpha_\infty-\frac{n}{p'_\infty}+\frac{n}{s}} \right\|_{L^q((2,\infty);dt/t)} \|f_0\|_{L^{p(\cdot)}(\mathbb{R}^n)} \\
&\lesssim \|f_0\|_{L^{p(\cdot)}(\mathbb{R}^n)}
\end{aligned}$$

where we have used the estimate (6.9); for the finiteness of the q-norm of the power function, we use the fact $\alpha_\infty - \frac{n}{p'_\infty} + \frac{n}{s} < 0$ and the relation (6.17) in the last line.
ESTIMATION OF $\mu_\Omega f_t$. For $x \in \mathsf{R}_{t,2t}$, using Minkowski's inequality, we have

$$|\mu_\Omega f_t(x)| \lesssim \int\limits_{R_{1,t/2}} \frac{|\Omega(x-y)||f(y)|}{|x-y|^n}dy.$$

Since $|x-y| \geqslant |x|-|y| \geqslant t/2$, using relation (6.10), Hölder's inequality, and (6.9), we obtain

$$\begin{aligned}
|\mu_\Omega f_t(x)| &\lesssim t^{-n} \int\limits_{1<|y|<t/2} |\Omega(x-y)||f(y)|dy \\
&\lesssim t^{-n}\int_1^t \frac{d\tau}{\tau} \int\limits_{\frac{\tau}{2}<|y|<\tau} |\Omega(x-y)||f(y)|dy \\
&\lesssim t^{-n}\int_1^t \|f\chi_{\mathsf{R}_{\frac{\tau}{2},\tau}}\|_{L^{p(\cdot)}}\|\Omega(x-\cdot)\chi_{\mathsf{R}_{\frac{\tau}{2},\tau}}\|_{L^{p'(\cdot)}}\frac{d\tau}{\tau}.
\end{aligned} \tag{6.20}$$

By Lemma 6.1, estimate (6.9) and taking $\frac{1}{p'(\cdot)} = \frac{1}{P'(\cdot)} + \frac{1}{s}$, we have

$$\|\Omega(x-\cdot)\chi_{\mathsf{R}_{\frac{\tau}{2},\tau}}\|_{L^{p'(\cdot)}} \lesssim \|\Omega(x-\cdot)\chi_{\mathsf{R}_{\frac{\tau}{2},\tau}}\|_{L^s(\mathbb{R}^n)}\|\chi_{\mathsf{R}_{\frac{\tau}{2},\tau}}\|_{L^{P'(\cdot)}}$$

$$\lesssim \left(\int_{\frac{\tau}{2}<|y|<\tau} |\Omega(x-y)|^s dy \right)^{\frac{1}{s}} \tau^{\frac{n}{p'_\infty}}.$$

Now by change of variables $z = x - y$ and noting that $|z| \leqslant |x| + |y| \leqslant 2t + \tau \leqslant 3t$, we have

$$\|\Omega(x-\cdot)\chi_{\mathrm{R}_{\frac{\tau}{2},\tau}}\|_{L^{p'(\cdot)}} \lesssim \left(\int_{|z|<3t} |\Omega(z)|^s dz \right)^{\frac{1}{s}} \tau^{\frac{n}{p'_\infty}}.$$

By change of variable in polar coordinates, we have

$$\|\Omega(x-\cdot)\chi_{\mathrm{R}_{\frac{\tau}{2},\tau}}\|_{L^{p'(\cdot)}} \lesssim t^{\frac{n}{s}} \|\Omega\|_{L^s(\mathbb{S}^{n-1})} \tau^{\frac{n}{p'_\infty}-\frac{n}{s}}.$$

Gathering all the previous estimates, we obtain

$$\begin{aligned}
t^{\alpha_\infty} \|\chi_{\mathrm{R}_{t,2t}} \mu_\Omega f_t\|_{L^{p(\cdot)}} &\lesssim t^{\alpha_\infty - n + \frac{n}{s}} \|\chi_{\mathrm{R}_{t,2t}}\|_{L^{p(\cdot)}} \int_1^t \|f\chi_{\mathrm{R}_{\frac{\tau}{2},\tau}}\|_{L^{p(\cdot)}} \tau^{\frac{n}{p'_\infty}-\frac{n}{s}-1} d\tau \\
&\lesssim t^{\alpha_\infty - n + \frac{n}{s} + \frac{n}{p_\infty}} \int_1^t \|f\chi_{\mathrm{R}_{\frac{\tau}{2},\tau}}\|_{L^{p(\cdot)}} \tau^{\frac{n}{p'_\infty}-\frac{n}{s}-1} d\tau \\
&\lesssim t^{\alpha_\infty + \frac{n}{s} - \frac{n}{p'_\infty}} \int_1^t \|f\chi_{\mathrm{R}_{\frac{\tau}{2},\tau}}\|_{L^{p(\cdot)}} \tau^{\frac{n}{p'_\infty}-\frac{n}{s}-1} d\tau \\
&\lesssim \int_1^t \left(\frac{t}{\tau}\right)^{\alpha_\infty + \frac{n}{s} - \frac{n}{p'_\infty}} \varphi(\tau) \frac{d\tau}{\tau}
\end{aligned} \tag{6.21}$$

where $\varphi(\tau) = \tau^{\alpha_\infty} \|f\chi_{\mathrm{R}_{\frac{\tau}{2},\tau}}\|_{L^{p(\cdot)}}$ and the right-hand side expression in the last inequality of (6.21) is a Hardy type operator. Taking $\mathcal{K}(t)$ as

$$\mathcal{K}(t) = \begin{cases} t^{\alpha_\infty + \frac{n}{s} - \frac{n}{p'_\infty}}, & t > 1, \\ 0, & 0 < t < 1, \end{cases} \tag{6.22}$$

we define the operator $K\varphi(t) = \int\limits_0^\infty \mathcal{K}(t/\tau)\varphi(\tau)\frac{d\tau}{\tau}$. With all the above taken into account, we have

$$\begin{aligned}
\mathcal{N}(\mu_\Omega f_t)_{p,q,\alpha} &\lesssim \|K\varphi\|_{L^q((2,\infty);dt/t)} \\
&\lesssim \|\varphi\|_{L^q((2,\infty);dt/t)} \\
&\lesssim \left\| t^{\alpha_\infty} \|f\chi_{\mathsf{R}_{t,2t}}\|_{L^{p(\cdot)}} \right\|_{L^q((1,2);dt/t)} \\
&\qquad + \left\| t^{\alpha_\infty} \|f\chi_{\mathsf{R}_{t,2t}}\|_{L^{p(\cdot)}} \right\|_{L^q((2,\infty);dt/t)} \\
&\lesssim \|f\|_{L^{p(\cdot)}(B(0,4)\setminus B(0,1))} + \left\| t^{\alpha_\infty} \|f\chi_{\mathsf{R}_{t,2t}}\|_{L^{p(\cdot)}} \right\|_{L^q((2,\infty);dt/t)} \\
&\lesssim \|f\|_{H_2^{p(\cdot),q,\alpha(\cdot)}}
\end{aligned}$$

where we have used the Lemma 6.2, the fact that $\alpha_\infty + \frac{n}{s} - \frac{n}{p'_\infty} < 0$ and for the last inequality we used the embedding

$$B(0,4)\setminus B(0,1) \subset B(0,2+\varepsilon) \cup B(0,4)\setminus B(0,2+\varepsilon)$$

and Lemma 6.6.

ESTIMATION OF $\mu_\Omega g_t$. By the $(L^{p(\cdot)} \to L^{p(\cdot)})$-boundedness of μ_Ω, we obtain

$$\|(\mu_\Omega g_t)\chi_{\mathsf{R}_{t,2t}}\|_{L^{p(\cdot)}} \lesssim \|g_t\|_{L^{p(\cdot)}} \lesssim \sum_{j=-1}^{2} \|f\chi_{\mathsf{R}_{2^j t,2^{j+1}t}}\|_{L^{p(\cdot)}}.$$

Then

$$\begin{aligned}
&\mathcal{N}(\mu_\Omega g_t)_{p,q,\alpha} \\
&\quad\lesssim \left\| t^{\alpha_\infty} \|f\chi_{\mathsf{R}_{\frac{t}{2},t}}\|_{L^{p(\cdot)}} \right\|_{L^q((2,\infty);dt/t)} + \left\| t^{\alpha_\infty} \|f\chi_{\mathsf{R}_{t,2t}}\|_{L^{p(\cdot)}} \right\|_{L^q((2,\infty);dt/t)} \\
&\quad\lesssim \left\| t^{\alpha_\infty} \|f\chi_{\mathsf{R}_{t,2t}}\|_{L^{p(\cdot)}} \right\|_{L^q((1,2);dt/t)} + \left\| t^{\alpha_\infty} \|f\chi_{\mathsf{R}_{t,2t}}\|_{p(\cdot)} \right\|_{L^q((2,\infty);dt/t)} \\
&\quad\lesssim \|f\|_{H_2^{p(\cdot),q,\alpha(\cdot)}}
\end{aligned}$$

where the first inequality comes from the fact that

$$\left\| t^{\alpha_\infty} \|f\chi_{\mathsf{R}_{2^j t,2^{j+1}t}}\|_{L^{p(\cdot)}} \right\|_{L^q((2,\infty);dt/t)} \lesssim \left\| t^{\alpha_\infty} \|f\chi_{\mathsf{R}_{t,2t}}\|_{L^{p(\cdot)}} \right\|_{L^q((2,\infty);dt/t)},$$

for $j = 1, 2$, which follows from a dilation change of variables in t (thanks to the fact that q is constant) and in the last inequality, we used Lemma 6.6 and the obvious inequality $\|f\chi_{\mathsf{R}_{t/2,t}}\|_{L^{p(\cdot)}} \leqslant \|f\chi_{\mathsf{B}(0,4)}\|_{L^{p(\cdot)}}$ for $1 < t < 2$.

ESTIMATION OF $\mu_\Omega h_t$. By the integral Minkowski's inequality, we have

$$|\mu_\Omega h_t(x)| \lesssim \int\limits_{|y|>8t} \frac{|\Omega(x-y)||f(y)|}{|x-y|^n} dy,$$

which is estimated as follows:

$$\int\limits_{|y|>8t} \frac{|\Omega(x-y)||f(y)|}{|x-y|^n} dy \lesssim \int\limits_{4t}^{\infty} \frac{d\tau}{\tau} \int\limits_{\tau<|y|<2\tau} |\Omega(x-y)| \frac{|f(y)|}{|x-y|^n} dy$$

$$\lesssim \int\limits_{4t}^{\infty} \tau^{-n} \|\Omega(x-\cdot)\chi_{\mathsf{R}_{\tau,2\tau}}\|_{L^{p'(\cdot)}} \|f\chi_{\mathsf{R}_{\tau,2\tau}}\|_{L^{p(\cdot)}} \frac{d\tau}{\tau}$$

where the first inequality comes from relation (6.11), the second by Hölder's inequality and the fact that $|x-y| \geqslant \frac{\tau}{2}$ (since $x \in \mathsf{R}_{t,2t}$ there), and the last one by relation (6.9).

Introduce the variable exponent $\frac{1}{p'(\cdot)} = \frac{1}{P'(\cdot)} + \frac{1}{s}$. By Lemma 6.1 and estimate (6.9) we have

$$\|\Omega(x-\cdot)\chi_{\mathsf{R}_{\tau,2\tau}}\|_{L^{p'(\cdot)}} \lesssim \|\Omega(x-\cdot)\chi_{\mathsf{R}_{\tau,2\tau}}\|_{L^s(\mathbb{R}^n)} \|\chi_{\mathsf{R}_{\tau,2\tau}}\|_{L^{P'(\cdot)}}$$

$$\lesssim \left(\int\limits_{\tau<|y|<2\tau} |\Omega(x-y)|^s dy \right)^{\frac{1}{s}} \tau^{\frac{n}{P'_\infty}}.$$

Now by change of variables $z = x - y$ and noting that

$$|z| \leqslant |x| + |y| \leqslant 2t + \tau \leqslant \frac{\tau}{2} + \tau \leqslant \frac{3}{2}\tau$$

we have

$$\|\Omega(x-\cdot)\chi_{\mathsf{R}_{\tau,2\tau}}\|_{L^{p'(\cdot)}} \lesssim \left(\int\limits_{|z|<\frac{3\tau}{2}} |\Omega(z)|^s dz \right)^{\frac{1}{s}} \tau^{\frac{n}{P'_\infty}}.$$

By change of variable in polar coordinates, we have

$$\|\Omega(x-\cdot)\chi_{\mathsf{R}_{\tau,2\tau}}\|_{L^{p'(\cdot)}} \lesssim \tau^{\frac{n}{s}}\|\Omega\|_{L^s(\mathbb{S}^{n-1})}\tau^{\frac{n}{p'_\infty}-\frac{n}{s}},$$

from which we obtain

$$\|\Omega(x-\cdot)\chi_{\mathsf{R}_{\tau,2\tau}}\|_{L^{p'(\cdot)}} \lesssim \|\Omega\|_{L^s(\mathbb{S}^{n-1})}\tau^{\frac{n}{p'_\infty}}.$$

By previous estimates and Lemma 6.3, we have

$$\begin{aligned} t^{\alpha_\infty}\|\chi_{\mathsf{R}_{t,2t}}\mu_\Omega h_t\|_{L^{p(\cdot)}} &\lesssim t^{\alpha_\infty+\frac{n}{p_\infty}}\int_{4t}^{\infty}\tau^{-n+\frac{n}{p'_\infty}}\|f\chi_{\mathsf{R}_{\tau,2\tau}}\|_{L^{p(\cdot)}}\frac{d\tau}{\tau} \\ &\lesssim \int_t^\infty\left(\frac{t}{\tau}\right)^{\alpha_\infty+\frac{n}{p_\infty}}\varphi(\tau)\frac{d\tau}{\tau} \end{aligned}$$

where $\varphi(\tau)=\tau^{\alpha_\infty}\|f\chi_{\mathsf{R}_{\tau,2\tau}}\|_{p(\cdot)}\chi_{(2,\infty)}(\tau)$. We note that we arrive at a Hardy-type inequality, and since $\alpha_\infty+\frac{n}{p_\infty}>0$, using Lemma 6.2, we obtain

$$\left\|t^{\alpha_\infty}\|\chi_{\mathsf{R}_{\tau,2\tau}}\mu_\Omega h_t\|_{p(\cdot)}\right\|_{L^q((2,\infty);dt/t)} \lesssim \|\varphi\|_{L^q((2,\infty);dt/t)} \lesssim \|f\|_{H_{2,2}^{p(\cdot),q,\alpha(\cdot)}}.$$

□

Proof of Theorem 6.14—The Case of Variable q The principal reason why we cannot act as in the case of constant q is the fact that, when q is variable, we cannot make a dilation change of variables in t as was done in the estimation of the term $\mu_\Omega g_t$. However, the proof follows on the whole the same arguments, so we omit details but dwell on changes.

We again choose $\nu=2$ and $\gamma=1,\delta=2$ for simplicity but stress once again that the norms in the Herz space with variable $q(\cdot)$ are not necessarily equivalent for different values of these parameters. We want to show that

$$\|\mu_\Omega f\|_{L^{p(\cdot)}(B(0,2+\varepsilon))}+\mathcal{N}_{1,2}^{p,q,\alpha}(\mu_\Omega f)\lesssim \|f\|_{L^{p(\cdot)}(B(0,2+\varepsilon))}+\mathcal{N}_{\gamma',\delta'}^{p,q,\alpha}(f)$$

with $\gamma'<1,\delta'>2$ and

$$\mathcal{N}_{\gamma,\delta}^{p,q,\alpha}(g):=\left\|t^{\alpha_\infty}\|g\chi_{\mathsf{R}_{\gamma t,\delta t}}\|_{L^{p(\cdot)}}\right\|_{L^{q(\cdot)}((2,\infty),dt/t)}.$$

The estimation for $\|\mu_\Omega f\|_{L^{p(\cdot)}(B(0,2+\varepsilon))}$ is obtained in a similar way as in the case of the constant q.

To estimate now the $\mathcal{N}(\mu_\Omega f)_{1,2}^{p,q,\alpha}$ term, we split the function $f(x)$ as

$$f(x) = f_0(x) + f_t(x) + g_t(x) + h_t(x)$$

where

$$f_0(x) = f(x)\chi_{B(0,1/2)}(x), \qquad f_t(x) = f(x)\chi_{B(0,\gamma' t)\setminus B(0,1/2)}(x)$$
$$g_t(x) = f(x)\chi_{B(0,\delta' t)\setminus B(0,\gamma' t)}(x), \qquad h_t(x) = f(x)\chi_{\mathbb{R}^n\setminus B(0,\delta' t)}.$$

Then

$$|\mu_\Omega f(x)| \leqslant |\mu_\Omega(f_0)(x)| + |\mu_\Omega(f_t)(x)| + |\mu_\Omega(g_t)(x)| + |\mu_\Omega(h_t)(x)|.$$

ESTIMATION OF $\mu_\Omega f_0$. Follows, mutatis mutandis, the one given in the constant q. ESTIMATION OF $\mu_\Omega f_t$. By the same estimations of (6.20) and taking into account that $|x-y| > (1-\gamma')t$ and doing the estimate of $\|\Omega(x-\cdot)\|_{L^{p'(\cdot)}}$, we obtain

$$|\mu_\Omega f_t(x)| \lesssim t^{-n+\frac{n}{s}} \int_{1/2}^{t} \|f\chi_{\mathsf{R}_{\frac{\tau}{2},\tau}}\|_{L^{p(\cdot)}} \tau^{\frac{n}{p'_\infty}-\frac{n}{s}-1} d\tau,$$

from which we get

$$\begin{aligned} t^{\alpha_\infty}\|\chi_{\mathsf{R}_{t,2t}}\mu_\Omega f_t\|_{L^{p(\cdot)}} &\lesssim \int_{1/2}^{t} \left(\frac{t}{\tau}\right)^{\alpha_\infty-\frac{n}{p'_\infty}+\frac{n}{s}} \tau^{\alpha_\infty}\|f\chi_{\mathsf{R}_{\tau/2,\tau}}\|_{L^{p(\cdot)}}\frac{d\tau}{\tau} \\ &\lesssim \int_{1/4}^{t} \left(\frac{t}{\tau}\right)^{\alpha_\infty-\frac{n}{p'_\infty}+\frac{n}{s}} \psi(\tau)\frac{d\tau}{\tau} \end{aligned} \tag{6.23}$$

where $\psi(\tau) = \tau^{\alpha_\infty}\|f\chi_{\mathsf{R}_{\tau,2\tau}}\|_{L^{p(\cdot)}}$. Observing, as in (6.21), that the right-hand side of (6.23) is a Hardy type operator and using Lemma 6.2 together with (6.22) we get

$$\mathcal{N}_{1,2}^{p,q,\alpha}(\mu_\Omega f_t) \lesssim \|K\psi\|_{L^{q(\cdot)}((2,\infty);dt/t)} \lesssim \|\psi\|_{L^{q(\cdot)}((2,\infty);dt/t)} \lesssim \|f\|_{H_{2;(\gamma',\delta')}^{p(\cdot),q(\cdot),\alpha(\cdot)}(\mathbb{R}^n)},$$

where we followed the same reasoning as in the case of constant q.
ESTIMATION OF $\mu_\Omega g_t$. By the $(L^{p(\cdot)} \to L^{p(\cdot)})$-boundedness of μ_Ω, we get

$$\left\|(\mu_\Omega g_t)\chi_{\mathsf{R}_{t,2t}}\right\|_{L^{p(\cdot)}} \lesssim \|g_t\|_{L^{p(\cdot)}} = \|f\chi_{\mathsf{R}_{\gamma' t,\delta' t}}\|_{L^{p(\cdot)}}$$

which implies

$$\mathcal{N}_{1,2}^{p,q,\alpha}(\mu_\Omega g_t) \lesssim \|f\|_{H_{2;(\gamma',\delta')}^{p(\cdot),q(\cdot),\alpha(\cdot)}(\mathbb{R}^n)}.$$

ESTIMATION OF $\mu_\Omega h_t$. Similarly adapting the estimate of $\mu_\Omega h_t$ from the proof of the constant case q to the case of variable q, we get

$$\left\| t^{\alpha_\infty} \| \chi_{\mathsf{R}_{\tau,2\tau}} \mu_\Omega h_t \|_{p(\cdot)} \right\|_{L^{q(\cdot)}((2,\infty);dt/t)} \lesssim \| f \|_{H^{p(\cdot),q(\cdot),\alpha(\cdot)}_{2;(\gamma',\delta')}(\mathbb{R}^n)}.$$

Taking all the estimates into account, we obtain the result. □

Homogeneous Herz Spaces (The Case $\nu = 0$)

The main difference in the proof is that now we have to arrange the corresponding splitting of f with respect to not only large values of t, but also near the origin. Correspondingly, the assumptions on all the variable exponents now should include the decay condition not only at infinity but also at the origin. But in the whole the proofs for this case follow the same lines as in the nonhomogeneous case, so we give only the formulation of the result.

Theorem 6.15 *Let* $0 < \gamma < \delta < \infty$, $0 < \lambda < n$, $p \in \mathbb{P}_{0,\infty}(\mathbb{R}^n)$ *with* $1 < p^- \leqslant p^+ < \infty$, $q \in \mathbb{P}^{\log}_{0,\infty}(\mathbb{R}_{\nu+})$ *with* $1 \leqslant q_- \leqslant q_+ < \infty$ *and* $\alpha \in \mathscr{M}_{0,\infty}(\mathbb{R}_+)$. *Let* Ω *be homogeneous of degree zero and* $\Omega \in L^s(\mathbb{S}^{n-1})$ $(s > p'_-)$. *If*

$$-\frac{n}{p(0)} < \alpha(0) < \frac{n}{p'(0)} - \frac{n}{s} \quad \text{and} \quad -\frac{n}{p(\infty)} < \alpha(\infty) < \frac{n}{p'(\infty)} - \frac{n}{s}, \tag{6.24}$$

then μ_Ω *is bounded from* $H^{p(\cdot),q(\cdot),\alpha(\cdot)}_{0;(\gamma',\delta')}(\mathbb{R}^n)$ *to* $H^{p(\cdot),q(\cdot),\alpha(\cdot)}_{0;(\gamma,\delta)}(\mathbb{R}^n)$ *for any* $0 < \gamma' < \gamma$ *and* $\delta < \delta' < \infty$. *When* q *is constant, then the same is true for* $\gamma' = \gamma$ *and* $\delta' = \delta$.

Finally, we remark that by checking the proofs of Theorems 6.14 and 6.15, we can obtain the same result for the boundedness of rough sublinear operators T_Ω which are bounded in $L^{p(\cdot)}(\mathbb{R}^n)$ and satisfy the size condition

$$|T_\Omega f(x)| \lesssim \int_{\mathbb{R}^n} \frac{|\Omega(x-y) f(y)|}{|x-y|^n} dy, \qquad x \notin \operatorname{supp} f. \tag{6.25}$$

where Ω is homogeneous function of degree zero, for integrable and compactly supported function f. Condition (6.25) is satisfied by many interesting operators in harmonic analysis, such as the Calderón–Zygmund operators, Carleson maximal operators, Hardy–Littlewood maximal operators, Bochner–Riesz means, etc. For more examples of such operators in the classical case, we refer, e.g., to Soria and Weiss [312] and Yang and Hu [339] (see details).

6.3 Central Calderón–Zygmund Operators on Herz-Type Hardy Spaces of Variable Smoothness and Integrability

We obtain the boundedness of central Calderón–Zygmund operators on Herz-type Hardy spaces $HK^{\alpha(\cdot)}_{p(\cdot),q(\cdot)}$ and $H\dot{K}^{\alpha(\cdot)}_{p(\cdot),q(\cdot)}$ by using the atomic decomposition of Herz–Hardy space obtained in Drihem and Seghiri [80]. In the proofs of the main theorems, we follow the ideas of Almeida and Drihem [12]. The results for the case $q = const.$ were obtained in Xu and Yang [338]. For the boundedness of such operators in the classical case, we refer to Yang and Hu [339].

6.3.1 *Variable Herz-Type Hardy Spaces*

We introduce variable Herz and Herz-type Hardy spaces. For the following definitions and statements, we refer to Drihem and Seghiri [80]. Set $B_k := B(0, 2^k)$, $R_k := B_k \backslash B_{k-1}$ and $\chi_k = \chi_{R_k}$ for $k \in \mathbb{Z}$. For the definition of the space $l^{q(\cdot)}(L^{p(\cdot)})$, see (6.5).

Definition 6.16 Let $p, q \in \mathcal{P}_0(\mathbb{R}^n)$ and $\alpha : \mathbb{R}^n \to \mathbb{R}$ with $\alpha \in L^\infty(\mathbb{R}^n)$. The nonhomogeneous Herz space $K^{\alpha(\cdot)}_{p(\cdot),q(\cdot)}(\mathbb{R}^n)$ consists of all $f \in L^{p(\cdot)}_{\text{loc}}(\mathbb{R}^n)$ such that

$$\|f\|_{K^{\alpha(\cdot)}_{p(\cdot),q(\cdot)}(\mathbb{R}^n)} := \|f\chi_{B_0}\|_{L^{p(\cdot)}(\mathbb{R}^n)} + \left\|\left(2^{k\alpha(\cdot)} f\chi_k\right)_{k\geqslant 1}\right\|_{l^{q(\cdot)}(L^{p(\cdot)})} < \infty. \tag{6.26}$$

The homogeneous Herz space $\dot{K}^{\alpha(\cdot)}_{p(\cdot),q(\cdot)}(\mathbb{R}^n)$ consists of all $f \in L^{p(\cdot)}_{\text{loc}}(\mathbb{R}^n\backslash\{0\})$ such that

$$\|f\|_{\dot{K}^{\alpha(\cdot)}_{p(\cdot),q(\cdot)}(\mathbb{R}^n)} := \left\|\left(2^{k\alpha(\cdot)} f\chi_k\right)_{k\in\mathbb{Z}}\right\|_{l^{q(\cdot)}(L^{p(\cdot)})} < \infty. \tag{6.27}$$

If α, p, q are constant, then these spaces coincide with the classical Herz spaces. Let us denote

$$\|\{g_k\}\|_{l^q_>(L^{p(\cdot)})} := \left(\sum_{k=0}^{\infty} \|g_k\|^q_{L^{p(\cdot)}}\right)^{1/q}$$

and

$$\|\{g_k\}\|_{l^q_<(L^{p(\cdot)})} := \left(\sum_{k=-\infty}^{-1} \|g_k\|^q_{L^{p(\cdot)}}\right)^{1/q}$$

for sequence $\{g_k\}_{k\in\mathbb{Z}}$ of measurable function. For the following proposition, we refer to Drihem and Seghiri [80].

Proposition 6.17 *Let $\alpha \in L^\infty(\mathbb{R}^n)$, $p, q \in \mathcal{P}_0(\mathbb{R}^n)$. If $\alpha, q \in \mathscr{M}_\infty(\mathbb{R}^n)$ then*

$$K^{\alpha(\cdot)}_{p(\cdot),q(\cdot)}(\mathbb{R}^n) = K^{\alpha_\infty}_{p(\cdot),q_\infty}(\mathbb{R}^n).$$

Furthermore, if α, q also satisfy the log-Hölder condition at origin, then

$$\|f\|_{\dot{K}^{\alpha(\cdot)}_{p(\cdot),q(\cdot)}(\mathbb{R}^n)} = \left\|\{2^{k\alpha(0)} f\chi_k\}\right\|_{l^{q(0)}_<(L^{p(\cdot)})} + \left\|\{2^{k\alpha_\infty} f\chi_k\}\right\|_{l^{q_\infty}_>(L^{p(\cdot)})} < \infty.$$

Let $G_N f$ be the grand maximal function of f defined by

$$G_N f(x) = \sup_{\phi\in\mathcal{A}_N} |\phi^*_N(f)(x)|$$

where

$$\mathcal{A}_N = \left\{\phi \in \mathscr{S}(\mathbb{R}^n) : \sup_{|\alpha|\leqslant N, |\beta|\leqslant N} |x^\alpha \partial^\beta \phi(x)| \leqslant 1\right\},$$

and

$$\phi^*_N(f)(x) = \sup_{t>0} |\phi_t * f(x)|,$$

with $\phi_t(\cdot) = t^{-n}\phi(\frac{\cdot}{t})$ and $\mathscr{S}(\mathbb{R}^n)$ denotes the class of Schwartz functions.

Definition 6.18 Let $p, q \in \mathcal{P}_0(\mathbb{R}^n)$, $\alpha : \mathbb{R}^n \to \mathbb{R}$ with $\alpha \in L^\infty(\mathbb{R}^n)$ and $N > n+1$. The nonhomogeneous Herz-type Hardy space $HK^{\alpha(\cdot)}_{p(\cdot),q(\cdot)}(\mathbb{R}^n)$ consists of all $f \in \mathscr{S}'(\mathbb{R}^n)$ such that $G_N f \in K^{\alpha(\cdot)}_{p(\cdot),q(\cdot)}(\mathbb{R}^n)$ and we define

$$\|f\|_{HK^{\alpha(\cdot)}_{p(\cdot),q(\cdot)}(\mathbb{R}^n)} = \|G_N f\|_{K^{\alpha(\cdot)}_{p(\cdot),q(\cdot)}(\mathbb{R}^n)}.$$

The homogeneous Herz-type Hardy space $H\dot{K}^{\alpha(\cdot)}_{p(\cdot),q(\cdot)}(\mathbb{R}^n)$ consists of all $f \in \mathscr{S}'(\mathbb{R}^n)$ such that $G_N f \in \dot{K}^{\alpha(\cdot)}_{p(\cdot),q(\cdot)}(\mathbb{R}^n)$ and we define

$$\|f\|_{H\dot{K}^{\alpha(\cdot)}_{p(\cdot),q(\cdot)}(\mathbb{R}^n)} = \|G_N f\|_{\dot{K}^{\alpha(\cdot)}_{p(\cdot),q(\cdot)}(\mathbb{R}^n)}.$$

Definition 6.19 Let $\alpha \in L^\infty(\mathbb{R}^n)$, $q \in \mathcal{P}_0(\mathbb{R}^n)$, $p \in \mathcal{P}(\mathbb{R}^n)$, and $s \in \mathbb{N}_0$. A function a is said to be a central $(\alpha(\cdot), p(\cdot))$-atom, if

(i) $\operatorname{supp} a \subset \overline{B(0,r)} = \{x \in \mathbb{R}^n : |x| \leqslant r\}$, $r > 0$.
(ii) $\|a\|_{L^{p(\cdot)}(\mathbb{R}^n)} \leqslant |B(0,r)|^{-\alpha(0)/n}$, $0 < r < 1$.

(iii) $\|a\|_{L^{p(\cdot)}(\mathbb{R}^n)} \leqslant |B(0,r)|^{-\alpha_\infty/n}, r \geqslant 1$.
(iv) $\int_{\mathbb{R}^n} x^\beta a(x)dx = 0, |\beta| \leqslant s$.

A function a is said to be a central $(\alpha(\cdot), p(\cdot))$–atom of restricted type if it satisfies the conditions (iii), (iv) and supp $a \subset \overline{B(0,r)}, r \geqslant 1$.

Now we present the atomic decomposition theorems. For these statements, we refer to Wang and Liu [332] in the case of α and q constant and Drihem and Seghiri [80] in the variable case.

Proposition 6.20 *Let $p \in \mathcal{P}(\mathbb{R}^n)$, $q \in \mathcal{P}_0(\mathbb{R}^n)$, and $\alpha \in L^\infty(\mathbb{R}^n)$. Let α, $q \in \mathscr{M}_\infty(\mathbb{R}^n)$ and $p \in \mathbb{P}^{\log}_\infty(\mathbb{R}^n)$. If $f \in HK^{\alpha(\cdot)}_{p(\cdot),q(\cdot)}(\mathbb{R}^n)$, then*

$$f = \sum_{k=0}^{\infty} \lambda_k a_k, \tag{6.28}$$

where the series converges in the sense of distribution, $\lambda_k \geqslant 0$, each a_k is a central $(\alpha(\cdot), p(\cdot))$-atom of restricted type with supp $a_k \subset B_k$ *and*

$$\left(\sum_{k=0}^{\infty} |\lambda_k|^{q_\infty}\right)^{1/q_\infty} \lesssim \|f\|_{HK^{\alpha(\cdot)}_{p(\cdot),q(\cdot)}(\mathbb{R}^n)}.$$

Proposition 6.21 *Let $p \in \mathcal{P}(\mathbb{R}^n)$, $q \in \mathcal{P}_0(\mathbb{R}^n)$ and $\alpha : \mathbb{R}^n \to \mathbb{R}$ with $\alpha \in L^\infty(\mathbb{R}^n)$. Let α, $q \in \mathscr{M}_{0,\infty}(\mathbb{R}^n)$ and $p \in \mathbb{P}^{\log}_{0,\infty}(\mathbb{R}^n)$. If $f \in H\dot{K}^{\alpha(\cdot)}_{p(\cdot),q(\cdot)}(\mathbb{R}^n)$, then*

$$f = \sum_{k=-\infty}^{\infty} \lambda_k a_k, \tag{6.29}$$

where the series converges in the sense of distribution, $\lambda_k \geqslant 0$, each a_k is a central $(\alpha(\cdot), p(\cdot))$-atom with supp $a_k \subset B_k$ *and*

$$\left(\sum_{k=-\infty}^{-1} |\lambda_k|^{q(0)}\right)^{1/q(0)} + \left(\sum_{k=0}^{\infty} |\lambda_k|^{q_\infty}\right)^{1/q_\infty} \lesssim \|f\|_{H\dot{K}^{\alpha(\cdot)}_{p(\cdot),q(\cdot)}(\mathbb{R}^n)}. \tag{6.30}$$

6.3.2 Boundedness of Central Calderón–Zygmund Operators

We prove that the central Calderón–Zygmund operator is bounded from the space $H\dot{K}^{\alpha(\cdot)}_{p(\cdot),q(\cdot)}(\mathbb{R}^n)$ to $\dot{K}^{\alpha(\cdot)}_{p(\cdot),q(\cdot)}(\mathbb{R}^n)$. Similar conclusion holds for nonhomogeneous case. Our main result follows from a general result of a larger class of operators. We begin this section with the following definition.

Definition 6.22 Let $p \in \mathcal{P}(\mathbb{R}^n)$, and $\alpha : \mathbb{R}^n \to \mathbb{R}$ with $\alpha \in L^\infty(\mathbb{R}^n)$. Let $\alpha \in \mathscr{M}_{0,\infty}(\mathbb{R}^n)$ and $p \in \mathbb{P}^{\log}_{0,\infty}(\mathbb{R}^n)$. Let s be the nonnegative integer given by $s = \max\{\lfloor \alpha_\infty + n(1/p_\infty - 1)\rfloor, \lfloor \alpha(0) + n(1/p(0) - 1)\rfloor\}$. We say that a sublinear operator T belongs to the class $\mathcal{I}_{0,\infty}$ ($T \in \mathcal{I}_{0,\infty}$) if the following conditions hold:

(1) T is bounded in $L^{p(\cdot)}(\mathbb{R}^n)$.
(2) there exists a constant $\delta > 0$ such that $s + \delta > \max\{\alpha_\infty + n(1/p_\infty - 1), \alpha(0) + n(1/p(0) - 1)\}$, and for any compactly supported function f with

$$\int_{\mathbb{R}^n} f(x)x^\beta dx = 0, \qquad |\beta| \leqslant s,$$

Tf satisfies the size condition

$$|Tf(x)| \lesssim (\mathrm{diam}(\mathrm{supp}\ f))^{s+\delta}\, |x|^{-(n+s+\delta)} \int_{\mathbb{R}^n} |f(y)|dy, \tag{6.31}$$

if $\mathrm{dist}(x, \mathrm{supp}\ f) \geqslant |x|/2$.

Analogously, we can define the class $\mathcal{I}_\infty$, where all exponents satisfy the log-Hölder condition near infinity and replace the condition in s as above by $s = \lfloor \alpha_\infty + n(1/p_\infty - 1)\rfloor$. The next statements are the main results of this section.

Theorem 6.23 *Assume that p, q, and α are as in Proposition 6.20, $T \in \mathcal{I}_\infty$, and δ be as in Definition 6.22. If*

$$\frac{n}{p'_\infty} < \alpha_\infty < \frac{n}{p'_\infty} + \delta \tag{6.32}$$

holds, then T is bounded from $HK^{\alpha(\cdot)}_{p(\cdot),q(\cdot)}(\mathbb{R}^n)$ to $K^{\alpha(\cdot)}_{p(\cdot),q(\cdot)}(\mathbb{R}^n)$.

Theorem 6.24 *Assume that p, q, and α are as in Proposition 6.21, $T \in \mathcal{I}_{0,\infty}$, and δ be as in Definition 6.22. If*

$$\frac{n}{p'(0)} < \alpha(0) < \frac{n}{p'(0)} + \delta, \qquad \frac{n}{p'_\infty} < \alpha_\infty < \frac{n}{p'_\infty} + \delta \tag{6.33}$$

holds, then T is bounded from $H\dot{K}^{\alpha(\cdot)}_{p(\cdot),q(\cdot)}(\mathbb{R}^n)$ to $\dot{K}^{\alpha(\cdot)}_{p(\cdot),q(\cdot)}(\mathbb{R}^n)$.

The proof of Theorem 6.24 is given below, as Theorem 6.23 follows by similar reasoning.

Proof of Theorem 6.24 From Proposition 6.17, we have

$$\|T(f)\|_{\dot{K}^{\alpha(\cdot)}_{p(\cdot),q(\cdot)}} \lesssim \left(\sum_{k=-\infty}^{-1} 2^{k\alpha(0)q(0)} \|T(f)\chi_k\|^{q(0)}_{L^{p(\cdot)}} \right)^{\frac{1}{q(0)}} +$$

$$
+\left(\sum_{k=0}^{\infty} 2^{k\alpha_\infty q_\infty} \|T(f)\chi_k\|_{L^{p(\cdot)}}^{q_\infty}\right)^{\frac{1}{q_\infty}}
$$

$$
=: I_< + I_>.
$$

Let $f \in H\dot{K}^{\alpha(\cdot)}_{p(\cdot),q(\cdot)}(\mathbb{R}^n)$. By Proposition 6.21 the function f can be represented as $f = \sum_{j=-\infty}^{\infty} \lambda_j a_j$, where a_j is a central $(\alpha(\cdot), p(\cdot))$-atom with $\operatorname{supp} a_j \subset B_j$. Hence,

$$
\begin{aligned}
I_< &\lesssim \left(\sum_{k=-\infty}^{-1} 2^{k\alpha(0)q(0)} \left(\sum_{j=-\infty}^{k-2} |\lambda_j| \left\|T(a_j)\chi_k\right\|_{L^{p(\cdot)}}\right)^{q(0)}\right)^{\frac{1}{q(0)}} \\
&\quad + \left(\sum_{k=-\infty}^{-1} 2^{k\alpha(0)q(0)} \left(\sum_{j=k-1}^{\infty} |\lambda_j| \left\|T(a_j)\chi_k\right\|_{L^{p(\cdot)}}\right)^{q(0)}\right)^{\frac{1}{q(0)}} \\
&=: I_<^{(1)} + I_<^{(2)}
\end{aligned}
$$

and

$$
\begin{aligned}
I_> &\lesssim \left(\sum_{k=0}^{\infty} 2^{k\alpha_\infty q_\infty} \left(\sum_{j=-\infty}^{k-2} |\lambda_j| \left\|T(a_j)\chi_k\right\|_{L^{p(\cdot)}}\right)^{q_\infty}\right)^{\frac{1}{q_\infty}} \\
&\quad + \left(\sum_{k=0}^{\infty} 2^{k\alpha_\infty q_\infty} \left(\sum_{j=k-1}^{\infty} |\lambda_j| \left\|T(a_j)\chi_k\right\|_{L^{p(\cdot)}}\right)^{q_\infty}\right)^{\frac{1}{q_\infty}} \\
&=: I_>^{(1)} + I_>^{(2)}.
\end{aligned}
$$

Estimation of $I_<^{(2)}$: By the $(L^{p(\cdot)}, L^{p(\cdot)})$ boundedness of T and Minkowski's inequality, we have

$$
\begin{aligned}
I_<^{(2)} &\lesssim \left(\sum_{k=-\infty}^{-1} 2^{k\alpha(0)q(0)} \left(\sum_{j=k-1}^{\infty} |\lambda_j| \left\|a_j\right\|_{L^{p(\cdot)}}\right)^{q(0)}\right)^{\frac{1}{q(0)}} \\
&\lesssim \left(\sum_{k=-\infty}^{-1} 2^{k\alpha(0)q(0)} \left(\sum_{j=k-1}^{-1} |\lambda_j| \left\|a_j\right\|_{L^{p(\cdot)}} + \sum_{j=0}^{\infty} |\lambda_j| \left\|a_j\right\|_{L^{p(\cdot)}}\right)^{q(0)}\right)^{\frac{1}{q(0)}}
\end{aligned}
$$

$$\lesssim \left(\sum_{k=-\infty}^{-1} 2^{k\alpha(0)q(0)} \left(\sum_{j=k-1}^{-1} |\lambda_j| \|a_j\|_{L^{p(\cdot)}} \right)^{q(0)} \right)^{\frac{1}{q(0)}} +$$

$$+ \left(\sum_{k=-\infty}^{-1} 2^{k\alpha(0)q(0)} \left(\sum_{j=0}^{\infty} |\lambda_j| \|a_j\|_{L^{p(\cdot)}} \right)^{q(0)} \right)^{\frac{1}{q(0)}}$$

$$\lesssim \left(\sum_{k=-\infty}^{-1} \left(\sum_{j=k-1}^{-1} |\lambda_j| 2^{(k-j)\alpha(0)} \right)^{q(0)} \right)^{\frac{1}{q(0)}} +$$

$$+ \left(\sum_{k=-\infty}^{-1} 2^{k\alpha(0)q(0)} \left(\sum_{j=0}^{\infty} |\lambda_j| 2^{-j\alpha_\infty} \right)^{q(0)} \right)^{\frac{1}{q(0)}}$$

$$=: H_1 + H_2.$$

Now for H_1, by Hölder's inequality with $(q(0), q'(0))$ and changing the order of summation, we have

$$H_1 \lesssim \left(\sum_{k=-\infty}^{-1} \sum_{j=k-1}^{-1} |\lambda_j|^{q(0)} 2^{(k-j)\frac{\alpha(0)q(0)}{2}} \right)^{\frac{1}{q(0)}}$$

$$\lesssim \left(\sum_{j=-\infty}^{-1} |\lambda_j|^{q(0)} \sum_{k=-\infty}^{j+1} 2^{(k-j)\frac{\alpha(0)q(0)}{2}} \right)^{\frac{1}{q(0)}}$$

$$\lesssim \left(\sum_{j=-\infty}^{-1} |\lambda_j|^{q(0)} \right)^{\frac{1}{q(0)}}.$$

$$\lesssim \|f\|_{H\dot{K}^{\alpha(\cdot)}_{p(\cdot),q(\cdot)}(\mathbb{R}^n)}.$$

For H_2, firstly let $q_\infty > 1$ then by Hölder's inequality with exponents (q_∞, q'_∞) and Proposition 6.21 (the fact that $\alpha_\infty > 0$ and inequality (6.30)), we have

$$H_2 = \left(\sum_{k=-\infty}^{-1} 2^{k\alpha(0)q(0)} \right)^{\frac{1}{q(0)}} \left(\sum_{j=0}^{\infty} |\lambda_j| 2^{-j\alpha_\infty} \right)$$

$$\lesssim \left(\sum_{j=0}^{\infty}|\lambda_j|^{q_\infty}\right)^{1/q_\infty}\left(\sum_{j=0}^{\infty}2^{-j\alpha_\infty q'_\infty}\right)^{1/q'_\infty}$$
$$\lesssim \left(\sum_{j=0}^{\infty}|\lambda_j|^{q_\infty}\right)^{1/q_\infty}$$
$$\lesssim \|f\|_{H\dot{K}^{\alpha(\cdot)}_{p(\cdot),q(\cdot)}(\mathbb{R}^n)}.$$

Estimation of $I_>^{(2)}$*:* The estimation is similar to the one for H_1 with $q(0)$ replaced by q_∞, since $j \geqslant k-1>0$. Thus, we have

$$I_>^{(2)} \lesssim \|f\|_{H\dot{K}^{\alpha(\cdot)}_{p(\cdot),q(\cdot)}(\mathbb{R}^n)}.$$

Estimation of $I_>^{(1)}$*:* Firstly, note that since $x \in R_k$, $y \in B_j$ and $j \leqslant k-2$ and $|x-y| \geqslant 2^k/2 \geqslant |x|/2$. By the size condition (6.31) and Hölder's inequality, we have

$$|T(a_j)(x)| \lesssim \begin{cases} 2^{(j-k)(s+\delta)}2^{-kn}2^{\frac{nj}{p'_\infty}}2^{-j\alpha_\infty} & j\geqslant 0,\\ 2^{(j-k)(s+\delta)}2^{-kn}2^{\frac{nj}{p'(0)}}2^{-j\alpha(0)} & j<0. \end{cases} \tag{6.34}$$

Hence,

$$\left\|T(a_j)\chi_k\right\|_{L^{p(\cdot)}(\mathbb{R}^n)} \lesssim \begin{cases} 2^{\frac{-nk}{p'_\infty}}2^{(j-k)(s+\delta)}2^{\frac{nj}{p'_\infty}}2^{-j\alpha_\infty} & j\geqslant 0,\\ 2^{\frac{-nk}{p'_\infty}}2^{(j-k)(s+\delta)}2^{\frac{nj}{p'(0)}}2^{-j\alpha(0)} & j<0, \end{cases} \tag{6.35}$$

thus

$$\begin{aligned} I_>^{(1)} &\lesssim \left(\sum_{k=0}^{\infty}2^{k\alpha_\infty q_\infty}\left(\sum_{j=-\infty}^{k-2}|\lambda_j|\left\|T(a_j)\chi_k\right\|_{L^{p(\cdot)}}\right)^{q_\infty}\right)^{\frac{1}{q_\infty}}\\ &\lesssim \left(\sum_{k=0}^{\infty}2^{k\alpha_\infty q_\infty}\left(\sum_{j=-\infty}^{-1}|\lambda_j|\left\|T(a_j)\chi_k\right\|_{L^{p(\cdot)}}\right)^{q_\infty}\right)^{\frac{1}{q_\infty}}\\ &\qquad + \left(\sum_{k=0}^{\infty}2^{k\alpha_\infty q_\infty}\left(\sum_{j=0}^{k-2}|\lambda_j|\left\|T(a_j)\chi_k\right\|_{L^{p(\cdot)}}\right)^{q_\infty}\right)^{\frac{1}{q_\infty}}\\ &\leqslant J_1+J_2, \end{aligned}$$

where $J_2 = 0$ for $k = 0, 1, 2$. Let $q(0) > 1$. By using the estimate (6.35) for J_1 and Hölder's inequality with exponents $(q(0), q'(0))$ and the fact $s + \delta > \max\{\alpha_\infty + n(1/p_\infty - 1), \alpha(0) + n(1/p(0) - 1)\}$, we have

$$
\begin{aligned}
J_1 &\lesssim \left(\sum_{k=0}^{\infty} 2^{k\alpha_\infty q_\infty}\left(\sum_{j=-\infty}^{-1} |\lambda_j| 2^{\frac{-nk}{p'_\infty}} 2^{(j-k)(s+\delta)} 2^{\frac{nj}{p'(0)}} 2^{-j\alpha(0)}\right)^{q_\infty}\right)^{\frac{1}{q_\infty}} \\
&\lesssim \left(\sum_{k=0}^{\infty} 2^{-kq_\infty(\frac{n}{p'_\infty}+s+\delta-\alpha_\infty)}\right)^{\frac{1}{q_\infty}} \left(\sum_{j=-\infty}^{-1} |\lambda_j| 2^{j(s+\delta+\frac{n}{p'(0)}-\alpha(0))}\right) \\
&\lesssim \left(\sum_{j=-\infty}^{-1} |\lambda_j|^{q(0)}\right)^{\frac{1}{q(0)}} \left(\sum_{j=-\infty}^{-1} 2^{j(s+\delta+\frac{n}{p'(0)}-\alpha(0))q'(0)}\right)^{\frac{1}{q'(0)}} \\
&\lesssim \|f\|_{H\dot{K}^{\alpha(\cdot)}_{p(\cdot),q(\cdot)}(\mathbb{R}^n)}.
\end{aligned}
$$

Similarly for J_2 inserting the appropriate estimate from (6.35) it follows from Hölder's inequality with exponents (q_∞, q'_∞) and the condition $s + \delta > \max\{\alpha_\infty + n(1/p_\infty - 1), \alpha(0) + n(1/p(0) - 1)\}$, we have

$$
\begin{aligned}
J_2 &\lesssim \left(\sum_{k=0}^{\infty} 2^{k\alpha_\infty q_\infty}\left(\sum_{j=0}^{k-2} |\lambda_j| 2^{(j-k)(s+\delta)} 2^{\frac{-kn}{p'_\infty}} 2^{\frac{nj}{p'_\infty}} 2^{-j\alpha_\infty}\right)^{q_\infty}\right)^{\frac{1}{q_\infty}} \\
&\lesssim \left(\sum_{k=0}^{\infty}\left(\sum_{j=0}^{k-2} |\lambda_j| 2^{(j-k)(s+\delta+\frac{n}{p'_\infty}-\alpha_\infty)}\right)^{q_\infty}\right)^{\frac{1}{q_\infty}} \\
&\lesssim \left(\sum_{k=0}^{\infty}\left(\sum_{j=0}^{k-2} |\lambda_j|^{q_\infty} 2^{(j-k)(s+\delta+\frac{n}{p'_\infty}-\alpha_\infty)\frac{q_\infty}{2}}\right) \times\right. \\
&\qquad\qquad \left.\times \left(\sum_{j=0}^{k-2} 2^{(j-k)(s+\delta+\frac{n}{p'_\infty}-\alpha_\infty)\frac{q'_\infty}{2}}\right)^{\frac{q_\infty}{q'_\infty}}\right)^{\frac{1}{q_\infty}} \\
&\lesssim \left(\sum_{j=0}^{\infty} |\lambda_j|^{q_\infty}\left(\sum_{k=j+2}^{\infty} 2^{(j-k)(s+\delta+\frac{n}{p'_\infty}-\alpha_\infty)\frac{q_\infty}{2}}\right)\right)^{\frac{1}{q_\infty}}
\end{aligned}
$$

$$\lesssim \left(\sum_{j=0}^{\infty} |\lambda_j|^{q_\infty} \right)^{\frac{1}{q_\infty}}$$

$$\lesssim \|f\|_{H\dot{K}^{\alpha(\cdot)}_{p(\cdot),q(\cdot)}(\mathbb{R}^n)}.$$

Estimation of $I^{(1)}_{<}$: The estimation is similar to the one for J_2 with q_∞ replaced by $q(0)$ since $j \leqslant k-2 < 0$. The case where $0 < q < 1$ can be obtained in a similar way using the inequality $(\sum x_i)^q \leqslant (\sum x_i^q)$. □

Application

We apply our main result to obtain the boundedness of central Calderón–Zygmund operators which are more general than the standard Calderón–Zygmund operators.

Definition 6.25 Let K be a locally integrable function on $\mathbb{R}^n \times \mathbb{R}^n \backslash \{x = y\}$. Then K is called a central kernel, if there exist $\delta \in (0, 1]$ and $C > 0$, such that

$$|K(x, y) - K(x, 0)| \lesssim \frac{|y|^\delta}{|x|^{n+\delta}}, \qquad |x| \geqslant 2|y|, \tag{6.36}$$

and

$$|K(x, 0)| \lesssim \frac{1}{|x|^n}.$$

Definition 6.26 A linear operator $T_K : \mathscr{S}(\mathbb{R}^n) \to \mathscr{S}'(\mathbb{R}^n)$ is said to be a Calderón–Zygmund operator associated with a central kernel K if:

(i) T_K can be extended to a bounded operator on $L^2(\mathbb{R}^n)$.
(ii) For any $f \in L^2(\mathbb{R}^n)$ with compact support and almost every $x \notin \operatorname{supp} f$,

$$T_K f(x) = \int_{\mathbb{R}^n} K(x, y) f(y) dy. \tag{6.37}$$

Now we present the boundedness of central Calderón–Zygmund operators. We note that the case for $q(\cdot) = const.$ was proved in Xu and Yang [338].

Theorem 6.27 *Assume that p, q, and α are as in Theorem 6.23. Let T_K be the Calderón–Zygmund operator bounded in $L^{p(\cdot)}(\mathbb{R}^n)$ and associated with the central kernel K in the sense (6.37). Then T_K is bounded from $HK^{\alpha(\cdot)}_{p(\cdot),q(\cdot)}(\mathbb{R}^n)$ to $K^{\alpha(\cdot)}_{p(\cdot),q(\cdot)}(\mathbb{R}^n)$.*

Theorem 6.28 *Assume that p, q, and α are as in Theorem 6.24. Let T_K be the Calderón–Zygmund operator bounded in $L^{p(\cdot)}(\mathbb{R}^n)$ and associated with the central kernel K in the sense* (6.37). *Then T_K is bounded from $H\dot{K}^{\alpha(\cdot)}_{p(\cdot),q(\cdot)}(\mathbb{R}^n)$ to $\dot{K}^{\alpha(\cdot)}_{p(\cdot),q(\cdot)}(\mathbb{R}^n)$.*

Proof We restrict ourselves to proving Theorem 6.27, since the proof of Theorem 6.28 proceeds in a similar manner. Firstly, the condition $n/p'_\infty \leqslant \alpha_\infty < n/p'_\infty + \delta$ implies that $s = \lfloor \alpha_\infty - n/p'_\infty \rfloor = 0$. Now, letting $f \in H\dot{K}^{\alpha(\cdot)}_{p(\cdot),q(\cdot)}(\mathbb{R}^n)$, then by Proposition 6.20, we have $f(x) = \sum_{k=0}^{\infty} \lambda_k a_k(x)$ where a_k is the central $(\alpha(\cdot), p(\cdot))$-atom. By the sublinearity of T_K, we have $|T_K f(x)| \leqslant \sum_{k=0}^{\infty} |\lambda_k||T_K(a_k)(x)|$. Following the proof of Theorem 6.23, we have that, for each $x \in A_k$ and supp $(f) \subset B_j$ $(j \leqslant k-2)$ together with the condition of vanishing moments (i.e., $\int\limits_{\mathbb{R}^n} a_k = 0$) and condition (6.36), the operator T_K satisfies condition (6.31) with $s = 0$. Hence, the result follows from the same arguments as presented in Theorem 6.23. □

6.4 On the Boundedness of Sublinear Operators on Grand Variable Herz Spaces

We introduce the grand variable Herz spaces $\dot{K}^{\alpha,p),\theta}_q(\mathbb{R}^n)$ and obtain the boundedness of sublinear operators on $\dot{K}^{\alpha,p),\theta}_q(\mathbb{R}^n)$.

6.4.1 Grand Lebesgue Sequence Space

We recall the notion of grand Lebesgue sequence space. For the following definitions and statements, see Kokilashvili et al. [190], Rafeiro et al. [281]. The letter $\mathbb{X}$ represents one of the sets $\mathbb{Z}^n$, $\mathbb{Z}$, $\mathbb{N}$ and $\mathbb{N}_0$.

Definition 6.29 Let $1 \leqslant p < \infty$ and $\theta > 0$. The grand Lebesgue sequence space $l^{p),\theta}$ is defined by the norm

$$\|\{x_k\}_{k\in\mathbb{X}}\|_{l^{p),\theta}(\mathbb{X})} = \|\boldsymbol{x}\|_{l^{p),\theta}(\mathbb{X})}$$

$$:= \sup_{\varepsilon>0} \left(\varepsilon^\theta \sum_{k\in\mathbb{X}} |x_k|^{p(1+\varepsilon)}\right)^{\frac{1}{p(1+\varepsilon)}} = \sup_{\varepsilon>0} \varepsilon^{\frac{\theta}{p(1+\varepsilon)}} \|x\|_{l^{p(1+\varepsilon)}(\mathbb{X})},$$

where $\boldsymbol{x} = \{x_k\}_{k\in\mathbb{X}}$.

Note that the following nesting properties hold:

$$l^{p(1-\varepsilon)} \hookrightarrow l^{p} \hookrightarrow l^{p),\theta_1} \hookrightarrow l^{p),\theta_2} \hookrightarrow l^{p(1+\delta)} \tag{6.38}$$

for $0 < \varepsilon < \dfrac{1}{p}$, $\delta > 0$ and $0 < \theta_1 \leqslant \theta_2$.

6.4.2 *Grand Variable Herz Space*

We introduce grand variable Herz space in a natural way using the discrete space of Definition 6.29.

Definition 6.30 Let $\alpha \in \mathbb{R}$, $1 \leqslant p < \infty$, $q \in \mathcal{P}(\mathbb{R}^n)$, and $\theta > 0$. We define the grand variable Herz space by

$$\dot{K}^{\alpha,p),\theta}_{q(\cdot)}(\mathbb{R}^n) = \{f \in L^{q(\cdot)}_{\mathrm{loc}}(\mathbb{R}^n \setminus \{0\}) : \|f\|_{\dot{K}^{\alpha,p),\theta}_{q(\cdot)}(\mathbb{R}^n)} < \infty\}$$

where

$$\begin{aligned}\|f\|_{\dot{K}^{\alpha,p),\theta}_{q(\cdot)}(\mathbb{R}^n)} &= \sup_{\varepsilon>0}\left(\varepsilon^{\theta}\sum_{k\in\mathbb{Z}} 2^{k\alpha p(1+\varepsilon)}\|f\chi_k\|^{p(1+\varepsilon)}_{L^{q(\cdot)}}\right)^{\frac{1}{p(1+\varepsilon)}} \\ &= \sup_{\varepsilon>0}\varepsilon^{\frac{1}{p(1+\varepsilon)}}\|f\|_{\dot{K}^{\alpha,p(1+\varepsilon)}_{q(\cdot)}(\mathbb{R}^n)}.\end{aligned}$$

In the following theorem, we prove that the Herz space is contained in the grand variable Herz space.

Theorem 6.31 *For $p > 1$, we have $\dot{K}^{\alpha,p}_{q(\cdot)}(\mathbb{R}^n) \subset \dot{K}^{\alpha,p),\theta}_{q(\cdot)}(\mathbb{R}^n)$, $\theta > 0$.*

Proof Let $f \in \dot{K}^{\alpha,p}_{q(\cdot)}(\mathbb{R}^n)$. Then

$$\begin{aligned}\|f\|_{\dot{K}^{\alpha,p),\theta}_{q(\cdot)}(\mathbb{R}^n)} &= \sup_{\varepsilon>0}\left(\varepsilon^{\theta}\sum_{k\in\mathbb{Z}} 2^{k\alpha p(1+\varepsilon)}\|f\chi_k\|^{p(1+\varepsilon)}_{L^{q(\cdot)}}\right)^{\frac{1}{p(1+\varepsilon)}} \\ &= \left\|2^{k\alpha}\|f\chi_k\|_{L^{q(\cdot)}}\right\|_{l^{p),\theta}} \\ &\lesssim \left\|2^{k\alpha}\|f\chi_k\|_{L^{q(\cdot)}}\right\|_{l^{p}} = C\left(\sum_{k\in\mathbb{Z}} 2^{k\alpha p}\|f\chi_k\|^{p}_{L^{q(\cdot)}}\right)^{\frac{1}{p}} \\ &= C\|f\|_{\dot{K}^{\alpha,p}_{q(\cdot)}(\mathbb{R}^n)}\end{aligned}$$

where we used (6.38). □

Now we prove the Hölder inequality for grand variable Herz space.

Theorem 6.32 *If* $0 < p_i \leqslant \infty$, $1 \leqslant q_- \leqslant q_+ < \infty$, $-\infty < \alpha_i < \infty$, $i = 1, 2$, $\frac{1}{p} = \frac{1}{p_1} + \frac{1}{p_2}$, $1 = \frac{1}{q_-} + \frac{1}{(q_+)'}$ *and* $\alpha = \alpha_1 + \alpha_2$, *then*

$$\|fg\|_{\dot{K}_1^{\alpha,p),\theta}(\mathbb{R}^n)} \leqslant \|f\|_{\dot{K}_{q(\cdot)}^{\alpha_1,p_1),\theta}(\mathbb{R}^n)} \|g\|_{\dot{K}_{q'(\cdot)}^{\alpha_2,p_2),\theta}(\mathbb{R}^n)}.$$

Proof We have

$$\begin{aligned}
\|fg\|_{\dot{K}_1^{\alpha,p),\theta}(\mathbb{R}^n)} &= \sup_{\varepsilon>0} \left(\varepsilon^\theta \sum_{k\in\mathbb{Z}} 2^{k\alpha p(1+\varepsilon)} \|fg\chi_k\|_{L^1(\mathbb{R}^n)}^{p(1+\varepsilon)} \right)^{\frac{1}{p(1+\varepsilon)}} \\
&= \sup_{\varepsilon>0} \left(\varepsilon^\theta \sum_{k\in\mathbb{Z}} 2^{k\alpha p(1+\varepsilon)} \left(\int_{2^k}^{2^{k+1}} |fg| \right)^{p(1+\varepsilon)} \right)^{\frac{1}{p(1+\varepsilon)}}.
\end{aligned}$$

By Hölder's inequality

$$\begin{aligned}
&\leqslant \sup_{\varepsilon>0} \left(\varepsilon^\theta \sum_{k\in\mathbb{Z}} 2^{k(\alpha_1+\alpha_2)p(1+\varepsilon)} \left(\|f\chi_k\|_{L^{q(\cdot)}}^{p(1+\varepsilon)} \right) \left(\|g\chi_k\|_{L^{q'(\cdot)}}^{p(1+\varepsilon)} \right) \right)^{\frac{1}{p(1+\varepsilon)}} \\
&= \sup_{\varepsilon>0} \left(\varepsilon^\theta \sum_{k\in\mathbb{Z}} \left(2^{k\alpha_1} \|f\chi_k\|_{L^{q(\cdot)}} \right)^{p(1+\varepsilon)} \left(2^{k\alpha_2} \|g\chi_k\|_{L^{q'(\cdot)}} \right)^{p(1+\varepsilon)} \right)^{\frac{1}{p(1+\varepsilon)}} \\
&\leqslant \sup_{\varepsilon>0} \left(\varepsilon^\theta \left(\sum_{k\in\mathbb{Z}} \left(2^{k\alpha_1} \|f\chi_k\|_{L^{q(\cdot)}} \right)^{p_1(1+\varepsilon)} \right)^{1/p_1(1+\varepsilon)} \times \right. \\
&\qquad \left. \times \sup_{\varepsilon>0} \left(\varepsilon^\theta \sum_{k\in\mathbb{Z}} \left(2^{k\alpha_2} \left(\|g\chi_k\|_{L^{q'(\cdot)}} \right) \right)^{p_2(1+\varepsilon)} \right)^{1/p_2(1+\varepsilon)} \right) \\
&= \|f\|_{\dot{K}_{q(\cdot)}^{\alpha_1,p_1),\theta}(\mathbb{R}^n)} \|g\|_{\dot{K}_{q'(\cdot)}^{\alpha_2,p_2),\theta}(\mathbb{R}^n)}.
\end{aligned}$$

□

6.4.3 Boundedness of Sublinear Operators

We show that sublinear operators are bounded on $\dot{K}^{\alpha,p),\theta}_{q(\cdot)}(\mathbb{R}^n)$. Hernández and Yang [138], Li and Yang [210], Lu and Yang [216] have proved that if a sublinear operator T is bounded on $L^p(\mathbb{R}^n)$ and satisfies the size condition

$$|Tf(x)| \lesssim \int\limits_{\mathbb{R}^n} |x-y|^{-n}|f(y)|\,dy, \quad x \notin \operatorname{supp} f \tag{6.39}$$

for all $f \in L^1(\mathbb{R}^n)$ with compact support, then T is bounded on the homogeneous Herz space $\dot{K}^{\alpha,p}_q$ and on the nonhomogeneous Herz space $K^{\alpha,p}_q$. The condition (6.39) is satisfied by many interesting operators in harmonic analysis, such as Calderón–Zygmund operators, Carleson's maximal operator, Hardy–Littlewood maximal operator, and the Bochner–Riesz means (for details, see Lu et al. [215]).

Theorem 6.33 *Let $1 < p < \infty$, $q \in \mathbb{P}_{0,\infty}(\mathbb{R}^n)$ and α such that $-n/q(0) < \alpha < n/q'(0)$ and $-n/q_\infty < \alpha < n/q'_\infty$. Suppose that T is a sublinear operator and bounded on $L^{q(\cdot)}(\mathbb{R}^n)$ satisfying the size condition* (6.39). *Then T is bounded on $\dot{K}^{\alpha,p),\theta}_{q(\cdot)}(\mathbb{R}^n)$.*

Proof Since T is sublinear, for every $f \in \dot{K}^{\alpha,p),\theta}_{q(\cdot)}(\mathbb{R}^n)$ we have

$$\begin{aligned}
\|Tf\|_{\dot{K}^{\alpha,p),\theta}_{q(\cdot)}(\mathbb{R}^n)} &= \sup_{\varepsilon>0}\left(\varepsilon^\theta \sum_{k\in\mathbb{Z}} 2^{k\alpha p(1+\varepsilon)} \|\chi_k Tf\|^{p(1+\varepsilon)}_{L^{q(\cdot)}(\mathbb{R}^n)}\right)^{\frac{1}{p(1+\varepsilon)}} \\
&\leqslant \sup_{\varepsilon>0}\left(\varepsilon^\theta \sum_{k\in\mathbb{Z}} 2^{k\alpha p(1+\varepsilon)} \left(\sum_{l=-\infty}^{\infty} \|\chi_k T(f\chi_l)\|^{p(1+\varepsilon)}_{L^{q(\cdot)}(\mathbb{R}^n)}\right)\right)^{\frac{1}{p(1+\varepsilon)}} \\
&\lesssim \sup_{\varepsilon>0}\left(\varepsilon^\theta \sum_{k\in\mathbb{Z}} 2^{k\alpha p(1+\varepsilon)} \left(\sum_{l=-\infty}^{k-2} \|\chi_k T(f\chi_l)\|_{L^{q(\cdot)}(\mathbb{R}^n)}\right)^{p(1+\varepsilon)}\right)^{\frac{1}{p(1+\varepsilon)}} \\
&\quad + \sup_{\varepsilon>0}\left(\varepsilon^\theta \sum_{k\in\mathbb{Z}} 2^{k\alpha p(1+\varepsilon)} \left(\sum_{l=k-1}^{k+1} \|\chi_k T(f\chi_l)\|_{L^{q(\cdot)}(\mathbb{R}^n)}\right)^{p(1+\varepsilon)}\right)^{\frac{1}{p(1+\varepsilon)}} \\
&\quad + \sup_{\varepsilon>0}\left(\varepsilon^\theta \sum_{k\in\mathbb{Z}} 2^{k\alpha p(1+\varepsilon)} \left(\sum_{l=k+2}^{\infty} \|\chi_k T(f\chi_l)\|_{L^{q(\cdot)}(\mathbb{R}^n)}\right)^{p(1+\varepsilon)}\right)^{\frac{1}{p(1+\varepsilon)}} \\
&=: E_1 + E_2 + E_3.
\end{aligned}$$

For E_2, using the $L^{q(\cdot)}(\mathbb{R}^n)$ boundedness of T, we obtain

$$\begin{aligned}
E_2 &\lesssim \sup_{\varepsilon>0}\left(\varepsilon^\theta\sum_{k\in\mathbb{Z}}2^{k\alpha p(1+\varepsilon)}\left(\sum_{l=k-1}^{k+1}\|T(f\chi_l)\|_{L^{q(\cdot)}(\mathbb{R}^n)}\right)^{p(1+\varepsilon)}\right)^{\frac{1}{p(1+\varepsilon)}}\\
&\lesssim \sup_{\varepsilon>0}\left(\varepsilon^\theta\sum_{k\in\mathbb{Z}}2^{k\alpha p(1+\varepsilon)}\left(\sum_{l=k-1}^{k+1}\|f\chi_l\|_{L^{q(\cdot)}(\mathbb{R}^n)}\right)^{p(1+\varepsilon)}\right)^{\frac{1}{p(1+\varepsilon)}}\\
&\lesssim \sup_{\varepsilon>0}\left(\varepsilon^\theta\sum_{k\in\mathbb{Z}}2^{k\alpha p(1+\varepsilon)}\|f\chi_k\|^{p(1+\varepsilon)}_{L^{q(\cdot)}(\mathbb{R}^n)}\right)^{\frac{1}{p(1+\varepsilon)}}\\
&= \|f\|_{\dot{K}^{\alpha,p),\theta}_{q(\cdot)}(\mathbb{R}^n)}.
\end{aligned}$$

For E_1, we use the facts that, for each $k\in\mathbb{Z}$ and $l\leqslant k-2$ and a.e. $x\in R_k$; size condition (6.39) and Hölder's inequality imply

$$\begin{aligned}
|T(f\chi_l)(x)| &\lesssim \int_{R_l}|x-y|^{-n}|f(y)|dy\\
&\lesssim 2^{-kn}\int_{R_l}|f(y)|dy \lesssim 2^{-kn}\|f\chi_l\|_{L^{q(\cdot)}(\mathbb{R}^n)}\|\chi_l\|_{L^{q'(\cdot)}(\mathbb{R}^n)}.
\end{aligned}\tag{6.40}$$

Moreover, splitting E_1 by means of Minkowski's inequality, we have

$$\begin{aligned}
E_1 &\lesssim \sup_{\varepsilon>0}\left(\varepsilon^\theta\sum_{k=-\infty}^{-1}2^{k\alpha p(1+\varepsilon)}\left(\sum_{l=-\infty}^{k-2}\|\chi_kT(f\chi_l)\|_{L^{q(\cdot)}(\mathbb{R}^n)}\right)^{p(1+\varepsilon)}\right)^{\frac{1}{p(1+\varepsilon)}}\\
&\quad+\sup_{\varepsilon>0}\left(\varepsilon^\theta\sum_{k=0}^{\infty}2^{k\alpha p(1+\varepsilon)}\left(\sum_{l=-\infty}^{k-2}\|\chi_kT(f\chi_l)\|_{L^{q(\cdot)}(\mathbb{R}^n)}\right)^{p(1+\varepsilon)}\right)^{\frac{1}{p(1+\varepsilon)}}\\
&:= E_{11}+E_{12}.
\end{aligned}$$

For E_{11} by Lemma 6.3, we have

$$2^{-kn}\|\chi_k\|_{L^{q(\cdot)}(\mathbb{R}^n)}\|\chi_l\|_{L^{q'(\cdot)}(\mathbb{R}^n)} \lesssim 2^{-kn}2^{\left(\frac{kn}{q(0)}\right)}2^{\left(\frac{ln}{q'(0)}\right)} \lesssim 2^{\frac{(l-k)n}{q'(0)}}.\tag{6.41}$$

Applying (6.40) and (6.41) to E_1, we get

$$E_{11} \lesssim \sup_{\varepsilon>0}\left(\varepsilon^{\theta}\sum_{k=-\infty}^{-1}2^{k\alpha p(1+\varepsilon)}\left(\sum_{l=-\infty}^{k-2}2^{-kn}\|\chi_k\|_{L^{q(\cdot)}(\mathbb{R}^n)}\times\right.\right.$$
$$\left.\left.\times\|f\chi_l\|_{L^{q(\cdot)}(\mathbb{R}^n)}\|\chi_l\|_{L^{q'(\cdot)}(\mathbb{R}^n)}\right)^{p(1+\varepsilon)}\right)^{\frac{1}{p(1+\varepsilon)}}$$
$$\lesssim \sup_{\varepsilon>0}\left(\varepsilon^{\theta}\sum_{k=-\infty}^{-1}\left(\sum_{l=-\infty}^{k-2}2^{\alpha l}\|f\chi_l\|_{L^{q(\cdot)}(\mathbb{R}^n)}2^{b(l-k)}\right)^{p(1+\varepsilon)}\right)^{\frac{1}{p(1+\varepsilon)}}$$

where $b := \frac{n}{q'(0)} - \alpha > 0$. Then we use Hölder's inequality, Fubini's theorem for series, and $2^{-p(1+\varepsilon)} < 2^{-p}$ to obtain

$$\lesssim \sup_{\varepsilon>0}\left(\varepsilon^{\theta}\sum_{k=-\infty}^{-1}\left(\sum_{l=-\infty}^{k-2}2^{\alpha p(1+\varepsilon)l}\|f\chi_l\|^{p(1+\varepsilon)}_{L^{q(\cdot)}(\mathbb{R}^n)}2^{bp(1+\varepsilon)(l-k)/2}\right)\times\right.$$
$$\left.\times\left(\sum_{l=-\infty}^{k-2}2^{b(p(1+\varepsilon))'(l-k)/2}\right)^{p(1+\varepsilon)/(p(1+\varepsilon))'}\right)^{\frac{1}{p(1+\varepsilon)}}$$
$$= \sup_{\varepsilon>0}\left(\varepsilon^{\theta}\sum_{k=-\infty}^{-1}\sum_{l=-\infty}^{k-2}2^{\alpha p(1+\varepsilon)l}\|f\chi_l\|^{p(1+\varepsilon)}_{L^{q(\cdot)}(\mathbb{R}^n)}2^{bp(1+\varepsilon)(l-k)/2}\right)^{\frac{1}{p(1+\varepsilon)}}$$
$$= \sup_{\varepsilon>0}\left(\varepsilon^{\theta}\sum_{l=-\infty}^{-1}2^{\alpha p(1+\varepsilon)l}\|f\chi_l\|^{p(1+\varepsilon)}_{L^{q(\cdot)}(\mathbb{R}^n)}\sum_{k=l+2}^{-1}2^{bp(1+\varepsilon)(l-k)/2}\right)^{\frac{1}{p(1+\varepsilon)}}$$
$$\lesssim \sup_{\varepsilon>0}\left(\varepsilon^{\theta}\sum_{l=-\infty}^{-1}2^{\alpha p(1+\varepsilon)l}\|f\chi_l\|^{p(1+\varepsilon)}_{L^{q(\cdot)}(\mathbb{R}^n)}\sum_{k=l+2}^{-1}2^{bp(l-k)/2}\right)^{\frac{1}{p(1+\varepsilon)}}$$
$$= \sup_{\varepsilon>0}\left(\varepsilon^{\theta}\sum_{l\in\mathbb{Z}}2^{\alpha p(1+\varepsilon)l}\|f\chi_l\|^{p(1+\varepsilon)}_{L^{q(\cdot)}(\mathbb{R}^n)}\right)^{\frac{1}{p(1+\varepsilon)}}$$
$$\lesssim \|f\|_{\dot{K}^{\alpha,p),\theta}_{q(\cdot)}(\mathbb{R}^n)}.$$

Now for E_{12} using Minkowski's inequality, we have

$$\begin{aligned}E_{12} &\lesssim \sup_{\varepsilon>0}\left(\varepsilon^{\theta}\sum_{k=0}^{\infty}2^{k\alpha p(1+\varepsilon)}\left(\sum_{l=-\infty}^{-1}\|\chi_k T(f\chi_l)\|_{L^{q(\cdot)}(\mathbb{R}^n)}\right)^{p(1+\varepsilon)}\right)^{\frac{1}{p(1+\varepsilon)}}\\&+\sup_{\varepsilon>0}\left(\varepsilon^{\theta}\sum_{k=0}^{\infty}2^{k\alpha p(1+\varepsilon)}\left(\sum_{l=0}^{k-2}\|\chi_k T(f\chi_l)\|_{L^{q(\cdot)}(\mathbb{R}^n)}\right)^{p(1+\varepsilon)}\right)^{\frac{1}{p(1+\varepsilon)}}\\&:=A_1+A_2\end{aligned}$$

The estimate for A_2 follows in a similar way to E_{11} with $q'(0)$ replaced by q'_∞ and using the fact that $\frac{n}{q'_\infty}-\alpha>0$. For A_1 using Lemma 6.3, we have

$$2^{-kn}\|\chi_k\|_{L^{q(\cdot)}(\mathbb{R}^n)}\|\chi_l\|_{L^{q'(\cdot)}(\mathbb{R}^n)}\lesssim 2^{-kn}2^{\left(\frac{kn}{q_\infty}\right)}2^{\left(\frac{ln}{q'_\infty}\right)}\lesssim 2^{\left(\frac{-kn}{q'_\infty}\right)}2^{\left(\frac{ln}{q(0)}\right)}. \tag{6.42}$$

Now using (6.40) and (6.42) and the fact that $\alpha-\frac{n}{q'_\infty}<0$, we have

$$\begin{aligned}A_1 &\leqslant \sup_{\varepsilon>0}\left(\varepsilon^{\theta}\sum_{k=0}^{\infty}2^{k\alpha p(1+\varepsilon)}\left(\sum_{l=-\infty}^{-1}\|\chi_k Tf(\chi_l)\|_{L^{q(\cdot)}(\mathbb{R}^n)}\right)^{p(1+\varepsilon)}\right)^{\frac{1}{p(1+\varepsilon)}}\\&\lesssim \sup_{\varepsilon>0}\left(\varepsilon^{\theta}\sum_{k=0}^{\infty}2^{k\alpha p(1+\varepsilon)}\left(\sum_{l=-\infty}^{-1}2^{-kn}2^{(kn/q_\infty)}2^{(ln/q'(0))}\times\right.\right.\\&\qquad\left.\left.\times\|f(\chi_l)\|_{L^{q(\cdot)}(\mathbb{R}^n)}\right)^{p(1+\varepsilon)}\right)^{\frac{1}{p(1+\varepsilon)}}\\&\lesssim \sup_{\varepsilon>0}\left(\varepsilon^{\theta}\sum_{k=0}^{\infty}2^{k\alpha p(1+\varepsilon)}\left(\sum_{l=-\infty}^{-1}2^{(-kn/q'_\infty)}2^{(ln/q'(0))}\times\right.\right.\\&\qquad\left.\left.\times\|f(\chi_l)\|_{L^{q(\cdot)}(\mathbb{R}^n)}\right)^{p(1+\varepsilon)}\right)^{\frac{1}{p(1+\varepsilon)}}\\&\lesssim \sup_{\varepsilon>0}\left(\varepsilon^{\theta}\sum_{k=0}^{\infty}2^{(k\alpha-kn/q'_\infty)p(1+\varepsilon)}\left(\sum_{l=-\infty}^{-1}2^{(ln/q'(0))}\times\right.\right.\\&\qquad\left.\left.\times\|f(\chi_l)\|_{L^{q(\cdot)}(\mathbb{R}^n)}\right)^{p(1+\varepsilon)}\right)^{\frac{1}{p(1+\varepsilon)}}\end{aligned}$$

$$\lesssim \sup_{\varepsilon>0}\left(\varepsilon^{\theta}\left(\sum_{l=-\infty}^{-1} 2^{(ln/q'(0))}\|f(\chi_l)\|_{L^{q(\cdot)}(\mathbb{R}^n)}\right)^{p(1+\varepsilon)}\right)^{\frac{1}{p(1+\varepsilon)}}$$

$$\lesssim \sup_{\varepsilon>0}\left(\varepsilon^{\theta}\left(\sum_{l=-\infty}^{-1} 2^{l\alpha}\|f(\chi_l)\|_{L^{q(\cdot)}(\mathbb{R}^n)}2^{(ln/q'(0)-l\alpha)}\right)^{p(1+\varepsilon)}\right)^{\frac{1}{p(1+\varepsilon)}}.$$

Now applying Hölder's inequality and using the inequality $\frac{n}{q'(0)} - \alpha > 0$, we have

$$A_1 \lesssim \sup_{\varepsilon>0}\left(\varepsilon^{\theta}\left(\sum_{l=-\infty}^{-1} 2^{l\alpha p(1+\varepsilon)}\|f(\chi_l)\|_{L^{q(\cdot)}(\mathbb{R}^n)}\right)\times\right.$$
$$\left.\times\left(\sum_{l=-\infty}^{-1} 2^{l(n/q'(0)-\alpha)}\right)^{p(1+\varepsilon)/p(1+\varepsilon)'}\right)^{\frac{1}{p(1+\varepsilon)}}$$
$$\lesssim \sup_{\varepsilon>0}\left(\varepsilon^{\theta}\left(\sum_{l\in\mathbb{Z}} 2^{l\alpha p(1+\varepsilon)}\|f(\chi_l)\|_{L^{q(\cdot)}(\mathbb{R}^n)}\right)\right)^{\frac{1}{p(1+\varepsilon)}}$$
$$\lesssim \|f\|_{\dot{K}^{\alpha,p),\theta}_{q(\cdot)}(\mathbb{R}^n)}.$$

Next, we estimate E_3. For each $k \in \mathbb{Z}$ and $l \geqslant k+2$ and a.e. $x \in R_k$; size condition (6.39) and Hölder's inequality imply

$$\begin{aligned}|T(f\chi_l)(x)| &\lesssim \int_{R_l} |x-y|^{-n}|f(y)|dy\\ &\lesssim 2^{-ln}\int_{R_l}|f(y)|dy \lesssim 2^{-ln}\|f\chi_l\|_{L^{q(\cdot)}(\mathbb{R}^n)}\|\chi_l\|_{L^{q'(\cdot)}(\mathbb{R}^n)}.\end{aligned} \tag{6.43}$$

Similar to E_1, splitting E_3 by means of Minkowski's inequality, we have

$$E_3 \lesssim \sup_{\varepsilon>0}\left(\varepsilon^{\theta}\sum_{k=-\infty}^{-1} 2^{k\alpha p(1+\varepsilon)}\left(\sum_{l=k+2}^{\infty}\|\chi_k T(f\chi_l)\|_{L^{q(\cdot)}(\mathbb{R}^n)}\right)^{p(1+\varepsilon)}\right)^{\frac{1}{p(1+\varepsilon)}}$$
$$+\sup_{\varepsilon>0}\left(\varepsilon^{\theta}\sum_{k=0}^{\infty} 2^{k\alpha p(1+\varepsilon)}\left(\sum_{l=k+2}^{\infty}\|\chi_k T(f\chi_l)\|_{L^{q(\cdot)}(\mathbb{R}^n)}\right)^{p(1+\varepsilon)}\right)^{\frac{1}{p(1+\varepsilon)}}$$
$$:= E_{31} + E_{32}.$$

For E_{32}, Lemma 6.3 yields

$$2^{-ln}\|\chi_k\|_{L^{q(\cdot)}(\mathbb{R}^n)}\|\chi_l\|_{L^{q'(\cdot)}(\mathbb{R}^n)} \lesssim 2^{-ln}2^{\left(\frac{kn}{q_\infty}\right)}2^{\left(\frac{ln}{q'_\infty}\right)} \lesssim 2^{\frac{(k-l)n}{q_\infty}}. \tag{6.44}$$

Using (6.43) and (6.44) for E_{32}, we get

$$E_{32} \lesssim \sup_{\varepsilon>0}\left(\varepsilon^\theta \sum_{k=0}^{\infty} 2^{k\alpha p(1+\varepsilon)}\left(\sum_{l=k+2}^{\infty} 2^{-ln}\|\chi_k\|_{L^{q(\cdot)}(\mathbb{R}^n)} \times \right.\right. \tag{6.45}$$

$$\left.\left.\times \|f\chi_l\|_{L^{q(\cdot)}(\mathbb{R}^n)}\|\chi_l\|_{L^{q'(\cdot)}(\mathbb{R}^n)}\right)^{p(1+\varepsilon)}\right)^{\frac{1}{p(1+\varepsilon)}} \tag{6.46}$$

$$\lesssim \sup_{\varepsilon>0}\left(\varepsilon^\theta \sum_{k=0}^{\infty}\left(\sum_{l=k+2}^{\infty} 2^{\alpha l}\|f\chi_l\|_{L^{q(\cdot)}(\mathbb{R}^n)}2^{d(k-l)}\right)^{p(1+\varepsilon)}\right)^{\frac{1}{p(1+\varepsilon)}}$$

where $d := \frac{n}{q_\infty} + \alpha > 0$. Then we use Hölder's inequality, Fubini's theorem for series, and inequality $2^{-p(1+\varepsilon)} < 2^{-p}$ to obtain

$$\lesssim \sup_{\varepsilon>0}\left(\varepsilon^\theta \sum_{k=0}^{\infty}\left(\sum_{l=k+2}^{\infty} 2^{\alpha p(1+\varepsilon)l}\|f\chi_l\|^{p(1+\varepsilon)}_{L^{q(\cdot)}(\mathbb{R}^n)}2^{dp(1+\varepsilon)(k-l)/2}\right) \times \right.$$

$$\left.\times\left(\sum_{l=k+2}^{\infty} 2^{d(p(1+\varepsilon))'(k-l)/2}\right)^{p(1+\varepsilon)/(p(1+\varepsilon))'}\right)^{\frac{1}{p(1+\varepsilon)}}$$

$$= \sup_{\varepsilon>0}\left(\varepsilon^\theta \sum_{k=0}^{\infty}\sum_{l=k+2}^{\infty} 2^{\alpha p(1+\varepsilon)l}\|f\chi_l\|^{p(1+\varepsilon)}_{L^{q(\cdot)}(\mathbb{R}^n)}2^{dp(1+\varepsilon)(k-l)/2}\right)^{\frac{1}{p(1+\varepsilon)}}$$

$$= \sup_{\varepsilon>0}\left(\varepsilon^\theta \sum_{l=0}^{\infty} 2^{\alpha p(1+\varepsilon)l}\|f\chi_l\|^{p(1+\varepsilon)}_{L^{q(\cdot)}(\mathbb{R}^n)}\sum_{k=0}^{l-2}2^{dp(1+\varepsilon)(k-l)/2}\right)^{\frac{1}{p(1+\varepsilon)}}$$

$$\lesssim \sup_{\varepsilon>0}\left(\varepsilon^\theta \sum_{l\in\mathbb{Z}} 2^{\alpha p(1+\varepsilon)l}\|f\chi_l\|^{p(1+\varepsilon)}_{L^{q(\cdot)}(\mathbb{R}^n)}\sum_{k=-\infty}^{l-2}2^{dp(k-l)/2}\right)^{\frac{1}{p(1+\varepsilon)}}$$

$$= \sup_{\varepsilon>0}\left(\varepsilon^\theta \sum_{l\in\mathbb{Z}} 2^{\alpha p(1+\varepsilon)l}\|f\chi_l\|^{p(1+\varepsilon)}_{L^{q(\cdot)}(\mathbb{R}^n)}\right)^{\frac{1}{p(1+\varepsilon)}}$$

$$\lesssim \|f\|_{\dot{K}^{\alpha,p),\theta}_{q(\cdot)}(\mathbb{R}^n)}.$$

Now for E_{31}, using Minkowski's inequality, we have

$$E_{31} \lesssim \sup_{\varepsilon>0}\left(\varepsilon^{\theta}\sum_{k=-\infty}^{-1}2^{k\alpha p(1+\varepsilon)}\left(\sum_{l=k+2}^{-1}\|\chi_k T(f\chi_l)\|_{L^{q(\cdot)}(\mathbb{R}^n)}\right)^{p(1+\varepsilon)}\right)^{\frac{1}{p(1+\varepsilon)}}$$
$$+c\sup_{\varepsilon>0}\left(\varepsilon^{\theta}\sum_{k=-\infty}^{-1}2^{k\alpha p(1+\varepsilon)}\left(\sum_{l=0}^{\infty}\|\chi_k T(f\chi_l)\|_{L^{q(\cdot)}(\mathbb{R}^n)}\right)^{p(1+\varepsilon)}\right)^{\frac{1}{p(1+\varepsilon)}}$$
$$:= B_1 + B_2.$$

The estimate for B_1 follows in similar manner as in E_{32} with q_∞ replaced by $q(0)$ and using the fact that $\frac{n}{q(0)}+\alpha>0$. For B_2, using Lemma 6.3, we have

$$2^{-ln}\|\chi_k\|_{L^{q(\cdot)}(\mathbb{R}^n)}\|\chi_l\|_{L^{q'(\cdot)}(\mathbb{R}^n)} \lesssim 2^{-ln}2^{\left(\frac{kn}{q(0)}\right)}2^{\left(\frac{ln}{q'_\infty}\right)} \lesssim 2^{\left(\frac{kn}{q(0)}\right)}2^{\left(\frac{ln}{q_\infty}\right)}. \tag{6.47}$$

Now using (6.43) and (6.47) and the estimate $\alpha+\frac{n}{q(0)}>0$, we have

$$B_2 \leqslant \sup_{\varepsilon>0}\left(\varepsilon^{\theta}\sum_{k=-\infty}^{-1}2^{k\alpha p(1+\varepsilon)}\left(\sum_{l=0}^{\infty}\|\chi_k Tf(\chi_l)\|_{L^{q(\cdot)}(\mathbb{R}^n)}\right)^{p(1+\varepsilon)}\right)^{\frac{1}{p(1+\varepsilon)}}$$
$$\lesssim \sup_{\varepsilon>0}\left(\varepsilon^{\theta}\sum_{k=-\infty}^{-1}2^{k\alpha p(1+\varepsilon)}\left(\sum_{l=0}^{\infty}2^{-ln}2^{(kn/q(0))}2^{(ln/q'_\infty))}\times\right.\right.$$
$$\left.\left.\times\|f(\chi_l)\|_{L^{q(\cdot)}(\mathbb{R}^n)}\right)^{p(1+\varepsilon)}\right)^{\frac{1}{p(1+\varepsilon)}}$$
$$\lesssim \sup_{\varepsilon>0}\left(\varepsilon^{\theta}\sum_{k=-\infty}^{-1}2^{k\alpha p(1+\varepsilon)}\left(\sum_{l=0}^{\infty}2^{(kn/q(0))}2^{-(ln/q_\infty)}\|f(\chi_l)\|_{L^{q(\cdot)}(\mathbb{R}^n)}\right)^{p(1+\varepsilon)}\right)^{\frac{1}{p(1+\varepsilon)}}$$
$$\lesssim \sup_{\varepsilon>0}\left(\varepsilon^{\theta}\sum_{k=-\infty}^{-1}2^{k(\alpha+n/q(0))p(1+\varepsilon)}\left(\sum_{l=0}^{\infty}2^{-(ln/q_\infty)}\|f(\chi_l)\|_{L^{q(\cdot)}(\mathbb{R}^n)}\right)^{p(1+\varepsilon)}\right)^{\frac{1}{p(1+\varepsilon)}}$$
$$\lesssim \sup_{\varepsilon>0}\left(\varepsilon^{\theta}\left(\sum_{l=0}^{\infty}2^{-(ln/q_\infty)}\|f(\chi_l)\|_{L^{q(\cdot)}(\mathbb{R}^n)}\right)^{p(1+\varepsilon)}\right)^{\frac{1}{p(1+\varepsilon)}}$$
$$\lesssim \sup_{\varepsilon>0}\left(\varepsilon^{\theta}\left(\sum_{l=0}^{\infty}2^{l\alpha}\|f(\chi_l)\|_{L^{q(\cdot)}(\mathbb{R}^n)}2^{-l(n/q_\infty+\alpha)}\right)^{p(1+\varepsilon)}\right)^{\frac{1}{p(1+\varepsilon)}}.$$

Now applying Hölder's inequality and using the fact that $\frac{n}{q_\infty}+\alpha>0$, we have

$$
\begin{aligned}
B_2 &\lesssim \sup_{\varepsilon>0}\left(\varepsilon^{\theta}\left(\sum_{l=0}^{\infty}2^{l\alpha p(1+\varepsilon)}\|f(\chi_l)\|_{L^{q(\cdot)}(\mathbb{R}^n)}\right)\times\right.\\
&\qquad\qquad\left.\times\left(\sum_{l=0}^{\infty}2^{-l(n/q_\infty+\alpha)}\right)^{p(1+\varepsilon)/p(1+\varepsilon)'}\right)^{\frac{1}{p(1+\varepsilon)}}\\
&\lesssim \sup_{\varepsilon>0}\left(\varepsilon^{\theta}\left(\sum_{l\in\mathbb{Z}}2^{l\alpha p(1+\varepsilon)}\|f(\chi_l)\|_{L^{q(\cdot)}(\mathbb{R}^n)}\right)\right)^{\frac{1}{p(1+\varepsilon)}}\\
&\lesssim \|f\|_{\dot{K}^{\alpha,p),\theta}_{q(\cdot)}(\mathbb{R}^n)}.
\end{aligned}
$$

Combining the estimates for E_1, E_2 and E_3, yields

$$
\|Tf\|_{\dot{K}^{\alpha,p),\theta}_{q(\cdot)}(\mathbb{R}^n)}\lesssim\|f\|_{\dot{K}^{\alpha,p),\theta}_{q(\cdot)}(\mathbb{R}^n)},
$$

which ends the proof. □

6.5 Boundedness of the Marcinkiewicz Integral on the Grand Variable Herz Space

This section aims to establish the boundedness of the Marcinkiewicz integral operator on the space $\dot{K}^{\alpha,p),\theta}_{q(\cdot)}(\mathbb{R}^n)$, introduced in Sect. 6.4.2.

6.5.1 Marcinkiewicz Integral on Grand Variable Herz Space

For definitions and results regarding the Marcinkiewicz integral, see Sect. 6.2.1.

To obtain Theorem 6.35, we need the following lemma.

Lemma 6.34 (Muckenhoupt and Wheeden [255]) *If $a>0, 1\leqslant s\leqslant\infty, 0<d\leqslant s$ and $-n+(n-1)d/s<\nu<\infty$, then*

$$
\left(\int\limits_{|y|\leqslant a|x|}|y|^{\nu}|\Omega(x-y)|^{d}dy\right)^{1/d}\lesssim|x|^{(\nu+n)/d}\|\Omega\|_{L^s(\mathbb{S}^{n-1})}.
$$

Theorem 6.35 *Let $0 < \nu \leqslant 1$, $\alpha, q \in \mathbb{P}_{0,\infty}(\mathbb{R}^n)$ with $1 < q^- \leqslant q^+ < \infty$, $1 \leqslant p < \infty$. Let Ω be homogeneous of degree zero and $\Omega \in L^s(\mathbb{S}^{n-1}), s > q'^-$. Let α be such that*

(a) $\dfrac{-n}{q(0)} - \nu - \dfrac{n}{s} < \alpha(0) < \dfrac{n}{q'(0)} - \nu - \dfrac{n}{s}$,

(b) $\dfrac{-n}{q_\infty} - \nu - \dfrac{n}{s} < \alpha_\infty < \dfrac{n}{q'_\infty} - \nu - \dfrac{n}{s}$.

Then the operator μ_Ω is bounded on $\dot{K}_{q(\cdot)}^{\alpha(\cdot),p),\theta}(\mathbb{R}^n)$.

Proof We decompose f as $f(x) = \sum_{i=-\infty}^{\infty} f(x)\chi_i(x) = \sum_{i=-\infty}^{\infty} f_i(x)$. We have

$$\begin{aligned}
&\|\mu_\Omega(f)\|_{\dot{K}_{q(\cdot)}^{\alpha(\cdot),p),\theta}(\mathbb{R}^n)} \\
&= \sup_{\varepsilon>0}\left(\varepsilon^\theta \sum_{k\in\mathbb{Z}} \|2^{k\alpha(\cdot)}\mu_\Omega(f)\chi_k\|_{L^{q(\cdot)}(\mathbb{R}^n)}^{p(1+\varepsilon)}\right)^{\frac{1}{p(1+\varepsilon)}} \\
&\lesssim \sup_{\varepsilon>0}\left(\varepsilon^\theta \sum_{k=-\infty}^{\infty}\left(\sum_{l=-\infty}^{k-2} \|2^{k\alpha(\cdot)}\chi_k\mu_\Omega(f\chi_l)\|_{L^{q(\cdot)}(\mathbb{R}^n)}\right)^{p(1+\varepsilon)}\right)^{\frac{1}{p(1+\varepsilon)}} \\
&\quad + \sup_{\varepsilon>0}\left(\varepsilon^\theta \sum_{k=-\infty}^{\infty}\left(\sum_{l=k-1}^{k+1} \|2^{k\alpha(\cdot)}\chi_k\mu_\Omega(f\chi_l)\|_{L^{q(\cdot)}(\mathbb{R}^n)}\right)^{p(1+\varepsilon)}\right)^{\frac{1}{p(1+\varepsilon)}} \\
&\quad + \sup_{\varepsilon>0}\left(\varepsilon^\theta \sum_{k=-\infty}^{\infty}\left(\sum_{l=k+2}^{\infty} \|2^{k\alpha(\cdot)}\chi_k\mu_\Omega(f\chi_l)\|_{L^{q(\cdot)}(\mathbb{R}^n)}\right)^{p(1+\varepsilon)}\right)^{\frac{1}{p(1+\varepsilon)}} \\
&=: I_1 + I_2 + I_3.
\end{aligned} \tag{6.48}$$

For I_2, we have

$$\begin{aligned}
I_2 &\lesssim \sup_{\varepsilon>0}\left(\varepsilon^\theta \sum_{k\in\mathbb{Z}}\left(\sum_{l=k-1}^{k+1} \|2^{k\alpha(\cdot)}\chi_k\mu_\Omega(f\chi_l)\|_{L^{q(\cdot)}(\mathbb{R}^n)}\right)^{p(1+\varepsilon)}\right)^{\frac{1}{p(1+\varepsilon)}} \\
&\lesssim \sup_{\varepsilon>0}\left(\varepsilon^\theta \sum_{k=-\infty}^{-1}\left(\sum_{l=k-1}^{k+1} \|2^{k\alpha(\cdot)}\chi_k\mu_\Omega(f\chi_l)\|_{L^{q(\cdot)}(\mathbb{R}^n)}\right)^{p(1+\varepsilon)}\right)^{\frac{1}{p(1+\varepsilon)}} \\
&\quad + \sup_{\varepsilon>0}\left(\varepsilon^\theta \sum_{k=0}^{\infty}\left(\sum_{l=k-1}^{k+1} \left\|2^{k\alpha(\cdot)}\chi_k\mu_\Omega(f\chi_l)\right\|_{L^{q(\cdot)}(\mathbb{R}^n)}\right)^{p(1+\varepsilon)}\right)^{\frac{1}{p(1+\varepsilon)}}
\end{aligned}$$

$$=: I_{21} + I_{22}.$$

Using the fact that $2^{k\alpha(x)} \approx 2^{k\alpha(0)}$, $k < 0$, $x \in R_k$ and considering that μ_Ω is a bounded operator on $L^{q(\cdot)}(\mathbb{R}^n)$, we get

$$\begin{aligned}
I_{21} &\lesssim \sup_{\varepsilon>0}\left(\varepsilon^\theta \sum_{k=-\infty}^{-1}\left(\sum_{l=k-1}^{k+1} \|2^{k\alpha(\cdot)}\chi_k\mu_\Omega(f\chi_l)\|_{L^{q(\cdot)}(\mathbb{R}^n)}\right)^{p(1+\varepsilon)}\right)^{\frac{1}{p(1+\varepsilon)}}\\
&\asymp \sup_{\varepsilon>0}\left(\varepsilon^\theta \sum_{k=-\infty}^{-1} 2^{k\alpha(0)p(1+\varepsilon)}\left(\sum_{l=k-1}^{k+1} \|\chi_k\mu_\Omega(f\chi_l)\|_{L^{q(\cdot)}(\mathbb{R}^n)}\right)^{p(1+\varepsilon)}\right)^{\frac{1}{p(1+\varepsilon)}}\\
&\lesssim \sup_{\varepsilon>0}\left(\varepsilon^\theta \sum_{k=-\infty}^{-1} 2^{k\alpha(0)p(1+\varepsilon)}\left(\sum_{l=k-1}^{k+1} \|\mu_\Omega(f\chi_l)\|_{L^{q(\cdot)}(\mathbb{R}^n)}\right)^{p(1+\varepsilon)}\right)^{\frac{1}{p(1+\varepsilon)}}\\
&\lesssim \sup_{\varepsilon>0}\left(\varepsilon^\theta \sum_{k=-\infty}^{-1} 2^{k\alpha(0)p(1+\varepsilon)}\left(\sum_{l=k-1}^{k+1} \|f\chi_l\|_{L^{q(\cdot)}(\mathbb{R}^n)}\right)^{p(1+\varepsilon)}\right)^{\frac{1}{p(1+\varepsilon)}}\\
&\lesssim \sup_{\varepsilon>0}\left(\varepsilon^\theta \sum_{k=-\infty}^{-1} 2^{k\alpha(0)p(1+\varepsilon)}\|f\chi_k\|_{L^{q(\cdot)}(\mathbb{R}^n)}^{p(1+\varepsilon)}\right)^{\frac{1}{p(1+\varepsilon)}}\\
&\lesssim \sup_{\varepsilon>0}\left(\varepsilon^\theta \sum_{k\in\mathbb{Z}} 2^{\alpha(0)p(1+\varepsilon)k}\|f\chi_k\|_{L^{q(\cdot)}(\mathbb{R}^n)}^{p(1+\varepsilon)}\right)^{\frac{1}{p(1+\varepsilon)}}\\
&\lesssim \|f\|_{\dot{K}_{q(\cdot)}^{\alpha(\cdot),p),\theta}(\mathbb{R}^n)}.
\end{aligned}$$

For estimation I_{22}, since $2^{k\alpha(x)} \approx 2^{k\alpha_\infty}, k \geqslant 0, x \in R_k$ and μ_Ω is a bounded operator on $L^{q(\cdot)}(\mathbb{R}^n)$, we have

$$\begin{aligned}
I_{22} &\lesssim \sup_{\varepsilon>0}\left(\varepsilon^\theta \sum_{k=0}^{\infty}\left(\sum_{l=k-1}^{k+1} \|2^{k\alpha(\cdot)}\chi_k\mu_\Omega(f\chi_l)\|_{L^{q(\cdot)}(\mathbb{R}^n)}\right)^{p(1+\varepsilon)}\right)^{\frac{1}{p(1+\varepsilon)}}\\
&\asymp \sup_{\varepsilon>0}\left(\varepsilon^\theta \sum_{k=0}^{\infty} 2^{k\alpha_\infty p(1+\varepsilon)}\left(\sum_{l=k-1}^{k+1} \|\chi_k\mu_\Omega(f\chi_l)\|_{L^{q(\cdot)}(\mathbb{R}^n)}\right)^{p(1+\varepsilon)}\right)^{\frac{1}{p(1+\varepsilon)}}
\end{aligned}$$

$$\lesssim \sup_{\varepsilon>0}\left(\varepsilon^{\theta}\sum_{k=0}^{\infty}2^{k\alpha_\infty p(1+\varepsilon)}\left(\sum_{l=k-1}^{k+1}\|\mu_\Omega(f\chi_l)\|_{L^{q(\cdot)}(\mathbb{R}^n)}\right)^{p(1+\varepsilon)}\right)^{\frac{1}{p(1+\varepsilon)}}$$

$$\lesssim \sup_{\varepsilon>0}\left(\varepsilon^{\theta}\sum_{k=0}^{\infty}2^{k\alpha_\infty p(1+\varepsilon)}\left(\sum_{l=k-1}^{k+1}\|f\chi_l\|_{L^{q(\cdot)}(\mathbb{R}^n)}\right)^{p(1+\varepsilon)}\right)^{\frac{1}{p(1+\varepsilon)}}$$

$$\lesssim \sup_{\varepsilon>0}\left(\varepsilon^{\theta}\sum_{k=0}^{\infty}2^{k\alpha_\infty p(1+\varepsilon)}\|f\chi_k\|_{L^{q(\cdot)}(\mathbb{R}^n)}^{p(1+\varepsilon)}\right)^{\frac{1}{p(1+\varepsilon)}}$$

$$\lesssim \sup_{\varepsilon>0}\left(\varepsilon^{\theta}\sum_{k\in\mathbb{Z}}2^{\alpha_\infty p(1+\varepsilon)k}\|f\chi_k\|_{L^{q(\cdot)}(\mathbb{R}^n)}^{p(1+\varepsilon)}\right)^{\frac{1}{p(1+\varepsilon)}}$$

$$\lesssim \|f\|_{\dot{K}_{q(\cdot)}^{\alpha(\cdot),p),\theta}(\mathbb{R}^n)}.$$

Combining these estimates, we get $I_2 \lesssim \|f\|_{\dot{K}_{q(\cdot)}^{\alpha(\cdot),p),\theta}(\mathbb{R}^n)}$.

Now we turn to estimate I_1. We consider

$$|\mu_\Omega(f\chi_l)(x)| \leqslant \left(\int_0^{|x|}\left|\int_{|x-y|\leqslant t}\frac{\Omega(x-y)}{|x-y|^{n-1}}f_l(y)dy\right|^2\frac{dt}{t^3}\right)^{1/2} + \left(\int_{|x|}^{\infty}\left|\int_{|x-y|\leqslant t}\frac{\Omega(x-y)}{|x-y|^{n-1}}f_l(y)dy\right|^2\frac{dt}{t^3}\right)^{1/2} =: I_{11}+I_{12}.$$

Let $x \in R_k$, $y \in R_l$ and $l \leqslant k-2$. We know that $|x-y| \asymp |x| \asymp 2^k$ and by the mean value theorem, we have

$$\left|\frac{1}{|x-y|^2}-\frac{1}{|x|^2}\right| \lesssim \frac{|y|}{|x-y|^3}. \tag{6.49}$$

By the integral Minkowski's inequality, inequality (6.49), and the generalized Hölder inequality, we have

$$I_{11} \lesssim \int_{\mathbb{R}^n}\frac{|\Omega(x-y)|}{|x-y|^{n-1}}|f_l(y)|\left(\int_{|x-y|}^{|x|}\frac{dt}{t^3}\right)^{1/2}dy$$

$$
\begin{aligned}
&\lesssim \int\limits_{\mathbb{R}^n} \frac{|\Omega(x-y)|}{|x-y|^{n-1}} |f_l(y)| \left| \frac{1}{|x-y|^2} - \frac{1}{|x|^2} \right|^{1/2} dy \\
&\lesssim \int\limits_{\mathbb{R}^n} \frac{|\Omega(x-y)|}{|x-y|^{n-1}} |f_l(y)| \frac{|y|^{1/2}}{|x-y|^{3/2}} dy \\
&\lesssim \frac{2^{l/2}}{|x|^{n+1/2}} \int\limits_{R_l} |\Omega(x-y)||f(y)| dy \\
&\lesssim 2^{(l-k)/2} 2^{-kn} \|f\chi_l\|_{L^{q(\cdot)}} \|\Omega(x-\cdot)\chi_l(\cdot)\|_{L^{q'(\cdot)}}.
\end{aligned}
$$

Similarly, we consider I_{12}. Noting that $|x-y| \asymp |x| \asymp 2^k$, by Minkowski and the generalized Hölder's inequality, we have

$$
\begin{aligned}
I_{12} &\lesssim \int\limits_{\mathbb{R}^n} \frac{|\Omega(x-y)|}{|x-y|^{n-1}} |f_l(y)| \left(\int\limits_{|x|}^{\infty} \frac{dt}{t^3} \right)^{1/2} dy \\
&\lesssim \int\limits_{\mathbb{R}^n} \frac{|\Omega(x-y)|}{|x-y|^{n}} |f_l(y)| dy \\
&\lesssim \frac{1}{|x|^n} \int\limits_{R_l} |\Omega(x-y)||f(y)| dy \\
&\lesssim 2^{-kn} \|f\chi_l\|_{L^{q(\cdot)}} \|\Omega(x-\cdot)\chi_l(\cdot)\|_{L^{q'(\cdot)}}.
\end{aligned}
$$

So, we have

$$
\begin{aligned}
|\mu_\Omega(f\chi_l)(x)| &\lesssim 2^{(l-k)/2} 2^{-kn} \|f\chi_l\|_{L^{q(\cdot)}} \|\Omega(x-\cdot)\chi_l(\cdot)\|_{L^{q'(\cdot)}} \\
&\qquad + 2^{-kn} \|f\chi_l\|_{L^{q(\cdot)}} \|\Omega(x-\cdot)\chi_l(\cdot)\|_{L^{q'(\cdot)}} \\
&\lesssim 2^{-kn} \|f\chi_l\|_{L^{q(\cdot)}} \|\Omega(x-\cdot)\chi_l(\cdot)\|_{L^{q'(\cdot)}}.
\end{aligned} \tag{6.50}
$$

Noting that $s > q'^{-}$, we define $\mathfrak{q}(\cdot)$ by the relations $\frac{1}{q'(x)} = \frac{1}{\mathfrak{q}(x)} + \frac{1}{s}$. By Lemma 6.34 and generalized Hölder's inequality, we have

$$
\begin{aligned}
\|\Omega(x-\cdot)\chi_l(\cdot)\|_{L^{q'(\cdot)}} &\lesssim \|\Omega(x-\cdot)\chi_l(\cdot)\|_{L^s} \|\chi_l(\cdot)\|_{L^{\mathfrak{q}(\cdot)}} \\
&\lesssim 2^{-lv} \left(\int\limits_{2^{l-1}<|y|<2^l} |\Omega(x-y)|^s |y|^{sv} dy \right)^{1/s} \|\chi_l\|_{L^{\mathfrak{q}(\cdot)}}
\end{aligned}
$$

$$\lesssim 2^{-l\nu}2^{k(\nu+\frac{n}{s})}\|\Omega\|_{L^s(\mathbb{S}^{n-1})}\|\chi_l\|_{L^{q(\cdot)}}. \tag{6.51}$$

Moreover, splitting I_1 by means of Minkowski's inequality, we have

$$\begin{aligned} I_1 &\lesssim \sup_{\varepsilon>0}\left(\varepsilon^{\theta}\sum_{k=-\infty}^{-1}\left(\sum_{l=-\infty}^{k-2}\|2^{k\alpha(\cdot)}\chi_k\mu_\Omega(f\chi_l)\|_{L^{q(\cdot)}(\mathbb{R}^n)}\right)^{p(1+\varepsilon)}\right)^{\frac{1}{p(1+\varepsilon)}} \\ &\quad + \sup_{\varepsilon>0}\left(\varepsilon^{\theta}\sum_{k=0}^{\infty}\left(\sum_{l=-\infty}^{k-2}\|2^{k\alpha(\cdot)}\chi_k\mu_\Omega(f\chi_l)\|_{L^{q(\cdot)}(\mathbb{R}^n)}\right)^{p(1+\varepsilon)}\right)^{\frac{1}{p(1+\varepsilon)}} \\ &:= I_{11}' + I_{12}'. \end{aligned}$$

For I_{11}', via (6.51) and (6.8), we have

$$\begin{aligned} &I_{11}' \\ &\lesssim \sup_{\varepsilon>0}\left(\varepsilon^{\theta}\sum_{k=-\infty}^{-1}2^{k\alpha(0)p(1+\varepsilon)}\left(\sum_{l=-\infty}^{k-2}2^{(l-k)(n/q'(0)-\nu-\frac{n}{s})}\|f\chi_l\|_{L^{q(\cdot)}}\right)^{p(1+\varepsilon)}\right)^{\frac{1}{p(1+\varepsilon)}} \\ &\lesssim \sup_{\varepsilon>0}\left(\varepsilon^{\theta}\sum_{k=-\infty}^{-1}\left(\sum_{l=-\infty}^{k-2}2^{\alpha(0)l}\|f\chi_l\|_{L^{q(\cdot)}(\mathbb{R}^n)}2^{b(l-k)}\right)^{p(1+\varepsilon)}\right)^{\frac{1}{p(1+\varepsilon)}} \end{aligned}$$

where $b := \frac{n}{q'(0)} - \nu - \frac{n}{s} - \alpha(0) > 0$. Then we use Hölder's inequality, Fubini's theorem for series, and the inequality $2^{-p(1+\varepsilon)} < 2^{-p}$ to obtain

$$\begin{aligned} I_{11}' &\lesssim \sup_{\varepsilon>0}\left(\varepsilon^{\theta}\sum_{k=-\infty}^{-1}\left(\sum_{l=-\infty}^{k-2}2^{\alpha(0)p(1+\varepsilon)l}\|f\chi_l\|_{L^{q(\cdot)}(\mathbb{R}^n)}^{p(1+\varepsilon)}2^{bp(1+\varepsilon)(l-k)/2}\right)\times\right. \\ &\qquad\left.\times\left(\sum_{l=-\infty}^{k-2}2^{b(p(1+\varepsilon))'(l-k)/2}\right)^{p(1+\varepsilon)/(p(1+\varepsilon))'}\right)^{\frac{1}{p(1+\varepsilon)}} \\ &\lesssim \sup_{\varepsilon>0}\left(\varepsilon^{\theta}\sum_{k=-\infty}^{-1}\sum_{l=-\infty}^{k-2}2^{\alpha(0)p(1+\varepsilon)l}\|f\chi_l\|_{L^{q(\cdot)}(\mathbb{R}^n)}^{p(1+\varepsilon)}2^{bp(1+\varepsilon)(l-k)/2}\right)^{\frac{1}{p(1+\varepsilon)}} \\ &\lesssim c\sup_{\varepsilon>0}\left(\varepsilon^{\theta}\sum_{l=-\infty}^{-1}2^{\alpha(0)p(1+\varepsilon)l}\|f\chi_l\|_{L^{q(\cdot)}(\mathbb{R}^n)}^{p(1+\varepsilon)}\sum_{k=l+2}^{-1}2^{bp(1+\varepsilon)(l-k)/2}\right)^{\frac{1}{p(1+\varepsilon)}} \end{aligned}$$

$$\lesssim \sup_{\varepsilon>0}\left(\varepsilon^{\theta}\sum_{l=-\infty}^{-1}2^{\alpha(0)p(1+\varepsilon)l}\|f\chi_l\|_{L^{q(\cdot)}(\mathbb{R}^n)}^{p(1+\varepsilon)}\sum_{k=l+2}^{-1}2^{bp(l-k)/2}\right)^{\frac{1}{p(1+\varepsilon)}}$$

$$\lesssim \sup_{\varepsilon>0}\left(\varepsilon^{\theta}\sum_{l\in\mathbb{Z}}2^{\alpha(0)p(1+\varepsilon)l}\|f\chi_l\|_{L^{q(\cdot)}(\mathbb{R}^n)}^{p(1+\varepsilon)}\right)^{\frac{1}{p(1+\varepsilon)}}$$

$$\lesssim \|f\|_{\dot{K}_{q(\cdot)}^{\alpha(\cdot),p),\theta}(\mathbb{R}^n)}.$$

Now for I'_{12} using Minkowski's inequality, we have

$$I'_{12}\lesssim \sup_{\varepsilon>0}\left(\varepsilon^{\theta}\sum_{k=0}^{\infty}\left(\sum_{l=-\infty}^{-1}\|2^{k\alpha(\cdot)}\chi_k\mu_{\Omega}(f\chi_l)\|_{L^{q(\cdot)}(\mathbb{R}^n)}\right)^{p(1+\varepsilon)}\right)^{\frac{1}{p(1+\varepsilon)}}$$

$$+\sup_{\varepsilon>0}\left(\varepsilon^{\theta}\sum_{k=0}^{\infty}\left(\sum_{l=0}^{k-2}\|2^{k\alpha(\cdot)}\chi_k\mu_{\Omega}(f\chi_l)\|_{L^{q(\cdot)}(\mathbb{R}^n)}\right)^{p(1+\varepsilon)}\right)^{\frac{1}{p(1+\varepsilon)}}$$

$$:= A_1 + A_2.$$

The estimate for A_2 follows in a similar manner as in I'_{11} with $q'(0)$ replaced by q'_∞ and using the fact that $b := \frac{n}{q'_\infty} - \nu - \frac{n}{s} - \alpha_\infty > 0$. For A_1 using (6.8), (6.9), (6.50), and (6.51), we have

$$\begin{aligned}\|\mu_{\Omega}(f\chi_l)\chi_k\|_{L^{q(\cdot)}} &\lesssim 2^{-kn}\|f\chi_l\|_{L^{q(\cdot)}}\|\Omega(x-\cdot)\chi_l(\cdot)\|_{L^{q'(\cdot)}}\|\chi_k\|_{L^{q(\cdot)}}\\ &\lesssim 2^{-kn}2^{-l\nu}2^{k(\nu+\frac{n}{s})}\|f\chi_l\|_{L^{q(\cdot)}}\|\Omega\|_{L^s(\mathbb{S}^{n-1})}\|\chi_l\|_{L^{q(\cdot)}}\|\chi_k\|_{L^{q(\cdot)}}\\ &\lesssim 2^{-kn}2^{-l\nu}2^{k(\nu+\frac{n}{s})}2^{ln/q(0)}2^{kn/q_\infty}\|f\chi_l\|_{L^{q(\cdot)}}\|\Omega\|_{L^s(\mathbb{S}^{n-1})}\|\\ &\lesssim 2^{l(\frac{n}{q(0)}-\nu)}2^{k(\nu+\frac{n}{s}-\frac{n}{q'_\infty})}\|f\chi_l\|_{L^{q(\cdot)}}\|\Omega\|_{L^s(\mathbb{S}^{n-1})}.\end{aligned} \tag{6.52}$$

Now using (6.52) and the fact that $\alpha_\infty + \nu + \frac{n}{s} - \frac{n}{q'_\infty} < 0$, we have

$$A_1 \leqslant \sup_{\varepsilon>0}\left(\varepsilon^{\theta}\sum_{k=0}^{\infty}2^{k\alpha_\infty p(1+\varepsilon)}\left(\sum_{l=-\infty}^{-1}\|\chi_k\mu_{\Omega}(f\chi_l)\|_{L^{q(\cdot)}(\mathbb{R}^n)}\right)^{p(1+\varepsilon)}\right)^{\frac{1}{p(1+\varepsilon)}}$$

$$\lesssim \sup_{\varepsilon>0}\left(\varepsilon^{\theta}\sum_{k=0}^{\infty}2^{k\alpha_\infty p(1+\varepsilon)}\left(\sum_{l=-\infty}^{-1}2^{l(\frac{n}{q(0)}-\nu)}2^{k(\nu+\frac{n}{s}-\frac{n}{q'_\infty})}\|f\chi_l\|_{L^{q(\cdot)}}\right)^{p(1+\varepsilon)}\right)^{\frac{1}{p(1+\varepsilon)}}$$

$$\lesssim \sup_{\varepsilon>0}\left(\varepsilon^{\theta}\sum_{k=0}^{\infty}2^{k(\alpha_\infty+\nu+\frac{n}{s}-\frac{n}{q'_\infty})p(1+\varepsilon)}\times\right.$$

$$\left.\times\left(\sum_{l=-\infty}^{-1}2^{l(\frac{n}{q(0)}-\nu)}\|f(\chi_l)\|_{L^{q(\cdot)}(\mathbb{R}^n)}\right)^{p(1+\varepsilon)}\right)^{\frac{1}{p(1+\varepsilon)}}$$

$$\lesssim \sup_{\varepsilon>0}\left(\varepsilon^{\theta}\left(\sum_{l=-\infty}^{-1}2^{l(\frac{n}{q(0)}-\nu)}\|f(\chi_l)\|_{L^{q(\cdot)}(\mathbb{R}^n)}\right)^{p(1+\varepsilon)}\right)^{\frac{1}{p(1+\varepsilon)}}$$

$$\lesssim \sup_{\varepsilon>0}\left(\varepsilon^{\theta}\left(\sum_{l=-\infty}^{-1}2^{l\alpha(0)}\|f(\chi_l)\|_{L^{q(\cdot)}(\mathbb{R}^n)}2^{l(\frac{n}{q'(0)}-\frac{n}{s}-\nu-\alpha(0))}\right)^{p(1+\varepsilon)}\right)^{\frac{1}{p(1+\varepsilon)}}.$$

Now applying Hölder's inequality and using the fact that $\frac{n}{q'(0)}-\frac{n}{s}-\nu-\alpha(0)>0$, we have

$$A_1\lesssim \sup_{\varepsilon>0}\left(\varepsilon^{\theta}\left(\sum_{l=-\infty}^{-1}2^{l\alpha(0)p(1+\varepsilon)}\|f(\chi_l)\|^{p(1+\varepsilon)}_{L^{q(\cdot)}(\mathbb{R}^n)}\right)\times\right.$$

$$\left.\times\left(\sum_{l=-\infty}^{-1}2^{l(\frac{n}{q'(0)}-\frac{n}{s}-\nu-\alpha(0))}\right)^{p(1+\varepsilon)/p(1+\varepsilon)'}\right)^{\frac{1}{p(1+\varepsilon)}}$$

$$\lesssim \sup_{\varepsilon>0}\left(\varepsilon^{\theta}\left(\sum_{l\in\mathbb{Z}}\|2^{l\alpha(\cdot)}f(\chi_l)\|^{p(1+\varepsilon)}_{L^{q(\cdot)}(\mathbb{R}^n)}\right)\right)^{\frac{1}{p(1+\varepsilon)}}$$

$$\lesssim \|f\|_{\dot K^{\alpha(\cdot),p),\theta}_{q(\cdot)}(\mathbb{R}^n)}.$$

Let us now estimate I_3. Note that $x\in R_k$, $y\in R_l$ and $l\geqslant k+2$, so we know that $|x-y|\asymp|y|\asymp 2^l$. We consider

$$|\mu_\Omega(f_l)(x)|\leqslant\left(\int_0^{|y|}\left|\int_{|x-y|\leqslant t}\frac{\Omega(x-y)}{|x-y|^{n-1}}f_l(y)dy\right|^2\frac{dt}{t^3}\right)^{1/2}$$

$$+\left(\int_{|y|}^{\infty}\left|\int_{|x-y|\leqslant t}\frac{\Omega(x-y)}{|x-y|^{n-1}}f_l(y)dy\right|^2\frac{dt}{t^3}\right)^{1/2}$$

$$=: I_{31}+I_{32}.$$

Similarly to the estimate for I_{11}, we get

$$I_{31} \lesssim 2^{(k-l)/2} 2^{-ln} \|f_l\|_{L^{q(\cdot)}} \|\Omega(x - \cdot)\chi_l(\cdot)\|_{L^{q'(\cdot)}}.$$

As was done for I_{12}, we get

$$I_{32} \lesssim 2^{-ln} \|f_l\|_{L^{q(\cdot)}} \|\Omega(x - \cdot)\chi_l(\cdot)\|_{L^{q'(\cdot)}}.$$

So, we have

$$|\mu_\Omega(f_l)(x)| \lesssim 2^{-ln} \|f_l\|_{L^{q(\cdot)}} \|\Omega(x - \cdot)\chi_l(\cdot)\|_{L^{q'(\cdot)}}.$$

Analogous to I_1, splitting I_3, we have

$$\begin{aligned} I_3 \lesssim & \sup_{\varepsilon>0} \left(\varepsilon^\theta \sum_{k=-\infty}^{-1} \left(\sum_{l=k+2}^{\infty} \|2^{k\alpha(\cdot)} \mu_\Omega(f\chi_l)\chi_k\|_{L^{q(\cdot)}(\mathbb{R}^n)} \right)^{p(1+\varepsilon)} \right)^{\dfrac{1}{p(1+\varepsilon)}} + \\ & + \sup_{\varepsilon>0} \left(\varepsilon^\theta \sum_{k=0}^{\infty} \left(\sum_{l=k+2}^{\infty} \|2^{k\alpha(\cdot)} \mu_\Omega(f\chi_l)\chi_k\|_{L^{q(\cdot)}(\mathbb{R}^n)} \right)^{p(1+\varepsilon)} \right)^{\frac{1}{p(1+\varepsilon)}} \\ := & I'_{31} + I'_{32}. \end{aligned}$$

For I'_{32} by condition (6.9), we obtain

$$\begin{aligned} \|\mu_\Omega(f\chi_l)\chi_k\|_{L^{q(\cdot)}} & \lesssim 2^{-ln} \|f\chi_l\|_{L^{q(\cdot)}} \|\Omega(x - \cdot)\chi_l(\cdot)\|_{L^{q'(\cdot)}} \|\chi_k\|_{L^{q(\cdot)}} \\ & \lesssim 2^{-ln} 2^{-l\nu} 2^{k(\nu+\frac{n}{s})} \|f\chi_l\|_{L^{q(\cdot)}} \|\Omega\|_{L^s(\mathbb{S}^{n-1})} \|\chi_l\|_{L^{q(\cdot)}} \|\chi_k\|_{L^{q(\cdot)}} \\ & \lesssim 2^{-ln} 2^{-l\nu} 2^{k(\nu+\frac{n}{s})} 2^{ln/q_\infty} 2^{kn/q_\infty} \|f\chi_l\|_{L^{q(\cdot)}} \|\Omega\|_{L^s(\mathbb{S}^{n-1})} \| \\ & \lesssim 2^{(k-l)} 2^{\left(\frac{n}{q_\infty}+\nu+\frac{n}{s}\right)} \|f\chi_l\|_{L^{q(\cdot)}} \|\Omega\|_{L^s(\mathbb{S}^{n-1})}. \end{aligned} \tag{6.53}$$

Hence,

$$\begin{aligned} I'_{32} & \lesssim \sup_{\varepsilon>0} \left(\varepsilon^\theta \sum_{k=0}^{\infty} 2^{k\alpha_\infty p(1+\varepsilon)} \left(\sum_{l=k+2}^{\infty} 2^{(k-l)} 2^{(\frac{n}{q_\infty}+\nu+\frac{n}{s})} \|f\chi_l\|_{L^{q(\cdot)}} \right)^{p(1+\varepsilon)} \right)^{\frac{1}{p(1+\varepsilon)}} \\ & \lesssim \sup_{\varepsilon>0} \left(\varepsilon^\theta \sum_{k=0}^{\infty} \left(\sum_{l=k+2}^{\infty} 2^{\alpha_\infty l} \|f\chi_l\|_{L^{q(\cdot)}(\mathbb{R}^n)} 2^{d(k-l)} \right)^{p(1+\varepsilon)} \right)^{\frac{1}{p(1+\varepsilon)}} \end{aligned}$$

where $d := \frac{n}{q_\infty} + \nu + \frac{n}{s} + \alpha_\infty > 0$. Then we use Hölder's inequality, Fubini's theorem for series, and the inequality $2^{-p(1+\varepsilon)} < 2^{-p}$ to obtain

$$I_{32}' \lesssim \sup_{\varepsilon>0}\left(\varepsilon^\theta \sum_{k=0}^{\infty}\left(\sum_{l=k+2}^{\infty} 2^{\alpha_\infty p(1+\varepsilon)l}\|f\chi_l\|_{L^{q(\cdot)}(\mathbb{R}^n)}^{p(1+\varepsilon)} 2^{dp(1+\varepsilon)(k-l)/2}\right)\times\right.$$
$$\left.\times\left(\sum_{l=k+2}^{\infty} 2^{d(p(1+\varepsilon))'(k-l)/2}\right)^{p(1+\varepsilon)/(p(1+\varepsilon))'}\right)^{\frac{1}{p(1+\varepsilon)}}$$
$$\lesssim \sup_{\varepsilon>0}\left(\varepsilon^\theta \sum_{k=0}^{\infty}\sum_{l=k+2}^{\infty} 2^{\alpha_\infty p(1+\varepsilon)l}\|f\chi_l\|_{L^{q(\cdot)}(\mathbb{R}^n)}^{p(1+\varepsilon)} 2^{dp(1+\varepsilon)(k-l)/2}\right)^{\frac{1}{p(1+\varepsilon)}}$$
$$\lesssim \sup_{\varepsilon>0}\left(\varepsilon^\theta \sum_{l=0}^{\infty} 2^{\alpha_\infty p(1+\varepsilon)l}\|f\chi_l\|_{L^{q(\cdot)}(\mathbb{R}^n)}^{p(1+\varepsilon)} \sum_{k=0}^{l-2} 2^{dp(1+\varepsilon)(k-l)/2}\right)^{\frac{1}{p(1+\varepsilon)}}$$
$$\lesssim \sup_{\varepsilon>0}\left(\varepsilon^\theta \sum_{l\in\mathbb{Z}} \|2^{\alpha(\cdot)}f\chi_l\|_{L^{q(\cdot)}(\mathbb{R}^n)}^{p(1+\varepsilon)}\right)^{\frac{1}{p(1+\varepsilon)}}$$
$$\lesssim \|f\|_{\dot{K}_{q(\cdot)}^{\alpha(\cdot),p),\theta}(\mathbb{R}^n)}.$$

Now for I_{31}, we have

$$E_{31} \lesssim \sup_{\varepsilon>0}\left(\varepsilon^\theta \sum_{k=-\infty}^{-1}\left(\sum_{l=k+2}^{-1} \|2^{k\alpha(\cdot)}\chi_k\mu_\Omega(f\chi_l)\|_{L^{q(\cdot)}(\mathbb{R}^n)}\right)^{p(1+\varepsilon)}\right)^{\frac{1}{p(1+\varepsilon)}}$$
$$+ \sup_{\varepsilon>0}\left(\varepsilon^\theta \sum_{k=-\infty}^{-1}\left(\sum_{l=0}^{\infty} \|2^{k\alpha(\cdot)}\chi_k\mu_\Omega(f\chi_l)\|_{L^{q(\cdot)}(\mathbb{R}^n)}\right)^{p(1+\varepsilon)}\right)^{\frac{1}{p(1+\varepsilon)}}$$
$$:= I_{31}^* + I_{32}^*.$$

For I_{32}^* by condition (6.8) and (6.9), we have

$$\begin{aligned}\|\mu_\Omega(f\chi_l)\chi_k\|_{L^{q(\cdot)}} &\lesssim 2^{-ln}\|f\chi_l\|_{L^{q(\cdot)}}\|\Omega(x-\cdot)\chi_l(\cdot)\|_{L^{q'(\cdot)}}\|\chi_k\|_{L^{q(\cdot)}}\\ &\lesssim 2^{-ln}2^{-l\nu}2^{k(\nu+\frac{n}{s})}\|f\chi_l\|_{L^{q(\cdot)}}\|\Omega\|_{L^s(\mathbb{S}^{n-1})}\|\chi_l\|_{L^{q(\cdot)}}\|\chi_k\|_{L^{q(\cdot)}}\\ &\lesssim 2^{-ln}2^{-l\nu}2^{k(\nu+\frac{n}{s})}2^{ln/q_\infty}2^{kn/q(0)}\|f\chi_l\|_{L^{q(\cdot)}}\|\Omega\|_{L^s(\mathbb{S}^{n-1})}\\ &\lesssim 2^{-l(\frac{n}{q_\infty}+\nu+\frac{n}{s})}2^{k(\frac{n}{q(0)}+\nu+\frac{n}{s})}\|f\chi_l\|_{L^{q(\cdot)}}\|\Omega\|_{L^s(\mathbb{S}^{n-1})}.\end{aligned} \tag{6.54}$$

Now using (6.54) and the fact that $\frac{n}{q(0)} + \nu + \frac{n}{s} + \alpha(0) > 0$, we have

$$
\begin{aligned}
I_{32}^* &\leqslant \sup_{\varepsilon>0}\left(\varepsilon^{\theta}\sum_{k=-\infty}^{-1}2^{k\alpha(0)p(1+\varepsilon)}\left(\sum_{l=0}^{\infty}\|\chi_k\mu_{\Omega}(f\chi_l)\|_{L^{q(\cdot)}(\mathbb{R}^n)}\right)^{p(1+\varepsilon)}\right)^{\frac{1}{p(1+\varepsilon)}}\\
&\lesssim \sup_{\varepsilon>0}\left(\varepsilon^{\theta}\sum_{k=-\infty}^{-1}2^{k\alpha(0)p(1+\varepsilon)}\left(\sum_{l=0}^{\infty}2^{-l(\frac{n}{q_\infty}+\nu+\frac{n}{s})}\times\right.\right.\\
&\qquad\left.\left.\times\ 2^{k(\frac{n}{q(0)}+\nu+\frac{n}{s})}\|f\chi_l\|_{L^{q(\cdot)}}\right)^{p(1+\varepsilon)}\right)^{\frac{1}{p(1+\varepsilon)}}\\
&\lesssim \sup_{\varepsilon>0}\left(\varepsilon^{\theta}\sum_{k=-\infty}^{-1}2^{k(\frac{n}{q(0)}+\nu+\frac{n}{s}+\alpha(0))p(1+\varepsilon)}\times\right.\\
&\qquad\left.\times\left(\sum_{l=0}^{\infty}2^{-l(\frac{n}{q_\infty}+\nu+\frac{n}{s})}\|f(\chi_l)\|_{L^{q(\cdot)}(\mathbb{R}^n)}\right)^{p(1+\varepsilon)}\right)^{\frac{1}{p(1+\varepsilon)}}\\
&\lesssim \sup_{\varepsilon>0}\left(\varepsilon^{\theta}\left(\sum_{l=0}^{\infty}2^{-l(\frac{n}{q_\infty}+\nu+\frac{n}{s})}\|f(\chi_l)\|_{L^{q(\cdot)}(\mathbb{R}^n)}\right)^{p(1+\varepsilon)}\right)^{\frac{1}{p(1+\varepsilon)}}\\
&\lesssim \sup_{\varepsilon>0}\left(\varepsilon^{\theta}\left(\sum_{l=0}^{\infty}2^{l\alpha_\infty}\|f(\chi_l)\|_{L^{q(\cdot)}(\mathbb{R}^n)}2^{-l(\frac{n}{q_\infty}+\nu+\frac{n}{s}+\alpha_\infty)}\right)^{p(1+\varepsilon)}\right)^{\frac{1}{p(1+\varepsilon)}}.
\end{aligned}
$$

Now applying Hölder's inequality and the fact that $\frac{n}{q_\infty} + \nu + \frac{n}{s} + \alpha_\infty > 0$, we have

$$
\begin{aligned}
I_{32}^* &\lesssim \sup_{\varepsilon>0}\left(\varepsilon^{\theta}\left(\sum_{l=0}^{\infty}2^{l\alpha_\infty p(1+\varepsilon)}\|f(\chi_l)\|^{p(1+\varepsilon)}_{L^{q(\cdot)}(\mathbb{R}^n)}\right)\times\right.\\
&\qquad\left.\times\left(\sum_{l=0}^{\infty}2^{-l(\frac{n}{q_\infty}+\nu+\frac{n}{s}+\alpha_\infty)(p(1+\varepsilon))'}\right)^{p(1+\varepsilon)/p(1+\varepsilon)'}\right)^{\frac{1}{p(1+\varepsilon)}}\\
&\lesssim \sup_{\varepsilon>0}\left(\varepsilon^{\theta}\left(\sum_{l\in\mathbb{Z}}\|2^{l\alpha(\cdot)}f(\chi_l)\|^{p(1+\varepsilon)}_{L^{q(\cdot)}(\mathbb{R}^n)}\right)\right)^{\frac{1}{p(1+\varepsilon)}}\\
&\lesssim \|f\|_{\dot{K}^{\alpha(\cdot),p),\theta}_{q(\cdot)}(\mathbb{R}^n)}.
\end{aligned}
$$

The estimate for I_{31}^* follows in a manner similar to that of I_{32}^* with q_∞ replaced by $q(0)$ and using the fact that $\frac{n}{q(0)} + \nu + \frac{n}{s} + \alpha(0) > 0$. Combining the estimates for I_1, I_2 and I_3, yields

$$\|\mu_\Omega(f)\|_{\dot{K}_{q(\cdot)}^{\alpha(\cdot),p),\theta}(\mathbb{R}^n)} \lesssim \|f\|_{\dot{K}_{q(\cdot)}^{\alpha(\cdot),p),\theta}(\mathbb{R}^n)},$$

which ends the proof. □

6.6 Boundedness of Multilinear Calderón–Zygmund Operators on Grand Variable Herz Spaces

We prove the boundedness of multilinear Calderón–Zygmund operators on grand variable Herz spaces.

For integers $n, m \in \mathbb{Z}$ with $n < m$, we define the set $\overline{n,m} := \{n, n+1, n+2, \ldots, m\}$.

6.6.1 *Boundedness of Multilinear Calderón–Zygmund Operators*

Consider the multilinear operator T of the form

$$T(f_1, \cdots, f_m)(x) = \int\limits_{(\mathbb{R}^n)^m} K(x, y_1, \cdots, y_m) f_1(y_1) \cdots f_m(y_m) dy_1 \cdots dy_m, \tag{6.55}$$

where $x \notin \bigcap_{j=1}^m \operatorname{supp}(f_j)$ and $f_1, \cdots, f_m \in L_c^\infty(\mathbb{R}^n)$, the space of compactly supported functions.

Let $K(x, y_1, \cdots, y_m)$ be a locally integrable function defined away from the diagonal $x = y_1 = \cdots = y_m \in (\mathbb{R}^n)^{m+1}$, which satisfies the size estimate

$$|K(x, y_1, \cdots, y_m)| \leqslant \frac{c}{(|x - y_1| + |x - y_2| + \cdots + |x - y_m|)^{mn}}, \tag{6.56}$$

for some $c > 0$ and all $(x, y_1, \cdots, y_m) \in (\mathbb{R}^n)^{m+1}$ with $x \neq y_j$ for some j. For smoothness, assume that for some $\varepsilon > 0$,

$$|K(x, y_1, \cdots, y_m) - K(x', y_1, \cdots, y_m)| \leqslant \frac{c|x - x'|^\varepsilon}{(|x - y_1| + \cdots + |x - y_m|)^{mn+\varepsilon}}, \tag{6.57}$$

provided that $|x - x'| \leqslant \frac{1}{2}\max\{|x - y_1|, \cdots, |x - y_m|\}$ and

$$|K(x, y_1, \cdots, y_j, \cdots, y_m) - K(x, y_1, \cdots, y_j', \cdots, y_m)| \leqslant \frac{c|y_j - y_j'|^{\varepsilon}}{(|x - y_1| + \cdots + |x - y_m|)^{mn+\varepsilon}}, \tag{6.58}$$

whenever $|y_j - y_j'| \leqslant \frac{1}{2}\max\{|x - y_1|, \cdots, |x - y_m|\}$ for all $j \in \overline{1, m}$.

Such kernels are called Calderón–Zygmund kernels, and the class of all functions satisfying (6.56), (6.57), and (6.58) with parameters m, c, and ε will be denoted by m-CZK(c, ε), cf. Grafakos and Torres [122]. We say that T, as defined in (6.55), is an m-linear Calderón–Zygmund operator, if

(i) the related kernel belongs to the m-CZK(c, ε) class;
(ii) T is bounded from $L^{q_1} \times \cdots \times L^{q_m}$ to L^q for some $1 < q_1, q_2, \cdots, q_m < \infty$, and $\frac{1}{q} = \frac{1}{q_1} + \cdots + \frac{1}{q_m}$.

Grafakos and Torres [122] proved the boundedness of T from $L^{q_1} \times \cdots \times L^{q_m}$ to $L^{q,\infty}$ for some $1 < q_1, q_2, \cdots, q_m < \infty$, and $\frac{1}{q} = \frac{1}{q_1} + \cdots + \frac{1}{q_m}$, and from $L^1 \times \cdots \times L^1$ to $L^{1/m,\infty}$.

The boundedness of the multilinear Calderón–Zygmund operator T on variable exponent Lebesgue spaces was proved in Tao et al. [319], as stated below.

Lemma 6.36 *Let $q_i \in \mathcal{B}(\mathbb{R}^n)$, $i \in \overline{1, m}$, $q \in \mathcal{P}_0(\mathbb{R}^n)$ with $\frac{1}{q(x)} = \frac{1}{q_1(x)} + \cdots + \frac{1}{q_m(x)}$, and $(q/s)' \in \mathcal{B}(\mathbb{R}^n)$ for some $0 < s < q^-$. Then the m-linear Calderón–Zygmund operator T is bounded on the product of variable exponent Lebesgue spaces. Moreover,*

$$\left\| T(\vec{f}) \right\|_{q(\cdot)} \lesssim \prod_{i=1}^{m} \| f_i \|_{q_i(\cdot)},$$

with the constant C independent of $\vec{f} = (f_1, \cdots, f_m)$.

We now state and prove the boundedness of multilinear Calderón–Zygmund operator on grand variable Herz spaces.

Theorem 6.37 *Let $q_i \in \mathcal{B}(\mathbb{R}^n)$, $i \in \overline{1, m}$, $q, q_i \in \mathbb{P}_{0,\infty}(\mathbb{R}^n)$ such that $q_i(0) \leqslant q_i(\infty)$, $\frac{1}{q(x)} = \frac{1}{q_1(x)} + \cdots + \frac{1}{q_m(x)}$, $1 < q_i^- \leqslant q_i^+ < \infty$, and $(q/s)' \in \mathcal{B}(\mathbb{R}^n)$ for some $0 < s < q^-$. Let $\theta > 0$, $1 < p_i < \infty$, and $\alpha_i \in L^\infty(\mathbb{R})$ be log-Hölder continuous both at the origin and at infinity for $i \in \overline{1, m}$ with*

$$\frac{-n}{q_i(\infty)} < \alpha_i(0) \leqslant \alpha_i(\infty) < n\left(1 - \frac{1}{q_i(0)}\right).$$

Suppose that $\alpha(x) = \sum_{i=1}^m \alpha_i(x)$ and $\frac{1}{p} = \sum_{i=1}^m \frac{1}{p_i}$. Then the m-linear Calderón–Zygmund operator T is bounded on the product of grand variable Herz spaces.

Moreover,

$$\left\|T(\vec{f})\right\|_{\dot{K}_{q(\cdot)}^{\alpha(\cdot),p),\theta}(\mathbb{R}^n)} \lesssim \prod_{i=1}^{m} \|f_i\|_{\dot{K}_{q_i(\cdot)}^{\alpha_i(\cdot),p_i),\theta}(\mathbb{R}^n)},$$

with the constant $C > 0$ independent of $\vec{f} = (f_1, \dots, f_m)$.

Proof We restrict ourselves to $m = 2$, the general case following in a similar manner. Defining $f_\varphi(x) := f_1(x)\chi_{R_\varphi}(x)$ and $f_\sigma(x) := f_2(x)\chi_{R_\sigma}(x)$, we decompose the component functions of $\vec{f}$ as

$$f_1(x) = \sum_{\varphi\in\mathbb{Z}} f_\varphi(x), \quad f_2(x) = \sum_{\sigma\in\mathbb{Z}} f_\sigma(x). \tag{6.59}$$

For future usage, we divide $\mathbb{Z}$ into the following sets:

$$\begin{aligned} L_k &:= \{n \in \mathbb{Z} \mid n \leqslant k-2\}, \\ M_k &:= \{n \in \mathbb{Z} \mid k-1 \leqslant n \leqslant k+1\}, \\ N_k &:= \{n \in \mathbb{Z} \mid n \geqslant k+2\}, \end{aligned}$$

and, for X and Y arbitrary subsets of $\mathbb{Z}$, we define

$$\nu_k^\alpha(X, Y) := \left\|\sum_{\varphi\in X}\sum_{\sigma\in Y} 2^{k\alpha(\cdot)}T(f_\varphi, f_\sigma)\chi_k\right\|_{q(\cdot)}, \text{ and } \nu_k(X, Y) := \nu_k^0(X, Y). \tag{6.60}$$

From Definition 6.30 and (6.59), we have

$$\begin{aligned} \|T(f_1, f_2)\|_{\dot{K}_{q(\cdot)}^{\alpha(\cdot),p),\theta}(\mathbb{R}^n)} &= \sup_{\varepsilon>0}\left(\varepsilon^\theta\sum_{k\in\mathbb{Z}}\left\|2^{k\alpha(\cdot)}T(f_1, f_2)\chi_k\right\|_{q(\cdot)}^{p(1+\varepsilon)}\right)^{\frac{1}{p(1+\varepsilon)}} \\ &\leqslant \sup_{\varepsilon>0}\left(\varepsilon^\theta\sum_{k\in\mathbb{Z}}\left\|\sum_{\varphi\in\mathbb{Z}}\sum_{\sigma\in\mathbb{Z}} 2^{k\alpha(\cdot)}T\left(f_\varphi, f_\sigma\right)\chi_k\right\|_{q(\cdot)}^{p(1+\varepsilon)}\right)^{\frac{1}{p(1+\varepsilon)}} \\ &\lesssim \sum_{j=1}^{9} I_j, \end{aligned}$$

where

$$I_j := \sup_{\varepsilon>0}\left(\varepsilon^\theta\sum_{k\in\mathbb{Z}}\lambda_j^{p(1+\varepsilon)}\right)^{\frac{1}{p(1+\varepsilon)}},$$

and

$$\lambda_1 = \nu_k^\alpha(L_k, L_k), \quad \lambda_2 = \nu_k^\alpha(L_k, M_k), \quad \lambda_3 = \nu_k^\alpha(L_k, N_k),$$

$$\lambda_4 = \nu_k^\alpha(M_k, L_k), \quad \lambda_5 = \nu_k^\alpha(M_k, M_k), \quad \lambda_6 = \nu_k^\alpha(M_k, N_k),$$

$$\lambda_7 = \nu_k^\alpha(N_k, L_k), \quad \lambda_8 = \nu_k^\alpha(N_k, M_k), \quad \lambda_9 = \nu_k^\alpha(N_k, N_k).$$

It is necessary to estimate I_1, I_2, I_3, I_5, I_6, and I_9, since I_4, I_7, and I_8 can be obtained in a similar manner as I_2, I_3, and I_6, respectively.

ESTIMATION FOR I_1: Splitting $\mathbb{Z} = \mathbb{Z}_- \cup \mathbb{N}_0$ and by the asymptotic $2^{k\alpha(x)} \simeq 2^{k\alpha(0)}$ ($x \in R_k$ and $k < 0$) and $2^{k\alpha(x)} \simeq 2^{k\alpha(\infty)}$ ($x \in R_k$ and $k \geqslant 0$), we get

$$\begin{aligned} I_1 \lesssim \sup_{\varepsilon>0} \left(\varepsilon^\theta \sum_{k\in\mathbb{Z}_-} 2^{k\alpha(0)p(1+\varepsilon)} \nu_k(L_k, L_k)^{p(1+\varepsilon)} \right)^{\frac{1}{p(1+\varepsilon)}} \\ + \sup_{\varepsilon>0} \left(\varepsilon^\theta \sum_{k\in\mathbb{N}_0} 2^{k\alpha(\infty)p(1+\varepsilon)} \nu_k(L_k, L_k)^{p(1+\varepsilon)} \right)^{\frac{1}{p(1+\varepsilon)}} \\ =: I_{11} + I_{12}. \end{aligned}$$

For $\varphi, \sigma \in L_k$, $x \in R_k$, $y_1 \in R_\varphi$, and $y_2 \in R_\sigma$, we have

$$|x - y_1| \geqslant ||x| - |y_1|| \geqslant 2^{k-1} - 2^\varphi \geqslant 2^{k-1} - 2^{k-2} \gtrsim 2^k,$$

and

$$|x - y_2| \geqslant ||x| - |y_2|| \geqslant 2^{k-1} - 2^\sigma \geqslant 2^{k-1} - 2^{k-2} \gtrsim 2^k,$$

yielding

$$\begin{aligned} \left|T(f_\varphi, f_\sigma)(x)\right| &\leqslant \int_{\mathbb{R}^n}\int_{\mathbb{R}^n} \frac{\left|f_\varphi(y_1)\right| |f_\sigma(y_2)|}{(|x-y_1|+|x-y_2|)^{2n}} dy_1 dy_2 \\ &\lesssim 2^{-2kn} \int_{\mathbb{R}^n} \left|f_\varphi(y_1)\right| dy_1 \int_{\mathbb{R}^n} |f_\sigma(y_2)| dy_2. \end{aligned} \tag{6.61}$$

From estimate (6.61), Hölder's inequality, $1/q(x) = 1/q_1(x) + 1/q_2(x)$, and Lemma 6.3, we obtain

$$\nu_k(L_k, L_k) \lesssim \sum_{\varphi\in L_k}\sum_{\sigma\in L_k} 2^{-2kn} \|f_\varphi\|_{q_1(\cdot)} \|\chi_\varphi\|_{q_1'(\cdot)} \|f_\sigma\|_{q_2(\cdot)} \|\chi_\sigma\|_{q_2'(\cdot)} \|\chi_k\|_{q(\cdot)}$$

$$\lesssim \sum_{\varphi\in L_k}\sum_{\sigma\in L_k} 2^{-2kn}\|f_\varphi\|_{q_1(\cdot)} 2^{\frac{\varphi n}{q_1'(0)}} \|f_\sigma\|_{q_2(\cdot)} 2^{\frac{\sigma n}{q_2'(0)}} 2^{\frac{kn}{q(0)}}$$

$$\lesssim \left(\sum_{\varphi\in L_k} 2^{(k-\varphi)\left(\frac{n}{q_1(0)}-n\right)}\|f_\varphi\|_{q_1(\cdot)}\right)\left(\sum_{\sigma\in L_k} 2^{(k-\sigma)\left(\frac{n}{q_2(0)}-n\right)}\|f_\sigma\|_{q_2(\cdot)}\right).$$

Taking into account the previous estimate for $\nu_k(L_k, L_k)$, the equality $1/p = 1/p_1 + 1/p_2$, and Hölder's inequality, we have

$$I_{11} \lesssim \sup_{\varepsilon>0}\left[\varepsilon^\theta \sum_{k\in\mathbb{Z}_-}\left(\sum_{\varphi\in L_k} 2^{(k-\varphi)\left(\frac{n}{q_1(0)}-n\right)+k\alpha_1(0)}\|f_\varphi\|_{q_1(\cdot)}\right)^{p_1(1+\varepsilon)}\right]^{\frac{1}{p_1(1+\varepsilon)}}$$

$$\times \sup_{\varepsilon>0}\left[\varepsilon^\theta \sum_{k\in\mathbb{Z}_-}\left(\sum_{\sigma\in L_k} 2^{(k-\sigma)\left(\frac{n}{q_2(0)}-n\right)+k\alpha_2(0)}\|f_\sigma\|_{q_2(\cdot)}\right)^{p_2(1+\varepsilon)}\right]^{\frac{1}{p_2(1+\varepsilon)}}$$

$$=: I_{111}I_{112}.$$

By Hölder's inequality, Fubini's theorem for series, the inequality $2^{-p_1(1+\varepsilon)} < 2^{-p_1}$, and defining $b_1 := n/q_1(0) - n + \alpha_1(0) < 0$, we obtain

$$I_{111} \lesssim \sup_{\varepsilon>0}\left(\varepsilon^\theta \sum_{k\in\mathbb{Z}_-}\left(\sum_{\varphi\in L_k} 2^{b_1(k-\varphi)\frac{p_1(1+\varepsilon)}{2}} 2^{\varphi\alpha_1(0)p_1(1+\varepsilon)}\|f_\varphi\|_{q_1(\cdot)}^{p_1(1+\varepsilon)}\right)\right.$$

$$\left.\times\left(\sum_{\varphi\in L_k} 2^{b_1(k-\varphi)\frac{(p_1(1+\varepsilon))'}{2}}\right)^{\frac{p_1(1+\varepsilon)}{(p_1(1+\varepsilon))'}}\right)^{\frac{1}{p_1(1+\varepsilon)}}$$

$$\lesssim \sup_{\varepsilon>0}\left(\varepsilon^\theta \sum_{\varphi=-\infty}^{-1} 2^{\varphi\alpha_1(0)p_1(1+\varepsilon)}\|f_\varphi\|_{q_1(\cdot)}^{p_1(1+\varepsilon)} \sum_{k=\varphi+2}^{-1} 2^{b_1(k-\varphi)\frac{p_1(1+\varepsilon)}{2}}\right)^{\frac{1}{p_1(1+\varepsilon)}}$$

$$\lesssim \sup_{\varepsilon>0}\left(\varepsilon^\theta \sum_{\varphi\in L_k} 2^{\varphi\alpha_1(0)p_1(1+\varepsilon)}\|f_\varphi\|_{q_1(\cdot)}^{p_1(1+\varepsilon)}\right)^{\frac{1}{p_1(1+\varepsilon)}}$$

$$\lesssim \|f_1\|_{\dot{K}^{\alpha_1(\cdot),p_1),\theta}_{q_1(\cdot)}(\mathbb{R}^n)}.$$

The estimate $I_{112} \lesssim \|f_2\|_{\dot{K}_{q_2(\cdot)}^{\alpha_2(\cdot),p_2),\theta}(\mathbb{R}^n)}$ is obtained, mutatis mutandis, via the estimation for I_{111}. With this estimates at hand, we obtain

$$I_{11} \lesssim \|f_1\|_{\dot{K}_{q_1(\cdot)}^{\alpha_1(\cdot),p_1),\theta}(\mathbb{R}^n)} \|f_2\|_{\dot{K}_{q_2(\cdot)}^{\alpha_2(\cdot),p_2),\theta}(\mathbb{R}^n)}.$$

To estimate I_{12}, we split as follows:

$$\begin{aligned} I_{12} \lesssim & \sup_{\varepsilon>0}\left(\varepsilon^\theta \sum_{k\in\mathbb{N}_0} 2^{k\alpha(\infty)p(1+\varepsilon)} \left\| \sum_{\varphi\in\mathbb{Z}_-}\sum_{\sigma\in\mathbb{Z}_-} T(f_\varphi, f_\sigma)\chi_k \right\|_{q(\cdot)}^{p(1+\varepsilon)}\right)^{\frac{1}{p(1+\varepsilon)}} \\ & + \sup_{\varepsilon>0}\left(\varepsilon^\theta \sum_{k\in\mathbb{N}_0} 2^{k\alpha(\infty)p(1+\varepsilon)} \left\| \sum_{\varphi=0}^{k-2}\sum_{\sigma=0}^{k-2} T(f_\varphi, f_\sigma)\chi_k \right\|_{q(\cdot)}^{p(1+\varepsilon)}\right)^{\frac{1}{p(1+\varepsilon)}} \\ =: & I_{121} + I_{122}. \end{aligned}$$

The estimate I_{122} follows in similar manner as in I_{11} with simply replacing $\alpha_i(0)$ by $\alpha_i(\infty)$ and the fact $q_i(0) \leqslant q_i(\infty)$.

For estimation I_{121}, by Hölder's inequality, Lemma 6.3, and the inequality $q_i(0) \leqslant q_i(\infty)$, we obtain

$$\begin{aligned} \nu_k(\mathbb{Z}_-,\mathbb{Z}_-) &\lesssim \sum_{\varphi\in\mathbb{Z}_-}\sum_{\sigma\in\mathbb{Z}_-} 2^{-2kn}\|f_\varphi\|_{q_1(\cdot)}\|\chi_\varphi\|_{q_1'(\cdot)}\|f_\sigma\|_{q_2(\cdot)}\|\chi_\sigma\|_{q_2'(\cdot)}\|\chi_k\|_{q(\cdot)} \\ &\lesssim \sum_{\varphi\in\mathbb{Z}_-}\sum_{\sigma\in\mathbb{Z}_-} 2^{-2kn}\|f_\varphi\|_{q_1(\cdot)} 2^{\frac{\varphi n}{q_1'(0)}}\|f_\sigma\|_{q_2(\cdot)} 2^{\frac{\sigma n}{q_2'(0)}} 2^{\frac{kn}{q(\infty)}} \\ &\lesssim \left(\sum_{\varphi\in\mathbb{Z}_-} 2^{(k-\varphi)\left(\frac{n}{q_1(0)}-n\right)}\|f_\varphi\|_{q_1(\cdot)}\right)\left(\sum_{\sigma\in\mathbb{Z}_-} 2^{(k-\sigma)\left(\frac{n}{q_2(0)}-n\right)}\|f_\sigma\|_{q_2(\cdot)}\right). \end{aligned}$$

From the estimate of $\nu_k(\mathbb{Z}_-,\mathbb{Z}_-)$, the equality $1/p = 1/p_1 + 1/p_2$, and Hölder's inequality, we have

$$\begin{aligned} I_{121} \lesssim & \sup_{\varepsilon>0}\left[\varepsilon^\theta \sum_{k\in\mathbb{N}_0}\left(\sum_{\varphi\in\mathbb{Z}_-} 2^{(k-\varphi)\left(\frac{n}{q_1(0)}-n\right)+k\alpha_1(\infty)}\|f_\varphi\|_{q_1(\cdot)}\right)^{p_1(1+\varepsilon)}\right]^{\frac{1}{p_1(1+\varepsilon)}} \\ & \times \sup_{\varepsilon>0}\left[\varepsilon^\theta \sum_{k\in\mathbb{N}_0}\left(\sum_{\sigma\in\mathbb{Z}_-} 2^{(k-\sigma)\left(\frac{n}{q_2(0)}-n\right)+k\alpha_2(\infty)}\|f_\sigma\|_{q_2(\cdot)}\right)^{p_2(1+\varepsilon)}\right]^{\frac{1}{p_2(1+\varepsilon)}} \\ =: & A_1A_2. \end{aligned}$$

Invoking the Hölder inequality and defining $\xi_1 := \frac{n}{q_1(0)} - n + \alpha_1(\infty) < 0$, we have

$$A_1 \lesssim \sup_{\varepsilon>0} \left[\varepsilon^{\theta} \sum_{k\in\mathbb{N}_0} \left(\sum_{\varphi\in\mathbb{Z}_-} 2^{(k-\varphi)\xi_1 \frac{p_1(1+\varepsilon)}{2}} 2^{\varphi\alpha_1(\infty)p_1(1+\varepsilon)} \|f_\varphi\|_{q_1(\cdot)}^{p_1(1+\varepsilon)} \right) \right.$$

$$\left. \times \left(\sum_{\varphi\in\mathbb{Z}_-} 2^{(k-\varphi)\xi_1 \frac{(p_1(1+\varepsilon))'}{2}} \right)^{\frac{p_1(1+\varepsilon)}{(p_1(1+\varepsilon))'}} \right]^{\frac{1}{p_1(1+\varepsilon)}}$$

$$\lesssim \sup_{\varepsilon>0} \left[\varepsilon^{\theta} \sum_{k\in\mathbb{N}_0} 2^{k\xi_1 \frac{p_1(1+\varepsilon)}{2}} \left(\sum_{\varphi\in\mathbb{Z}_-} 2^{-\varphi\xi_1 \frac{p_1(1+\varepsilon)}{2}} 2^{\varphi\alpha_1(\infty)p_1(1+\varepsilon)} \|f_\varphi\|_{q_1(\cdot)}^{p_1(1+\varepsilon)} \right) \right]^{\frac{1}{p_1(1+\varepsilon)}}$$

$$\lesssim \sup_{\varepsilon>0} \left(\varepsilon^{\theta} \sum_{\varphi\in\mathbb{Z}_-} 2^{\varphi\alpha_1(\infty)p_1(1+\varepsilon)} \|f_\varphi\|_{q_1(\cdot)}^{p_1(1+\varepsilon)} \right)^{\frac{1}{p_1(1+\varepsilon)}}$$

$$\lesssim \|f_1\|_{\dot{K}_{q_1(\cdot)}^{\alpha_1(\cdot),p_1),\theta}(\mathbb{R}^n)}.$$

Similar estimate, with the corresponding changes, is obtained for A_2, from which we obtain $I_{12} \lesssim \|f_1\|_{\dot{K}_{q_1(\cdot)}^{\alpha_1(\cdot),p_1),\theta}(\mathbb{R}^n)} \|f_2\|_{\dot{K}_{q_2(\cdot)}^{\alpha_2(\cdot),p_2),\theta}(\mathbb{R}^n)}$. Hence

$$I_1 \lesssim \|f_1\|_{\dot{K}_{q_1(\cdot)}^{\alpha_1(\cdot),p_1),\theta}(\mathbb{R}^n)} \|f_2\|_{\dot{K}_{q_2(\cdot)}^{\alpha_2(\cdot),p_2),\theta}(\mathbb{R}^n)}. \tag{6.62}$$

ESTIMATION FOR I_2: As in the case of I_1, we obtain the following estimate

$$I_2 \lesssim \sup_{\varepsilon>0} \left(\varepsilon^{\theta} \sum_{k\in\mathbb{Z}_-} 2^{k\alpha(0)p(1+\varepsilon)} \nu_k(L_k, M_k)^{p(1+\varepsilon)} \right)^{\frac{1}{p(1+\varepsilon)}}$$

$$+ \sup_{\varepsilon>0} \left(\varepsilon^{\theta} \sum_{k\in\mathbb{N}_0} 2^{k\alpha(\infty)p(1+\varepsilon)} \nu_k(L_k, M_k)^{p(1+\varepsilon)} \right)^{\frac{1}{p(1+\varepsilon)}}$$

$$=: I_{21} + I_{22}.$$

Notice that, for $x \in R_k$, $y_1 \in R_\varphi$, $y_2 \in R_\sigma$, $\varphi \in L_k$, and $\sigma \in M_k$, we have

$$|x - y_1| + |x - y_2| \geqslant |x - y_1| \geqslant ||x| - |y_1|| \gtrsim 2^k, \tag{6.63}$$

from which, taking Lemma 6.3 into consideration and elementary computations, we obtain

$$\nu_k(L_k, M_k) \lesssim \sum_{\varphi\in L_k}\sum_{\sigma\in M_k} 2^{(k-\varphi)\left(\frac{n}{q_1(0)}-n\right)}\|f_\varphi\|_{q_1(\cdot)}2^{(\sigma-k)n}2^{\frac{(k-\sigma)n}{q_2(0)}}\|f_\sigma\|_{q_2(\cdot)}, \quad (k\in\mathbb{Z}_-).$$

From the estimate for $\nu_k(L_k, M_k)$ and Hölder's inequality, we get

$$\begin{aligned} I_{21} &\lesssim \sup_{\varepsilon>0}\left(\varepsilon^\theta\sum_{k\in\mathbb{Z}_-}\left(\sum_{\varphi\in L_k}2^{(k-\varphi)\left(\frac{n}{q_1(0)}-n\right)+k\alpha_1(0)}\|f_\varphi\|_{q_1(\cdot)}\right)^{p_1(1+\varepsilon)}\right)^{\frac{1}{p_1(1+\varepsilon)}} \\ &\quad\times \sup_{\varepsilon>0}\left(\varepsilon^\theta\sum_{k\in\mathbb{Z}}\left(\sum_{\sigma\in M_k}2^{k\alpha_2(0)}2^{(\sigma-k)n}\|f_\sigma\|_{q_2(\cdot)}\right)^{p_2(1+\varepsilon)}\right)^{\frac{1}{p_2(1+\varepsilon)}} \\ &=: I_{211}I_{212}. \end{aligned}$$

Notice that $I_{211} = I_{111}$. For the estimate I_{212}, we reason as follows:

$$\begin{aligned} I_{212} &\lesssim \sup_{\varepsilon>0}\left(\varepsilon^\theta\sum_{k\in\mathbb{Z}_-}2^{k\alpha_2(0)p_2(1+\varepsilon)}\left(\sum_{\sigma\in M_k}2^{(\sigma-k)n}\|f_\sigma\|_{q_2(\cdot)}\right)^{p_2(1+\varepsilon)}\right)^{\frac{1}{p_2(1+\varepsilon)}} \\ &\lesssim \sup_{\varepsilon>0}\left(\varepsilon^\theta\sum_{k\in\mathbb{Z}_-}2^{k\alpha_2(0)p_2(1+\varepsilon)}\|f_2\chi_k\|_{q_2(\cdot)}^{p_2(1+\varepsilon)}\right)^{\frac{1}{p_2(1+\varepsilon)}} \\ &\lesssim \|f_2\|_{\dot{K}^{\alpha_2(\cdot),p_2),\theta}_{q_2(\cdot)}(\mathbb{R}^n)}. \end{aligned}$$

The term I_{22} is estimated by

$$\begin{aligned} I_{22} &\lesssim \sup_{\varepsilon>0}\left(\varepsilon^\theta\sum_{k\in\mathbb{N}_0}2^{k\alpha(\infty)p(1+\varepsilon)}\nu_k(\mathbb{Z}_-, M_k)^{p(1+\varepsilon)}\right)^{\frac{1}{p(1+\varepsilon)}} \\ &\quad+ \sup_{\varepsilon>0}\left(\varepsilon^\theta\sum_{k\in\mathbb{N}_0}2^{k\alpha(\infty)p(1+\varepsilon)}\nu_k(\overline{0, k-2}, M_k)^{p(1+\varepsilon)}\right)^{\frac{1}{p(1+\varepsilon)}} \\ &=: I_{221} + I_{222}. \end{aligned}$$

To estimate I_{221}, by Hölder's inequality, Lemma 6.3, and the inequality $q_1(0) \leqslant q_1(\infty)$, we have

$$\nu_k(\mathbb{Z}_-, M_k) \lesssim \sum_{\varphi\in\mathbb{Z}_-}\sum_{\sigma\in M_k} 2^{(k-\varphi)\left(\frac{n}{q_1(0)}-n\right)} \| f_\varphi \|_{q_1(\cdot)} 2^{(\sigma-k)n} 2^{\frac{(k-\sigma)n}{q_2(\infty)}} \| f_\sigma \|_{q_2(\cdot)}. \tag{6.64}$$

Thus using (6.64) and by Hölder's inequality, we have

$$\begin{aligned} I_{221} &\lesssim \sup_{\varepsilon>0}\left(\varepsilon^\theta \sum_{k\in\mathbb{N}_0}\left(\sum_{\varphi\in\mathbb{Z}_-} 2^{(k-\varphi)\left(\frac{n}{q_1(0)}-n\right)+k\alpha_1(\infty)} \| f_\varphi \|_{q_1(\cdot)}\right)^{p_1(1+\varepsilon)}\right)^{\frac{1}{p_1(1+\varepsilon)}} \\ &\quad\times \sup_{\varepsilon>0}\left(\varepsilon^\theta \sum_{k\in\mathbb{N}_0}\left(\sum_{\sigma\in M_k} 2^{k\alpha_2(\infty)} 2^{(\sigma-k)n} \| f_\sigma \|_{q_2(\cdot)}\right)^{p_2(1+\varepsilon)}\right)^{\frac{1}{p_2(1+\varepsilon)}} \\ &=: B_1 B_2. \end{aligned}$$

The term B_1 is equal to A_1 and for B_2 we use similar arguments as for I_{212}, replacing $\alpha_2(0)$ with $\alpha_2(\infty)$.

For the term I_{222}, by Hölder's inequality and Lemma 6.3, we have

$$\nu_k(\overline{0, k-2}, M_k) \lesssim \sum_{\varphi=0}^{k-2}\sum_{\sigma\in M_k} 2^{(k-\varphi)\left(\frac{n}{q_1(\infty)}-n\right)} \| f_\varphi \|_{q_1(\cdot)} 2^{(\sigma-k)n} \| f_\sigma \|_{q_2(\cdot)}. \tag{6.65}$$

Taking into consideration (6.65) and applying Hölder's inequality, we get

$$\begin{aligned} I_{222} &\lesssim \sup_{\varepsilon>0}\left(\varepsilon^\theta \sum_{k\in\mathbb{N}_0}\left(\sum_{\varphi=0}^{k-2} 2^{(k-\varphi)\left(\frac{n}{q_1(\infty)}-n\right)+k\alpha_1(\infty)} \| f_\varphi \|_{q_1(\cdot)}\right)^{p_1(1+\varepsilon)}\right)^{\frac{1}{p_1(1+\varepsilon)}} \\ &\quad\times \sup_{\varepsilon>0}\left(\varepsilon^\theta \sum_{k\in\mathbb{N}_0}\left(\sum_{\sigma\in M_k} 2^{k\alpha_2(\infty)} 2^{(\sigma-k)n} \| f_\sigma \|_{q_2(\cdot)}\right)^{p_2(1+\varepsilon)}\right)^{\frac{1}{p_2(1+\varepsilon)}} \\ &=: B_1' B_2'. \end{aligned}$$

Noting that $B_2' = B_2$ and estimating B_1' in a similar fashion as I_{212}, we obtain $I_{22} \lesssim \|f_1\|_{\dot{K}^{\alpha_1(\cdot),p_1),\theta}_{q_1(\cdot)}(\mathbb{R}^n)} \|f_2\|_{\dot{K}^{\alpha_2(\cdot),p_2),\theta}_{q_2(\cdot)}(\mathbb{R}^n)}$, from which we get

$$I_2 \lesssim \|f_1\|_{\dot{K}^{\alpha_1(\cdot),p_1),\theta}_{q_1(\cdot)}(\mathbb{R}^n)} \|f_2\|_{\dot{K}^{\alpha_2(\cdot),p_2),\theta}_{q_2(\cdot)}(\mathbb{R}^n)}. \tag{6.66}$$

ESTIMATION FOR I_3: We have

$$\begin{aligned} I_3 &\lesssim \sup_{\varepsilon>0} \left(\varepsilon^\theta \sum_{k\in\mathbb{Z}_-} 2^{k\alpha(0)p(1+\varepsilon)} v_k(L_k, N_k)^{p(1+\varepsilon)} \right)^{\frac{1}{p(1+\varepsilon)}} \\ &\quad + \sup_{\varepsilon>0} \left(\varepsilon^\theta \sum_{k\in\mathbb{N}_0} 2^{k\alpha(\infty)p(1+\varepsilon)} v_k(L_k, N_k)^{p(1+\varepsilon)} \right)^{\frac{1}{p(1+\varepsilon)}} \\ &=: I_{31} + I_{32}. \end{aligned}$$

For $\varphi \in L_k$, $\sigma \in N_k$, $x \in R_k$, $y_1 \in R_\varphi$, and $y_2 \in R_\sigma$, we get

$$|x - y_1| \geqslant |x| - |y_1| \gtrsim 2^k, \quad |x - y_2| \geqslant |y_2| - |x| \gtrsim 2^\sigma,$$

from which, taking into account the elementary inequality $(2^k + 2^\sigma)^{2n} \gtrsim 2^{kn+\sigma n}$, it follows

$$|T(f_\varphi, f_\sigma)(x)| \lesssim 2^{-kn} \int\limits_{\mathbb{R}^n} |f_\varphi(y_1)| dy_1 2^{-\sigma n} \int\limits_{\mathbb{R}^n} |f_\sigma(y_2)| dy_2, \quad (x \in R_k).$$

By Hölder's inequality, Lemma 6.3, $q_2(0) \leqslant q_2(\infty)$, and $k \in \mathbb{Z}_-$, we obtain

$$v_k(L_k, N_k) \lesssim \sum_{\varphi\in L_k} 2^{(k-\varphi)\left(\frac{n}{q_1(0)} - n\right)} \|f_\varphi\|_{q_1(\cdot)} \sum_{\sigma\in N_k} 2^{\frac{(k-\sigma)n}{q_2(\infty)}} \|f_\sigma\|_{q_2(\cdot)}. \tag{6.67}$$

To estimate the term I_{31}, using (6.67) and the Hölder inequality, we obtain

$$I_{31} \lesssim \sup_{\varepsilon>0} \left(\varepsilon^\theta \sum_{k\in\mathbb{Z}_-} \left(\sum_{\varphi\in L_k} 2^{(k-\varphi)\left(\frac{n}{q_1(0)} - n + \alpha_1(0)\right)} 2^{\varphi\alpha_1(0)} \|f_\varphi\|_{q_1(\cdot)} \right)^{p_1(1+\varepsilon)} \right)^{\frac{1}{p_1(1+\varepsilon)}}$$

$$\times \sup_{\varepsilon>0}\left(\varepsilon^{\theta}\sum_{k\in\mathbb{Z}_-}\left(\sum_{\sigma\in N_k}2^{(k-\sigma)\left(\frac{n}{q_2(\infty)}+\alpha_2(0)\right)}2^{\sigma\alpha_2(0)}\|f_\sigma\|_{q_2(\cdot)}\right)^{p_2(1+\varepsilon)}\right)^{\frac{1}{p_2(1+\varepsilon)}}$$

$$=: I_{311}I_{312}.$$

We have $I_{311}\lesssim\|f_1\|_{\dot{K}^{\alpha_1(\cdot),p_1),\theta}_{q_1(\cdot)}(\mathbb{R}^n)}$, since $I_{311}=I_{111}$.

As for I_{312}, we split as follows:

$$\begin{aligned}I_{312}&\lesssim \sup_{\varepsilon>0}\left(\varepsilon^{\theta}\sum_{k\in\mathbb{Z}_-}\left(\sum_{\sigma=k+2}^{-1}2^{(k-\sigma)\left(\frac{n}{q_2(\infty)}+\alpha_2(0)\right)}2^{\sigma\alpha_2(0)}\|f_\sigma\|_{q_2(\cdot)}\right)^{p_2(1+\varepsilon)}\right)^{\frac{1}{p_2(1+\varepsilon)}}\\&\quad+\sup_{\varepsilon>0}\left(\varepsilon^{\theta}\sum_{k\in\mathbb{Z}_-}\left(\sum_{\sigma\in\mathbb{N}_0}2^{(k-\sigma)\left(\frac{n}{q_2(\infty)}+\alpha_2(0)\right)}2^{\sigma\alpha_2(0)}\|f_\sigma\|_{q_2(\cdot)}\right)^{p_2(1+\varepsilon)}\right)^{\frac{1}{p_2(1+\varepsilon)}}\\&=: D_1+D_2.\end{aligned}$$

By Hölder's inequality, Fubini's theorem for series, the inequality $\frac{n}{q_2(\infty)}+\alpha_2(0)>0$ and $2^{-p_2(1+\varepsilon)}<2^{-p_2}$, we obtain

$$\begin{aligned}D_1&\lesssim \sup_{\varepsilon>0}\left[\varepsilon^{\theta}\sum_{k\in\mathbb{Z}_-}\left(\sum_{\sigma=k+2}^{-1}2^{\sigma\alpha_2(0)p_2(1+\varepsilon)}\|f_\sigma\|_{q_2(\cdot)}^{p_2(1+\varepsilon)}2^{\frac{(k-\sigma)\left(\frac{n}{q_2(\infty)}+\alpha_2(0)\right)p_2(1+\varepsilon)}{2}}\right)\right.\\&\qquad\left.\times\left(\sum_{\sigma=k+2}^{-1}2^{\frac{(k-\sigma)\left(\frac{n}{q_2(\infty)}+\alpha_2(0)\right)(p_2(1+\varepsilon))'}{2}}\right)^{\frac{p_2(1+\varepsilon)}{(p_2(1+\varepsilon))'}}\right]^{\frac{1}{p_2(1+\varepsilon)}}\\&\lesssim \sup_{\varepsilon>0}\left(\varepsilon^{\theta}\sum_{k\in\mathbb{Z}_-}\sum_{\sigma=k+2}^{-1}2^{\sigma\alpha_2(0)}2^{\frac{(k-\sigma)\left(\frac{n}{q_2(\infty)}+\alpha_2(0)\right)p_2(1+\varepsilon)}{2}}\|f_\sigma\|_{q_2(\cdot)}^{p_2(1+\varepsilon)}\right)^{\frac{1}{p_2(1+\varepsilon)}}\\&\lesssim \sup_{\varepsilon>0}\left(\varepsilon^{\theta}\sum_{\sigma=-\infty}^{-1}2^{\sigma\alpha_2(0)p_2(1+\varepsilon)}\|f_\sigma\|_{q_2(\cdot)}^{p_2(1+\varepsilon)}\sum_{k=-\infty}^{\sigma-2}2^{\frac{(k-\sigma)\left(\frac{n}{q_2(\infty)}+\alpha_2(0)\right)}{2}}\right)^{\frac{1}{p_2(1+\varepsilon)}}\\&\lesssim\|f_2\|_{\dot{K}^{\alpha_2(\cdot),p_2),\theta}_{q_2(\cdot)}(\mathbb{R}^n)}.\end{aligned}$$

For D_2, using the same argument as above and the inequality $\alpha_2(0) \leqslant \alpha_2(\infty)$, we obtain

$$D_2 \lesssim \sup_{\varepsilon>0}\left(\varepsilon^{\theta}\sum_{k\in\mathbb{Z}_-}\sum_{\sigma\in\mathbb{N}_0} 2^{\sigma\alpha_2(0)p_2(1+\varepsilon)}\|f_\sigma\|_{q_2(\cdot)}^{p_2(1+\varepsilon)} 2^{\frac{(k-\sigma)\left(\frac{n}{q_2(\infty)}+\alpha_2(0)\right)p_2(1+\varepsilon)}{2}}\right)^{\frac{1}{p_2(1+\varepsilon)}}$$

$$\lesssim \sup_{\varepsilon>0}\left(\varepsilon^{\theta}\sum_{k\in\mathbb{Z}_-}\sum_{\sigma\in\mathbb{N}_0} 2^{\frac{(k-\sigma)\left(\frac{n}{q_2(\infty)}+\alpha_2(0)\right)p_2(1+\varepsilon)}{2}}\times\right.$$

$$\left.\times\sum_{j=\sigma}^{\infty} 2^{j\alpha_2(\infty)p_2(1+\varepsilon)}\|f_2\chi_{R_j}\|_{q_2(\cdot)}^{p_2(1+\varepsilon)}\right)^{\frac{1}{p_2(1+\varepsilon)}}$$

$$\lesssim \sup_{\varepsilon>0}\left(\sum_{k\in\mathbb{Z}_-}\sum_{\sigma\in\mathbb{N}_0} 2^{\frac{(k-\sigma)\left(\frac{n}{q_2(\infty)}+\alpha_2(0)\right)p_2(1+\varepsilon)}{2}}\|f_2\|_{\dot{K}_{q_2(\cdot)}^{\alpha_2(\cdot),p_2),\theta}(\mathbb{R}^n)}^{p_2(1+\varepsilon)}\right)^{\frac{1}{p_2(1+\varepsilon)}}$$

$$\lesssim \|f_2\|_{\dot{K}_{q_2(\cdot)}^{\alpha_2(\cdot),p_2),\theta}(\mathbb{R}^n)}.$$

So $I_{31} \lesssim \|f_1\|_{\dot{K}_{q_1(\cdot)}^{\alpha_1,p_1),\theta}(\mathbb{R}^n)}\|f_2\|_{\dot{K}_{q_2(\cdot)}^{\alpha_2,p_2),\theta}(\mathbb{R}^n)}$.

The term I_{32} can be estimated as

$$I_{32} \lesssim \sup_{\varepsilon>0}\left(\varepsilon^{\theta}\sum_{k\in\mathbb{N}_0} 2^{k\alpha(\infty)p(1+\varepsilon)}\nu_k(\mathbb{Z}_-,N_k)^{p(1+\varepsilon)}\right)^{\frac{1}{p(1+\varepsilon)}}$$

$$+\sup_{\varepsilon>0}\left(\varepsilon^{\theta}\sum_{k\in\mathbb{N}_0} 2^{k\alpha(\infty)p(1+\varepsilon)}\nu_k(\overline{0,k-2},N_k)^{p(1+\varepsilon)}\right)^{\frac{1}{p(1+\varepsilon)}}$$

$$=: I_{321}+I_{322}.$$

For $k \in \mathbb{N}_0$, $\varphi \in \mathbb{Z}_-$, $\sigma \in N_k$, $q_1(0) \leqslant q_1(\infty)$, and Lemma 6.3, we obtain

$$\nu_k(\mathbb{Z}_-,N_k) \lesssim \sum_{\varphi\in\mathbb{Z}_-} 2^{(k-\varphi)\left(\frac{n}{q_1(0)}-n\right)}\|f_\varphi\|_{q_1(\cdot)}\sum_{\sigma\in N_k} 2^{\frac{(k-\sigma)n}{q_2(\infty)}}\|f_\sigma\|_{q_2(\cdot)}. \tag{6.68}$$

By (6.68) and Hölder's inequality, we get

$$
\begin{aligned}
&I_{321}\\
&\lesssim \sup_{\varepsilon>0}\left(\varepsilon^{\theta}\sum_{k\in\mathbb{N}_0}\left(\sum_{\varphi\in\mathbb{Z}_-}2^{(k-\varphi)\left(\frac{n}{q_1(0)}-n+\alpha_1(\infty)\right)}2^{\varphi\alpha_1(\infty)}\|f_\varphi\|_{q_1(\cdot)}\right)^{p_1(1+\varepsilon)}\right)^{\frac{1}{p_1(1+\varepsilon)}}\\
&\times \sup_{\varepsilon>0}\left(\varepsilon^{\theta}\sum_{k\in\mathbb{N}_0}\left(\sum_{\sigma\in N_k}2^{(k-\sigma)\left(\frac{n}{q_2(\infty)}+\alpha_2(\infty)\right)}2^{\sigma\alpha_2(\infty)}\|f_\sigma\|_{q_2(\cdot)}\right)^{p_2(1+\varepsilon)}\right)^{\frac{1}{p_2(1+\varepsilon)}}\\
&=: D_1'D_2'.
\end{aligned}
$$

Note that $D_1' = A_1$. The estimate for D_2' can be obtained in a similar way as for I_{312}, by replacing $\alpha_2(0)$ with $\alpha_2(\infty)$.

For estimation I_{322}, by Hölder's inequality and Lemma 6.3, we have

$$
\nu_k(\overline{0,k-2},N_k)\lesssim\sum_{\varphi=0}^{k-2}\sum_{\sigma\in N_k}2^{(k-\varphi)\left(\frac{n}{q_1(\infty)}-n\right)}\|f_\varphi\|_{q_1(\cdot)}2^{\frac{(k-\sigma)n}{q_2(\infty)}}\|f_\sigma\|_{q_2(\cdot)}.
\tag{6.69}
$$

Using (6.69) and Hölder's inequality, we obtain

$$
\begin{aligned}
I_{322}&\lesssim \sup_{\varepsilon>0}\left(\varepsilon^{\theta}\sum_{k\in\mathbb{N}_0}\left(\sum_{\varphi=0}^{k-2}2^{(k-\varphi)\left(\frac{n}{q_1(\infty)}-n+\alpha_1(\infty)\right)}2^{\varphi\alpha_1(\infty)}\|f_\varphi\|_{q_1(\cdot)}\right)^{p_1(1+\varepsilon)}\right)^{\frac{1}{p_1(1+\varepsilon)}}\\
&\quad\times \sup_{\varepsilon>0}\left(\varepsilon^{\theta}\sum_{k\in\mathbb{N}_0}\left(\sum_{\sigma\in N_k}2^{(k-\sigma)\left(\frac{n}{q_2(\infty)}+\alpha_2(\infty)\right)}2^{\sigma\alpha_2(\infty)}\|f_\sigma\|_{q_2(\cdot)}\right)^{p_2(1+\varepsilon)}\right)^{\frac{1}{p_2(1+\varepsilon)}}\\
&=: D_1''D_2''.
\end{aligned}
$$

We have $D_2'' = D_2'$ and the estimate for the term D_1'' is similar to B_1'. Taking all the estimates into account yields

$$
I_3\lesssim \|f_1\|_{\dot{K}^{\alpha_1(\cdot),p_1),\theta}_{q_1(\cdot)}(\mathbb{R}^n)}\|f_2\|_{\dot{K}^{\alpha_2(\cdot),p_2),\theta}_{q_2(\cdot)}(\mathbb{R}^n)}.
\tag{6.70}
$$

ESTIMATION FOR I_5: we have

$$I_5 \lesssim \sup_{\varepsilon>0}\left(\varepsilon^{\theta}\sum_{k\in\mathbb{Z}_-}2^{k\alpha(0)p(1+\varepsilon)}\nu_k(M_k, M_k)^{p(1+\varepsilon)}\right)^{\frac{1}{p(1+\varepsilon)}}$$

$$+\sup_{\varepsilon>0}\left(\varepsilon^{\theta}\sum_{k\in\mathbb{N}_0}2^{k\alpha(\infty)p(1+\varepsilon)}\nu_k(M_k, M_k)^{p(1+\varepsilon)}\right)^{\frac{1}{p(1+\varepsilon)}}$$

$$=: I_{51}+I_{52}.$$

By the $L^{q(\cdot)}$-boundedness of T (see Lemma 6.36), we obtain that

$$\nu_k(M_k, M_k)\lesssim \sum_{\varphi\in M_k}\sum_{\sigma\in M_k}\|f_\varphi\|_{q_1(\cdot)}\|f_\sigma\|_{q_2(\cdot)}.$$

By Hölder's inequality, we have

$$I_{51}\lesssim \sup_{\varepsilon>0}\left(\varepsilon^{\theta}\sum_{k\in\mathbb{Z}_-}\left(\sum_{\varphi\in M_k}2^{k\alpha_1(0)}\|f_\varphi\|_{q_1(\cdot)}\right)^{p_1(1+\varepsilon)}\right)^{\frac{1}{p_1(1+\varepsilon)}}$$

$$\times \sup_{\varepsilon>0}\left(\varepsilon^{\theta}\sum_{k\in\mathbb{Z}_-}\left(\sum_{\sigma\in M_k}2^{k\alpha_2(0)}\|f_\sigma\|_{q_2(\cdot)}\right)^{p_2(1+\varepsilon)}\right)^{\frac{1}{p_2(1+\varepsilon)}}$$

$$\lesssim \|f_1\|_{\dot{K}^{\alpha_1(\cdot),p_1),\theta}_{q_1(\cdot)}(\mathbb{R}^n)}\|f_2\|_{\dot{K}^{\alpha_2(\cdot),p_2),\theta}_{q_2(\cdot)}(\mathbb{R}^n)}.$$

We can obtain similar estimate for I_{52}, replacing $\alpha_i(0)$ by $\alpha_i(\infty)$ with $k \in \mathbb{N}_0$. Therefore

$$I_5 \lesssim \|f_1\|_{\dot{K}^{\alpha_1(\cdot),p_1),\theta}_{q_1(\cdot)}(\mathbb{R}^n)}\|f_2\|_{\dot{K}^{\alpha_2(\cdot),p_2),\theta}_{q_2(\cdot)}(\mathbb{R}^n)}. \tag{6.71}$$

ESTIMATION FOR I_6: we have

$$I_6 \lesssim \sup_{\varepsilon>0}\left(\varepsilon^{\theta}\sum_{k\in\mathbb{Z}_-}2^{k\alpha(0)p(1+\varepsilon)}\nu_k(M_k, N_k)^{p(1+\varepsilon)}\right)^{\frac{1}{p(1+\varepsilon)}}$$

$$+\sup_{\varepsilon>0}\left(\varepsilon^{\theta}\sum_{k\in\mathbb{N}_0}2^{k\alpha(\infty)p(1+\varepsilon)}\nu_k(M_k,N_k)^{p(1+\varepsilon)}\right)^{\frac{1}{p(1+\varepsilon)}}$$

$$=: I_{61}+I_{62}.$$

For $k\in\mathbb{Z}_-$, $\varphi\in M_k$, $\sigma\in N_k$, and $x\in R_k$, we have

$$|x-y_2|\geqslant|y_2|-|x|\gtrsim 2^{\sigma},$$

yielding

$$|T(f_\varphi,f_\sigma)(x)|\lesssim 2^{-kn}\int_{\mathbb{R}^n}|f_\varphi(y_1)|dy_1 2^{-\sigma n}\int_{\mathbb{R}^n}|f_\sigma(y_2)|dy_2.$$

By Hölder's inequality, Lemma 6.3, $k\in\mathbb{Z}_-$, and inequality $q_2(0)\leqslant q_2(\infty)$, we obtain

$$\nu_k(M_k,N_k)\lesssim\sum_{\varphi\in M_k}\sum_{\sigma\in N_k}2^{(\varphi-k)n}\|f_\varphi\|_{q_1(\cdot)}2^{\frac{(k-\sigma)n}{q_2(\infty)}}\|f_\sigma\|_{q_2(\cdot)}. \tag{6.72}$$

By (6.72) and invoking Hölder's inequality, we get

$$I_{61}\lesssim\sup_{\varepsilon>0}\left(\varepsilon^{\theta}\sum_{k\in\mathbb{Z}_-}\left(\sum_{\varphi\in M_k}2^{(\varphi-k)n}2^{k\alpha_1(0)}\|f_\varphi\|_{q_1(\cdot)}\right)^{p_1(1+\varepsilon)}\right)^{\frac{1}{p_1(1+\varepsilon)}}$$

$$\times\sup_{\varepsilon>0}\left(\varepsilon^{\theta}\sum_{k\in\mathbb{Z}_-}\left(\sum_{\sigma\in N_k}2^{(k-\sigma)\left(\frac{n}{q_2(\infty)}+\alpha_2(0)\right)}2^{\sigma\alpha_2(0)}\|f_\sigma\|_{q_2(\cdot)}\right)^{p_2(1+\varepsilon)}\right)^{\frac{1}{p_2(1+\varepsilon)}}$$

$$=: I_{611}I_{612}.$$

Note that the estimate I_{611} is equal to that of I_{51} and I_{612} is obtained by similar argument used in I_{312}.

For $k\in\mathbb{N}_0$, we have

$$\nu_k(M_k,N_k)\lesssim\sum_{\varphi\in M_k}\sum_{\sigma\in N_k}2^{(\varphi-k)n}\|f_\varphi\|_{q_1(\cdot)}2^{\frac{(k-\sigma)n}{q_2(\infty)}}\|f_\sigma\|_{q_2(\cdot)}. \tag{6.73}$$

Using (6.73) and Hölder's inequality, we get

$$I_{62} \lesssim \sup_{\varepsilon>0}\left(\varepsilon^{\theta}\sum_{k\in\mathbb{N}_0}\left(\sum_{\varphi\in M_k}2^{(\varphi-k)n}2^{k\alpha_1(\infty)}\|f_\varphi\|_{q_1(\cdot)}\right)^{p_1(1+\varepsilon)}\right)^{\frac{1}{p_1(1+\varepsilon)}}$$

$$\times \sup_{\varepsilon>0}\left(\varepsilon^{\theta}\sum_{k\in\mathbb{N}_0}\left(\sum_{\sigma\in N_k}2^{(k-\sigma)\left(\frac{n}{q_2(\infty)}+\alpha_2(\infty)\right)}2^{\sigma\alpha_2(\infty)}\|f_\sigma\|_{q_2(\cdot)}\right)^{p_2(1+\varepsilon)}\right)^{\frac{1}{p_2(1+\varepsilon)}}$$

$$=: I_{621}I_{622}.$$

The estimate I_{621} is similar to I_{52} and $I_{622} = D_2''$. Hence

$$I_6 \lesssim \|f_1\|_{\dot{K}^{\alpha_1,p_1),\theta}_{q_1(\cdot)}(\mathbb{R}^n)}\|f_2\|_{\dot{K}^{\alpha_2,p_2),\theta}_{q_2(\cdot)}(\mathbb{R}^n)}. \tag{6.74}$$

Estimation for I_9: We estimate I_9 as usual

$$I_9 \lesssim \sup_{\varepsilon>0}\left(\varepsilon^{\theta}\sum_{k\in\mathbb{Z}_-}2^{k\alpha(0)p(1+\varepsilon)}\nu_k(N_k,N_k)^{p(1+\varepsilon)}\right)^{\frac{1}{p(1+\varepsilon)}}$$

$$+\sup_{\varepsilon>0}\left(\varepsilon^{\theta}\sum_{k\in\mathbb{N}_0}2^{k\alpha(\infty)p(1+\varepsilon)}\nu_k(N_k,N_k)^{p(1+\varepsilon)}\right)^{\frac{1}{p(1+\varepsilon)}}$$

$$=: I_{91}+I_{92}.$$

For $\varphi \in N_k$, $y_1 \in R_\varphi$, and $x \in R_k$, we have $|x-y_1| \gtrsim 2^{\varphi}$, whereas for $\sigma \in N_k$, $y_2 \in R_\sigma$, and $x \in R_k$, we have $|x-y_2| \gtrsim 2^{\sigma}$. Under the assumptions for the obtained inequalities, we have

$$|T(f_\varphi, f_\sigma)(x)| \lesssim 2^{-\varphi n}\int_{\mathbb{R}^n}|f_\varphi(y_1)|dy_1 2^{-\sigma n}\int_{\mathbb{R}^n}|f_\sigma(y_2)|dy_2.$$

When $k \in \mathbb{Z}_-$, by Hölder's inequality, Lemma 6.3, and $q_j(0) \leqslant q_j(\infty)$, we obtain

$$\nu_k(N_k,N_k) \lesssim \sum_{\varphi\in N_k}\sum_{\sigma\in N_k}2^{(k-\varphi)\frac{n}{q_1(\infty)}}2^{(k-\sigma)\frac{n}{q_2(\infty)}}\|f_\varphi\|_{q_1(\cdot)}\|f_\sigma\|_{q_2(\cdot)}.$$

Thus

$$I_{91} \lesssim \sup_{\varepsilon>0}\left(\varepsilon^{\theta}\sum_{k\in\mathbb{Z}_-}\left(\sum_{\varphi\in N_k}2^{(k-\varphi)\left(\frac{n}{q_1(\infty)}+\alpha_1(0)\right)}2^{\varphi\alpha_1(0)}\|f_\varphi\|_{q_1(\cdot)}\right)^{p_1(1+\varepsilon)}\right)^{\frac{1}{p_1(1+\varepsilon)}}$$
$$\times \sup_{\varepsilon>0}\left(\varepsilon^{\theta}\sum_{k\in\mathbb{Z}_-}\left(\sum_{\sigma\in N_k}2^{(k-\sigma)\left(\frac{n}{q_1(\infty)}+\alpha_1(0)\right)}2^{\sigma\alpha_2(0)}\|f_\sigma\|_{q_2(\cdot)}\right)^{p_2(1+\varepsilon)}\right)^{\frac{1}{p_2(1+\varepsilon)}}$$
$$=: I_{911}I_{912}.$$

The estimates for I_{911} and I_{912} are obtained in the same way as for I_{312}. Therefore

$$I_{91} \leqslant C\|f_1\|_{\dot{K}^{\alpha_1(\cdot),p_1),\theta}_{q_1(\cdot)}(\mathbb{R}^n)}\|f_2\|_{\dot{K}^{\alpha_2(\cdot),p_2),\theta}_{q_2(\cdot)}(\mathbb{R}^n)}.$$

Finally, for the term I_{92}, we have

$$I_{92} \lesssim \sup_{\varepsilon>0}\left(\varepsilon^{\theta}\sum_{k\in\mathbb{N}_0}\left(\sum_{\varphi\in N_k}2^{(k-\varphi)\left(\frac{n}{q_1(\infty)}+\alpha_1(\infty)\right)}2^{\varphi\alpha_1(\infty)}\|f_\varphi\|_{q_1(\cdot)}\right)^{p_1(1+\varepsilon)}\right)^{\frac{1}{p_1(1+\varepsilon)}}$$
$$\times \sup_{\varepsilon>0}\left(\varepsilon^{\theta}\sum_{k\in\mathbb{N}_0}\left(\sum_{\sigma\in N_k}2^{(k-\sigma)\left(\frac{n}{q_2(\infty)}+\alpha_2(\infty)\right)}2^{\sigma\alpha_2(\infty)}\|f_\sigma\|_{q_2(\cdot)}\right)^{p_2(1+\varepsilon)}\right)^{\frac{1}{p_2(1+\varepsilon)}}$$
$$=: I_{921}I_{922}.$$

The terms I_{921} and I_{922} can be estimated in a similar manner as for that I_{322}. Thus

$$I_9 \lesssim \|f_1\|_{\dot{K}^{\alpha_1(\cdot),p_1),\theta}_{q_1(\cdot)}(\mathbb{R}^n)}\|f_2\|_{\dot{K}^{\alpha_2(\cdot),p_2),\theta}_{q_2(\cdot)}(\mathbb{R}^n)}. \tag{6.75}$$

Taking into account the estimates (6.62), (6.66), (6.70), (6.71), (6.74), and (6.75), we get

$$\|T(f_1, f_2)\|_{\dot{K}^{\alpha(\cdot),p),\theta}_{q(\cdot)}(\mathbb{R}^n)} \lesssim \|f_1\|_{\dot{K}^{\alpha_1(\cdot),p_1),\theta}_{q_1(\cdot)}(\mathbb{R}^n)}\|f_2\|_{\dot{K}^{\alpha_2(\cdot),p_2),\theta}_{q_2(\cdot)}(\mathbb{R}^n)},$$

which completes the proof. □

6.7 Comments to Chap. 6

The classical versions of Herz spaces $K^{\alpha}_{p,q}(\mathbb{R}^n)$ and $\dot{K}^{\alpha}_{p,q}(\mathbb{R}^n)$, introduced in Herz [139], known under the names of *nonhomogeneous* and *homogeneous* Herz spaces, are defined by the norms

$$\|f\|_{K^{\alpha}_{p,q}} := \|f\|_{L^p(B(0,1))} + \left\{ \sum_{k\in\mathbb{N}} 2^{k\alpha q} \left(\int\limits_{\mathsf{R}_{2^{k-1},2^k}} |f(x)|^p dx \right)^{\frac{q}{p}} \right\}^{\frac{1}{q}} \tag{6.76}$$

$$\|f\|_{\dot{K}^{\alpha}_{p,q}} := \left\{ \sum_{k\in\mathbb{Z}} 2^{k\alpha q} \left(\int\limits_{\mathsf{R}_{2^{k-1},2^k}} |f(x)|^p dx \right)^{\frac{q}{p}} \right\}^{\frac{1}{q}}, \tag{6.77}$$

respectively, where $\mathsf{R}_{t,\tau}$ stands for the annulus $\mathsf{R}_{t,\tau} := B(0,\tau)\backslash B(0,t)$. These spaces were studied in many papers; see, for instance, Feichtinger and Weisz [102], Grafakos et al. [123], Hernández and Yang [138], Izuki [153, 154], Johnson [156, 157], Komori [192], and references therein.

Herz spaces with variable exponents have recently been introduced in Almeida and Drihem [12], Izuki [153, 154]. In the last two articles, the exponent p was variable, and the remaining exponents α and q were kept constant. The most general results were obtained in Almeida and Drihem [12], where the variability of α was allowed. The main results obtained, for example, in Almeida and Drihem [12] concern the boundedness of sublinear operators (including the maximal function and Calderón–Zygmund singular operators) and a Spanne-type result for the Riesz potential operator. The approach used in Almeida and Drihem [12] allowed us to cover the case where p and α are variable and depend on the point x of the underlying set, keeping the exponent q constant.

In Samko [294, 295] (see also Kokilashvili et al. [183]), there was used another approach, based on the "continual" interpretation of Herz spaces via the norm

$$k^{\alpha}_{p,q}(f) := \|f\|_{L^p(B(0,\gamma\nu+\varepsilon))} + \left\{ \int\limits_{\nu}^{\infty} t^{\alpha q} \left(\int\limits_{\gamma t<|x|<\delta t} |f(x)|^p \, dx \right)^{\frac{q}{p}} \frac{dt}{t} \right\}^{\frac{1}{q}} \tag{6.78}$$

where $1 \leqslant p < \infty$, $1 \leqslant q < \infty$, $\alpha \in \mathbb{R}$, $\varepsilon > 0$, $\nu \geqslant 0$ and $0 < \gamma < \delta < \infty$, the cases $\nu = 0$ and $\nu > 0$ corresponding to homogeneous and nonhomogeneous Herz spaces (the first term in (6.78) should be omitted in the case $\nu = 0$). This norm is equivalent to the classical Herz norm given by discrete ℓ^q-norm; see Kokilashvili et al. [183], Samko [294] for more details. The Sobolev-type theorem on these spaces was studied by Rafeiro and Samko [279]. A different generalization to include the case of the variable q was also given in Izuki and Noi [155].

Drihem and Seghiri [80] studied Herz and Herz-type Hardy spaces with variable exponents, which were introduced in Izuki and Noi [155] and obtained several results regarding equivalence of norms and the atomic decomposition of Herz-type Hardy spaces. Furthermore, they studied the boundedness of some sublinear operators on Herz spaces with variable exponent.

Grand Lebesgue spaces on bounded sets were introduced in Iwaniec and Sbordone [152]. Grand spaces proved to be useful in application to partial differential equations. Various operators of harmonic analysis have been intensively studied in recent years; cf. Fiorenza et al. [103], Kokilashvili [169], Kokilashvili and Meskhi [172, 173], Kokilashvili et al. [182], Meskhi [241, 242], Soria and Weiss [312], and the references therein. Grand Lebesgue sequence spaces were introduced recently in Rafeiro et al. [281] (see also Kokilashvili et al. [190]), where various operators of harmonic analysis were studied in these spaces, e.g., maximal, convolutions, Hardy, Hilbert, and fractional operators, among others.

There has been increased interest in the study of multilinear singular integral operators in recent years. The class of multilinear singular integrals with standard Calderón–Zygmund kernels provides the foundation and starting point of research of the theory. This class of multilinear Calderón–Zygmund operators was introduced and first studied by Coifman and Meyer [54, 55, 56] and later by Grafakos and Torres [122]. For the boundedness and other properties of multilinear fractional integrals, we refer to, e.g., Kokilashvili et al. [179, 181, 183, 187].

The content of Sects. 6.2 and 6.3 is based on Meskhi et al. [248, 249], respectively, while Sects. 6.4, 6.5, and 6.6 contain the results obtained in Nafis et al. [258, 259, 260], respectively.

Chapter 7
Approximation in Subspaces of Morrey Spaces

We start by introducing a new subspace of Morrey spaces whose elements can be approximated by infinitely differentiable compactly supported functions. Consequently, we give an explicit description of the closure of the set of such functions in Morrey spaces. A generalization of known embeddings of Morrey spaces into weighted Lebesgue spaces is also obtained.

Furthermore, we study the approximation of functions from generalized Morrey spaces by nice functions. We introduce a new subspace whose elements can be also approximated by infinitely differentiable compactly supported functions. This provides, in particular, an explicit description of the closure of the set of such functions in generalized Morrey spaces.

Moreover, we show that the aforementioned introduced subspaces are invariant with respect to some classical operators of harmonic analysis, such as the Hardy–Littlewood maximal operator, singular-type operators, and Hardy operators. We also show that the vanishing properties defining those subspaces are preserved under the action of Riesz potential operators and fractional maximal operators.

7.1 Approximation in Morrey Spaces

In this section, we introduce some subspaces of Morrey spaces by adding some "vanishing"-type conditions. One of them is similar to the already known vanishing property, but related to the behavior at infinity instead of that at the origin. Another is connected with the truncation of functions to the exterior of large balls. These two additional properties, together with the vanishing property at the origin, allow us to show that all elements in this new subspace, denoted in the sequel by $V_{0,\infty}^{(*)}L^{p,\lambda}$, may be approximated by C_0^∞ functions in Morrey norm. In particular, dilation-

A. Meskhi et al., *Integral Operators in Non-Standard Function Spaces*,
Operator Theory: Advances and Applications 310,
https://doi.org/10.1007/978-3-032-03691-9_7

type identity approximations with integrable kernels strongly converges for all the functions in $V_{0,\infty}^{(*)}L^{p,\lambda}$.

As we shall see, the set $V_{0,\infty}^{(*)}L^{p,\lambda}$ is strictly smaller than the Zorko class. An example of a Morrey function belonging to Zorko class but not belonging to this new subspace is given below. Moreover, $V_{0,\infty}^{(*)}L^{p,\lambda}$ is closed in the Morrey space. Consequently, we give an explicit description of the closure of C_0^∞ in Morrey spaces. This closure plays an important role in harmonic analysis on Morrey spaces, including Calderón–Zygmund theory, since its dual constitutes a predual of Morrey spaces (cf. Adams and Xiao [8]). Preduals of Morrey spaces have been studied by many authors; see the recent book by Adams [6] and the paper Rosenthal and Triebel [287] for further details and references.

In addition to consideration of smaller subspaces of Morrey spaces, we also generalize some known embeddings of these space into weighted Lebesgue spaces. Various examples are presented showing the difference between all those spaces.

7.1.1 Preliminaries on Morrey Spaces

By supp f we denote the support of the function f, while $X \hookrightarrow Y$ stands for continuous embeddings from the normed space X into the normed space Y. The set $\mathcal{S}(\mathbb{R}^n)$ denotes the usual Schwartz class of complex-valued rapidly decreasing infinitely differentiable functions on $\mathbb{R}^n$, and $\mathcal{S}'(\mathbb{R}^n)$ is the set of all tempered distributions in $\mathbb{R}^n$. The class $C_0^\infty(\mathbb{R}^n)$ consists of all those functions in $\mathcal{S}(\mathbb{R}^n)$ with compact support. As usual, $L^p(E)$ with $1 \leqslant p < \infty$ and E a measurable subset of $\mathbb{R}^n$ is the standard Lebesgue space normed by

$$\|f\|_{L^p(E)} = \left(\int\limits_E |f(x)|^p \, dx \right)^{1/p} .$$

When $E = \mathbb{R}^n$, we shall write only $\|\cdot\|_p$ instead of $\|\cdot\|_{L^p(\mathbb{R}^n)}$.

In the sequel $\Omega \subseteq \mathbb{R}^n$ is an open set, $\widetilde{B}(x,r) := B(x,r) \cap \Omega$ for $x \in \Omega$, $r > 0$, and $L^p_{\text{loc}}(\Omega)$ stands for the space of all locally p-integrable functions on Ω. Moreover, we use the notation

$$\mathfrak{M}_{p,\lambda}(f; x, r) := \frac{1}{r^\lambda} \int\limits_{\widetilde{B}(x,r)} |f(y)|^p \, dy , \qquad x \in \Omega, \quad r > 0, \quad f \in L^1_{\text{loc}}(\Omega)$$

(as in Samko [291]).

Classical Morrey Spaces

Let $1 \leqslant p < \infty$ and $0 \leqslant \lambda \leqslant n$. The *homogeneous Morrey space* $L^{p,\lambda}(\Omega)$ (sometimes also called *global Morrey space*) is defined as

$$L^{p,\lambda}(\Omega) = \left\{ f \in L^p_{\text{loc}}(\Omega) : \sup_{x\in\Omega,\, r>0} \mathfrak{M}_{p,\lambda}(f; x, r) < \infty \right\}.$$

Its *nonhomogeneous* counterpart $\mathcal{L}^{p,\lambda}(\Omega)$ (sometimes also called *local Morrey space*) is the space

$$\mathcal{L}^{p,\lambda}(\Omega) = \left\{ f \in L^p_{\text{loc}}(\Omega) : \sup_{x\in\Omega,\, r\in(0,1]} \mathfrak{M}_{p,\lambda}(f; x, r) < \infty \right\}.$$

They are both Banach spaces equipped with the corresponding norms

$$\|f\|_{L^{p,\lambda}(\Omega)} := \sup_{x\in\Omega,\, r>0} \mathfrak{M}_{p,\lambda}(f; x, r)^{1/p} \tag{7.1}$$

and

$$\|f\|_{\mathcal{L}^{p,\lambda}(\Omega)} := \sup_{x\in\Omega,\, r\in(0,1]} \mathfrak{M}_{p,\lambda}(f; x, r)^{1/p}. \tag{7.2}$$

In general, we write $\|\cdot\|_{p,\lambda}$ for brevity to denote the Morrey norms above, when it is clear from the context which version is being considered. Moreover, we shall omit the reference to the domain when it coincides with the whole $\mathbb{R}^n$; e.g., we write $L^{p,\lambda}$ instead of $L^{p,\lambda}(\mathbb{R}^n)$ for short.

Morrey spaces can be seen as a refinement of L^p spaces, and they are part of a larger scale, called Morrey–Campanato spaces, which also includes Hölder spaces and the BMO space.

It can be shown that both spaces $L^{p,\lambda}(\Omega)$ and $\mathcal{L}^{p,\lambda}(\Omega)$ would reduce to $\{0\}$ if $\lambda > n$. Moreover, we have

$$L^{p,0}(\Omega) = L^p(\Omega), \quad L^{p,n}(\Omega) = L^\infty(\Omega) = \mathcal{L}^{p,n}(\Omega) \quad \text{and} \quad \mathcal{L}^{p,0}(\Omega) = \mathcal{L}^p(\Omega),$$

where $\mathcal{L}^p(\Omega)$ stands for the *uniform Lebesgue space* normed by

$$\|f\|_{\mathcal{L}^p(\Omega)} := \sup_{x\in\Omega} \|f\|_{L^p(\widetilde{B}(x,1))}. \tag{7.3}$$

It is clear that we always have

$$L^{p,\lambda}(\Omega) \hookrightarrow \mathcal{L}^{p,\lambda}(\Omega) \quad \text{and} \quad L^\infty(\Omega) \hookrightarrow \mathcal{L}^{p,\lambda}(\Omega).$$

Given $1 \leqslant p \leqslant q < \infty$ and $0 \leqslant \lambda, \mu \leqslant n$, an application of Hölder's inequality yields the embedding

$$L^{q,\mu}(\Omega) \hookrightarrow L^{p,\lambda}(\Omega)$$

under the condition

$$\frac{\lambda - n}{p} = \frac{\mu - n}{q}$$

if $|\Omega| = \infty$, and the condition

$$\frac{\lambda - n}{p} \leqslant \frac{\mu - n}{q}$$

if $|\Omega| < \infty$. In particular, for bounded domains Ω, we have

$$L^{\infty}(\Omega) \hookrightarrow L^{p,\lambda}(\Omega) \hookrightarrow L^{p}(\Omega).$$

Easy calculations show that the homogeneous Morrey norm has the property

$$\|f(t\cdot)\|_{L^{p,\lambda}(\Omega)} = t^{\frac{\lambda-n}{p}} \|f\|_{L^{p,\lambda}(\Omega)}, \quad t > 0,$$

which reveals the homogeneous nature of the spaces $L^{p,\lambda}(\Omega)$. It is not hard to check that a Young's convolution inequality holds true for homogeneous Morrey spaces $L^{p,\lambda}(\mathbb{R}^n)$:

$$\|f * g\|_{p,\lambda} \leqslant \|g\|_1 \, \|f\|_{p,\lambda}, \quad 0 \leqslant \lambda \leqslant n, \quad 1 \leqslant p < \infty. \tag{7.4}$$

The following chain of embeddings is known (cf. Rosenthal and Triebel [287, Theorem 3.1]):

$$\mathcal{S} \hookrightarrow B^{\frac{\lambda}{p}}_{p,p} \hookrightarrow L^{\frac{pn}{n-\lambda}} \hookrightarrow L^{p,\lambda} \hookrightarrow \mathcal{L}^{p,\lambda} \hookrightarrow L^{p}(w_\alpha) \hookrightarrow \mathcal{S}',$$

with $1 < p < \infty$, $0 < \lambda < n$ and $\alpha < -n/p$, where $B^{s}_{q,r}$ stands for classical Besov spaces on $\mathbb{R}^n$ (see, e.g., Triebel [323]) and $L^p(w_\alpha)$ denotes the weighted Lebesgue space on $\mathbb{R}^n$ normed by

$$\|f\|_{L^p(w_\alpha)} := \|w_\alpha f\|_p ,$$

with the notation $w_\alpha(x) := (1 + |x|^2)^{\alpha/2}$. All these embeddings are strict. For example,

$$|x|^{-\frac{n-\lambda}{p}} \in L^{p,\lambda} \quad \text{but} \quad |x|^{-\frac{n-\lambda}{p}} \notin L^{\frac{pn}{n-\lambda}}. \tag{7.5}$$

As regards the embeddings of Morrey spaces into weighted Lebesgue spaces, the weight above can be improved when we consider the homogeneous space $L^{p,\lambda}$. In fact, it also holds

$$L^{p,\lambda} \hookrightarrow L^p(w_\alpha) \quad \text{if} \quad \alpha < -\lambda/p$$

(cf. Kato [162, p. 132]). Later, in Sect. 7.1.5, we shall return to such embeddings into weighted Lebesgue spaces and generalize the corresponding results given in Kato [162] and Rosenthal and Triebel [287].

Morrey spaces $L^{p,\lambda}$ and $\mathcal{L}^{p,\lambda}$ are non-separable when $\lambda > 0$. Moreover, none of the usual classes of smooth functions (e.g., C_0^∞ or $\mathcal{S}$) is dense in $L^{p,\lambda}$ and $\mathcal{L}^{p,\lambda}$; see Rosenthal and Triebel [287, Proposition 3.7]. Nevertheless, C_0^∞ is dense in the dual $(\overset{\circ}{L}{}^{p,\lambda})'$, where $\overset{\circ}{L}{}^{p,\lambda}$ denotes the closure of C_0^∞ in Morrey norm. This fact, observed in Adams and Xiao [8], plays a crucial role in the study of harmonic analysis on Morrey spaces. One has

$$\big(\overset{\circ}{L}{}^{p,\lambda}\big)'' = \Big(\big(\overset{\circ}{L}{}^{p,\lambda}\big)'\Big)' = L^{p,\lambda}, \qquad 1 < p < \infty, \quad 0 \leqslant \lambda < n \tag{7.6}$$

(cf. Rosenthal and Triebel [287, Corollary 4.2 and Remark 4.3]), which extends to Morrey spaces the well-known assertion (for Lebesgue spaces)

$$\big(\overset{\circ}{L}{}^{p,0}\big)'' = \big(L^p\big)'' = L^p.$$

Preduals of Morrey spaces have been investigated by many researchers starting in the paper Zorko [342]. In Adams and Xiao [8], it was described by means of Hausdorff capacities and Muckenhoupt weights. We refer to Rosenthal and Triebel [287] for a study in the framework of tempered distributions and further discussions including historical remarks.

Known Subspaces

Approximations to the identity do not behave well in general Morrey spaces, since these spaces may contain functions with singularities like homogeneous functions (cf. (7.5)). This is one reason supporting the need of finding appropriate subspaces of Morrey spaces where certain nice properties still hold.

The so-called vanishing Morrey space (at the origin) $V_0L^{p,\lambda}(\Omega)$ is the subspace of $L^{p,\lambda}(\Omega)$ consisting of all those functions f such that

$$\lim_{r\to 0} \sup_{x\in\Omega} \mathfrak{M}_{p,\lambda}(f;x,r) = 0. \tag{V^0}$$

This is a closed subset of $L^{p,\lambda}$ (cf. Lemma 7.7 below), and it was introduced by Chiarenza and Franciosi [52] in the study of elliptic equations in nondivergence form; see also Vitanza [327, 328] for regularity results for elliptic equations with coefficients in such subspace. Note that

$$\lim_{r\to 0}\sup_{x\in\Omega}\mathfrak{M}_{p,\lambda}(f;x,r)=0 \iff \lim_{r\to 0}\sup_{\substack{x\in\Omega\\ t\in(0,r]}}\mathfrak{M}_{p,\lambda}(f;x,t)=0.$$

Since all L^p functions have this vanishing property, we have $V_0L^{p,0}(\Omega) = L^p(\Omega)$. However, if $\lambda > 0$, then $V_0L^{p,\lambda}(\Omega)$ is a proper subset of $L^{p,\lambda}(\Omega)$ (see Example 7.2).

An interesting feature of $V_0L^{p,\lambda}(\Omega)$, not shared by the whole Morrey space $L^{p,\lambda}(\Omega)$, is the possibility of approximation by bounded functions on bounded domains, as observed in Chiarenza and Franciosi [52, Lemma 1.1].

Another important subset of $L^{p,\lambda}(\Omega)$ is the so-called Zorko subspace $\mathbb{L}^{p,\lambda}(\Omega)$, introduced in Zorko [342], consisting of all functions $f \in L^{p,\lambda}(\Omega)$ on which the translation operator is continuous, i.e.,

$$\mathbb{L}^{p,\lambda}(\Omega) := \{f \in L^{p,\lambda}(\Omega) : \ \|\tau_\xi f - f\|_{p,\lambda} \to 0 \ \text{ as } \ \xi \to 0\}, \tag{7.7}$$

where $\tau_\xi f = f(\cdot - \xi)$, $\xi \in \mathbb{R}^n$ (setting $f(x) = 0$ for $x \in \mathbb{R}^n \setminus \Omega$). This is also a closed set in Morrey spaces, and it is known that approximation via mollifiers is possible in such subspace; see Zorko [342, Proposition 3].

In the limiting case $\lambda = 0$, we have $\mathbb{L}^{p,0}(\Omega) = V_0L^{p,0}(\Omega) = L^p(\Omega)$. For $\lambda > 0$, in general, the previous subspaces are related as follows:

$$\mathbb{L}^{p,\lambda}(\Omega) \subset V_0L^{p,\lambda}(\Omega) \subset L^{p,\lambda}(\Omega) \tag{7.8}$$

(cf. Kato [162, Corollary 3.3]). Nevertheless, taking into account Chiarenza and Franciosi [52, Lemma 1.2], we have

$$\mathbb{L}^{p,\lambda}(\Omega) = V_0L^{p,\lambda}(\Omega) \quad \text{if} \quad |\Omega| < \infty.$$

For unbounded domains, e.g., $\Omega = \mathbb{R}^n$, the inclusions in (7.8) are strict.

7.1.2 New Subspaces of Morrey Spaces

In the sequel, we introduce new subspaces of Morrey spaces by means of vanishing-type conditions at infinity and show how they are related with the already known subspaces mentioned above. The context now is the whole $\mathbb{R}^n$, so that we omit the domain in the notations as agreed before.

Definition 7.1 Let $0 \leqslant \lambda < n$ and $1 \leqslant p < \infty$. We define $V_\infty L^{p,\lambda}$ as the subset of $L^{p,\lambda}$ consisting of all those functions f such that

$$\lim_{r\to\infty} \sup_{x\in\mathbb{R}^n} \mathfrak{M}_{p,\lambda}(f;x,r) = 0. \tag{V^∞}$$

Note that the subset $V_\infty L^{p,\lambda}$ is defined similarly to $V_0 L^{p,\lambda}$, but now in terms of a corresponding vanishing property at infinity.

Example 7.2 For $\alpha, \beta > 0$ let

$$f_{\alpha,\beta}(x) := \begin{cases} \frac{1}{|x|^\alpha}, & |x| \leqslant 1, \\ \frac{1}{|x|^\beta}, & |x| > 1. \end{cases}$$

Then

$$f_{\alpha,\beta} \in V_0 L^{p,\lambda} \cap V_\infty L^{p,\lambda} \quad \text{if} \quad \alpha < \frac{n-\lambda}{p} < \beta.$$

In the limiting case $\alpha = \frac{n-\lambda}{p} = \beta$, we have

$$f_{\alpha,\beta} \in L^{p,\lambda} \quad \text{but} \quad f_{\alpha,\beta} \notin V_0 L^{p,\lambda} \cup V_\infty L^{p,\lambda}.$$

We introduce another subset by distinguishing a vanishing property after truncation in large balls. For convenience, we use the notation

$$\mathcal{A}_{N,p}(f) := \sup_{x\in\mathbb{R}^n} \int_{B(x,1)} |f(y)|^p \, \chi_N(y)\, dy$$

with

$$\chi_N := \chi_{\mathbb{R}^n \setminus B(0,N)}, \qquad N \in \mathbb{N}.$$

These truncations make sense, for example, for functions f in the uniform Lebesgue space $\mathcal{L}^p$ for $1 \leqslant p < \infty$.

Definition 7.3 For $0 \leqslant \lambda < n$ and $1 \leqslant p < \infty$, we define $V^{(*)} L^{p,\lambda}$ as the set of all functions $f \in L^{p,\lambda}$ having the vanishing property

$$\lim_{N\to\infty} \mathcal{A}_{N,p}(f) = 0. \tag{V^*}$$

By the Lebesgue dominated convergence theorem, we can see that every L^p function has property (V^*) and hence $V^{(*)} L^{p,0} = L^p$. In a sense, property (V^*)

allows us to overcome difficulties in applying such a theorem when the functions are not necessarily in L^p.

Lemma 7.4 *A function $f \in L^{p,\lambda}$ satisfies property (V^*) if and only if*

$$\lim_{N\to\infty} \sup_{x\in\mathbb{R}^n} \int_{B(x,r)} |f(y)|^p \chi_N(y)\,dy = 0 \tag{7.9}$$

uniformly in $r \in (0, R_0]$ for any fixed $R_0 > 0$.

Proof It is clear that condition (7.9) immediately implies property (V^*). Conversely, suppose that f satisfies (V^*). Let $R_0 > 0$ be arbitrary fixed and let $x \in \mathbb{R}^n$. Then there exist $K_0 \in \mathbb{N}$ (depending only on R_0 and n) and $x_j \in B(x, R_0)$, $j = 1, \ldots, K_0$, such that

$$B(x,r) \subset \bigcup_{j=1}^{K_0} B(x_j, 1)$$

for any $r \in (0, R_0]$. Since

$$\int_{B(x,r)} |f(y)|^p \chi_N(y)\,dy \leqslant \sum_{j=1}^{K_0} \int_{B(x_j,1)} |f(y)|^p \chi_N(y)\,dy$$

$$\leqslant c \sup_{z\in\mathbb{R}^n} \int_{B(z,1)} |f(y)|^p \chi_N(y)\,dy$$

with $c > 0$ not depending on x, r, and N, we get the estimate

$$\sup_{x\in\mathbb{R}^n} \int_{B(x,r)} |f(y)|^p \chi_N(y)\,dy \leqslant c\,\mathcal{A}_{N,p}(f),$$

from which (7.9) follows, uniformly in $r \in (0, R_0]$. □

Definition 7.5 For $0 \leqslant \lambda < n$ and $1 \leqslant p < \infty$, one defines the subset $V_{0,\infty}^{(*)} L^{p,\lambda}$ of $L^{p,\lambda}$ as

$$V_{0,\infty}^{(*)} L^{p,\lambda} = V_0 L^{p,\lambda} \cap V_\infty L^{p,\lambda} \cap V^{(*)} L^{p,\lambda}.$$

Example 7.6 Consider the function

$$\varphi(x) = \sum_{k=2}^{\infty} \chi_{B_k}(x)$$

where $B_k = B(2^k e_1, 1)$, $k \in \mathbb{N}$, with $e_1 = (1, 0, \ldots, 0)$. Then

$$\varphi \in V_0 L^{p,\lambda} \cap V_\infty L^{p,\lambda} \quad \text{but} \quad \varphi \notin V^{(*)} L^{p,\lambda}$$

(cf. proof of Theorem 7.10).

Lemma 7.7 *All the vanishing subsets $V_0 L^{p,\lambda}$, $V_\infty L^{p,\lambda}$ and $V^{(*)} L^{p,\lambda}$ (and consequently $V^{(*)}_{0,\infty} L^{p,\lambda}$) are closed in $L^{p,\lambda}$.*

Proof The proof uses standard arguments which can be adapted to all the cases. By this reason, we give details for the subset $V^{(*)} L^{p,\lambda}$ only.

Let $(f_k)_k$ be a sequence of functions in $V^{(*)} L^{p,\lambda}$ converging to f in $L^{p,\lambda}$. We want to show that $f \in V^{(*)} L^{p,\lambda}$. For any $\varepsilon > 0$ there exists $\overline{k} \in \mathbb{N}$ such that

$$\mathcal{A}_{N,p}(f - f_{\overline{k}}) \leqslant \|f - f_{\overline{k}}\|^p_{p,\lambda} < \frac{\varepsilon}{2^{p+1}}.$$

On the other hand, for such fixed $\overline{k}$, by the vanishing property (V^*) of $f_{\overline{k}}$ there exists $\overline{N} \in \mathbb{N}$ such that

$$\mathcal{A}_{N,p}(f_{\overline{k}}) < \frac{\varepsilon}{2^{p+1}}$$

for every $N \geqslant \overline{N}$. Hence, we have

$$\mathcal{A}_{N,p}(f) \leqslant 2^p \mathcal{A}_{N,p}(f - f_{\overline{k}}) + 2^p \mathcal{A}_{N,p}(f_{\overline{k}}) < \varepsilon$$

for arbitrary large $N \in \mathbb{N}$, which completes the proof. □

We end this section with a discussion on the preservation of the vanishing properties introduced above by convolution operators with integrable kernels.

Theorem 7.8 *Let $0 \leqslant \lambda \leqslant n$, $1 \leqslant p < \infty$. Then the Morrey subspaces $V_0 L^{p,\lambda}$, $V_\infty L^{p,\lambda}$ and $V^{(*)} L^{p,\lambda}$ are invariant with respect to convolutions with integrable kernels.*

Proof Let $\phi \in L^1$. By (7.4) it is clear that $f * \phi \in L^{p,\lambda}$ when $f \in L^{p,\lambda}$. Let $x \in \mathbb{R}^n$ and $r > 0$. By Minkowski's integral inequality and a change of variables, we get

$$\begin{aligned} \frac{1}{r^\lambda} \int\limits_{B(x,r)} |(f * \phi)(y)|^p \, dy &\leqslant \frac{1}{r^\lambda} \left(\int\limits_{\mathbb{R}^n} \left(\int\limits_{B(x,r)} |f(y - z)\, \phi(z)|^p \, dy \right)^{1/p} dz \right)^p \\ &= \frac{1}{r^\lambda} \left(\int\limits_{\mathbb{R}^n} |\phi(z)| \left(\int\limits_{B(x-z,r)} |f(u)|^p \, du \right)^{1/p} dz \right)^p \\ &\leqslant \|\phi\|_1^p \sup_{v \in \mathbb{R}^n} \mathfrak{M}_{p,\lambda}(f; v, r). \end{aligned}$$

From this it follows that

$$f \in V_0 L^{p,\lambda} \;\to\; f * \phi \in V_0 L^{p,\lambda} \quad \text{and} \quad f \in V_\infty L^{p,\lambda} \;\to\; f * \phi \in V_\infty L^{p,\lambda}.$$

It remains to prove that the convolution also preserves property (V^*). As above, for $x \in \mathbb{R}^n$ and $N \in \mathbb{N}$, we get

$$\left(\int_{B(x,1)} |(f * \phi)(y)|^p \chi_N(y)\, dy \right)^{1/p} \leqslant \int_{\mathbb{R}^n} |\phi(z)| \left(\int_{B(x,1)} |f(y-z)|^p \chi_N(y)\, dy \right)^{1/p} dz.$$

We extend the notation from Definition 7.3 as follows:

$$\chi_a := \chi_{\mathbb{R}^n \setminus B(0,a)} \text{ if } a > 0 \quad \text{and} \quad \chi_a \equiv 1 \text{ if } a \leqslant 0. \tag{7.10}$$

So we have $\chi_N(y) \leqslant \chi_{N-|z|}(y-z)$. Using this fact and changing variables on the right-hand side of the inequality above, we obtain the estimate

$$\left(\int_{B(x,1)} |(f * \phi)(y)|^p \chi_N(y)\, dy \right)^{1/p}$$
$$\leqslant \int_{\mathbb{R}^n} |\phi(z)| \left(\sup_{v \in \mathbb{R}^n} \int_{B(v,1)} |f(u)|^p \chi_{N-|z|}(u)\, du \right)^{1/p} dz.$$

Since this is uniform with respect to $x \in \mathbb{R}^n$, we get

$$\left[\mathcal{A}_{N,p}(f * \phi) \right]^{1/p} \leqslant \int_{\mathbb{R}^n} |\phi(z)| \left[\mathcal{A}_{N-|z|,p}(f) \right]^{1/p} dz,$$

with the interpretation

$$\mathcal{A}_{a,p}(f) := \sup_{x \in \mathbb{R}^n} \int_{B(x,1)} |f(y)|^p \chi_a(y)\, dy\,, \qquad a \in \mathbb{R}$$

(according to (7.10)). Therefore we have

$$\mathcal{A}_{N,p}(f * \phi) \to 0 \quad \text{as} \quad N \to \infty,$$

by the Lebesgue dominated convergence theorem. □

Remark 7.9 An inspection of the last part of the previous proof shows that the convolution operator (with integrable kernel) preserves property (V^*) of functions belonging to the uniform Lebesgue space $\mathcal{L}^p$ (recall (7.3)).

The preservation of the vanishing properties above is also true for other operators from harmonic analysis. We shall discuss such topic in Sect. 7.3.

7.1.3 On Strict Embeddings Between Morrey Subspaces

Theorem 7.10 *For any $0 < \lambda < n$ and $1 \leqslant p < \infty$, there are functions in $V_0L^{p,\lambda} \cap V_\infty L^{p,\lambda}$ which do not have property (V^*).*

Proof We take the function φ from Example 7.6,

$$\varphi(x) = \sum_{k=2}^{\infty} \chi_{B_k}(x),$$

keeping the notation $B_k = B(2^k e_1, 1)$. For every $N \in \mathbb{N}$, we have

$$\begin{aligned} \mathcal{A}_{N,p}(\varphi) &= \sup_{x\in\mathbb{R}^n} \sum_{k=2}^{\infty} \int_{B(x,1)} \chi_{B_k}(y)\, \chi_N(y)\, dy \\ &\geqslant \int_{B(2^{\bar{k}}e_1,1)} \chi_{B_{\bar{k}}}(y)\, \chi_N(y)\, dy \\ &= |B(2^{\bar{k}}e_1, 1)| = |B(0,1)|, \end{aligned}$$

where $\bar{k} \geqslant 2$ is any integer chosen such that $2^{\bar{k}} > N+1$. Hence φ fails to have property (V^*). It remains to show that $\varphi \in L^{p,\lambda}$ and that φ has both the vanishing properties (V_0) and (V_∞).

For any $x \in \mathbb{R}^n$ and $r \leqslant 1$, there exists at most one ball B_{k_0} intersecting $B(x,r)$. Then

$$\mathfrak{M}_{p,\lambda}(\varphi; x, r) = r^{-\lambda} \int_{B(x,r)} \chi_{B_{k_0}}(y)\, dy \lesssim r^{n-\lambda},$$

with the implicit constant not depending on x and r. Therefore, noting also that $\lambda \in (0,n)$,

$$\sup_{x\in\mathbb{R}^n} \mathfrak{M}_{p,\lambda}(\varphi; x, r) \to 0 \quad \text{as} \quad r \to 0,$$

which shows that φ has the vanishing property at the origin.

Suppose now that $r > 1$. We need to count the number of balls intersecting $B(x,r)$. In the case $|x| \leqslant 2r$, we calculate

$$r+1 \geqslant |x-2^k e_1| \geqslant |2^k e_1| - |x| \geqslant 2^k - 2r,$$

so that

$$2^k \leqslant 3r+1.$$

Therefore, for $|x| \leqslant 2r$, we get the estimate

$$\mathfrak{M}_{p,\lambda}(\varphi;x,r) \leqslant r^{-\lambda} \sum_{k=2}^{[\log(3r)]+1} |B(x,r)\cap B_k| \leqslant \frac{\log(3r)}{r^\lambda}, \tag{7.11}$$

where $[a]$ stands for the integer part of $a \in \mathbb{R}$. As before, we should have

$$|x| - r - 1 \leqslant 2^k \leqslant |x| + r + 1.$$

Hence, in the case $|x| > 2r$, the number of balls B_k intersecting $B(x,r)$ can be estimated by

$$\log(|x|+r+1) - \log(|x|-r) = \log\frac{|x|+r+1}{|x|-r} = \log\left(1+\frac{2r+1}{|x|-r}\right) \leqslant 2.$$

We then obtain

$$\mathfrak{M}_{p,\lambda}(\varphi;x,r) \leqslant \frac{2}{r^\lambda}. \tag{7.12}$$

Taking into account (7.11) and (7.12), we conclude that

$$\sup_{x\in\mathbb{R}^n} \mathfrak{M}_{p,\lambda}(\varphi;x,r) \leqslant \frac{\log(4r)}{r^\lambda} \to 0 \quad \text{as} \quad r\to\infty,$$

from which the vanishing property at infinity follows (note that $\lambda > 0$). □

Corollary 7.11 *For any* $0<\lambda<n$ *and* $1 \leqslant p < \infty$, *we have*

$$V_{0,\infty}^{(*)}L^{p,\lambda} \subsetneqq V_0 L^{p,\lambda} \cap V_\infty L^{p,\lambda} \subsetneqq V_0 L^{p,\lambda} \subsetneqq L^{p,\lambda}.$$

The next result (and its corollary) shows that the new subspace $V_{0,\infty}^{(*)}L^{p,\lambda}$ is strictly smaller than Zorko subspace $\mathbb{L}^{p,\lambda}$ (from (7.7)). Its proof uses the fact that Morrey functions having all the vanishing properties above can always be approximated by compactly supported functions. However, for convenience, this claim will be proved later in Sect. 7.1.4.

Theorem 7.12 *For any $0 \leqslant \lambda < n$ and $1 \leqslant p < \infty$, we have $V_{0,\infty}^{(*)} L^{p,\lambda} \subset \mathbb{L}^{p,\lambda}$.*

Proof Let $f \in V_{0,\infty}^{(*)} L^{p,\lambda}$. For any $\varepsilon > 0$, there exists $g \in L^{p,\lambda}$ with compact support such that

$$\|f - g\|_{p,\lambda} < \varepsilon/4$$

(cf. Step 2 in the proof of Theorem 7.17). Since, for any $\xi \in \mathbb{R}^n$, we have

$$\|\tau_\xi f - f\|_{p,\lambda} \leqslant 2\,\|f - g\|_{p,\lambda} + \|\tau_\xi g - g\|_{p,\lambda}$$

it suffices to show that the second norm is less that $\varepsilon/2$ for small values of $|\xi|$. By the (V_0) and (V_∞) properties, there exist r_0, $R > 0$ such that

$$S_1 := \sup_{\substack{x\in\mathbb{R}^n \\ 0<r<r_0}} \mathfrak{M}(\tau_\xi g - g; x, r) \leqslant 2^p \sup_{\substack{x\in\mathbb{R}^n \\ 0<r<r_0}} \mathfrak{M}(g; x, r) < (\varepsilon/2)^p$$

and

$$S_2 := \sup_{\substack{x\in\mathbb{R}^n \\ r>R}} \mathfrak{M}(\tau_\xi g - g; x, r) \leqslant 2^p \sup_{\substack{x\in\mathbb{R}^n \\ r>R}} \mathfrak{M}(g; x, r) < (\varepsilon/2)^p.$$

For such fixed r_0 and R, we estimate

$$\|\tau_\xi g - g\|_{p,\lambda}^p \leqslant \max\{S_1, S_2, S_3\}$$

with

$$S_3 := \sup_{\substack{x\in\mathbb{R}^n \\ r_0\leqslant r\leqslant R}} \mathfrak{M}(\tau_\xi g - g; x, r).$$

We have

$$S_3 \lesssim \sup_{x\in\mathbb{R}^n} \int\limits_{B(x,R)} |g(y-\xi) - g(y)|^p\, dy \leqslant \max\left\{ \sup_{|x|<M} (\cdots),\ \sup_{|x|>M} (\cdots) \right\}.$$

where $M > 0$ is chosen below. Since g has compact support, there exists $K > 0$ such that $g(u) = 0$ if $|u| > K$. In the case $|x| > M$, we have

$$\int\limits_{|y-x|<R} |g(y-\xi) - g(y)|^p\, dy = \int\limits_{|z|<R} |g(z+x-\xi) - g(z+x)|^p\, dz.$$

Hence, if we choose $M > R + K + 1$, then $g(z + x) = 0$ and $g(z + x - \xi) = 0$ for small values of $|\xi|$, say $|\xi| < 1$, since

$$|z + x| > M - R > K \quad \text{and} \quad |z + x - \xi| \geqslant |z + x| - |\xi| > K.$$

Let then $M > R+K+1$ be fixed and let us now estimate the integral when $|x| < M$. In this case we are just taking the L^p-norm on a ball centered at the origin with fixed radius, precisely $B(0, R + M + 1)$, again for $|\xi| < 1$. Therefore we also obtain

$$S_3 < (\varepsilon/2)^p$$

by the continuity of the L^p-norm with respect to translations. □

Remark 7.13 From the previous proof, we see that for any $\varepsilon > 0$ and $f \in V_0 L^{p,\lambda} \cap V_\infty L^{p,\lambda}$, there are $r_0, R > 0$ such that

$$\|\tau_\xi f - f\|_{p,\lambda}^p \leqslant \max\{\varepsilon, S_{r_0,R}(\xi)\}$$

for all $\xi \in \mathbb{R}^n$, where

$$S_{r_0,R}(\xi) = r_0^{-\lambda} \sup_{x\in\mathbb{R}^n} \int_{B(x,R)} |f(y - \xi) - f(y)|^p \, dy.$$

Thus, for Morrey functions in the intersection $V_0 L^{p,\lambda} \cap V_\infty L^{p,\lambda}$, the Zorko property is reduced to the continuity of the translation operator in the uniform Lebesgue space $\mathcal{L}^p$. Note also that the vanishing property (V^*) ensures the continuity of the translation operator in $\mathcal{L}^p$ and consequently Zorko property.

Corollary 7.14 *For any* $0 < \lambda < n$ *and* $1 \leqslant p < \infty$, *there holds*

$$V_{0,\infty}^{(*)} L^{p,\lambda} \subsetneq \mathbb{L}^{p,\lambda} \subsetneq V_0 L^{p,\lambda} \subsetneq L^{p,\lambda}.$$

7.1.4 Approximation in Morrey Subspaces

As observed by Zorko [342], there are functions in $L^{p,\lambda}$ that cannot be approximated even by continuous functions. It is the case of functions with the form $f_{x_0}(x) = |x - x_0|^{\frac{\lambda-n}{p}}$ for fixed $x_0 \in \mathbb{R}^n$. It should be mentioned that the Morrey norm in Zorko [342] was taken as the sum $\|\cdot\|_p + \|\cdot\|_{p,\lambda}$. In particular, Morrey functions there were supposed to be in L^p which is not necessarily the case in the approach followed in this section. Anyway, it was the failure in approximation by nice functions that motivated the introduction of the subspace $\mathbb{L}^{p,\lambda}$.

In the intersection $L^{p,\lambda} \cap L^p$ standard approximations using mollifiers can be used in order to approximate functions with Zorko property by C_0^∞ functions (cf. Adams [6, Theorem 2.3]). However, things change when we deal with the Morrey space $L^{p,\lambda}$, $\lambda > 0$, with the norm defined by (7.1), since then we cannot approximate all Morrey functions by regular compactly supported functions.

Consider the usual dilations $\phi_t(x) = t^{-n}\phi(x/t)$, $t > 0$, where ϕ is an integrable function with $\|\phi\|_1 = 1$. Suppose that $f \in \mathbb{L}^{p,\lambda}$. Standard calculations and Minkowski's integral inequality yield

$$\|f * \phi_t - f\|_{p,\lambda} \leqslant \int\limits_{\mathbb{R}^n} \|\tau_{tz} f - f\|_{p,\lambda} \, |\phi(z)| \, dz.$$

Since $\|\tau_{tz} f - f\|_{p,\lambda} \to 0$ as $t \to 0$ and $\|\tau_{tz} f - f\|_{p,\lambda} \leqslant 2\|f\|_{p,\lambda}$ for any $t > 0$ and $z \in \mathbb{R}^n$, by the Lebesgue theorem, we get

$$\|f * \phi_t - f\|_{p,\lambda} \to 0 \quad \text{as} \quad t \to 0. \tag{7.13}$$

If we take smooth kernels, say $\phi \in \mathcal{S}$, then the mollifiers $f * \phi_t \in \mathbb{L}^{p,\lambda} \cap C^\infty$ for any $t > 0$. Consequently, we derive the following result.

Theorem 7.15 *Let $0 \leqslant \lambda < n$ and $1 \leqslant p < \infty$. Then every Morrey function with Zorko property can be approximated in Morrey norm by C^∞ functions. Moreover, we have*

$$\overline{\mathbb{L}^{p,\lambda} \cap C^\infty} = \mathbb{L}^{p,\lambda}.$$

We discuss now the approximation of Morrey functions having both vanishing properties at the origin and at infinity.

Theorem 7.16 *Let $0 \leqslant \lambda < n$ and $1 \leqslant p < \infty$. If $f \in V_0L^{p,\lambda} \cap V_\infty L^{p,\lambda}$ is uniformly continuous, then f can be approximated in Morrey norm by functions from $V_0L^{p,\lambda} \cap V_\infty L^{p,\lambda} \cap C^\infty$.*

Proof Let $\phi \in \mathcal{S}$ with $\|\phi\|_1 = 1$. Then $f * \phi_t \in V_0L^{p,\lambda} \cap V_\infty L^{p,\lambda} \cap C^\infty$ for any $t > 0$ by Theorem 7.8. Thus it remains to show that $f * \phi_t \to f$ in $L^{p,\lambda}$ as $t \to 0$.

Let $\varepsilon > 0$. For any $x \in \mathbb{R}^n$, $r > 0$ and $t > 0$, we have

$$\left(\int\limits_{B(x,r)} |(f * \phi_t)(y) - f(y)|^p \, dy \right)^{1/p}$$
$$\leqslant \int\limits_{\mathbb{R}^n} |\phi_t(z)| \left(\int\limits_{B(x,r)} |f(y-z) - f(y)|^p \, dy \right)^{1/p} dz.$$

Since $f \in V_0 L^{p,\lambda} \cap V_\infty L^{p,\lambda}$, there are $r_0, R > 0$ such that

$$\frac{1}{r^\lambda} \int\limits_{B(x,r)} |f(y-z) - f(y)|^p \, dy < \varepsilon$$

for every $r < r_0$ or $r > R$ (and all $x, z \in \mathbb{R}^n$). Thus we have

$$\sup_{r>0} \frac{1}{r^\lambda} \int\limits_{B(x,r)} |(f * \phi_t)(y) - f(y)|^p \, dy \leqslant \max\{\varepsilon, S_{r_0,R}(x,t)\} \tag{7.14}$$

where

$$S_{r_0,R}(x,t) := \int\limits_{\mathbb{R}^n} |\phi_t(z)| \left(\frac{1}{r_0^\lambda} \int\limits_{B(x,R)} |f(y-z) - f(y)|^p \, dy \right)^{1/p} dz.$$

By the uniform continuity of f, one can find $\delta > 0$ such that $|f(y-z) - f(y)| < \varepsilon$ for any y and z with $|z| < \delta$. For such fixed δ, we split the outer integral above into

$$S_{r_0,R}(x,t) = \int\limits_{|z|<\delta} (\cdots) \, dz + \int\limits_{|z|\geqslant\delta} (\cdots) \, dz. \tag{7.15}$$

For the first integral, we use the uniform continuity of f and get

$$\int\limits_{|z|<\delta} (\cdots) \, dz \leqslant \varepsilon \, r_0^{-\lambda} \, |B(x,R)| \int\limits_{|z|<\delta} |\phi_t(z)| \, dz \leqslant |B(0,1)| \, r_0^{-\lambda} \, R^n \, \varepsilon \tag{7.16}$$

for every $x \in \mathbb{R}^n$ and $t > 0$. In the second integral in (7.15), we use the fact that $f \in L^{p,\lambda}$ and $\phi \in \mathcal{S}$ to derive the inequality

$$\int\limits_{|z|\geqslant\delta} (\cdots) \, dz \leqslant c \, r_0^{-\lambda} \, R^\lambda \, t \, \|f\|_{p,\lambda} \tag{7.17}$$

with $c > 0$ depending only on ϕ, n, and δ. In particular, we used the estimate

$$\int\limits_{|z|\geqslant\delta} |\phi_t(z)| \, dz \lesssim \int\limits_{|z|\geqslant\delta} \frac{t}{|z|^{n+1}} \, dz \lesssim t\,,$$

where the implicit constant is independent of t. Using (7.16) and (7.17) in (7.15), from (7.14) we obtain

$$\sup_{\substack{x\in\mathbb{R}^n \\ r>0}} \mathfrak{M}_{p,\lambda}(f*\phi_t - f; x, r) \lesssim \varepsilon$$

for sufficiently small $t > 0$. This implies $\|f*\phi_t - f\|_{p,\lambda} \to 0$ as $t \to 0$, and hence the proof is complete. □

Finally we discuss the approximation of Morrey functions having all the vanishing properties.

Theorem 7.17 *Let* $0 \leqslant \lambda < n$ *and* $1 \leqslant p < \infty$. *Then every function in* $V_{0,\infty}^{(*)}L^{p,\lambda}$ *can be approximated in Morrey norm by* C_0^∞ *functions.*

Proof *Step 1*: The claim holds true for functions $f \in V_{0,\infty}^{(*)}L^{p,\lambda}$ with compact support. In fact, if the kernel $\phi \in C_0^\infty$, then the mollifiers $f*\phi_t$ have compact support and belong to $V_{0,\infty}^{(*)}L^{p,\lambda}$ (cf. 7.8). Moreover, they approximate f in Morrey norm (recall the discussion in the beginning of this section leading to (7.13)).

Step 2: We show now that functions from $V_{0,\infty}^{(*)}L^{p,\lambda}$ can be approximated by compactly supported functions in Morrey norm. Let $f \in V_{0,\infty}^{(*)}L^{p,\lambda}$. As before, let $\chi_k := \chi_{\mathbb{R}^n\setminus B(0,k)}$, $k \in \mathbb{N}$. For each k, set

$$f_k = f \quad \text{on the ball } B(0,k) \quad \text{and} \quad f_k = 0 \quad \text{otherwise.}$$

Let $\varepsilon > 0$ be arbitrary. By the vanishing properties (V_0) and (V^∞), one finds $r_0, R > 0$ such that

$$\sup_{x\in\mathbb{R}^n} \mathfrak{M}_{p,\lambda}\big(f - f_k; x, r\big) = \sup_{x\in\mathbb{R}^n} \mathfrak{M}_{p,\lambda}\big(f\chi_k; x, r\big) < \varepsilon,$$

for all $k \in \mathbb{N}$ and every $r < r_0$ or $r > R$. Hence

$$\|f - f_k\|_{p,\lambda}^p < \max\{\varepsilon, S_{r_0,R}\}$$

where

$$S_{r_0,R}(k) := \sup_{\substack{x\in\mathbb{R}^n \\ r\in[r_0,R]}} \mathfrak{M}_{p,\lambda}\big(f\chi_k; x, r\big).$$

Now, by the vanishing property (V^*) and Lemma 7.4, we get

$$S_{r_0,R}(k) \leqslant \sup_{x\in\mathbb{R}^n} r_0^{-\lambda} \int_{B(x,R)} |f(y)|^p \chi_k(y)\, dy < \varepsilon$$

for all k large enough. Therefore,

$$\|f - f_k\|_{p,\lambda} \to 0 \quad \text{as} \quad k \to \infty,$$

which ends the proof. □

From the previous result and the fact that $V_{0,\infty}^{(*)} L^{p,\lambda}$ is closed (cf. Lemma 7.7), we obtain an explicit description of $\overset{\circ}{L}{}^{p,\lambda}$, the closure of C_0^∞ in the homogeneous Morrey spaces.

Corollary 7.18 *For $0 \leqslant \lambda < n$ and $1 \leqslant p < \infty$, the set C_0^∞ is dense in $V_{0,\infty}^{(*)} L^{p,\lambda}$. Moreover,*

$$\overset{\circ}{L}{}^{p,\lambda} = V_{0,\infty}^{(*)} L^{p,\lambda}.$$

Recalling the known result (7.6), we see that $L^{p,\lambda}$ is precisely the bidual of the new subspace $V_{0,\infty}^{(*)} L^{p,\lambda}$.

7.1.5 A Generalization of Embeddings into Weighted L^p Spaces

The next result generalizes some known embeddings of Morrey spaces into weighted Lebesgue spaces (see, for instance, Kato [162], Rosenthal and Triebel [287]). Recall that a non-negative function g on $[0, \infty)$ is called *almost decreasing* if there exists a constant $c > 0$ such that $g(s) \leqslant c\, g(t)$ for all $s \geqslant t (\geqslant 0)$.

Theorem 7.19 *Let $1 \leqslant p < \infty$ and $0 < \lambda < n$. Let $w = w(|x|)$ be a radial weight such that $w(t) \approx 1$ for $0 \leqslant t < 1$. Then*

$$L^{p,\lambda}(\mathbb{R}^n) \hookrightarrow L^p(\mathbb{R}^n, w) \quad \text{and} \quad \mathcal{L}^{p,\lambda}(\mathbb{R}^n) \hookrightarrow L^p(\mathbb{R}^n, w)$$

if $w(t)^p\, t^\gamma$ is almost decreasing, and

$$\int_1^\infty w(t)^p\, t^{\gamma-1}\, dt < \infty \tag{7.18}$$

with $\gamma = \lambda$ in the homogeneous case and $\gamma = n$ in the nonhomogeneous case.

Proof We split the weighted norm of f as

$$\|f\|_{L^p(w)}^p = \int_{|y|<1} |f(y)|^p\, w(y)^p\, dy + \int_{|y|\geqslant 1} |f(y)|^p\, w(y)^p\, dy =: I_1(f) + I_2(f).$$

Since $w(|y|) \approx 1$ for $|y| < 1$ we have

$$I_1(f) \lesssim \|f\|^p_{L^{p,\lambda}} \quad \text{and} \quad I_1(f) \lesssim \|f\|^p_{\mathcal{L}^{p,\lambda}}.$$

As regards to $I_2(f)$ the estimation is different in each situation. In the homogeneous case, we have

$$I_2(f) = \sum_{k=0}^{\infty} \int_{2^k \leqslant |y| < 2^{k+1}} |f(y)|^p \, w(y)^p \, dy \lesssim \sum_{k=0}^{\infty} w(2^k)^p \, 2^{k\lambda} \, \|f\|^p_{L^{p,\lambda}} \lesssim \|f\|^p_{L^{p,\lambda}},$$

where in the first inequality we used the monotonicity of $w(t)^p \, t^\gamma$ and in the second we used the integral assumption (7.18) with $\gamma = \lambda$.

In the nonhomogeneous case, we use similar arguments (with $\gamma = n$ in the integral condition (7.18)) and estimate the integral as follows:

$$I_2(f) \leqslant \sum_{\substack{m \in \mathbb{Z}^n \\ |m| \neq 0}} \int_{B(m,1)} |f(y)|^p \, w(y)^p \, dy \lesssim \sum_{|m| \geqslant 1} w(|m| - 1)^p \, \|f\|^p_{\mathcal{L}^{p,\lambda}} \lesssim \|f\|^p_{\mathcal{L}^{p,\lambda}},$$

which completes the argument. □

Example 7.20 It is easy to see that the weights

$$w(x) = (1 + |x|)^\alpha \approx (1 + |x|^2)^{\alpha/2} =: w_\alpha(x)$$

satisfy the assumptions of Theorem 7.19 when $\alpha < -\gamma/p$. Hence we have

$$L^{p,\lambda}(\mathbb{R}^n) \hookrightarrow L^p(\mathbb{R}^n, w_\alpha) \quad \text{if} \quad \alpha < -\lambda/p$$

and

$$\mathcal{L}^{p,\lambda}(\mathbb{R}^n) \hookrightarrow L^p(\mathbb{R}^n, w_\alpha) \quad \text{if} \quad \alpha < -n/p$$

(cf. Kato [162, p.132] and Rosenthal and Triebel [287, Theorem 3.1], respectively).

Example 7.21 The weights

$$w(x) = (1 + |x|)^{-\frac{\gamma}{p}} \ln^{-\beta}(e + |x|), \quad \text{with} \quad \beta > 1/p,$$

are also admissible in the sense of Theorem 7.19. This corresponds to the limiting case $\alpha = -\gamma/p$ given in Example 7.20.

7.2 Approximation in Generalized Morrey Spaces

Recall that the classical Morrey spaces $L^{p,\lambda}(\mathbb{R}^n)$, $1 \leqslant p < \infty$, $0 \leqslant \lambda \leqslant n$, consist of all locally p-integrable functions f on $\mathbb{R}^n$ such that

$$\|f\|_{L^{p,\lambda}(\mathbb{R}^n)} := \sup_{x\in\mathbb{R}^n, r>0} \left(\frac{1}{r^\lambda} \int_{B(x,r)} |f(y)|^p \, dy \right)^{1/p} < \infty. \tag{7.19}$$

Straightforward calculations show that

$$\|f(t\cdot)\|_{L^{p,\lambda}(\mathbb{R}^n)} = t^{\frac{\lambda-n}{p}} \|f\|_{L^{p,\lambda}(\mathbb{R}^n)}, \quad t > 0,$$

which implies a modification of the scaling factor in comparison with L^p-spaces. This property reveals the homogeneous nature of the spaces $L^{p,\lambda}(\mathbb{R}^n)$, and it is very useful in the study of partial differential equations.

The main goal of this section is to extend the approximation scheme developed in Sect. 7.1 (appearing in Almeida and Samko [14]), for classical Morrey spaces to the case of generalized Morrey spaces. The latter are defined by replacing the power r^λ in (7.19) by a more general positive function $\varphi(r)$ (cf. (7.20), (7.21)), mainly satisfying monotonicity type conditions. Such spaces proved to be useful in the study of critical Sobolev type embeddings (cf. Sawano and Wadade [301], Wadade [331]).

Using appropriate vanishing properties, we identify a closed subspace of $L^{p,\varphi}(\mathbb{R}^n)$ whose elements may be approximated by C_0^∞ functions in the generalized Morrey norm. We also show how the various Morrey subspaces, including a generalized version of Zorko class, are related to each other by proving embeddings and presenting examples showing its strictness. Moreover, we prove that such vanishing properties are preserved by convolution operators with integrable kernels.

7.2.1 Preliminaries on Generalized Morrey Spaces

A function $\varphi : (0, \infty) \to (0, \infty)$ is said to be *almost increasing* (resp., *almost decreasing*) if $\varphi(s) \lesssim \varphi(t)$ (resp., $\varphi(s) \gtrsim \varphi(t)$) for $s \leqslant t$.

As usual the class $C_0^\infty(\mathbb{R}^n)$ consists of all complex-valued infinitely differentiable functions on $\mathbb{R}^n$ with compact support, and $L^p(\mathbb{R}^n)$ stands for the classical Lebesgue space normed by

$$\|f\|_p = \left(\int_{\mathbb{R}^n} |f(x)|^p \, dx \right)^{1/p}, \quad 1 \leqslant p < \infty.$$

In the sequel, we use the following notation for locally integrable functions f on $\mathbb{R}^n$:

$$\mathfrak{M}_{p,\varphi}(f;x,r) := \frac{1}{\varphi(r)} \int\limits_{B(x,r)} |f(y)|^p \, dy\,, \qquad x \in \mathbb{R}^n, \quad r > 0. \tag{7.20}$$

Definition 7.22 Let $1 \leqslant p < \infty$ and let $\varphi : (0,\infty) \to (0,\infty)$ be a measurable function. The generalized Morrey space $L^{p,\varphi}(\mathbb{R}^n)$ consists of all locally p-integrable functions f on $\mathbb{R}^n$ with finite norm

$$\|f\|_{p,\varphi} := \sup_{x\in\mathbb{R}^n,\, r>0} \mathfrak{M}_{p,\varphi}(f;x,r)^{1/p}. \tag{7.21}$$

If $\varphi(r) = r^\lambda$, $0 \leqslant \lambda \leqslant n$, then $L^{p,\varphi}(\mathbb{R}^n) = L^{p,\lambda}(\mathbb{R}^n)$ are the well-known classical Morrey spaces, which in turn coincide with $L^p(\mathbb{R}^n)$ when $\lambda = 0$.

It is not hard to see that

$$L^{p,\varphi}(\mathbb{R}^n) \hookrightarrow L^\infty(\mathbb{R}^n) \quad \text{if} \quad \sup_{r>0} \frac{\varphi(r)}{r^n} < \infty$$

and

$$L^\infty(\mathbb{R}^n) \hookrightarrow L^{p,\varphi}(\mathbb{R}^n) \quad \text{if} \quad \inf_{r>0} \frac{\varphi(r)}{r^n} > 0.$$

Consequently,

$$L^{p,\varphi}(\mathbb{R}^n) = L^\infty(\mathbb{R}^n) \quad \text{when} \quad \varphi(r) \approx r^n.$$

In general we have to require some assumptions on the function parameter φ in order to ensure good properties for the spaces $L^{p,\varphi}(\mathbb{R}^n)$. Let us consider the following class.

Definition 7.23 The class Φ consists of all measurable functions $\varphi : (0,\infty) \to (0,\infty)$ such that:

1. φ is almost increasing;
2. $\varphi(t)/t^n$ is almost decreasing.
3. $\inf\limits_{t>\delta} \varphi(t) > 0$ for every $\delta > 0$.

The conditions above defining the class Φ are widely used in papers on generalized Morrey spaces. As observed in Nakai [262], if $\varphi \in \Phi$ then the space $L^{p,\varphi}(\mathbb{R}^n)$ is non-trivial. For example, characteristic functions on balls belong to $L^{p,\varphi}(\mathbb{R}^n)$ under such assumptions on φ. Moreover, for any $x_0 \in \mathbb{R}^n$ and $r_0 > 0$ there holds

$$\|\chi_{B(x_0,r_0)}\|_{p,\varphi}^p \approx \frac{r_0^n}{\varphi(r_0)}, \qquad x_0 \in \mathbb{R}^n, \quad r_0 > 0 \tag{7.22}$$

(cf. Eridani and Sawano [96, Proposition A]). It is clear that if $\varphi \lesssim \psi$ then $L^{p,\varphi}(\mathbb{R}^n) \hookrightarrow L^{p,\psi}(\mathbb{R}^n)$. Moreover, using (7.22), we can see that, for $\varphi, \psi \in \Phi$, the condition $\varphi \lesssim \psi$ is also necessary for the above embedding.

There are Morrey functions on $\mathbb{R}^n$ which are not in $L^p(\mathbb{R}^n)$. Indeed, if $\varphi \in \Phi$ and, for some $\varepsilon > 0$, the function $\varphi(t)/t^\varepsilon$ is almost increasing, then it is known that

$$\left(\frac{\varphi(|x|)}{|x|^n}\right)^{1/p} \in L^{p,\varphi}(\mathbb{R}^n) \quad \text{but} \quad \left(\frac{\varphi(|x|)}{|x|^n}\right)^{1/p} \notin L^p(\mathbb{R}^n), \tag{7.23}$$

see, for example, Almeida and Samko [13, Lemma 2.1] and Eridani et al. [97, Lemma 2.4].

7.2.2 New and Known Morrey Subspaces

It is known that approximations to the identity do not behave well in Morrey spaces, since these spaces may contain functions with singularities like (7.23). The lack of such property has motivated the consideration of appropriate Morrey subspaces mainly in the case of power functions $\varphi(r) = r^\lambda$. For example, it was introduced in Zorko [342] the subset $\mathbb{L}^{p,\lambda}(\mathbb{R}^n)$ consisting of all Morrey functions for which the translation is continuous in Morrey norm. We introduce a generalized Zorko subspace as follows.

Definition 7.24 For $1 \leqslant p < \infty$ and $\varphi \in \Phi$, we consider

$$\mathbb{L}^{p,\varphi}(\mathbb{R}^n) := \left\{ f \in L^{p,\varphi}(\mathbb{R}^n) : \ \|\tau_\xi f - f\|_{p,\varphi} \to 0 \ \text{ as } \ \xi \to 0 \right\}, \tag{7.24}$$

where $\tau_\xi f := f(\cdot - \xi)$, $\xi \in \mathbb{R}^n$.

We also consider some Morrey classes by using vanishing-type properties related to the behavior of (7.20) at the origin and at infinity.

Definition 7.25 Let $1 \leqslant p < \infty$ and $\varphi \in \Phi$. The class $V_0 L^{p,\varphi}(\mathbb{R}^n)$ consists of all those functions $f \in L^{p,\varphi}(\mathbb{R}^n)$ such that

$$\lim_{r\to 0} \sup_{x\in\mathbb{R}^n} \mathfrak{M}_{p,\varphi}(f; x, r) = 0. \tag{V_0}$$

Similarly, $V_\infty L^{p,\varphi}(\mathbb{R}^n)$ is the set of all $f \in L^{p,\varphi}(\mathbb{R}^n)$ such that

$$\lim_{r\to \infty} \sup_{x\in\mathbb{R}^n} \mathfrak{M}_{p,\varphi}(f; x, r) = 0. \tag{V_∞}$$

We shall call $V_0 L^{p,\varphi}(\mathbb{R}^n)$ and $V_\infty L^{p,\varphi}(\mathbb{R}^n)$ *vanishing Morrey spaces*, at the origin and at infinity, respectively. As usual, in the classical case of $\varphi(r) = r^\lambda$, we write $V_0 L^{p,\lambda}(\mathbb{R}^n)$ and $V_\infty L^{p,\lambda}(\mathbb{R}^n)$; see Sect. 7.1.

Remark 7.26 The generalized vanishing space $V_0L^{p,\varphi}(\mathbb{R}^n)$ has been already considered in other papers (see, for instance, Burtseva and Samko [43], Persson et al. [270], Ragusa [282], Samko [291] for harmonic analysis in such spaces).

The space $V_0L^{p,\varphi}(\mathbb{R}^n)$ is non-trivial if $\varphi \in \Phi$ and

$$\lim_{r\to 0} \frac{r^n}{\varphi(r)} = 0 \tag{7.25}$$

since then it contains bounded functions with compact support. Note also that bounded Morrey functions always satisfy the vanishing property (V_0) when (7.25) holds. We state this property as a separate lemma.

Lemma 7.27 *Let $1 \leqslant p < \infty$ and $\varphi \in \Phi$. If, in addition, φ satisfies* (7.25), *then*

$$L^{p,\varphi}(\mathbb{R}^n) \cap L^\infty(\mathbb{R}^n) \subset V_0L^{p,\varphi}(\mathbb{R}^n).$$

As regards to $V_\infty L^{p,\varphi}(\mathbb{R}^n)$, it is not hard to see that it is non-trivial if $\frac{\varphi(r)}{r^\varepsilon}$ is almost increasing for some $\varepsilon > 0$ and $\inf_{r>\delta} \varphi(r) > 0$ for all $\delta > 0$. In fact, under such assumptions on φ, compactly supported bounded functions belong to this vanishing space. We also observe the obvious fact that

$$V_0L^{p,\varphi}(\mathbb{R}^n) = L^p(\mathbb{R}^n) \quad \text{when} \quad \varphi(r) \approx 1.$$

Following the classical case studied in Almeida and Samko [14] (which appears in Sect. 7.1), we also introduce a new subspace by using another vanishing property related to truncations in large balls. Below we use the notation

$$\mathcal{A}_{N,p}(f) := \sup_{x\in\mathbb{R}^n} \int\limits_{B(x,1)} |f(y)|^p\, \chi_N(y)\, dy\,, \quad \text{where } \chi_N := \chi_{\mathbb{R}^n\setminus B(0,N)}\,, \qquad N \in \mathbb{N}.$$

Definition 7.28 For $1 \leqslant p < \infty$ and $\varphi \in \Phi$, we define $V^{(*)}L^{p,\varphi}(\mathbb{R}^n)$ as the set of all functions $f \in L^{p,\varphi}(\mathbb{R}^n)$ having the vanishing property

$$\lim_{N\to\infty} \mathcal{A}_{N,p}(f) = 0. \tag{V^*}$$

Remark 7.29 As shown in Lemma 7.4, a Morrey function f satisfies property (V^*) if and only if

$$\lim_{N\to\infty} \sup_{x\in\mathbb{R}^n} \int\limits_{B(x,r)} |f(y)|^p\, \chi_N(y)\, dy = 0 \tag{7.26}$$

uniformly in $r \in (0, R_0]$, for any fixed $R_0 > 0$. Note also that the generalized parameter φ does not interfere in the vanishing property (V^*) itself.

By the Lebesgue dominated convergence theorem, we can see that every L^p-function has property (V^*) and hence $V^{(*)}L^{p,\varphi}(\mathbb{R}^n) = L^p(\mathbb{R}^n)$ when $\varphi(r) \approx 1$. Nevertheless, there are Morrey functions which fail to have this vanishing property. An example of such a function is given in Example 7.40 below.

The subspaces $V_0L^{p,\varphi}(\mathbb{R}^n)$ and $V^{(*)}L^{p,\varphi}(\mathbb{R}^n)$ make difference only when we are dealing with Morrey spaces different from Lebesgue spaces. Indeed, it is easy to check that all L^p-functions have the vanishing properties (V_0) and (V^*). On the other hand, the translation operator is continuous in L^p-norm. Therefore, for any $p \in [1, \infty)$ we have the coincidence

$$\mathbb{L}^{p,\varphi}(\mathbb{R}^n) = V_0L^{p,\varphi}(\mathbb{R}^n) = V^{(*)}L^{p,\varphi}(\mathbb{R}^n) = L^p(\mathbb{R}^n) \quad \text{if} \quad \varphi(r) \approx 1.$$

We end this section by defining a smaller Morrey class which will play an important role later on.

Definition 7.30 Let $1 \leqslant p < \infty$ and $\varphi \in \Phi$. We set

$$V_{0,\infty}^{(*)}L^{p,\varphi}(\mathbb{R}^n) := V_0L^{p,\varphi}(\mathbb{R}^n) \cap V_\infty L^{p,\varphi}(\mathbb{R}^n) \cap V^{(*)}L^{p,\varphi}(\mathbb{R}^n).$$

Remark 7.31 It can be shown that all the vanishing subsets $\mathbb{L}^{p,\varphi}(\mathbb{R}^n)$, $V_0L^{p,\varphi}(\mathbb{R}^n)$, $V_\infty L^{p,\varphi}(\mathbb{R}^n)$ and $V^{(*)}L^{p,\varphi}(\mathbb{R}^n)$ (and consequently, $V_{0,\infty}^{(*)}L^{p,\varphi}(\mathbb{R}^n)$) are closed in $L^{p,\varphi}(\mathbb{R}^n)$ (see Lemma 7.7 for a proof in the classical case).

7.2.3 *Convolution in Morrey Spaces*

The classical Young's inequality for convolutions with kernels in $L^1(\mathbb{R}^n)$, known for L^p-spaces, holds also for Morrey spaces. In a more general definition of Morrey spaces, Young's inequality was recently studied in Burenkov and Tararykova [42]. Here we are interested in the preservation of the vanishing properties (V_0), (V_∞) and (V^*) by convolution operators.

The following two lemmas can be proved using Minskowski's inequality and a simple change of variables. Further details can be found in the proof of Theorem 7.8.

Lemma 7.32 *Let $1 \leqslant p < \infty$ and $\varphi \in \Phi$. If $\mathcal{K} \in L^1(\mathbb{R}^n)$ and f is locally p-integrable on $\mathbb{R}^n$, then*

$$\mathfrak{M}_{p,\varphi}\big(\mathcal{K} * f; x, r\big) \leqslant \|\mathcal{K}\|_1^p \sup_{z\in\mathbb{R}^n} \mathfrak{M}_{p,\varphi}(f; z, r) \tag{7.27}$$

for every $x \in \mathbb{R}^n$ and $r > 0$. Consequently,

$$\|\mathcal{K} * f\|_{p,\varphi} \leqslant \|\mathcal{K}\|_1 \, \|f\|_{p,\varphi}.$$

In the next lemma we use the interpretation

$$\mathcal{A}_{a,p}(f) := \sup_{x\in\mathbb{R}^n} \int\limits_{B(x,1)} |f(y)|^p \chi_a(y)\, dy\,, \qquad a \in \mathbb{R},$$

where

$$\chi_a := \chi_{\mathbb{R}^n\setminus B(0,a)} \text{ if } a > 0 \quad \text{and} \quad \chi_a \equiv 1 \text{ if } a \leqslant 0.$$

Lemma 7.33 *Let $1 \leqslant p < \infty$. If $\mathcal{K} \in L^1(\mathbb{R}^n)$ and f is locally integrable on $\mathbb{R}^n$, then*

$$\left[\mathcal{A}_{N,p}(\mathcal{K} * f)\right]^{1/p} \leqslant \int\limits_{\mathbb{R}^n} |\mathcal{K}(z)| \left[\mathcal{A}_{N-|z|,p}(f)\right]^{1/p} dz, \tag{7.28}$$

for any $N \in \mathbb{N}$.

From (7.27) we easily see that both vanishing properties, at the origin and at infinity, are preserved by convolutions with integrable kernels. Moreover, the same holds for the vanishing property (V^*). Indeed, by (7.28) and the Lebesgue dominated convergence theorem, we conclude that

$$\mathcal{A}_{N,p}(\mathcal{K} * f) \to 0 \quad \text{as} \quad N \to \infty.$$

Corollary 7.34 *Let $1 \leqslant p < \infty$ and $\varphi \in \Phi$. Then the Morrey subspaces $V_0L^{p,\varphi}(\mathbb{R}^n)$, $V_\infty L^{p,\varphi}(\mathbb{R}^n)$ and $V^{(*)}L^{p,\varphi}(\mathbb{R}^n)$ are all invariant with respect to convolution operators with a kernel $\mathcal{K} \in L^1(\mathbb{R}^n)$:*

$$f \in V_0L^{p,\varphi}(\mathbb{R}^n) \;\to\; \mathcal{K} * f \in V_0L^{p,\varphi}(\mathbb{R}^n),$$

$$f \in V_\infty L^{p,\varphi}(\mathbb{R}^n) \;\to\; \mathcal{K} * f \in V_\infty L^{p,\varphi}(\mathbb{R}^n),$$

and

$$f \in V^{(*)}L^{p,\varphi}(\mathbb{R}^n) \;\to\; \mathcal{K} * f \in V^{(*)}L^{p,\varphi}(\mathbb{R}^n).$$

We already know (by Lemma 7.32) that the convolution operator with an integrable kernel is bounded on $L^{p,\varphi}(\mathbb{R}^n)$. The next result shows that the convolution of a Morrey function is bounded when the kernel possesses additional desirable properties.

Theorem 7.35 *Let $1 \leqslant p < \infty$ and $\varphi \in \Phi$. Let $\mathcal{K}(x) = \widetilde{k}(|x|)$ be an integrable kernel in $\mathbb{R}^n$ with $\widetilde{k}$ a non-negative C^1 function on $[0, \infty)$. Suppose that*

$$\widetilde{k}(r)\, r^{\frac{n}{p'}}\, \varphi(r)^{\frac{1}{p}} \to 0 \quad \text{as} \quad r \to \infty \tag{7.29}$$

and

$$C(\mathcal{K}, \varphi, p) := \int_0^\infty |\widetilde{k}'(r)|\, r^{\frac{n}{p'}}\, \varphi(r)^{\frac{1}{p}}\, dr < \infty. \tag{7.30}$$

Then the convolution operator with kernel $\mathcal{K}$ is bounded from $L^{p,\varphi}(\mathbb{R}^n)$ into $L^\infty(\mathbb{R}^n)$ and

$$\|\mathcal{K} * f\|_\infty \leqslant \omega_n^{\frac{1}{p'}}\, C(\mathcal{K}, \varphi, p)\, \|f\|_{p,\varphi}$$

for every $f \in L^{p,\varphi}(\mathbb{R}^n)$.

Proof Passing to polar coordinates, we get

$$\int_{\mathbb{R}^n} \mathcal{K}(y)\, |f(x-y)|\, dy = \int_0^\infty \int_{\mathbb{S}^{n-1}} \widetilde{k}(r)\, |f(x-r\sigma)|\, r^{n-1}\, d\sigma\, dr = \int_0^\infty \widetilde{k}(r)\, dF(r) \tag{7.31}$$

where

$$F(r) = \int_0^r G(s)\, ds \quad \text{with} \quad G(s) = \int_{\mathbb{S}^{n-1}} |f(x-s\sigma)|\, \mathbb{S}^{n-1}\, d\sigma.$$

Integrating by parts, we can write the right-hand side of (7.31) as

$$-\int_0^\infty \widetilde{k}'(r)\, F(r)\, dr = -\int_0^\infty \widetilde{k}'(r) \int_{B(0,r)} |f(x-y)|\, dy\, dr = -\int_0^\infty \widetilde{k}'(r) \int_{B(x,r)} |f(y)|\, dy\, dr,$$

where we noted (by Hölder's inequality) that

$$F(r) = \int_{B(x,r)} |f(y)|\, dy \leqslant (\omega_n r^n)^{\frac{1}{p'}}\, \varphi(r)^{\frac{1}{p}}\, \|f\|_{p,\varphi}$$

and used (7.29) in particular. Therefore, with the help of Hölder's inequality and condition (7.30), the convolution may be estimated, at any $x \in \mathbb{R}^n$, as follows:

$$|\mathcal{K} * f(x)| \leqslant \int_0^\infty |\widetilde{k}'(r)| \int_{B(x,r)} |f(y)|\, dy\, dr \leqslant \omega_n^{\frac{1}{p'}}\, C(\mathcal{K}, \varphi, p)\, \|f\|_{p,\varphi},$$

and the claim is proved. □

Remark 7.36 An example of a radial kernel satisfying the assumptions of Theorem 7.35 is the heat kernel

$$\mathcal{K}_t(x) = c_n\, t^{-\frac{n}{2}}\, e^{-\frac{|x|^2}{4t}}, \qquad t > 0, \tag{7.32}$$

where $c_n > 0$ is a certain normalization constant depending on n. This special case was considered in Kato [162] in the case of classical Morrey spaces.

We end this section with an observation on the approximation of Morrey functions with Zorko property by standard mollifiers. This will be useful later on.

Consider the usual dilations $\mathcal{K}_t(x) = t^{-n}\mathcal{K}(x/t), t > 0$, where $\mathcal{K}$ is an integrable function on $\mathbb{R}^n$ with $\|\mathcal{K}\|_1 = 1$. By straightforward calculations and Minkowski's integral inequality, we get

$$\|f * \mathcal{K}_t - f\|_{p,\varphi} \leqslant \int\limits_{\mathbb{R}^n} \|\tau_{tz} f - f\|_{p,\varphi}\, |\mathcal{K}(z)|\, dz.$$

If $f \in \mathbb{L}^{p,\varphi}(\mathbb{R}^n)$, we have $\|\tau_{tz} f - f\|_{p,\varphi} \to 0$ as $t \to 0$. Since $\|\tau_{tz} f - f\|_{p,\varphi} \leqslant 2\,\|f\|_{p,\varphi}$ for any $t > 0$ and $z \in \mathbb{R}^n$, we see that

$$\|f * \mathcal{K}_t - f\|_{p,\varphi} \to 0 \quad \text{as} \quad t \to 0 \tag{7.33}$$

by the Lebesgue theorem.

7.2.4 *Strict Embeddings Between Morrey Subspaces*

Examples and Embeddings Between Vanishing Morrey Spaces

We consider the families of functions given by

$$g_\alpha(x) := \left(\frac{\varphi(|x|)}{|x|^{n-\alpha}}\right)^{1/p} \chi_{B(0,1)}(x) \quad \text{and} \quad h_\beta(x) := \left(\frac{\varphi(|x|)}{|x|^{n+\beta}}\right)^{1/p} \chi_{\mathbb{R}^n\setminus B(0,1)}(x)$$

for $\alpha, \beta \geqslant 0$. In the limiting cases $\alpha = 0 = \beta$, we already know that $g_\alpha, h_\beta \in L^{p,\varphi}(\mathbb{R}^n)$ when $\varphi \in \Phi$ and $\frac{\varphi(t)}{t^\varepsilon}$ is almost increasing for some $\varepsilon > 0$. For the non-limiting cases, we have the following lemma.

Lemma 7.37 *Let $\varphi \in \Phi$ and $1 \leqslant p < \infty$.*

(i) If $\alpha > 0$ and $\varphi(t)/t^{n-\delta}$ is almost decreasing for some $\delta \in (0, \alpha]$, then $g_\alpha \in V_0L^{p,\varphi}(\mathbb{R}^n)$.

(ii) If $\beta > 0$ and $\varphi(t)/t^\varepsilon$ is almost increasing for some $\varepsilon > \beta$, then $h_\beta \in V_\infty L^{p,\varphi}(\mathbb{R}^n)$.

Proof *Step 1*: We prove (i). Let $x \in \mathbb{R}^n$ and $r > 0$.
If $|x| < 2r$ we have

$$\mathfrak{M}_{p,\varphi}(g_\alpha; x, r) \leqslant \frac{1}{\varphi(r)} \int\limits_{B(0,\min\{1,3r\})} \frac{\varphi(|y|)}{|y|^{n-\alpha}}\, dy$$

since $B(x, r) \subset B(0, 3r)$. Passing to polar coordinates in the last integral and using the fact that φ is almost increasing and $\varphi(t)/t^n$ is almost decreasing, we get

$$\mathfrak{M}_{p,\varphi}(g_\alpha; x, r) \lesssim \min\{1, r\}^\alpha, \tag{7.34}$$

with the implicit constant independent of x and r.
In the case $|x| \geqslant 2r$ we have $|y| > |x - y|$. Hence

$$\begin{aligned}\mathfrak{M}_{p,\varphi}(g_\alpha; x, r) &\leqslant \frac{1}{\varphi(r)} \int\limits_{B(x,r)\cap B(x,1)} \frac{\varphi(|y|)}{|y|^{n-\alpha}} \chi_{B(0,1)}(y)\, dy \\ &\leqslant \frac{1}{\varphi(r)} \int\limits_{B(x,\min\{1,r\})} \frac{\varphi(|y|)}{|y|^{n-\delta}}\, dy\end{aligned}$$

where we used that $\delta \leqslant \alpha$. Since $\varphi(t)/t^{n-\delta}$ is almost decreasing the last integral can be estimated from above by

$$\int\limits_{B(x,\min\{1,r\})} \frac{\varphi(|x-y|)}{|x-y|^{n-\delta}}\, dy = \int\limits_{B(0,\min\{1,r\})} \frac{\varphi(|z|)}{|z|^{n-\delta}}\, dz.$$

Passing again to polar coordinates and observing that φ is almost increasing, we obtain the final estimate

$$\mathfrak{M}_{p,\varphi}(g_\alpha; x, r) \lesssim \min\{1, r\}^\delta. \tag{7.35}$$

By (7.34) and (7.35) we see that $g_\alpha \in V_0 L^{p,\varphi}(\mathbb{R}^n)$.
Step 2: We prove (ii). If $|x| < 2r$ we use again that $B(x, r) \subset B(0, 3r)$ and obtain

$$\mathfrak{M}_{p,\varphi}(h_\beta; x, r) \leqslant \frac{1}{\varphi(r)} \int\limits_{1\leqslant|y|\leqslant 3r} \frac{\varphi(|y|)}{|y|^{n+\beta}}\, dy$$

when $3r \geqslant 1$ (otherwise the integral vanishes taking into account the definition of h_β). Writing the last integral in polar coordinates and noting that $\varphi(t)/t^\varepsilon$ is almost increasing and that $\varphi(t)/t^n$ is almost decreasing, we estimate the right-hand side as

$$\frac{1}{\varphi(r)} \int\limits_{1\leqslant|y|\leqslant 3r} \frac{\varphi(|y|)}{|y|^{n+\beta}}\,dy \lesssim \frac{\varphi(3r)}{\varphi(r)\,r^{\varepsilon}} \int\limits_{1}^{3r} \rho^{\varepsilon-\beta-1}d\rho \lesssim r^{-\beta}.$$

Hence, in the case $|x| < 2r$, there holds the estimate

$$\mathfrak{M}_{p,\varphi}(h_\beta; x, r) \lesssim \max\{1, r\}^{-\beta}, \tag{7.36}$$

with the implicit constant not depending on x and r. We can check that estimate (7.36) also holds in the remaining case $|x| > 2r$. This can be handled following the corresponding case in Step 1, but now using that $\varphi(t)/t^n$ is almost decreasing and $\varphi(t)/t^{\varepsilon}$ is almost increasing (with $\beta < \varepsilon$). □

The previous lemma and similar calculations used in its proof prompt us to exhibit the following example which summarizes the main conclusions on the preciseness of embeddings.

Example 7.38 Let $1 \leqslant p < \infty$ and $\varphi \in \Phi$. Define

$$f_{\alpha,\beta}(x) := g_\alpha(x) + h_\beta(x) = \begin{cases} \left(\frac{\varphi(|x|)}{|x|^{n-\alpha}}\right)^{1/p}, & |x| < 1, \\ \left(\frac{\varphi(|x|)}{|x|^{n+\beta}}\right)^{1/p}, & |x| \geqslant 1. \end{cases}$$

If $\alpha, \beta > 0$ and there exist $\delta \in (0, \alpha]$ and $\varepsilon > \beta$ such that $\varphi(r)/r^{n-\delta}$ is almost decreasing and $\varphi(r)/r^{\varepsilon}$ is almost increasing, then

$$f_{\alpha,\beta} \in V_0 L^{p,\varphi}(\mathbb{R}^n) \cap V_\infty L^{p,\varphi}(\mathbb{R}^n).$$

Moreover,

$$f_{\alpha,0} \in V_0 L^{p,\varphi}(\mathbb{R}^n) \quad \text{but} \quad f_{\alpha,0} \notin V_\infty L^{p,\varphi}(\mathbb{R}^n)$$

and

$$f_{0,\beta} \in V_\infty L^{p,\varphi}(\mathbb{R}^n) \quad \text{but} \quad f_{0,\beta} \notin V_0 L^{p,\varphi}(\mathbb{R}^n).$$

In the limiting case $\alpha = 0 = \beta$, there holds

$$f_{\alpha,\beta} \in L^{p,\varphi}(\mathbb{R}^n) \quad \text{but} \quad f_{\alpha,\beta} \notin V_0 L^{p,\varphi}(\mathbb{R}^n) \cup V_\infty L^{p,\varphi}(\mathbb{R}^n),$$

where the failure of the vanishing properties (V_0) and (V_∞) can be seen from the estimate

$$\mathfrak{M}_{p,\varphi}(f_{0,0}; 0, r) = \frac{1}{\varphi(r)} \int_{B(0,r)} \frac{\varphi(|y|)}{|y|^n}\, dy \gtrsim 1$$

(with the implicit constant independent of $r > 0$), which follows by the monotonicity of $\varphi(r)/r^n$.

Theorem 7.39 *Let* $1 \leqslant p < \infty$ *and* $\varphi \in \Phi$. *If, in addition,* φ *satisfies* (7.25) *and*

$$\lim_{r\to\infty} \frac{\log^p r}{\varphi(r)} = 0, \tag{7.37}$$

then there are functions in $V_0 L^{p,\varphi}(\mathbb{R}^n) \cap V_\infty L^{p,\varphi}(\mathbb{R}^n)$ *which do not have property* (V^*).

Proof Let f_0 be the function defined by

$$f_0(x) := \sum_{k=2}^{\infty} \chi_{B_k}(x)\,,$$

where $B_k = B(2^k e_1, 1)$, $k \in \mathbb{N}$, with $e_1 = (1, 0, \ldots, 0)$. The calculations involving the vanishing properties are quite similar to those presented in the proof of Theorem 7.10. Nevertheless, for reader's convenience, we present some steps to show where the assumptions on φ are used. Since $\mathcal{A}_{N,p}(f_0) \geqslant |B(0, 1)|$ for any $N \in \mathbb{N}$, then f_0 fails to have the vanishing property (V^*). In order to see that f_0 belongs to the other vanishing spaces, we need to count the number of balls B_k intersecting $B(x, r)$ (for any fixed $x \in \mathbb{R}^n$ and $r > 0$). If $r \leqslant 1$ there exists at most one such a ball. Hence

$$\mathfrak{M}_{p,\varphi}(f_0; x, r) \lesssim \frac{r^n}{\varphi(r)} \quad \text{for} \quad r \leqslant 1, \tag{7.38}$$

with the implicit constant independent of x and r. From (7.38) and (7.25) we conclude that f_0 satisfies property (V_0). The case of large values of r is technically more complicated. Following the aforementioned theorem, we get

$$\mathfrak{M}_{p,\varphi}(f_0; x, r) \leqslant \frac{\log^p(4r)}{\varphi(r)} \quad \text{for} \quad r > 1. \tag{7.39}$$

Therefore, by (7.39) and (7.37), we see that f_0 also satisfies property (V_∞). The proof is complete. □

Let us explicit the example contained in the proof above.

Example 7.40 Let f_0 be the function given by

$$f_0(x) := \sum_{k=2}^{\infty} \chi_{B_k}(x) \qquad (\text{with } B_k = B(2^k e_1, 1)).$$

If $1 \leqslant p < \infty$ and $\varphi \in \Phi$ satisfies the conditions (7.25) and (7.37), then we have

$$f_0 \in V_0 L^{p,\varphi}(\mathbb{R}^n) \cap V_\infty L^{p,\varphi}(\mathbb{R}^n) \quad \text{but} \quad f_0 \notin V^{(*)} L^{p,\varphi}(\mathbb{R}^n).$$

Corollary 7.41 *Under the same assumptions of Theorem 7.39, there holds*

$$V_{0,\infty}^{(*)} L^{p,\varphi}(\mathbb{R}^n) \subsetneqq V_0 L^{p,\varphi}(\mathbb{R}^n) \cap V_\infty L^{p,\varphi}(\mathbb{R}^n) \subsetneqq V_0 L^{p,\varphi}(\mathbb{R}^n) \subsetneqq L^{p,\varphi}(\mathbb{R}^n).$$

Embeddings Involving the Generalized Zorko Space

We have seen in the previous section that the new subspace $V_{0,\infty}^{(*)} L^{p,\varphi}(\mathbb{R}^n)$ is strictly smaller than the intersection $V_0 L^{p,\varphi}(\mathbb{R}^n) \cap V_\infty L^{p,\varphi}(\mathbb{R}^n)$. We will see that $V_{0,\infty}^{(*)} L^{p,\varphi}(\mathbb{R}^n)$ is also strictly smaller than the generalized Zorko subspace $\mathbb{L}^{p,\varphi}(\mathbb{R}^n)$ (recall Definition 7.24).

First we observe the connection of the Zorko space with the vanishing space $V_0 L^{p,\varphi}(\mathbb{R}^n)$. The next result is known in the classical case of $\varphi(r) = r^\lambda, 0 \leqslant \lambda \leqslant n$, see Kato [162, Corollary 3.3].

Theorem 7.42 *Let* $1 \leqslant p < \infty$ *and* $\varphi \in \Phi$. *If* φ *also satisfies* (7.25), *then*

$$\mathbb{L}^{p,\varphi}(\mathbb{R}^n) \subset V_0 L^{p,\varphi}(\mathbb{R}^n). \tag{7.40}$$

Proof Let $f \in \mathbb{L}^{p,\varphi}(\mathbb{R}^n)$ and consider the heat kernel $\mathcal{K}_t(x)$ given by (7.32). Recalling the discussion in the very end of Sect. 7.2.3, we have

$$\| f * \mathcal{K}_t - f \|_{p,\varphi} \to 0 \quad \text{as} \quad t \to 0.$$

On the other hand, by Lemma 7.32 and Theorem 7.35 (and also Remark 7.36), we have, for any $t > 0$,

$$f * \mathcal{K}_t \in L^{p,\varphi}(\mathbb{R}^n) \cap L^\infty(\mathbb{R}^n).$$

Therefore, $f * \mathcal{K}_t \in V_0 L^{p,\varphi}(\mathbb{R}^n)$ in view of Lemma 7.27. Since $V_0 L^{p,\varphi}(\mathbb{R}^n)$ is closed we conclude that $f \in V_0 L^{p,\varphi}(\mathbb{R}^n)$. □

The strictness of the embedding (7.40) was proved in Kato [162, Example 3.4] in the case of classical Morrey spaces. Below we use the same example from Kato [162] to show that there are functions satisfying both vanishing properties at the origin and at infinity but not having Zorko property, when, in addition to the conditions in Theorem 7.42, φ also satisfies (7.37). It shows, in particular, that embedding (7.40) is strict under appropriate assumptions.

Example 7.43 For simplicity, consider the case $n = 1$. Define

$$f_1(x) := \sum_{k=1}^{\infty} \eta_k(x - 2^k) \quad \text{where} \quad \eta_k(y) = \sin(2k\pi y)\, \chi_{(0,1)}(y), \quad k \in \mathbb{N}.$$

It is clear that f_1 is bounded and $\|f_1\|_\infty = 1$. Thus, for every x and r, the modular $\mathfrak{M}_{p,\varphi}(f_1; x, r)$ has the same bound as in (7.38). To control the behavior of the modular for large values of r, we need to count the number of intervals $(2^k, 1 + 2^k)$ contained in $(x - r, x + r)$. Using similar arguments to those used in the proof of Theorem 7.39, we can arrive at an estimate similar to (7.39). Taking into account the assumptions on φ, we conclude that $f_1 \in V_0L^{p,\varphi}(\mathbb{R}^n) \cap V_\infty L^{p,\varphi}(\mathbb{R}^n)$.

In order to show that $f_1 \notin \mathbb{L}^{p,\varphi}(\mathbb{R}^n)$, first we calculate $\tau_{\frac{1}{2k}}\eta_k - \eta_k$ in the intervals $(0, 1 + \frac{1}{2k})$, $k \in \mathbb{N}$. Following Kato [162, p. 137], after some calculations we get

$$\int_{B\left(2^k+\frac{1}{2},1\right)} \left|\tau_{\frac{1}{2k}} f_1(y) - f_1(y)\right|^p dy \geqslant 4^{-p}.$$

Therefore,

$$\left\|\tau_{\frac{1}{2k}} f_1 - f_1\right\|_{p,\varphi} \geqslant c_\varphi > 0$$

for some constant c_φ independent of k. Hence we have

$$f_1 \in V_0L^{p,\varphi}(\mathbb{R}^n) \cap V_\infty L^{p,\varphi}(\mathbb{R}^n) \quad \text{but} \quad f_1 \notin \mathbb{L}^{p,\varphi}(\mathbb{R}^n).$$

In particular, by Theorem 7.42, we have the strict embedding

$$\mathbb{L}^{p,\varphi}(\mathbb{R}^n) \subsetneqq V_0L^{p,\varphi}(\mathbb{R}^n).$$

Remark 7.44 It is known that the vanishing Morrey space at the origin and the Zorko space coincide on bounded domains. At least in the standard case of power functions $\varphi(t) = t^\lambda$, this follows from Chiarenza and Franciosi [52, Lemma 1.2] and Kato [162, Corollary 3.3]. In a sense the bounded domains setting is much better to deal with since then Morrey functions are always in L^p.

As mentioned above, we want to show that our space $V_{0,\infty}^{(*)}L^{p,\varphi}(\mathbb{R}^n)$ is even strictly smaller than the Zorko space. To this end we need an approximation argument which already touches the main topic of this section. Note that the approximation by nice functions in Morrey subspaces will be discussed in detail in Sect. 7.2.5 below. By convenience, the proof of the next lemma is postponed to Sect. 7.2.5 (see Step 2 in the proof of Theorem 7.50).

Lemma 7.45 *Let $1 \leqslant p < \infty$ and $\varphi \in \Phi$. Then every function from $V^{(*)}_{0,\infty}L^{p,\varphi}(\mathbb{R}^n)$ can be approximated in Morrey norm by compactly supported functions.*

Theorem 7.46 *For any $1 \leqslant p < \infty$ and $\varphi \in \Phi$, we have $V^{(*)}_{0,\infty}L^{p,\varphi}(\mathbb{R}^n) \subset \mathbb{L}^{p,\varphi}(\mathbb{R}^n)$.*

Proof Let $f \in V^{(*)}_{0,\infty}L^{p,\varphi}(\mathbb{R}^n)$. For any $\varepsilon > 0$ there exists $g \in L^{p,\varphi}(\mathbb{R}^n)$ with compact support such that

$$\|f - g\|_{p,\varphi} < \varepsilon/4$$

(cf. Lemma 7.45). Since for any $\xi \in \mathbb{R}^n$, we have

$$\|\tau_\xi f - f\|_{p,\varphi} \leqslant 2\,\|f - g\|_{p,\varphi} + \|\tau_\xi g - g\|_{p,\varphi}$$

it suffices to show that the second norm is less that $\varepsilon/2$ for small values of $|\xi|$. By the vanishing properties (V_0) and (V_∞), one can find $r_0, r_1 > 0$, with $r_0 < r_1$, such that

$$S_1 := \sup_{x\in\mathbb{R}^n, 0<r<r_0} \mathfrak{M}_{p,\varphi}(\tau_\xi g - g; x, r) \leqslant 2^p \sup_{x\in\mathbb{R}^n, 0<r<r_0} \mathfrak{M}_{p,\varphi}(g; x, r) < (\varepsilon/2)^p$$

and

$$S_2 := \sup_{x\in\mathbb{R}^n, r>r_1} \mathfrak{M}_{p,\varphi}(\tau_\xi g - g; x, r) \leqslant 2^p \sup_{x\in\mathbb{R}^n, r>r_1} \mathfrak{M}_{p,\varphi}(g; x, r) < (\varepsilon/2)^p.$$

For such fixed r_0 and r_1, we estimate

$$\|\tau_\xi g - g\|^p_{p,\varphi} \leqslant \max\{S_1, S_2, S_3\}$$

with

$$S_3 := \sup_{x\in\mathbb{R}^n, r_0\leqslant r\leqslant r_1} \mathfrak{M}_{p,\varphi}(\tau_\xi g - g; x, r).$$

Since $\inf_{r\geqslant r_0} \varphi(r) > 0$, we have

$$S_3 \lesssim \sup_{x\in\mathbb{R}^n} \int\limits_{B(x,r_1)} |g(y-\xi) - g(y)|^p\, dy \leqslant \max\left\{ \sup_{|x|<M} (\cdots),\ \sup_{|x|>M} (\cdots) \right\}.$$

where $M > 0$ is chosen below. Since g has compact support, there exists $K > 0$ such that $g(u) = 0$ if $|u| > K$. In the case $|x| > M$, we have

$$\int_{|y-x|<r_1} |g(y-\xi) - g(y)|^p \, dy = \int_{|z|<r_1} |g(z+x-\xi) - g(z+x)|^p \, dz.$$

Hence, if we choose $M > r_1 + K + 1$, then $g(z+x) = 0$ and $g(z+x-\xi) = 0$ for small values of $|\xi|$, say $|\xi| < 1$, since

$$|z+x| > M - r_1 > K \quad \text{and} \quad |z+x-\xi| \geqslant |z+x| - |\xi| > K.$$

Let then $M > r_1 + K + 1$ be fixed, and let us now estimate the integral when $|x| < M$. In this case we are just taking the L^p-norm on a ball centered at the origin with fixed radius, precisely $B(0, r_1 + M + 1)$, again for $|\xi| < 1$. Therefore we also obtain

$$S_3 < (\varepsilon/2)^p$$

by the continuity of the L^p-norm with respect to translations. □

We end this section with the following chain of strict embeddings.

Corollary 7.47 *Let* $1 \leqslant p < \infty$ *and* $\varphi \in \Phi$ *satisfying* (7.25) *and* (7.37). *Then*

$$V_{0,\infty}^{(*)} L^{p,\varphi}(\mathbb{R}^n) \subsetneq \mathbb{L}^{p,\varphi}(\mathbb{R}^n) \subsetneq V_0 L^{p,\varphi}(\mathbb{R}^n) \subsetneq L^{p,\varphi}(\mathbb{R}^n).$$

7.2.5 Approximation in Subspaces of Generalized Morrey Spaces

As observed by Zorko [342] in the classical case $\varphi(r) = r^\lambda$, $\lambda \in (0, n)$, there are Morrey functions that cannot be approximated even by continuous functions. It is the case of functions with the form (7.23). This fact has motivated the introduction of the subspace $\mathbb{L}^{p,\lambda}(\mathbb{R}^n)$. It turns out that even in this subspace we cannot approximate (in Morrey norm) by infinitely differentiable compactly supported functions.

Recalling the details given in the end of Sect. 7.2.3, one knows that if $f \in \mathbb{L}^{p,\varphi}(\mathbb{R}^n)$ and $\mathcal{K} \in L^1(\mathbb{R}^n)$, with $\|\mathcal{K}\|_1 = 1$, then $\|f * \mathcal{K}_t - f\|_{p,\varphi} \to 0$ as $t \to 0$ (cf. (7.33)). If, in addition, the kernel $\mathcal{K}$ is smooth, say $\mathcal{K}$ is a Schwartz function, then the mollifiers $f * \mathcal{K}_t \in \mathbb{L}^{p,\varphi}(\mathbb{R}^n) \cap C^\infty(\mathbb{R}^n)$ for any $t > 0$ (note that the convolution is invariant with respect to translations). Consequently, we derive the following result.

Theorem 7.48 *Let $\varphi \in \Phi$ and $1 \leqslant p < \infty$. Then every Morrey function with Zorko property can be approximated in Morrey norm by C^∞ functions. Moreover, we have*

$$\overline{\mathbb{L}^{p,\lambda}(\mathbb{R}^n) \cap C^\infty(\mathbb{R}^n)} = \mathbb{L}^{p,\varphi}(\mathbb{R}^n).$$

Now we discuss the approximation of Morrey functions having both vanishing properties at the origin and at infinity.

Theorem 7.49 *Let $\varphi \in \Phi$ and $1 \leqslant p < \infty$. If $f \in V_0 L^{p,\varphi}(\mathbb{R}^n) \cap V_\infty L^{p,\varphi}(\mathbb{R}^n)$ is uniformly continuous, then f can be approximated in Morrey norm by functions from $V_0 L^{p,\varphi}(\mathbb{R}^n) \cap V_\infty L^{p,\varphi}(\mathbb{R}^n) \cap C^\infty(\mathbb{R}^n)$.*

Proof Take a Schwartz kernel $\mathcal{K}$ with $\|\mathcal{K}\|_1 = 1$. Then, by Corollary 7.34, we have $f * \mathcal{K}_t \in V_0 L^{p,\varphi}(\mathbb{R}^n) \cap V_\infty L^{p,\varphi}(\mathbb{R}^n) \cap C^\infty(\mathbb{R}^n)$ for any $t > 0$.

It remains to show that $f * \mathcal{K}_t \to f$ in $L^{p,\varphi}(\mathbb{R}^n)$ as $t \to 0$. Let $\varepsilon > 0$. For any $x \in \mathbb{R}^n$, $r > 0$ and $t > 0$, we have

$$\left(\int_{B(x,r)} |(f * \mathcal{K}_t)(y) - f(y)|^p \, dy \right)^{1/p}$$

$$\leqslant \int_{\mathbb{R}^n} |\mathcal{K}_t(z)| \left(\int_{B(x,r)} |f(y-z) - f(y)|^p \, dy \right)^{1/p} dz.$$

Since $f \in V_0 L^{p,\varphi}(\mathbb{R}^n) \cap V_\infty L^{p,\varphi}(\mathbb{R}^n)$, there are $r_0, r_1 > 0$ such that

$$\frac{1}{\varphi(r)} \int_{B(x,r)} |f(y-z) - f(y)|^p \, dy < \varepsilon$$

for every $r < r_0$ or $r > r_1$ (and all $x, z \in \mathbb{R}^n$). We also have

$$\sup_{r>0} \frac{1}{\varphi(r)} \int_{B(x,r)} |(f * \phi_t)(y) - f(y)|^p \, dy \leqslant \frac{1}{\inf_{r \geqslant r_0} \varphi(r)} \max\{\varepsilon, S_{r_0,r_1}(x,t)^p\}, \tag{7.41}$$

where

$$S_{r_0,r_1}(x,t) := \int_{\mathbb{R}^n} |\mathcal{K}_t(z)| \left(\int_{B(x,r_1)} |f(y-z) - f(y)|^p \, dy \right)^{1/p} dz.$$

By the uniform continuity of f one can find $\delta > 0$ such that $|f(y-z) - f(y)| < \varepsilon$ for any y and z with $|z| < \delta$. For such fixed δ we split the outer integral above into

$$S_{r_0,r_1}(x,t) = \int\limits_{|z|<\delta} (\cdots)\,dz + \int\limits_{|z|\geqslant\delta} (\cdots)\,dz. \tag{7.42}$$

For the first integral we use the uniform continuity of f and get

$$\int\limits_{|z|<\delta} (\cdots)\,dz \leqslant \varepsilon\,|B(x,r_1)|^{1/p} \int\limits_{|z|<\delta} |\mathcal{K}_t(z)|\,dz \leqslant |B(0,1)|\,r_1^{n/p}\,\varepsilon \tag{7.43}$$

for every $x \in \mathbb{R}^n$ and $t > 0$. In the second integral in (7.42), we use the fact that $f \in L^{p,\varphi}(\mathbb{R}^n)$ and

$$\int\limits_{|z|\geqslant\delta} |\mathcal{K}_t(z)|\,dz \lesssim \int\limits_{|z|\geqslant\delta} \frac{t}{|z|^{n+1}}\,dz \lesssim t$$

(where the implicit constant is independent of t) to derive the inequality

$$\int\limits_{|z|\geqslant\delta} (\cdots)\,dz \lesssim \sup_{r\in[r_0,r_1]} \varphi(r)^{1/p}\,t\,\|f\|_{p,\varphi} \tag{7.44}$$

with the implicit constant depending only on $\mathcal{K}$, n and δ. Taking into account (7.43) and (7.44) in (7.42), from (7.41) we obtain

$$\sup_{x\in\mathbb{R}^n, r>0} \mathfrak{M}_{p,\varphi}(f * \mathcal{K}_t - f; x, r) \lesssim \varepsilon$$

for sufficiently small $t > 0$. This implies $\|f * \mathcal{K}_t - f\|_{p,\varphi} \to 0$ as $t \to 0$, and hence the proof is complete. □

Finally we discuss the approximation of Morrey functions having all the vanishing properties. In particular, property (V^*) allow us to approximate by compactly supported functions.

Theorem 7.50 *Let $\varphi \in \Phi$ and $1 \leqslant p < \infty$. Then every function in $V_{0,\infty}^{(*)}L^{p,\varphi}(\mathbb{R}^n)$ can be approximated in Morrey norm by C_0^∞ functions.*

Proof We split the proof into two steps.

Step 1: The claim holds true for functions $f \in V_{0,\infty}^{(*)}L^{p,\varphi}(\mathbb{R}^n)$ with compact support. In fact, if we take a C_0^∞ kernel $\mathcal{K}$, then the mollifiers $f * \mathcal{K}_t$ have compact support and belong to $V_{0,\infty}^{(*)}L^{p,\varphi}(\mathbb{R}^n)$ (cf. Corollary 7.34). Moreover, they approximate f in Morrey norm (recall again the discussion in the end of Sect. 7.2.3 leading to (7.33)).

Step 2: We show now that functions from $V_{0,\infty}^{(*)}L^{p,\varphi}(\mathbb{R}^n)$ can be approximated by compactly supported functions in Morrey norm. Let $f \in V_{0,\infty}^{(*)}L^{p,\varphi}(\mathbb{R}^n)$. As before let $\chi_k := \chi_{\mathbb{R}^n\setminus B(0,k)}$, $k \in \mathbb{N}$. For each k, set

$$f_k = f \quad \text{on the ball } B(0,k) \quad \text{and} \quad f_k = 0 \quad \text{otherwise.}$$

Let $\varepsilon > 0$ be arbitrary. Again by the vanishing properties (V_0) and (V_∞), there exist $r_0, r_1 > 0$ such that

$$\sup_{x\in\mathbb{R}^n} \mathfrak{M}_{p,\varphi}\big(f - f_k; x, r\big) = \sup_{x\in\mathbb{R}^n} \mathfrak{M}_{p,\varphi}\big(f\chi_k; x, r\big) < \varepsilon$$

for all $k \in \mathbb{N}$ and every $r < r_0$ or $r > r_1$. Hence

$$\|f - f_k\|_{p,\varphi}^p < \max\{\varepsilon, S_{r_0,r_1}(k)\}$$

where

$$S_{r_0,r_1}(k) := \sup_{x\in\mathbb{R}^n, r\in[r_0,r_1]} \mathfrak{M}_{p,\varphi}\big(f\chi_k; x, r\big).$$

Now, by the vanishing property (V^*) and Remark 7.29, we get

$$S_{r_0,r_1}(k) \leqslant \sup_{x\in\mathbb{R}^n} \frac{1}{\inf_{r\geqslant r_0}\varphi(r)} \int\limits_{B(x,r_1)} |f(y)|^p \chi_k(y)\, dy < \varepsilon$$

for all k large enough. Therefore,

$$\|f - f_k\|_{p,\varphi} \to 0 \quad \text{as} \quad k \to \infty,$$

which completes the proof. □

Since $V_{0,\infty}^{(*)}L^{p,\varphi}(\mathbb{R}^n)$ is closed (cf. Remark 7.31), from Theorem 7.50 we are ready to give an explicit description of the closure of $C_0^\infty(\mathbb{R}^n)$ in the generalized Morrey space.

Corollary 7.51 *Let $1 \leqslant p < \infty$ and let $\varphi \in \Phi$ satisfy (7.25) and (7.37). Then the class $C_0^\infty(\mathbb{R}^n)$ is dense in $V_{0,\infty}^{(*)}L^{p,\varphi}(\mathbb{R}^n)$. Moreover, $V_{0,\infty}^{(*)}L^{p,\varphi}(\mathbb{R}^n)$ coincides with the closure of $C_0^\infty(\mathbb{R}^n)$ in $L^{p,\varphi}(\mathbb{R}^n)$.*

7.3 On the Invariance of Certain Vanishing Subspaces of Morrey Spaces with Respect to Some Classical Operators

Recall that the classical Morrey space $L^{p,\lambda}$ consists of all locally p-integrable functions f on $\mathbb{R}^n$ with finite norm

$$\|f\|_{p,\lambda} := \sup_{x\in\mathbb{R}^n,\, r>0} \mathfrak{M}_{p,\lambda}(f;x,r)^{1/p}, \tag{7.45}$$

where

$$\mathfrak{M}_{p,\lambda}(f;x,r) := \frac{1}{r^\lambda} \int\limits_{B(x,r)} |f(y)|^p\, dy, \qquad x \in \mathbb{R}^n, \quad r > 0. \tag{7.46}$$

Many classical operators from harmonic analysis, such as maximal operators, singular operators, potential operators, and Hardy operators, are known to be bounded in Morrey spaces. In this section, we are interested in studying the behavior of those classical operators in certain subspaces of Morrey spaces, as introduced an discussed in Sect. 7.1.

7.3.1 *Preliminaries*

Some Classical Operators

We shall consider the following class of sublinear operators which was introduced in Samko [291]. Recall that an operator T is said sublinear if $|T(\alpha f)| = |\alpha||Tf|$ and $|T(f+g)| \leqslant |Tf| + |Tg|$ for any scalar α and every functions f, g.

Definition 7.52 Let $1 < p < \infty$. A sublinear operator T is called p-admissible singular-type operator if it is bounded in $L^p(\mathbb{R}^n)$ and it satisfies a "size condition" of the form

$$\chi_{B(x,r)}(z) \left|T\left(f\, \chi_{\mathbb{R}^n\setminus B(x,2r)}\right)(z)\right| \lesssim \chi_{B(x,r)}(z) \int\limits_{\mathbb{R}^n\setminus B(x,2r)} \frac{|f(y)|}{|y-z|^n} dy$$

for every $x \in \mathbb{R}^n$ and $r > 0$.

An example of p-admissible singular-type operators is the *Hardy–Littlewood maximal operator*

$$Mf(x) := \sup_{t>0} \frac{1}{|B(x,t)|} \int\limits_{B(x,t)} |f(y)| dy, \quad x \in \mathbb{R}^n.$$

It is well-known that the maximal operator controls various other important operators of harmonic analysis. This is the case of the sharp maximal function

$$M^{\sharp} f(x) := \sup_{t>0} \frac{1}{|B(x,t)|} \int\limits_{B(x,t)} |f(y) - f_{B(x,t)}|\, dy, \tag{7.47}$$

with $f_B = \frac{1}{|B|} \int\limits_B f(z)dz$. By straightforward calculations, we have

$$(M^{\sharp} f)(x) \leqslant 2(Mf)(x), \quad x \in \mathbb{R}^n. \tag{7.48}$$

The class above includes also singular integral operators S, defined by

$$Sf(x) := \int\limits_{\mathbb{R}^n} K(x,y) f(y)\, dy := \lim_{\varepsilon \to 0} \int\limits_{|x-y|>\varepsilon} K(x,y) f(y)\, dy, \tag{7.49}$$

which are bounded in $L^p(\mathbb{R}^n)$ and whose kernel satisfies

$$|K(x,y)| \lesssim |x-y|^{-n}, \quad \text{for all } x \neq y. \tag{7.50}$$

This is the case of Calderón–Zygmund operators with "standard kernels" (cf. Duoandikoetxea [81, p. 99]).

Other examples of p-admissible singular-type operators are the multidimensional *Hardy operators* H and $\mathcal{H}$, defined by

$$Hf(x) := \frac{1}{|x|^n} \int\limits_{|y|<|x|} f(y)\, dy \quad \text{and} \quad \mathcal{H} f(x) := \int\limits_{|y|>|x|} \frac{f(y)}{|y|^n}\, dy.$$

Using that $|x - y| < 2|x|$ in the integral defining H, we get the pointwise estimate

$$H\big(|f|\big)(x) \leqslant 2^n v_n\, Mf(x), \quad x \in \mathbb{R}^n. \tag{7.51}$$

We shall consider more general Hardy-type operators, H^{α} and $\mathcal{H}^{\alpha}$, $0 \leqslant \alpha < n$, defined for appropriate functions f by

$$H^{\alpha} f(x) := |x|^{\alpha-n} \int\limits_{|y|<|x|} f(y)\, dy \quad \text{and} \quad \mathcal{H}^{\alpha} f(x) := |x|^{\alpha} \int\limits_{|y|>|x|} \frac{f(y)}{|y|^n}\, dy.$$

It can be easily shown that the operator H^{α} is now dominated by the *fractional maximal operator*

$$M^{\alpha} f(x) := \sup_{t>0} \frac{1}{|B(x,t)|^{1-\frac{\alpha}{n}}} \int\limits_{B(x,t)} |f(y)| dy, \quad x \in \mathbb{R}^n,$$

which in turn can be estimated by the *Riesz potential operator*

$$I^{\alpha} f(x) := \int\limits_{\mathbb{R}^n} \frac{f(y)}{|x-y|^{n-\alpha}}\, dy, \quad x \in \mathbb{R}^n.$$

More precisely, for $0 < \alpha < n$ there holds

$$\left|H^{\alpha} f(x)\right| \leqslant v_n\, 2^{n-\alpha} \left(M^{\alpha} f\right)(x) \leqslant 2^{n-\alpha}\, I^{\alpha}\left(|f|\right)(x), \qquad x \in \mathbb{R}^n. \tag{7.52}$$

The Hardy operator $\mathcal{H}^{\alpha}$ is also dominated by the Riesz potential operator. The pointwise estimate below should be known, but since we did not find a reference in the literature, we take the opportunity to give a simple proof of it.

Lemma 7.53 *If* $0 < \alpha < n$, *then we have*

$$\left|\mathcal{H}^{\alpha} f(x)\right| \leqslant 2^{n-\alpha}\, I^{\alpha}\left(|f|\right)(x)\,, \quad x \in \mathbb{R}^n. \tag{7.53}$$

Proof The inequality follows from the estimate

$$\frac{|x|^{\alpha}}{|y|^n} \leqslant \frac{2^{n-\alpha}}{|x-y|^{n-\alpha}}, \qquad \text{for} \quad |y| > |x|.$$

Putting $t = \frac{|x|}{|y|}$ and $x' = \frac{x}{|x|}$, $y' = \frac{y}{|y|}$, the latter is equivalent to

$$t^{\alpha} \leqslant 2^{n-\alpha} \left(\frac{|y|}{|x-y|}\right)^{n-\alpha} \qquad \text{or} \qquad t^{\alpha} \leqslant \frac{2^{n-\alpha}}{|tx'-y'|^{n-\alpha}}$$

which is a consequence of having $t < 1$ and $|tx' - y'| \leqslant 2$. □

Pointwise Estimates for Modulars

The estimates given in the next two lemmas are taken from Samko [291, Theorems 4.1 and 4.3].

Lemma 7.54 *Let* $1 < p < \infty$ *and* $0 \leqslant \lambda < n$. *If* T *is a* p*-admissible sublinear singular-type operator, then*

$$\mathfrak{M}_{p,\lambda}(Tf; x, r) \lesssim r^{n-\lambda} \left(\int\limits_r^{\infty} t^{\frac{\lambda-n}{p}-1} \left(\mathfrak{M}_{p,\lambda}(f; x, t)\right)^{\frac{1}{p}}\, dt\right)^p$$

with the implicit constant independent of $x \in \mathbb{R}^n$, $r > 0$ *and* $f \in L^p_{\mathrm{loc}}(\mathbb{R}^n)$.

Although the previous lemma is formulated for functions f in $L^p_{\text{loc}}(\mathbb{R}^n)$, the finiteness of the right-hand side implies that the function f must have already some prescribed behavior at infinity which, together with the size condition, ensures the well-posedness of the operator T.

Remark 7.55 The estimate given in Lemma 7.54 implies that p-admissible sublinear singular-type operators are bounded on $L^{p,\lambda}$ and also on the vanishing space $V_0L^{p,\lambda}$, with $1 < p < \infty$ and $0 \leqslant \lambda < n$ (cf. Samko [291]). In particular, all the operators M, $M^\sharp$, S, H and $\mathcal{H}$ are bounded on both spaces $L^{p,\lambda}$ and $V_0L^{p,\lambda}$.

Lemma 7.56 *Let* $0 < \alpha < n$, $1 < p < n/\alpha$, $1/q = 1/p - \alpha/n$, *and* $0 \leqslant \lambda, \mu < n$. *Then*

$$\mathfrak{M}_{q,\mu}(I^\alpha f; x, r) \lesssim r^{n-\mu} \left(\int_r^\infty t^{\frac{\lambda}{p} - \frac{n}{q} - 1} \left(\mathfrak{M}_{p,\lambda}(f; x, t) \right)^{\frac{1}{p}} dt \right)^q$$

with the implicit constant not depending on $x \in \mathbb{R}^n$, $r > 0$ *and* $f \in L^p_{\text{loc}}(\mathbb{R}^n)$.

Remark 7.57 The estimate given in Lemma 7.56 implies a Sobolev–Spanne result on the $L^{p,\lambda} \to L^{q,\mu}$- boundedness of I^α, under the additional assumptions $0 \leqslant \lambda < n - \alpha p$ and $\lambda/p = \mu/q$. Moreover, the same estimate was used in Samko [291] to show that the Riesz potential operator I^α, and consequently the fractional maximal operator M^α, are also bounded from the vanishing space $V_0L^{p,\lambda}$ into the vanishing space $V_0L^{q,\mu}(\mathbb{R}^n)$.

7.3.2 Main Results

In this section we show that both subspaces $V_\infty L^{p,\lambda}$ and $V^{(*)}L^{p,\lambda}$ are invariant with respect to the operators mentioned above. Since the boundedness of such operators is already known in the whole Morrey space, we only have to show that the corresponding vanishing properties are preserved under the action of those operators.

Preservation of the Property (V_∞)

Theorem 7.58 *Let* $1 < p < \infty$ *and* $0 \leqslant \lambda < n$. *Then any p-admissible sublinear singular-type operator T is bounded in* $V_\infty L^{p,\lambda}$.

Proof The proof is based on the modular estimate given in Lemma 7.54. Let $f \in V_\infty L^{p,\lambda}$. For any $\varepsilon > 0$ there exists $R = R(\varepsilon) > 0$ such that $\mathfrak{M}_{p,\lambda}(f; x, t) < \varepsilon$ for every $t \geqslant R$ and all $x \in \mathbb{R}^n$. Thus, for $r \geqslant R$, we have

$$\mathfrak{M}_{p,\lambda}(Tf; x, r) \lesssim \varepsilon\, r^{n-\lambda}\Big(\int_r^\infty t^{\frac{\lambda-n}{p}-1}dt\Big)^p \lesssim \varepsilon$$

with the implicit constants independent of x and r. This shows that

$$\lim_{r\to\infty}\sup_{x\in\mathbb{R}^n}\mathfrak{M}_{p,\lambda}(Tf; x, r) = 0$$

and hence $Tf \in V_\infty L^{p,\lambda}$. □

Corollary 7.59 *If $1 < p < \infty$ and $0 \leqslant \lambda < n$, then the operators M, $M^\sharp$, H, $\mathcal{H}$ and S are bounded in $V_\infty L^{p,\lambda}$.*

In the sequel T^α, $0 < \alpha < n$, stands for any of the operators I^α, M^α, H^α and $\mathcal{H}^\alpha$ above.

Theorem 7.60 *Let $0 < \alpha < n$, $1 < p < n/\alpha$, $0 \leqslant \lambda < n - \alpha p$, $1/q = 1/p - \alpha/n$ and $\lambda/p = \mu/q$. Then T^α is bounded from $V_\infty L^{p,\lambda}$ into $V_\infty L^{q,\mu}(\mathbb{R}^n)$.*

Proof Since I^α is bounded from $L^{p,\lambda}$ into $L^{q,\mu}(\mathbb{R}^n)$ (this is a well-known result by Spanne and published in Peetre [269]), the same norm inequalities also hold for the operators M^α, H^α and $\mathcal{H}^\alpha$ in virtue of the estimates (7.52) and (7.53). Moreover, by the same estimates, the preservation of the vanishing property at infinity by the maximal and the Hardy operators follows from the corresponding preservation by the action of the Riesz potential operator. Hence, it remains to show that

$$\sup_{x\in\mathbb{R}^n}\mathfrak{M}_{p,\lambda}(f; x, r) \to 0 \quad\Longrightarrow\quad \sup_{x\in\mathbb{R}^n}\mathfrak{M}_{q,\mu}(I^\alpha f; x, r) \to 0 \qquad \text{as} \quad r\to\infty.$$

This can be done as in the proof of Theorem 7.58, but now using the modular estimate given in Lemma 7.56 instead of that in Lemma 7.54. Nevertheless, we write the proof in slightly different terms by applying the Lebesgue dominated convergence theorem. For any $x \in \mathbb{R}^n$ and $r > 0$, we have

$$\mathfrak{M}_{q,\mu}(I^\alpha f; x, r) \lesssim r^{n-\mu}\Big(\int_r^\infty t^{\frac{\lambda}{p}-\frac{n}{q}-1}\big(\mathfrak{M}_{p,\lambda}(f; x, t)\big)^{\frac{1}{p}}\,dt\Big)^q$$

$$= \Big(\int_1^\infty s^{\frac{\lambda}{p}-\frac{n}{q}-1}\big(\mathfrak{M}_{p,\lambda}(f; x, rs)\big)^{\frac{1}{p}}\,ds\Big)^q. \tag{7.54}$$

Since the implicit constant does not depend on $x \in \mathbb{R}^n$ and $r > 0$, and the integrand admits an integrable dominant (note that $f \in L^{p,\lambda}$ and $\lambda/p < n/q$), then the right-hand side of (7.54) tends to 0 as $r \to \infty$ (uniformly on x). This gives the desired result. □

Theorem 7.61 *Let $0 < \alpha < n$, $0 \leqslant \lambda < n$, $1 < p < (n-\lambda)/\alpha$, and $1/q = 1/p - \alpha/(n-\lambda)$. Then T^α is bounded from $V_\infty L^{p,\lambda}$ into $V_\infty L^{q,\lambda}(\mathbb{R}^n)$.*

Proof As in the proof of Theorem 7.60, we only have to prove the statement for the Riesz potential operator. The $(L^{p,\lambda} \to L^{q,\lambda})$- boundedness of I^α is well-known (cf. Adams [4]). To show the preservation of the vanishing property at infinity we make use of the pointwise estimate of the Riesz potential operator in terms of the maximal function. It is known that there exists $c > 0$ such that

$$\left|I^\alpha f(x)\right| \leqslant c \left(Mf(x)\right)^{\frac{p}{q}} \|f\|_{p,\lambda}^{1-\frac{p}{q}} \tag{7.55}$$

for all $f \in L^{p,\lambda}$ and $x \in \mathbb{R}^n$ (see, for instance, Burtseva and Samko [43], Adams [6, Chapter 7]). From (7.55) we get

$$\mathfrak{M}_{q,\lambda}\left(I^\alpha f; x, r\right) \lesssim \|f\|_{p,\lambda}^{\frac{q-p}{q}} \mathfrak{M}_{p,\lambda}\left(Mf; x, r\right)$$

for $r > 0$ and $x \in \mathbb{R}^n$. If $f \in V_\infty L^{p,\lambda}$ then $Mf \in V_\infty L^{p,\lambda}$ by Corollary 7.59. Consequently, we have $I^\alpha f \in V_\infty L^{q,\lambda}(\mathbb{R}^n)$ taking into account the previous estimate. □

Preservation of the Property (V^*)

First we show that the space $V^{(*)}L^{p,\lambda}$ is invariant under the action of the Hardy–Littlewood maximal operator.

Theorem 7.62 *Let $1 < p < \infty$ and $0 \leqslant \lambda < n$. Then the maximal operator M is bounded in $V^{(*)}L^{p,\lambda}$.*

Proof Since M is bounded in $L^{p,\lambda}$ (cf. Chiarenza and Frasca [53]) we only have to show that it preserves the vanishing property (V^*), that is

$$\lim_{N\to\infty} \mathcal{A}_{N,p}(f) = 0 \quad \Rightarrow \quad \lim_{N\to\infty} \mathcal{A}_{N,p}(Mf) = 0.$$

Given $x \in \mathbb{R}^n$ and $N \in \mathbb{N}$, we split f into

$$f = f_1 + f_2, \quad \text{with} \quad f_1 := f\,\chi_{\Omega_{x,N/2}}, \quad f_2 = f\,\chi_{\mathbb{R}^n \setminus \Omega_{x,N/2}},$$

where, for short, we use the notation

$$\Omega_{x,N} := B(x,2) \cap \left(\mathbb{R}^n \setminus B(0,N)\right).$$

Since M is sublinear, we have

$$\mathcal{A}_{N,p}(Mf) \lesssim \mathcal{A}_{N,p}(Mf_1) + \mathcal{A}_{N,p}(Mf_2). \tag{7.56}$$

We show next that both quantities in the right-hand side of (7.56) tend to zero as $N \to \infty$. The boundedness of M in $L^p(\mathbb{R}^n)$ gives

$$\begin{aligned} \int\limits_{B(x,1)} \big(M(f_1)(y)\big)^p \chi_N(y)\, dy &\leqslant \int\limits_{\mathbb{R}^n} \big(M(f_1)(y)\big)^p dy \\ &\lesssim \int\limits_{\mathbb{R}^n} |f_1(y)|^p dy \\ &= \int\limits_{\Omega_{x,N/2}} |f(y)|^p\, dy \end{aligned} \tag{7.57}$$

with the implicit constant independent of x, N, and f. Since $f \in V^{(*)}L^{p,\lambda}$, the right-hand side above tends to zero uniformly on x as $N \to \infty$ (note that the property (V^*) does not depend on the particular value of the radius taken in the balls centered at x, see Sect. 7.1). Therefore, $\lim_{N\to\infty} \mathcal{A}_{N,p}(Mf_1) = 0$.

Now we deal with the second term in the sum in (7.56). Let $\varepsilon > 0$ be arbitrary. Then there exists $t_1 > 1$ such that $t^{\lambda-n} < \varepsilon$ for all $t \geqslant t_1$. For such fixed t_1, we have

$$\int\limits_{B(x,1)} \big(M(f_2)(y)\big)^p \chi_N(y)\, dy \lesssim I_1(x, N) + I_2(x, N)$$

where

$$I_1(x, N) := \int\limits_{B(x,1)} \chi_N(y) \sup_{0<t<t_1} \left[\frac{1}{|B(y, t)|} \int\limits_{B(y,t)} |f(z)|\, \chi_{\mathbb{R}^n\setminus\Omega_{x,N/2}}(z)\, dz \right]^p dy$$

and

$$I_2(x, N) := \int\limits_{B(x,1)} \chi_N(y) \sup_{t\geqslant t_1} \left[\frac{1}{|B(y, t)|} \int\limits_{B(y,t)} |f(z)|\, \chi_{\mathbb{R}^n\setminus\Omega_{x,N/2}}(z)\, dz \right]^p dy.$$

First we estimate $I_2(x, N)$. By Hölder's inequality we have

$$\frac{1}{|B(y, t)|} \int\limits_{B(y,t)} |f(z)|\, dz \leqslant \frac{1}{|B(y, t)|^{1/p}} \|f\|_{L^p(B(y,t))}.$$

Hence

$$I_2(x,N) \leqslant \int\limits_{B(x,1)} \sup_{t \geqslant t_1} \frac{1}{|B(y,t)|} \int\limits_{B(y,t)} |f(z)|^p dz\, dy \lesssim \sup_{t \geqslant t_1} \frac{1}{t^{n-\lambda}} \|f\|_{p,\lambda}^p \leqslant \varepsilon \|f\|_{p,\lambda}^p.$$

As regards $I_1(x,N)$, we have two different cases to be analyzed for $z \in B(y,t)$ and $z \notin \Omega_{x,N/2}$. If $z \in B(0,N/2)$, then $t > |z-y| \geqslant |y| - |z| > N/2$. Thus, there is no contribution to the supremum on $t \in (0,t_1)$ for $N \geqslant 2t_1$. If $z \notin B(x,2)$, then $t > |z-y| \geqslant |z-x| - |y-x| \geqslant 1$. Hence, it remains to handle $I_1(x,N)$ when the supremum is taken over all $t \in (1,t_1)$. For such values of t, we have

$$t^{-n} \int\limits_{B(y,t)} |f(z)|\, \chi_{\mathbb{R}^n \setminus \Omega_{x,N/2}}(z)\, dz \leqslant \int\limits_{B(y,t_1)} |f(z)|\, dz = \int\limits_{B(0,t_1)} |f(y-z)|\, dz.$$

Using this, Minkowski's inequality and a simple change of variables, we get

$$\begin{aligned} I_1(x,N) &\leqslant \int\limits_{B(x,1)} \chi_N(y) \left[\int\limits_{\mathbb{R}^n} \chi_{B(0,t_1)}(z)\, |f(y-z)|\, dz \right]^p dy \\ &\leqslant \left(\int\limits_{\mathbb{R}^n} \chi_{B(0,t_1)}(z) \left[\sup_{v \in \mathbb{R}^n} \int\limits_{B(v,1)} |f(u)|^p\, \chi_{N-|z|}(u)\, du \right]^{1/p} dz \right)^p \\ &=: \left(\int\limits_{\mathbb{R}^n} g_N(z)\, dz \right)^p \end{aligned}$$

with the interpretation

$$\chi_a := 1 \quad \text{if } a \leqslant 0 \quad \text{and} \quad \chi_a := \chi_{\mathbb{R}^n \setminus B(0,a)} \text{ if } a > 0.$$

This gives an uniform bound for $I_1(x,N)$. Since $f \in L^{p,\lambda}$ and $g_N(z)$ has an integrable majorant (depending on t_1), an application of the Lebesgue convergence theorem shows that $\int\limits_{\mathbb{R}^n} g_N(z)\, dz \to 0$ as $N \to \infty$, which implies that

$$I_1(x,N) \to 0 \quad \text{uniformly on } x, \quad \text{as} \quad N \to \infty.$$

The proof is complete. □

Theorem 7.63 *Let $1 < p < \infty$ and $0 \leqslant \lambda < n$. Then the operators $M^\sharp$, H and $\mathcal{H}$ are bounded in $V^{(*)}L^{p,\lambda}$.*

Proof The boundedness of $M^\sharp$ and H in $V^{(*)}L^{p,\lambda}$ is a consequence of Theorem 7.62 and inequalities (7.48), (7.51). The case of the Hardy operator $\mathcal{H}$ requires a different approach since it cannot be estimated by the maximal function. Observing that

$$\lim_{|z|\to\infty} \mathcal{H}f(z) = 0,$$

then for any $\varepsilon > 0$ there exists $N_\varepsilon \in \mathbb{N}$ such that $|\mathcal{H}f(y)| \leqslant (\varepsilon/v_n)^{1/p}$ for all $|y| \geqslant N_\varepsilon$. Therefore

$$\int\limits_{\mathbb{R}^n} |\mathcal{H}f(y)|^p \chi_{B(x,1)\cap\{y:|y|>N\}}(y)\, dy \leqslant \varepsilon/v_n \int\limits_{\mathbb{R}^n} \chi_{B(x,1)\cap\{y:|y|>N\}}(y)\, dy \leqslant \varepsilon$$

for all $N \geqslant N_\varepsilon$, uniformly on $x \in \mathbb{R}^n$. This shows that $\mathcal{A}_{N,p}(\mathcal{H}f) \to 0$ as $N \to \infty$. □

Remark 7.64 As regards the preservation of the property (V^*) by the action of the Hardy operator $\mathcal{H}$, one can formulate a stronger result. Indeed, as we can see from the proof above, $\mathcal{H}$ is bounded from the whole space Morrey $L^{p,\lambda}$ into $V^{(*)}L^{p,\lambda}$.

As in Sect. 7.3.2, suppose that T^α, $0 < \alpha < n$, denotes any of the operators I^α, M^α, H^α and $\mathcal{H}^\alpha$.

Theorem 7.65 *Let* $0 < \alpha < n$, $0 \leqslant \lambda < n$, $1 < p < (n-\lambda)/\alpha$ *and* $1/q = 1/p - \alpha/(n-\lambda)$. *Then* T^α *is bounded from* $V^{(*)}L^{p,\lambda}$ *into* $V^{(*)}L^{q,\lambda}(\mathbb{R}^n)$.

Proof By (7.55) we get

$$\mathcal{A}_{N,q}(I^\alpha f) \lesssim \|f\|_{p,\lambda}^{q-p} \mathcal{A}_{N,p}(Mf)$$

with the implicit constant independent of f and $N \in \mathbb{N}$. If $f \in V^{(*)}L^{p,\lambda}$ then $Mf \in V^{(*)}L^{p,\lambda}$ by Theorem 7.62. Consequently, we also have $I^\alpha f \in V^{(*)}L^{q,\lambda}(\mathbb{R}^n)$ by the previous estimate. □

Invariance of the Closure of $C_0^\infty(\mathbb{R}^n)$

As shown in Sect. 7.1, the subspace

$$V_{0,\infty}^{(*)}L^{p,\lambda} := V_0L^{p,\lambda} \cap V_\infty L^{p,\lambda} \cap V^{(*)}L^{p,\lambda}$$

coincides with the closure of the class $C_0^\infty(\mathbb{R}^n)$ in Morrey norm. Note that this closure plays an important role in harmonic analysis on Morrey spaces since its dual provides a predual space for Morrey spaces (cf. Adams [6], Adams and Xiao [8], Rosenthal and Triebel [287], Triebel [325]). Moreover (cf. Almeida and Samko

[13], see also Corollary 7.11), we have the strict embeddings

$$V_{0,\infty}^{(*)}L^{p,\lambda} \subsetneqq V_0L^{p,\lambda} \cap V_\infty L^{p,\lambda} \subsetneqq V_0L^{p,\lambda} \subsetneqq L^{p,\lambda}.$$

The next corollaries are immediate consequences of the results obtained in Sects. 7.1 and 7.2 and the already known corresponding boundedness in the vanishing space $V_0L^{p,\lambda}$.

Corollary 7.66 *Let $1 < p < \infty$ and $0 \leqslant \lambda < n$. Then the operators M, $M^\sharp$, H, and $\mathcal{H}$ are bounded in $V_{0,\infty}^{(*)}L^{p,\lambda}$.*

Corollary 7.67 *Let $0 < \alpha < n$, $0 \leqslant \lambda < n$, $1 < p < (n-\lambda)/\alpha$ and $1/q = 1/p - \alpha/(n-\lambda)$. Then T^α is bounded from $V_{0,\infty}^{(*)}L^{p,\lambda}$ into $V_{0,\infty}^{(*)}L^{q,\lambda}(\mathbb{R}^n)$.*

We end this section with a further result for singular integral operators S defined by (7.49), with the kernel satisfying (7.50). One knows that S is bounded in $V_0L^{p,\lambda}$, for $1 < p < \infty$ and $0 \leqslant \lambda < n$ (cf. Samko [291, Theorem 5.1]). On the other hand, we have seen that S is also bounded in $V_\infty L^{p,\lambda}$ (cf. Corollary 7.59). Unfortunately, we do not know whether S preserves the vanishing property (V^*). Nevertheless, we have the following result.

Theorem 7.68 *Let $1 < p < \infty$ and $0 \leqslant \lambda < n$. Then S is bounded from $V_{0,\infty}^{(*)}L^{p,\lambda}$ into itself.*

Proof For bounded compactly supported functions f, there holds

$$|Sf(y)| \leqslant \frac{c}{|y|^n} \tag{7.58}$$

for sufficiently large $|y|$, with $c > 0$ not depending on y. In fact, suppose that $\operatorname{supp} f \subset B(0, M)$. We have

$$|y-z| \geqslant |y| - M \geqslant \frac{|y|}{2} \quad \text{for } |z| \leqslant M \quad \text{and} \quad |y| \geqslant 2M.$$

Hence (7.58) follows thanks to the size condition (7.50).

Let now $f \in V_{0,\infty}^{(*)}L^{p,\lambda}$. Then there exists a sequence $(f_k) \subset C_0^\infty(\mathbb{R}^n)$ such that $f_k \to f$ in $L^{p,\lambda}$ as $k \to \infty$. By the continuity of S in $L^{p,\lambda}$, we get

$$Sf = S(\lim_{k\to\infty} f_k) = \lim_{k\to\infty}(Sf_k).$$

Thus we have $Sf \in V_{0,\infty}^{(*)}L^{p,\lambda}$ since this subspace is closed in $L^{p,\lambda}$. □

7.3.3 Additional Results

In this section, we consider the following hybrids of Hardy and potential operators:

$$K_\beta f(x) := \frac{1}{|x|^\beta} \int\limits_{|y|<|x|} \frac{f(y)}{|x-y|^{n-\beta}}\, dy$$

and

$$\mathcal{K}_\beta f(x) := \int\limits_{|y|>|x|} \frac{f(y)}{|y|^\beta |x-y|^{n-\beta}}\, dy$$

(with $0 < \beta \leqslant n$). As observed in Persson et al. [271], K_β and $\mathcal{K}_\beta$ are integral operators bounded in $L^p(\mathbb{R}^n)$, $1 < p < \infty$, with the corresponding kernels satisfying the size condition (7.50). Consequently, by Theorem 7.58 we have the following result.

Theorem 7.69 *Let $1 < p < \infty$, $0 \leqslant \lambda < n$ and $0 < \beta \leqslant n$. Then both operators K_β and $\mathcal{K}_\beta$ are bounded on $V_\infty L^{p,\lambda}$.*

7.4 Preservation of Certain Vanishing Properties of Generalized Morrey Spaces by Some Classical Operators

In this section we consider generalized Morrey spaces $L^{p,\varphi}$ where in the corresponding norm the power r^λ is replaced by a suitable positive function $\varphi(r)$ on $(0, \infty)$:

$$\|f\|_{p,\varphi} := \sup_{x\in\mathbb{R}^n,\, r>0} \mathfrak{M}_{p,\varphi}(f; x, r)^{1/p}, \tag{7.59}$$

with

$$\mathfrak{M}_{p,\varphi}(f; x, r) := \frac{1}{\varphi(r)} \int\limits_{B(x,r)} |f(y)|^p\, dy\,, \qquad x \in \mathbb{R}^n, \quad r > 0. \tag{7.60}$$

Similarly to the classes in Sect. 7.3, we now recall the classes related to the generalized Morrey spaces, viz., $V_0 L^{p,\varphi}$ and $V_\infty L^{p,\varphi}$, which consist of all functions $f \in L^{p,\varphi}$ such that

$$\lim_{r\to 0} \sup_{x\in\mathbb{R}^n} \mathfrak{M}_{p,\varphi}(f; x, r) = 0 \tag{V_0}$$

and

$$\lim_{r\to\infty} \sup_{x\in\mathbb{R}^n} \mathfrak{M}_{p,\varphi}(f;x,r) = 0, \tag{V_∞}$$

respectively. We also consider the class $V^{(*)}L^{p,\varphi}$ consisting of all those functions $f \in L^{p,\varphi}$ having the vanishing property

$$\lim_{N\to\infty} \mathcal{A}_{N,p}(f) := \lim_{N\to\infty} \sup_{x\in\mathbb{R}^n} \int_{B(x,1)} |f(y)|^p \chi_N(y)\,dy = 0, \tag{V^*}$$

with the notation $\chi_N := \chi_{\mathbb{R}^n\setminus B(0,N)}$, $N \in \mathbb{N}$. Note that the three sets defined above are closed in $L^{p,\varphi}$ with respect to the norm (7.59).

The main goal of this section is to show that the vanishing properties defining the three generalized Morrey subspaces are preserved under the action of maximal, singular, potential, and Hardy operators. This extends the results obtained in Sect. 7.3 to the setting of generalized Morrey spaces.

7.4.1 Preliminaries

A function $\varphi : (0,\infty) \to (0,\infty)$ is said to be *almost increasing* (resp., *almost decreasing*) if $\varphi(s) \lesssim \varphi(t)$ (resp., $\varphi(s) \gtrsim \varphi(t)$) for $s \leqslant t$. As usual $C_0^\infty(\mathbb{R}^n)$ stands for the class of all complex-valued infinitely differentiable functions on $\mathbb{R}^n$ with compact support, and $L^p(\mathbb{R}^n)$ denotes the classical Lebesgue space equipped with the usual norm.

Generalized Morrey Spaces and Vanishing Subspaces

As we can see directly from (7.59), with (7.60), if $\varphi(r) = r^\lambda$, $0 \leqslant \lambda \leqslant n$, then $L^{p,\varphi}$ coincides with the classical Morrey space $L^{p,\lambda}$. We also have

$$L^{p,\varphi} = L^\infty(\mathbb{R}^n) \quad \text{when} \quad \varphi(r) \approx r^n.$$

and

$$L^{p,\varphi} = L^p(\mathbb{R}^n) \quad \text{if} \quad \varphi(r) \approx 1.$$

In general, we have to require some assumptions on the function parameter φ in order to ensure good properties for the spaces $L^{p,\varphi}$. The following class, already defined in Sect. 7.2, is similar to the class defined by Nakai [262].

Definition 7.70 The class Φ consists of all measurable functions $\varphi : (0, \infty) \to (0, \infty)$ such that:

1. φ is almost increasing;
2. $\varphi(t)/t^n$ is almost decreasing.
3. $\inf_{t>\delta} \varphi(t) > 0$ for every $\delta > 0$.

Typical examples of functions belonging to the class Φ are the power functions

$$\varphi(t) = t^\lambda, \quad 0 \leqslant \lambda \leqslant n,$$

and the more general functions

$$\varphi(t) = \begin{cases} t^{\lambda_1} \log^{\mu_1}\left(\frac{2}{t}\right), \ t < 1, \\ t^{\lambda_2} \log^{\mu_2}(2t), \ t \geqslant 1, \end{cases} \quad 0 \leqslant \lambda_1, \lambda_2 \leqslant n, \quad \mu_1, \mu_2 \in \mathbb{R}.$$

The conditions above defining the class Φ are widely used in papers on generalized Morrey spaces. As observed in Nakai [262], if $\varphi \in \Phi$ then the space $L^{p,\varphi}$ is non-trivial. For example, characteristic functions on balls belong to $L^{p,\varphi}$ under such assumptions on φ.

The vanishing subspace $V_0 L^{p,\varphi}$ is non-trivial if $\varphi \in \Phi$ satisfies the additional condition

$$\lim_{r \to 0} \frac{r^n}{\varphi(r)} = 0 \tag{7.61}$$

since then it contains bounded functions with compact support. Note also that

$$V_0 L^{p,\varphi} = L^p(\mathbb{R}^n) \quad \text{when} \quad \varphi(t) \approx 1$$

and

$$L^{p,\varphi} \cap L^\infty(\mathbb{R}^n) \subset V_0 L^{p,\varphi}$$

when (7.61) holds.

As for the subspace $V_\infty L^{p,\varphi}$, it is not hard to show that it is non-trivial if $\frac{\varphi(r)}{r^\varepsilon}$ is almost increasing for some $\varepsilon > 0$ and $\inf_{r>\delta} \varphi(r) > 0$ for all $\delta > 0$. In fact, under such assumptions on φ, compactly supported bounded functions belong to this vanishing space.

With appropriate assumptions on $\varphi \in \Phi$, the subspaces $V_0 L^{p,\varphi}$, $V_\infty L^{p,\varphi}$ and $V^{(*)} L^{p,\varphi}$ are proper closed subspaces of $L^{p,\varphi}$.

Known Boundedness Results

For definitions and notations of the operators used, see Sect. 7.3.1.

The next two lemmas are taken from Samko [291, Theorems 4.1 and 4.3], and they provide pointwise estimates for the modulars of p-admissible sublinear singular-type operators and fractional integral operators.

Lemma 7.71 *Let $1 < p < \infty$ and $\varphi \in \Phi$. If T is a p-admissible sublinear singular-type operator, then*

$$\mathfrak{M}_{p,\varphi}(Tf; x, r) \lesssim \frac{r^n}{\varphi(r)} \Big(\int_r^\infty \frac{\varphi^{\frac{1}{p}}(t)}{t^{1+\frac{n}{p}}} (\mathfrak{M}_{p,\varphi}(f; x, t))^{\frac{1}{p}} \, dt \Big)^p \tag{7.62}$$

with the implicit constant independent of $x \in \mathbb{R}^n$, $r > 0$ and $f \in L^p_{\mathrm{loc}}(\mathbb{R}^n)$.

Lemma 7.72 *Let $0 < \alpha < n$, $1 < p < n/\alpha$, $1/q = 1/p - \alpha/n$ and $\varphi, \psi \in \Phi$. Then*

$$\mathfrak{M}_{q,\psi}(I^\alpha f; x, r) \lesssim \frac{r^n}{\psi(r)} \Big(\int_r^\infty \frac{\varphi^{\frac{1}{p}}(t)}{t^{1+\frac{n}{q}}} (\mathfrak{M}_{p,\varphi}(f; x, t))^{\frac{1}{p}} \, dt \Big)^q \tag{7.63}$$

with the implicit constant not depending on $x \in \mathbb{R}^n$, $r > 0$ and $f \in L^p_{\mathrm{loc}}(\mathbb{R}^n)$.

The modular estimates (7.62) and (7.63), together with some additional assumptions on φ (cf. (7.64), (7.65) below), allow us to obtain corresponding ($L^{p,\varphi} \to L^{p,\varphi}$)- and ($L^{p,\varphi} \to L^{q,\psi}$)-boundedness results for the operators under consideration. However, the most remarkable point in Samko [291] is that those boundedness results hold true also in the vanishing subspaces at the origin. The next two theorems are taken from Samko [291, Theorems 5.1 and 5.4].

Theorem 7.73 *Let $1 < p < \infty$ and $\varphi \in \Phi$ satisfy the condition*

$$\int_r^\infty \frac{\varphi^{\frac{1}{p}}(t)}{t^{1+\frac{n}{p}}} dt \lesssim \frac{\varphi^{\frac{1}{p}}(r)}{r^{\frac{n}{p}}}, \tag{7.64}$$

with the implicit constant independent of $r > 0$. Then every p-admissible singular operator T is bounded in the vanishing subspace $V_0 L^{p,\varphi}$.

In particular, all the operators M, $M^\sharp$, S, H, and $\mathcal{H}$ are bounded on $L^{p,\varphi}$ and they are also bounded on $V_0 L^{p,\varphi}$.

In the sequel T^α, $0 < \alpha < n$, denotes any of the operators I^α, M^α, H^α, and $\mathcal{H}^\alpha$.

Theorem 7.74 *Let $0 < \alpha < n$, $1 < p < n/\alpha$, $1/q = 1/p - \alpha/n$ and let $\varphi, \psi \in \Phi$ satisfy the condition*

$$\int_r^\infty \frac{\varphi^{\frac{1}{p}}(t)}{t^{1+\frac{n}{q}}} dt \lesssim \frac{\psi^{\frac{1}{q}}(r)}{r^{\frac{n}{q}}}, \tag{7.65}$$

with the implicit constant not depending on $r > 0$. *Then* T^α *is bounded from* $V_0 L^{p,\varphi}$ *into* $V_0 L^{q,\psi}(\mathbb{R}^n)$.

The result formulated in Theorem 7.74 was given in Samko [291, Theorem 5.4] for the operators I^α and M^α. The corresponding result for the Hardy operators H^α and $\mathcal{H}^\alpha$ follows immediately from the boundedness of I^α in view of the estimates (7.52) and (7.53).

Remark 7.75 In the particular case of power functions $\varphi(r) = r^\lambda$ and $\psi(r) = r^\mu$, with $0 \leqslant \lambda, \mu < n$, Theorem 7.74 recovers the known Spanne's result on the $(L^{p,\lambda} \to L^{q,\mu})$-boundedness of I^α (published in Peetre [269]), under the additional assumptions $0 \leqslant \lambda < n - \alpha p$ and $\lambda/p = \mu/q$.

The following theorem gives an Adams type result (cf. Adams [4]) on the $(L^{p,\varphi} \to L^{q,\varphi})$-boundedness of the Riesz potential operator.

Theorem 7.76 *Let* $0 < \alpha < n$, $1 < p < \frac{n}{\alpha}$ *and* $q > p$. *Suppose that* $\varphi \in \Phi$ *satisfies the condition* (7.64) *and the condition*

$$\int_r^\infty \frac{\varphi^{\frac{1}{p}}(t)}{t^{1-\alpha+\frac{n}{p}}}\, dt \lesssim r^{-\frac{\alpha p}{q-p}}, \tag{7.66}$$

with the implicit constant independent of $r > 0$. *Then* I^α *is bounded from* $L^{p,\varphi}(\mathbb{R}^n)$ *into* $L^{q,\varphi}(\mathbb{R}^n)$.

Theorem 7.76 can be proved following the steps of the proof from Guliyev et al. [129, Theorem 5.5.]. In particular, it is worth noting that condition (7.66) is used to derive the pointwise estimate

$$\left|I^\alpha f(x)\right| \lesssim \left(Mf(x)\right)^{\frac{p}{q}} \|f\|_{p,\varphi}^{1-\frac{p}{q}}, \tag{7.67}$$

from which the $(L^{p,\varphi} \to L^{q,\varphi})$-boundedness of I^α follows, taking into account the boundedness of the maximal operator on $L^{p,\varphi}$ spaces under the assumption (7.64).

Remark 7.77 The integral condition (7.66) appears in Guliyev et al. [129, Theorem 5.5] to study the boundedness of the Riesz potential operator in generalized variable exponent Morrey spaces on bounded domains; see also Guliyev and Shukurov [128]. In the classical case $\varphi(r) = r^\lambda$, with $0 \leqslant \lambda < n$, Theorem 7.76 recovers the well-known Adams result on the $(L^{p,\lambda} \to L^{q,\lambda})$- boundedness of I^α (cf. Adams [4]), where the usual Sobolev-Adams exponent q, given by $\frac{1}{q} = \frac{1}{p} - \frac{\alpha}{n-\lambda}$, is implied by the corresponding assumption (7.66).

7.4.2 Preservation of the Property (V_∞)

In this section, we show that the subspace $V_\infty L^{p,\varphi}$ is invariant with respect to sublinear p-admissible singular-type operators. Moreover, we show that the vanishing property (V_∞) is preserved under the action of any of the operators I^α, M^α, H^α and $\mathcal{H}^\alpha$ above.

Theorem 7.78 *Let $1 < p < \infty$ and let $\varphi \in \Phi$ satisfy condition* (7.64). *Then any p-admissible sublinear singular-type operator T is bounded in $V_\infty L^{p,\varphi}$.*

Proof Since the norm inequality is known (cf. Theorem 7.73), it remains to show that

$$\sup_{x\in\mathbb{R}^n} \mathfrak{M}_{p,\varphi}(f;x,r) \to 0 \qquad \Longrightarrow \qquad \sup_{x\in\mathbb{R}^n} \mathfrak{M}_{p,\varphi}(Tf;x,r) \to 0 \qquad \text{as} \quad r \to \infty.$$

The proof is based on the modular estimate given in Lemma 7.71. Let $f \in V_\infty L^{p,\varphi}$. For any $\varepsilon > 0$ there exists $R = R(\varepsilon) > 0$ such that $\mathfrak{M}_{p,\varphi}(f;x,t) < \varepsilon$ for every $t \geqslant R$ and all $x \in \mathbb{R}^n$. Thus, for $r \geqslant R$, we have

$$\mathfrak{M}_{p,\varphi}(Tf;x,r) \lesssim \varepsilon \, \frac{r^n}{\varphi(r)} \Big(\int_r^\infty \frac{\varphi^{\frac{1}{p}}(t)}{t^{1+\frac{n}{p}}} dt \Big)^p \lesssim \varepsilon$$

with the implicit constants independent of x and r. This shows that

$$\lim_{r\to\infty} \sup_{x\in\mathbb{R}^n} \mathfrak{M}_{p,\varphi}(Tf;x,r) = 0$$

and hence $Tf \in V_\infty L^{p,\varphi}$. □

Corollary 7.79 *Let $1 < p < \infty$ and $\varphi \in \Phi$. If φ satisfies* (7.64), *then the operators M, $M^\sharp$, H, $\mathcal{H}$ and S are bounded in $V_\infty L^{p,\varphi}$.*

Recall that T^α, $0 < \alpha < n$, stands for any of the operators I^α, M^α, H^α and $\mathcal{H}^\alpha$ above.

Theorem 7.80 *Let $0 < \alpha < n$, $1 < p < n/\alpha$, $1/q = 1/p - \alpha/n$ and let $\varphi, \psi \in \Phi$ satisfy condition* (7.65). *Then T^α is bounded from $V_\infty L^{p,\varphi}$ into $V_\infty L^{q,\psi}(\mathbb{R}^n)$.*

Proof Since I^α is bounded from $V_\infty L^{p,\varphi}$ into $V_\infty L^{q,\psi}(\mathbb{R}^n)$ (cf. Theorem 7.74; see also Nakai [261]), the same norm inequalities also hold for the operators M^α, H^α and $\mathcal{H}^\alpha$. So we only need to check the action of the Riesz potential operator. Hence, it remains to show that

$$\sup_{x\in\mathbb{R}^n} \mathfrak{M}_{p,\varphi}(f;x,r) \to 0 \qquad \Longrightarrow \qquad \sup_{x\in\mathbb{R}^n} \mathfrak{M}_{q,\psi}(I^\alpha f;x,r) \to 0 \qquad \text{as} \quad r \to \infty.$$

This can be done as in the proof of Theorem 7.78, but now using the modular estimate (7.63) instead of (7.62). □

The theorem below provides an Adams-type result for the action of the operators T^α on vanishing subspaces at infinity.

Theorem 7.81 *Let* $0 < \alpha < n$, $1 < p < \frac{n}{\alpha}$ *and* $q > p$. *Suppose that* $\varphi \in \Phi$ *satisfies the conditions* (7.64) *and* (7.66). *Then* T^α *is bounded from* $V_\infty L^{p,\varphi}$ *into* $V_\infty L^{q,\varphi}(\mathbb{R}^n)$.

Proof As in the proof of Theorem 7.80, we only have to prove the statement for the Riesz potential operator. The $(L^{p,\varphi} \to L^{q,\varphi})$-boundedness of I^α follows from Theorem 7.76. To show the preservation of the vanishing property at infinity, we make use of the pointwise estimate (7.67). From that estimate, we get

$$\mathfrak{M}_{q,\varphi}\big(I^\alpha f; x, r\big) \lesssim \|f\|_{p,\varphi}^{q-p}\, \mathfrak{M}_{p,\varphi}\big(Mf; x, r\big)$$

for all $r > 0$ and $x \in \mathbb{R}^n$. If $f \in V_\infty L^{p,\varphi}$, then $Mf \in V_\infty L^{p,\varphi}$ by Corollary 7.79. Consequently, we have $I^\alpha f \in V_\infty L^{q,\varphi}(\mathbb{R}^n)$ taking into account the modular estimate above. □

7.4.3 Preservation of the Property (V^*)

In this section, we show invariance of the subspace $V^{(*)}L^{p,\varphi}$ with respect to the Hardy–Littlewood maximal operator and, consequently, also with respect to the sharp maximal function and the Hardy operators. Further, we show that the vanishing property (V^*) is also preserved under the action of any of the operators I^α, M^α, H^α and $\mathcal{H}^\alpha$.

Theorem 7.82 *Let* $1 < p < \infty$ *and* $\varphi \in \Phi$. *If* φ *satisfies* (7.64) *and the condition*

$$\lim_{t\to\infty} \frac{\varphi(t)}{t^n} = 0, \tag{7.68}$$

then the maximal operator M *is bounded in* $V^{(*)}L^{p,\varphi}$.

The proof of Theorem 7.82 essentially follows the same steps as the proof of Theorem 7.62, with the necessary modifications. However, since the arguments are quite different from those used to show the preservation of the other vanishing properties (at the origin and at infinity), we give the full proof here for the sake of completeness.

Proof Since M is bounded in $L^{p,\varphi}$ when φ satisfies (7.64) (cf. Theorem 7.73, see also Guliyev [127], Mizuhara [250], Nakai [261]), we only have to show that

$$\lim_{N\to\infty} \mathcal{A}_{N,p}(f) = 0 \quad \Longrightarrow \quad \lim_{N\to\infty} \mathcal{A}_{N,p}(Mf) = 0.$$

Given $x \in \mathbb{R}^n$ and $N \in \mathbb{N}$, we split f into

$$f = f_1 + f_2, \quad \text{with} \quad f_1 := f\, \chi_{\Omega_{x,N/2}}, \quad f_2 = f\, \chi_{\mathbb{R}^n \setminus \Omega_{x,N/2}},$$

where, for short, we use the notation

$$\Omega_{x,N} := B(x,2) \cap \big(\mathbb{R}^n \setminus B(0,N)\big).$$

Since M is sublinear, we have

$$\mathcal{A}_{N,p}(Mf) \lesssim \mathcal{A}_{N,p}(Mf_1) + \mathcal{A}_{N,p}(Mf_2). \tag{7.69}$$

We show next that both quantities in the right-hand side of (7.69) tend to zero as $N \to \infty$. From the boundedness of M in $L^p(\mathbb{R}^n)$, we get

$$\begin{aligned}\int_{B(x,1)} \big(M(f_1)(y)\big)^p \chi_N(y)\, dy &\leqslant \int_{\mathbb{R}^n} \big(M(f_1)(y)\big)^p dy \\ &\lesssim \int_{\mathbb{R}^n} |f_1(y)|^p dy = \int_{\Omega_{x,N/2}} |f(y)|^p\, dy\end{aligned}$$

with the implicit constant independent of x, N and f. Since $f \in V^{(*)}L^{p,\varphi}$, the right-hand side above tends to zero uniformly on x as $N \to \infty$ (note that the property (V^*) does not depend on the particular value of the radius taken in the balls centered at x; see Sect. 7.1). Therefore,

$$\lim_{N\to\infty} \mathcal{A}_{N,p}(Mf_1) = 0.$$

Now we deal with the second term of the sum in (7.69). Let $\varepsilon > 0$ be arbitrary. By condition (7.68) there exists $t_1 > 1$ such that $t^{-n}\varphi(t) < \varepsilon$ for all $t \geqslant t_1$. For such fixed t_1, we have

$$\int_{B(x,1)} \big(M(f_2)(y)\big)^p \chi_N(y)\, dy \lesssim I_1(x,N) + I_2(x,N),$$

where

$$I_1(x,N) := \int_{B(x,1)} \chi_N(y) \sup_{0<t<t_1} \left[\frac{1}{|B(y,t)|} \int_{B(y,t)} |f(z)|\, \chi_{\mathbb{R}^n\setminus\Omega_{x,N/2}}(z)\, dz \right]^p dy$$

and

$$I_2(x,N) := \int_{B(x,1)} \chi_N(y) \sup_{t\geqslant t_1} \left[\frac{1}{|B(y,t)|} \int_{B(y,t)} |f(z)|\, \chi_{\mathbb{R}^n\setminus\Omega_{x,N/2}}(z)\, dz \right]^p dy.$$

First we estimate $I_2(x,N)$. By Hölder's inequality, we have

$$\frac{1}{|B(y,t)|} \int_{B(y,t)} |f(z)|\, dz \leqslant \frac{1}{|B(y,t)|^{1/p}} \|f\|_{L^p(B(y,t))}.$$

Hence

$$\begin{aligned} I_2(x,N) &\leqslant \int_{B(x,1)} \sup_{t\geqslant t_1} \frac{1}{|B(y,t)|} \int_{B(y,t)} |f(z)|^p dz \\ &\lesssim \int_{B(x,1)} \sup_{t\geqslant t_1} \frac{\varphi(t)}{t^n} \|f\|_{p,\varphi}^p \\ &\leqslant \varepsilon\, \|f\|_{p,\varphi}^p. \end{aligned}$$

Now we estimate $I_1(x,N)$. There are two different cases to be analyzed for $z \in B(y,t)$ and $z \notin \Omega_{x,N/2}$. If $z \in B(0,N/2)$ then $t > |z-y| \geqslant |y|-|z| > N/2$. Thus there is no contribution to the supremum on $t \in (0,t_1)$ for $N \geqslant 2t_1$. If $z \notin B(x,2)$ then $t > |z-y| \geqslant |z-x|-|y-x| \geqslant 1$. Hence it remains to handle $I_1(x,N)$ when the supremum is taken over all $t \in (1,t_1)$. For such values of t we have

$$t^{-n} \int_{B(y,t)} |f(z)|\, \chi_{\mathbb{R}^n\setminus\Omega_{x,N/2}}(z)\, dz \leqslant \int_{B(y,t_1)} |f(z)|\, dz = \int_{B(0,t_1)} |f(y-z)|\, dz.$$

Then Minkowski's inequality and a simple change of variables yield

$$\begin{aligned} I_1(x,N) &\leqslant \int_{B(x,1)} \chi_N(y) \left[\int_{\mathbb{R}^n} \chi_{B(0,t_1)}(z)\, |f(y-z)|\, dz \right]^p dy \\ &\leqslant \left(\int_{\mathbb{R}^n} \chi_{B(0,t_1)}(z) \left[\sup_{v\in\mathbb{R}^n} \int_{B(v,1)} |f(u)|^p\, \chi_{N-|z|}(u)\, du \right]^{1/p} dz \right)^p \\ &=: \left(\int_{\mathbb{R}^n} g_N(z)\, dz \right)^p \end{aligned}$$

with the interpretation

$$\chi_a := 1 \quad \text{if } a \leqslant 0 \quad \text{and} \quad \chi_a := \chi_{\mathbb{R}^n \setminus B(0,a)} \text{ if } a > 0.$$

This gives a uniform bound for $I_1(x, N)$. Since $f \in L^{p,\varphi}$ and $g_N(z)$ has an integrable majorant (depending on t_1), by the Lebesgue convergence theorem we conclude that $\int\limits_{\mathbb{R}^n} g_N(z)\,dz \to 0$ as $N \to \infty$, which implies that

$$I_1(x, N) \to 0 \quad \text{uniformly on } x, \quad \text{as} \quad N \to \infty.$$

The proof is complete. □

In view of the inequalities (7.48) and (7.51), we get the following corollary.

Corollary 7.83 *Under the same assumptions of Theorem 7.82, the operators $M^\sharp$ and H are bounded in $V^{(*)}L^{p,\varphi}$.*

Regarding the behavior of the Hardy operator $\mathcal{H}$, as observed in Alabalik et al. [10, Remark 3.7], see also Sect. 7.3, we have the following stronger result.

Theorem 7.84 *Let $1 < p < \infty$ and $\varphi \in \Phi$ satisfy (7.64). Then the operator $\mathcal{H}$ is bounded from $L^{p,\varphi}$ into $V^{(*)}L^{p,\varphi}$.*

We end this section by showing that the vanishing property (V^*) is also preserved by the operators T^α (where again T^α stands for any of the operators I^α, M^α, H^α and $\mathcal{H}^\alpha$).

Theorem 7.85 *Let $0 < \alpha < n$, $1 < p < \frac{n}{\alpha}$ and $q > p$. Suppose that $\varphi \in \Phi$ satisfies the conditions (7.64) and (7.66). Then T^α is bounded from $V^{(*)}L^{p,\varphi}$ into $V^{(*)}L^{q,\varphi}(\mathbb{R}^n)$.*

Proof Since the operators M^α, H^α and $\mathcal{H}^\alpha$ are pointwise dominated by the Riesz potential operator I^α, it suffices to show the result for the latter operator. From (7.67) we get

$$\mathcal{A}_{N,q}\big(I^\alpha f\big) \lesssim \|f\|_{p,\varphi}^{q-p}\, \mathcal{A}_{N,p}(Mf)$$

with the implicit constant not depending on f and $N \in \mathbb{N}$. Then the result follows from the previous inequality and Theorem 7.82. □

7.4.4 Preservation of the Closure of $C_0^\infty(\mathbb{R}^n)$

Let us denote by $\overset{\circ}{L}{}^{p,\varphi}(\mathbb{R}^n)$ the closure of $C_0^\infty(\mathbb{R}^n)$ in the Morrey norm $\|\cdot\|_{p,\varphi}$. In Almeida and Samko [15, Theorem 6.3 and Corollary 6.4] (see also Sect. 7.2), it was shown that $\overset{\circ}{L}{}^{p,\varphi}(\mathbb{R}^n)$ can be described in an explicit way by combining the three vanishing properties, namely,

$$\overset{\circ}{L}^{p,\varphi}(\mathbb{R}^n) = V_0 L^{p,\varphi} \cap V_\infty L^{p,\varphi} \cap V^{(*)} L^{p,\varphi} =: V_{0,\infty}^{(*)} L^{p,\varphi}$$

when $\varphi \in \Phi$ satisfies the decay assumptions (7.61) and

$$\lim_{t\to\infty} \frac{\log^p t}{\varphi(t)} = 0. \tag{7.70}$$

Note that the closure of $C_0^\infty(\mathbb{R}^n)$ plays an important role in harmonic analysis on Morrey spaces since its dual provides a predual space for Morrey spaces (cf. Adams [6], Adams and Xiao [8], Rosenthal and Triebel [287], Triebel [325]). Moreover, we have the strict embeddings (see Sect. 7.2):

$$V_{0,\infty}^{(*)} L^{p,\varphi} \subsetneqq V_0 L^{p,\varphi} \cap V_\infty L^{p,\varphi} \subsetneqq V_0 L^{p,\varphi} \subsetneqq L^{p,\varphi}.$$

The next corollaries are immediate consequences of the results obtained in Sects. 7.4.2, 7.4.3 and the already known boundedness discussed in Sect. 7.4.1.

Corollary 7.86 *Let $1 < p < \infty$ and $\varphi \in \Phi$. If φ satisfies the Zygmund-type condition* (7.64) *and the decay conditions* (7.61), (7.68) *and* (7.70), *then the operators M, $M^\sharp$, H, and $\mathcal{H}$ are bounded from $\overset{\circ}{L}^{p,\varphi}(\mathbb{R}^n)$ into itself.*

Corollary 7.87 *Let $0 < \alpha < n$, $1 < p < \frac{n}{\alpha}$ and $q > p$. Suppose that $\varphi \in \Phi$ satisfies the integral conditions* (7.64), (7.66) *and the decay conditions* (7.61), (7.68) *and* (7.70). *Then T^α is bounded from $\overset{\circ}{L}^{p,\varphi}(\mathbb{R}^n)$ into $\overset{\circ}{L}^{q,\varphi}(\mathbb{R}^n)$.*

We end this section with a further result for singular integral operators S defined by (7.49), with the kernel satisfying (7.50). One knows that S is bounded in $V_0 L^{p,\varphi}$, for $1 < p < \infty$ and appropriate functions φ (cf. Samko [291, Theorem 5.1]). On the other hand, we have seen that S is also bounded in $V_\infty L^{p,\varphi}$, cf. Corollary 7.79. Unfortunately, we do not know whether S preserves the vanishing property (V^*). Nevertheless, we have the following result.

Theorem 7.88 *Let $1 < p < \infty$ and let $\varphi \in \Phi$ satisfy the conditions* (7.64), (7.61) *and* (7.70). *Then S is bounded from $\overset{\circ}{L}^{p,\varphi}(\mathbb{R}^n)$ into itself.*

Proof For bounded compactly supported functions f, there holds

$$|Sf(y)| \leqslant \frac{c}{|y|^n} \tag{7.71}$$

for sufficiently large $|y|$, with $c > 0$ not depending on y. In fact, suppose that $\operatorname{supp} f \subset B(0, R)$ for some $R > 0$. We have

$$|y - z| \geqslant |y| - R \geqslant \frac{|y|}{2} \qquad \text{for } |z| \leqslant R \quad \text{and} \quad |y| \geqslant 2R.$$

Hence (7.71) follows thanks to the size condition (7.50).

Let now $f \in \overset{\circ}{L}{}^{p,\varphi}(\mathbb{R}^n)$. Then there exists a sequence $(f_k) \subset C_0^\infty(\mathbb{R}^n)$ such that $f_k \to f$ in $L^{p,\varphi}$ as $k \to \infty$. By the continuity of S in $L^{p,\varphi}$, we get

$$Sf = S(\lim_{k\to\infty} f_k) = \lim_{k\to\infty} (Sf_k).$$

Thus we have $Sf \in \overset{\circ}{L}{}^{p,\varphi}(\mathbb{R}^n)$ since the intersection of the three vanishing subspaces is closed in $L^{p,\varphi}$. □

7.5 Comments to Chap. 7

The theory of Morrey spaces goes back to Morrey [252] who considered related integral inequalities in connection with regularity properties of solutions to nonlinear elliptic equations. In the form of Banach spaces of functions, called thereafter Morrey spaces, the ideas of Morrey [252] were further developed by Campanato [45]. A more systematic study of these (and even more general) spaces was carried out by Peetre [269] and Brudnyĭ [36]. We refer to the books Adams [6], Giaquinta [115], Pick et al. [274], Sawano et al. [305, 306], Taylor [320], Triebel [324] and the overview Rafeiro et al. [280] for additional references and basic properties and generalizations of Morrey spaces. We also refer to Adams and Xiao [8] for Harmonic Analysis in Morrey spaces and the special issue with Editorial Sawano et al. [304] for a discussion on related function spaces.

Morrey spaces are widely used in applications to regularity properties of solutions to PDE including the study of Navier–Stokes equations (see Triebel [324] and references therein). Although such spaces allow to describe local properties of functions better than Lebesgue spaces, they have some unpleasant properties. It is well-known that Morrey spaces are non-separable and that the usual classes of nice functions are not dense in such spaces.

In Zorko [342] it was observed that the set of functions in Morrey spaces for which the translation is continuous in Morrey norm plays an important role in approximation. This was sketched in Zorko [342, Proposition. 3] and also discussed in Chiarenza and Franciosi [52] and Kato [162]. In Zorko [342] and Kato [162] there were given some results on equivalence between belonging to Zorko space and approximation by mollifiers.

The space $V_0L^{p,\lambda}$, often called in the literature just by *vanishing Morrey space*, was already introduced in Chiarenza and Franciosi [52], Vitanza [327, 328] in connection with applications to PDE. The subspaces $V_\infty L^{p,\lambda}$ and $V^{(*)}L^{p,\lambda}$ were recently introduced in Almeida and Samko [13, 15] to study the approximation problem by nice functions in Morrey spaces. Note that $V_\infty L^{p,\lambda}$ was independently considered in Yuan et al. [341] in the study of interpolation problems.

The boundedness of classical operators in vanishing Morrey spaces at the origin was already studied in some papers, including the case of generalized parameters; see Persson et al. [270], Ragusa [282], Samko [290, 291]. In particular, in Samko [291] it was studied a class of sublinear singular-type operators which includes the Hardy–Littlewood maximal function and Calderón–Zygmund operators with standard kernels. The boundedness results in Samko [291] were given in terms of Zygmund-type integral conditions on the function parameter defining the Morrey space.

The results presented in Sects. 7.1 and 7.2 were obtained in Almeida and Samko [14, 15], respectively. Section 7.3 follows the paper Alabalik et al. [10] while Sect. 7.4 appears in Alabalik et al. [11].

Chapter 8
Bounded Variation Spaces and Related Topics

In the foundational paper of Riesz [286], it is proved (in modern terminology) that an absolutely continuous function $f : I \to \mathbb{R}$ belongs to the Sobolev space $W^{1,p}(I)$, $1 < p < \infty$ and $I \subset \mathbb{R}$ an interval, if and only if

$$\sup \sum_{j} \frac{|f(x_j) - f(x_{j-1})|^p}{|x_j - x_{j-1}|^{p-1}} < \infty, \tag{8.1}$$

where the supremum is taken over all finite partitions of I. The quantity in (8.1) is called the *Riesz p-variation of f on I*; see also Castillo et al. [48, 50], Cruz-Uribe et al. [65], Kakochashvili and Zviadadze [160] for generalizations of Riesz bounded variation spaces. The notion of Riesz bounded variation spaces has also been extended to functions on general domains in $\mathbb{R}^n$, $n > 1$: see, for instance, Angeloni [20], Barza and Lind [24], Bojarski [33], Malý [220]. We refer the interested readers to Appell et al. [21] for a comprehensive survey of the classical theory of bounded variation spaces.

In this chapter, we introduce variable exponent bounded variation spaces in the Riesz sense, prove some embeddings, and show that a type of Riesz representation lemma is valid in the newly introduced spaces. We also characterize the linear functionals on variable exponent Bochner–Lebesgue spaces in terms of the variable exponent Riesz bounded variation spaces for vector measures. Furthermore, we introduce a weighted version of the Riesz bounded variation space and establish that a generalized Riesz theorem, connecting this space with the Sobolev space $W^{1,p}(I)$, remains valid for weighted Riesz bounded variation spaces when the weight belongs to the Muckenhoupt class. Additionally, we characterize globally Lipschitz Nemytskii operators acting on these weighted Riesz bounded variation spaces. Finally, we introduce weighted Riesz bounded variation spaces defined on an open subset of n-dimensional Euclidean space and use them to characterize weighted Sobolev spaces when the weight belongs to the Muckenhoupt class. As an

A. Meskhi et al., *Integral Operators in Non-Standard Function Spaces*,
Operator Theory: Advances and Applications 310,
https://doi.org/10.1007/978-3-032-03691-9_8

application, using Rubio de Francia's extrapolation theory, a similar characterization of the variable exponent Sobolev spaces via variable exponent Riesz bounded variation spaces is obtained.

8.1 Preliminaries

8.1.1 Modular Spaces

Let X be a vector space over $\mathbb{K}$ ($\mathbb{K} = \mathbb{C}$ or $\mathbb{K} = \mathbb{R}$). A convex and left-continuous function $\rho : \mathsf{X} \to [0, +\infty]$ is called a *convex pseudomodular* on X if for arbitrary x and y, there holds:

(i) $\rho(0_{\mathsf{X}}) = 0$;
(ii) $\rho(\alpha x) = \rho(x)$ for every $\alpha \in \mathbb{K}$ such that $|a| = 1$;
(iii) $\rho(\alpha x + (1-\alpha)y) \leqslant \alpha\rho(x) + (1-\alpha)\rho(y)$ for every $\alpha \in [0, 1]$.

If, in addition to (i), $\rho(\lambda x) = 0$ for all $\lambda > 0$ implies $x = 0$, then ρ is called a semimodular. If, moreover, $\rho(x) = 0$ implies $x = 0$, then ρ is called a modular.

If ρ is a pseudomodular in X then the set defined by

$$\mathsf{X}_\rho = \left\{ x \in \mathsf{X} : \lim_{\lambda \to 0} \rho(\lambda x) = 0 \right\}$$

is called a *modular space*.

For a convex pseudomodular ρ in X_ρ, we define the *Luxemburg norm*, also known as Luxemburg–Nakano norm, as

$$\|f\|_\rho = \inf \left\{ \lambda > 0 : \rho\left(\frac{f}{\lambda}\right) \leqslant 1 \right\}.$$

It is also possible to introduce another norm, the so-called Amemiya norm

$$|||f|||_\rho = \inf_{\lambda > 0} \lambda \left(1 + \rho\left(\frac{f}{\lambda}\right)\right),$$

for details see Musielak [257].

The following propositions encapsulate the main properties that we need regarding modular spaces, see Diening et al. [74], Harjulehto and Hästö [132], Kozlowski [199] for the proofs.

Proposition 8.1 *Let X be a vector space, ρ be a semimodular on X, and $f \in X$. Then*

(i) $\rho(f) \leqslant 1$ if and only if $\|f\|_\rho \leqslant 1$;
(ii) if $\|f\|_\rho \leqslant 1$, then $\rho(f) \leqslant \|f\|_\rho$;

(iii) if $\|f\|_\rho > 1$*, then* $\rho(f) \geqslant \|f\|_\rho$*;*
(iv) for every $f \in \mathsf{X}$*,* $\|f\|_\rho \leqslant \rho(f) + 1$.

Definition 8.2 Let ρ be a semimodular on X and $x, x_k \in \mathsf{X}_\rho$. Then we say that x_k is modular convergent (ρ-convergent) to x if there exists $\lambda > 0$ such that $\rho(\lambda(x_k - x)) \to 0$.

Proposition 8.3 *Let* X_ρ *be a semimodular space. Then modular convergence and norm convergence are equivalent if and only if* $\rho(x_k) \to 0$ *implies* $\rho(2x_k) \to 0$.

If any of the equivalent conditions in the previous proposition hold, then we say that the semimodular satisfies the weak Δ_2-condition. From the above we get the following.

Proposition 8.4 *Let* ρ *be a semimodular on* X *that satisfies the weak* Δ_2*-condition. Then for every* $\varepsilon > 0$ *there exists* $\delta > 0$ *such that* $\rho(f) \leqslant \delta$ *implies* $\|f\|_\rho \leqslant \varepsilon$.

For more details about modular spaces, see Diening et al. [74], Kozlowski [199], Musielak [257].

8.1.2 Variable Exponent Spaces

The basics on variable exponent Lebesgue spaces may be found in Cruz-Uribe and Fiorenza [58], Diening et al. [74], Kokilashvili et al. [182], but we recall here some necessary definitions and propositions without explicitly citing the exact references from the aforementioned books. For $\Omega \subset \mathbb{R}^n$ we put $p_\Omega^+ := \operatorname{ess\,sup}_{x \in \Omega} p(x)$ and $p_\Omega^- := \operatorname{ess\,inf}_{x \in \Omega} p(x)$; we use the abbreviations $p^+ = p_\Omega^+$ and $p^- = p_\Omega^-$. For a measurable function $p : \Omega \to [1, \infty)$, we call it a *variable exponent* and define the set of all variable exponents with $p^+ < \infty$ as $\mathscr{P}(\Omega)$.

For a measurable function $\varphi : \Omega \to \mathbb{R}$, we define the *modular* $\varrho_{p(\cdot)}$ by

$$\varrho_{p(\cdot)}(f) := \int_\Omega |f(x)|^{p(x)} dx$$

and the *Luxemburg–Nakano norm* by

$$\|f\|_{L^{p(\cdot)}(\Omega)} := \inf\left\{\lambda > 0 : \varrho_{p(\cdot)}\left(\frac{f}{\lambda}\right) \leqslant 1\right\}. \tag{8.2}$$

Definition 8.5 Let $p \in \mathscr{P}(\Omega)$. The *variable exponent Lebesgue space* $L^{p(\cdot)}(\Omega)$ is introduced as the set of all real-valued measurable functions $f : \Omega \to \mathbb{R}$ for which the modular is finite, i.e., $\varrho_{p(\cdot)}(f) < \infty$. Equipped with the Luxemburg–Nakano norm (8.2) this is a Banach space.

We need to impose some regularity in the variable exponent in order to have some "fruitful" theory (e.g., the boundedness of the maximal operator).

Given $p \in \mathscr{P}(\Omega)$, we say that p satisfies the local log-Hölder continuity condition and denote this by $p \in \mathbb{P}^{\log}(\Omega)$, if there exists a constant C_0 such that

$$|p(x) - p(y)| \leqslant \frac{C_0}{-\ln(|x-y|)}, \quad x, y \in \Omega, \quad |x-y| < 1/2, \tag{8.3}$$

and we say that p is log-Hölder continuous at infinity or that it satisfies the decay condition, if there exist p_∞ and $C_\infty > 0$ such that

$$|p(x) - p_\infty| \leqslant \frac{C_\infty}{\ln(e + |x|)}, \quad x \in \Omega. \tag{8.4}$$

When p satisfies (8.3) and (8.4), we say that p is a log-Hölder continuous function and denote this by $p \in \mathbb{P}^{\log}_{\infty}(\Omega)$.

8.2 Variable Exponent Bounded Variation Spaces in the Riesz Sense

In this section, we introduce variable exponent bounded variation spaces in the Riesz sense, denoted by $RBV^{p(\cdot)}([a, b])$, and prove some embedding results, and finally we show that a type of Riesz representation lemma is valid in $RBV^{p(\cdot)}([a, b])$.

8.2.1 Bounded Variation in the Riesz Sense

We use the notions of modular and variable exponent Lebesgue space as introduced in Sect. 8.1.

We will need the following embedding result (see, e.g., Cruz-Uribe and Fiorenza [58], Diening et al. [74]).

Theorem 8.6 *Let $p, q \in \mathscr{P}([a, b])$ and $p(x) \leqslant q(x)$, $x \in [a, b]$. Then*

$$L^{q(\cdot)}[a, b] \hookrightarrow L^{p(\cdot)}[a, b]. \tag{8.5}$$

The following lemma is adapted from a more general result; see, e.g. Diening et al. [74].

Lemma 8.7 *Let $p \in \mathbb{P}^{\log}(I)$. Then*

$$|I|^{p_I^- - p_I^+} \leqslant C, \tag{8.6}$$

for some constant $C > 0$ and every finite interval $I \subset \mathbb{R}$.

Proof Since $p_I^- - p_I^+ < 0$, we find that $|I|^{p_I^- - p_I^+} < 1$ for every I satisfying $|I| > 1/4$. If $|I| \leqslant 1/4$ and $x, y \in I$, we estimate as follows:

$$
\begin{aligned}
|p_I^- - p_I^+| \ln \frac{1}{|I|} &\leqslant \ln \frac{1}{|I|} |p(x) - p(y)| \\
&\lesssim \frac{\ln (1/|I|)}{\ln (1/|x-y|)} \\
&\lesssim \frac{\ln (1/|I|)}{\ln (1/|I|)} \\
&\lesssim 1,
\end{aligned}
$$

which implies (8.6). □

Let us state the following lemma, which is tailored to our purposes. A similar result is already known in a slightly different formulation; see Diening [71], Diening and Schwarzacher [73].

Lemma 8.8 *Let $[a, b] \subset \mathbb{R}$ be a finite interval. If $p \in \mathbb{P}^{\log}([a, b])$, then*

$$
\left(\frac{1}{b-a} \int_a^b |f(y)| dy \right)^{p(x)} \lesssim \frac{1}{b-a} \int_a^b |f(y)|^{p(y)} dy + 1, \tag{8.7}
$$

for all $x \in [a, b]$ and $\|f\|_{L^{p(\cdot)}([a,b])} \leqslant 1$.

Proof For the sake of simplicity let us denote $I = [a, b]$ and, as usual, $|I| = b - a$. We define the radius of I by $r = |I|/2$. The proof is divided in two cases.
Case 1: $r > 1/2$. We estimate as follows:

$$
\begin{aligned}
\left(\frac{1}{|I|} \int_I |f(y)| dy \right)^{p(x)} &\leqslant \left[\frac{1}{|I|} \int_I \left(|f(y)|^{p(y)} + 1 \right) dy \right]^{p(x)} \\
&\leqslant \left(|I|^{-1} \varrho_{p(\cdot)}(f) + 1 \right)^{p(x)} \\
&\leqslant \left(|[-1/2, 1/2]|^{-1} + 1 \right)^{p_+} \\
&= C(p).
\end{aligned}
$$

Case 2: $0 < r \leqslant 1/2$. Observe that $|I| \leqslant 1$. In virtue of Proposition 8.1, we get

$$
\frac{1}{3} \int_I (|f(y)|^{p(y)} + 1) dy < \frac{1}{3} \int_I (|f(y)|^{p(y)} + 2) dy = \frac{1}{3} \varrho_{p(\cdot)}(f) + \frac{2}{3} |I| \leqslant 1. \tag{8.8}
$$

Hence, by Hölder's inequality, (8.8), and Lemma 8.7, we have

$$\begin{aligned}
&\left(\frac{1}{|I|}\int_I |f(y)|dy\right)^{p(x)}\\
&\leqslant \left[\left(\frac{1}{|I|}\int_I |f(y)|^{p_I^-}dy\right)^{1/p_I^-}\left(\frac{1}{|I|}\int_I \chi_I(y)^{(p_I^-)'}dy\right)^{1/(p_I^-)'}\right]^{p(x)}\\
&= \left(\frac{1}{|I|}\int_I |f(y)|^{p_I^-}dy\right)^{p(x)/p_I^-} \leqslant \left(\frac{1}{|I|}\int_I \left(|f(y)|^{p(y)}+1\right)dy\right)^{p(x)/p_I^-}\\
&= |I|^{\frac{-p(x)}{p_I^-}}3^{\frac{p(x)}{p_I^-}}\left[\frac{1}{3}\int_I \left(|f(y)|^{p(y)}+2\right)dy\right]^{p(x)/p_I^-}\\
&\leqslant |I|^{\frac{-p(x)}{p_I^-}}3^{\frac{p(x)}{p_I^-}}\left[\frac{1}{3}\int_I \left(|f(y)|^{p(y)}+2\right)dy\right]\\
&= |I|^{\frac{-p(x)}{p_I^-}}3^{p_+}\left(\frac{1}{3}\int_I |f(y)|^{p(y)}dy+\frac{2}{3}|I|\right)\\
&\leqslant |I|^{\frac{-p(x)}{p_I^-}+1}3^{p_+-1}\left(\frac{1}{|I|}\int_I |f(y)|^{p(y)}dy+1\right)\\
&\leqslant |I|^{\frac{p_I^- -p_I^+}{p_I^-}}3^{p_+-1}\left(\frac{1}{|I|}\int_I |f(y)|^{p(y)}dy+1\right)\\
&\lesssim \frac{1}{|I|}\int_I |f(y)|^{p(y)}dy+1,
\end{aligned}$$

which ends the proof. □

For further usage, we recall the notion of variable exponent Lebesgue sequence space $\ell_{\{p_n\}} = \ell_{\{p_n\}_{n\in\mathbb{N}}}$ given by

$$\ell_{\{p_n\}} := \left\{\mathbf{x} = (x_1, x_2, \ldots, x_n, \ldots) : \sigma_{\{p_n\}}(\mathbf{x}) := \sum_{i=1}^{\infty} |x_i|^{p_i} < \infty\right\},$$

and normed using the Luxemburg norm, viz.

$$\|\mathbf{x}\|_{\ell_{\{pn\}}} := \inf\left\{\lambda > 0 : \sigma_{\{p_n\}}\left(\frac{\mathbf{x}}{\lambda}\right) \leqslant 1\right\}.$$

By $\ell^m_{\{p_n\}}$ we denote the set of all $\mathbf{x} \in \ell_{\{p_n\}}$ such that the kth projection of $\mathbf{x}$ is zero for all $k > m$, i.e., $\mathrm{pr}_k(\mathbf{x}) = 0$ for all natural $k > m$; see, e.g., Nakano [263], Nekvinda [265]

Variable exponent Lebesgue sequence spaces satisfy the following Hölder inequality:

$$\sum_{i=1}^{\infty} x_i y_i \leqslant 2\|\mathbf{x}\|_{\ell_{\{pn\}}} \|\mathbf{y}\|_{\ell_{\{qn\}}} \tag{8.9}$$

where

$$\frac{1}{p_i} + \frac{1}{q_i} = 1,$$

for all $i \in \mathbb{N}$.

8.2.2 *Bounded Variation in the Riesz Sense*

In this section we introduce variable exponent bounded variation spaces in the Riesz sense. We also show some function modular properties and embedding results.

Definition 8.9 By an *admissible function* we mean a function whose range is in $(1, \infty)$ and such that $p_+ := \sup_{x\in[a,b]} p(x)$ is finite.

Definition 8.10 Let $\Pi_{[a,b]} = \{a = t_0 < t_1 < \ldots < t_n = b\}$ be a partition of $[a, b]$ and p an admissible function. We define the functionals $\sigma^{p(\cdot)}_{[a,b]}, V^{p(\cdot)}_{[a,b]} : \mathbb{R}^{[a,b]} \to \overline{\mathbb{R}}_+$ by

$$\sigma^{p(\cdot)}_{[a,b]}\left(f, \Pi^*_{[a,b]}\right) = \sum_{i=1}^{n} \frac{|f(t_i) - f(t_{i-1})|^{p(x_{i-1})}}{(t_i - t_{i-1})^{p(x_{i-1})-1}},$$

and

$$V^{p(\cdot)}_{[a,b]}(f) = \sup_{\Pi^*_{[a,b]}} \sigma^{p(\cdot)}_{[a,b]}\left(f, \Pi^*_{[a,b]}\right),$$

where $\Pi^*_{[a,b]}$ is a tagged partition of the interval $[a, b]$, i.e., a partition of the interval $[a, b]$ together with a finite sequence of numbers $x_0, \ldots, x_{n-1}$ subject to the conditions that for each i, $t_i \leqslant x_i \leqslant t_{i+1}$. We now introduce the space of bounded $p(\cdot)$ variation in Riesz sense as

$$\mathrm{RBV}^{p(\cdot)}([a,b]) := \left\{ f \in \mathbb{R}^{[a,b]} : V_{[a,b]}^{p(\cdot)}(f) < \infty \right\}.$$

Lemma 8.11 *The functional $V_{[a,b]}^{p(\cdot)}$ is a convex pseudomodular.*

Proof We have that $V_{[a,b]}^{p(\cdot)}(0) = 0$, or more generally, we have that $V_{[a,b]}^{p(\cdot)}(f) = 0$ if and only if $f = \text{const}$. To see this, observe that if $f = \text{const}$, then for every partition $\Pi_{[a,b]}$, we have that $|f(t_i) - f(t_{i-1})| = 0$, so $V_{[a,b]}^{p(\cdot)}(f) = 0$. Now, we suppose that f is not constant on $[a,b]$ then there exist t_{i-1} and t_i in $[a,b]$ such that $|f(t_{i-1}) - f(t_i)| \neq 0$. Let $\Pi_{[a,b]}$ be a partition of $[a,b]$ that contains t_{i-1} and t_i, we obtain

$$0 < \sum_{j=1}^{m} \frac{|f(t_i) - f(t_{i-1})|^{p(x_{i-1})}}{(t_i - t_{i-1})^{p(x_{i-1})-1}} \leqslant V_{[a,b]}^{p(\cdot)}(f).$$

The property that $V_{[a,b]}^{p(\cdot)}(\alpha f) = V_{[a,b]}^{p(\cdot)}(f)$ whenever $|\alpha| = 1$ is also clear. To show the convexity of $V_{[a,b]}^{p(\cdot)}$, we note that

$$\begin{aligned} &\frac{|\alpha(f(t_i) - f(t_{i-1})) + (1-\alpha)(g(t_i) - g(t_{i-1}))|^{p(x_{i-1})}}{(t_i - t_{i-1})^{p(x_{i-1})-1}} \\ &\qquad \leqslant \alpha \frac{|f(t_i) - f(t_{i-1})|^{p(x_{i-1})}}{(t_i - t_{i-1})^{p(x_{i-1})-1}} + (1-\alpha) \frac{|g(t_i) - g(t_{i-1})|^{p(x_{i-1})}}{(t_i - t_{i-1})^{p(x_{i-1})-1}} \end{aligned}$$

which follows from the convexity of $x \mapsto x^p$ on $[0, +\infty)$ when $p(x) \geqslant 1$ and the definition of $V_{[a,b]}^{p(\cdot)}$. The left-continuity follows from the fact that

$$\lambda^{p_+} V_{[a,b]}^{p(\cdot)}(f) \leqslant V_{[a,b]}^{p(\cdot)}(\lambda f) \leqslant \lambda^{p_-} V_{[a,b]}^{p(\cdot)}(f)$$

when $\lambda < 1$. □

We can now introduce a norm on the quotient space

$$RBV^{p(\cdot)}([a,b]) := \mathrm{RBV}^{p(\cdot)}([a,b])/\mathcal{C},$$

where $\mathcal{C}$ is the space of all constant functions, viz.

$$\|f\|_{RBV^{p(\cdot)}([a,b])} = \inf\left\{ \lambda > 0 : V_{[a,b]}^{p(\cdot)}\left(\frac{f}{\lambda}\right) \leqslant 1 \right\}.$$

From the definition of the $RBV^{p(\cdot)}$ norm we have that, if

$$V_{[a,b]}^{p(\cdot)}\left(\frac{f}{a}\right) \leqslant b,$$

then

$$\|f\|_{RBV^{p(\cdot)}([a,b])} \leqslant ab^{\nu}, \tag{8.10}$$

where

$$\nu = \begin{cases} \frac{1}{p_+}, & b \leqslant 1; \\ \frac{1}{p_-}, & b \geqslant 1. \end{cases}$$

8.2.3 *Embedding Results*

Theorem 8.12 *Let* $1 \leqslant p(x) \leqslant q(x)$, $x \in [a, b]$. *Then*

$$RBV^{q(\cdot)}([a, b]) \hookrightarrow RBV^{p(\cdot)}([a, b]).$$

Proof Let s and p' be defined by

$$\frac{1}{q(x)} + \frac{1}{s(x)} = \frac{1}{p(x)}, \qquad \frac{1}{p'(x)} + \frac{1}{p(x)} = 1, \quad x \in [a, b].$$

Using the inequality $(AB)^p \leqslant \frac{p}{q}A^q + \frac{p}{s}B^s$ and specifying A and B as

$$A = \frac{|f(t_i) - f(t_{i-1})|}{(t_i - t_{i-1})^{\frac{1}{p'(x_{i-1})} + \frac{1}{s(x_{i-1})}} \|f\|_{RBV^{q(\cdot)}([a,b])}}, \ B = (t_i - t_{i-i})^{\frac{1}{s(x_{i-1})}},$$

we obtain

$$\frac{|f(t_i) - f(t_{i-1})|^{p(x_{i-1})}}{(t_i - t_{i-1})^{p(x_{i-1})-1} \|f\|_{RBV^{q(\cdot)}([a,b])}^{p(x_{i-1})}}$$
$$\leqslant \left(\frac{p}{q}\right)^+ \frac{|f(t_i) - f(t_{i-1})|^{q(x_{i-1})}}{(t_i - t_{i-1})^{q(x_{i-1})-1} \|f\|_{RBV^{q(\cdot)}([a,b])}^{q(x_{i-1})}} + \left(\frac{p}{s}\right)^+ (t_i - t_{i-1}), \tag{8.11}$$

from which it follows that

$$V_{[a,b]}^{p(\cdot)}\left(\frac{f}{\|f\|_{RBV^{q(\cdot)}([a,b])}}\right) \leqslant \left(\frac{p}{q}\right)^{+} V_{[a,b]}^{q(\cdot)}\left(\frac{f}{\|f\|_{RBV^{q(\cdot)}([a,b])}}\right) + \left(\frac{p}{s}\right)^{+}(b-a)$$

which implies, from (8.10) and the definition of the RBV-norm, that

$$\|f\|_{RBV^{p(\cdot)}([a,b])} \lesssim \|f\|_{RBV^{q(\cdot)}([a,b])}$$

with the implicit constant independent of f, which ends the proof. □

Remark 8.13 To apply Proposition 8.3, it suffices to note that $V_{[a,b]}^{p(\cdot)}(f_j) \to 0$ implies $V_{[a,b]}^{p(\cdot)}(2f_j) \to 0$ for any sequence $\{f_j\} \subset RBV^{p(\cdot)}([a,b])$, since

$$V_{[a,b]}^{p(\cdot)}(2f_j) \leqslant 2^{p_+} V_{[a,b]}^{p(\cdot)}(f_j) \to 0.$$

Thus, modular and norm convergence are equivalent in $RBV^{p(\cdot)}([a,b])$.

Next, we show that the introduced $RBV^{p(\cdot)}([a,b])$ space contains all Lipschitz functions and is contained in the set of absolutely continuous functions.

Theorem 8.14 *Let* Lip$([a,b])$ *be the space of Lipschitz functions on* $[a,b]$ *and* $AC([a,b])$ *be the space of absolutely continuous functions in* $[a,b]$. *Then*

$$\mathrm{Lip}([a,b]) \subseteq RBV^{p(\cdot)}([a,b]) \subseteq AC([a,b]).$$

Proof Let $\Pi_{[a,b]}^*$ be a tagged partition of $[a,b]$ and f a Lipschitz function on $[a,b]$ with Lipschitz constant L. Then

$$\begin{aligned}\sum_{i=1}^{n} \frac{|f(t_i)-f(t_{i-1})|^{p(x_{i-1})}}{(t_i-t_{i-1})^{p(x_{i-1})-1}} &\leqslant \sum_{i=1}^{n} L^{p(x_{i-1})}(t_i-t_{i-1})^{p(x_{i-1})-p(x_{i-1})+1}\\ &\leqslant (1+L)^{p_+}\sum_{i=1}^{n}(t_i-t_{i-1})\\ &= (1+L)^{p_+}(b-a),\end{aligned}$$

from which it follows that $V_{[a,b]}^{p(\cdot)}(f) < \infty$ and therefore $f \in RBV^{p(\cdot)}([a,b])$.

Let us now prove that $RBV^{p(\cdot)}([a,b]) \subseteq AC([a,b])$. We have that

$$\sum_{i=1}^{n}|f(t_i)-f(t_{i-1})| = \sum_{i=1}^{n} \frac{|f(t_i)-f(t_{i-1})|}{(t_i-t_{i-1})^{\frac{1}{p'(x_{i-1})}}}(t_i-t_{i-1})^{\frac{1}{p'(x_{i-1})}}$$

$$\leqslant 2\|\mathbf{X}\|_{\ell^n_{\{p_i\}}}\|\mathbf{Y}\|_{\ell^n_{\{p'_i\}}}$$

$$\leqslant 2\left(1+V^{p(\cdot)}_{[a,b]}(f)\right)\|\mathbf{Y}\|_{\ell^n_{\{p'_i\}}}$$

where $\mathbf{X}=\{X_i\}_{i=1}^n$, $\mathbf{Y}=\{Y_i\}_{i=1}^n$ and

$$X_i=\frac{|f(t_i)-f(t_{i-1})|}{(t_i-t_{i-1})^{\frac{1}{p'(x_{i-1})}}},\quad Y_i=(t_i-t_{i-1})^{\frac{1}{p'(x_{i-1})}}.$$

From Proposition 8.4, for all $\varepsilon>0$ we can choose $\delta>0$ such that

$$\|\mathbf{Y}\|_{\ell^n_{\{p'_i\}}}\leqslant\varepsilon$$

whenever

$$\sigma_{\{p'_i\}}(Y)=\sum_{i=1}^n(t_i-t_{i-1})\leqslant\delta,$$

from which we get the desired embedding. □

We now prove the completeness of the $RBV^{p(\cdot)}([a,b])$ space.

Theorem 8.15 *The space $RBV^{p(\cdot)}([a,b])$ is a Banach space.*

Proof Let $\{f_m\}_{m\in\mathbb{N}}$ be a Cauchy sequence in $RBV^{p(\cdot)}([a,b])$, i.e.,

$$\|f_m-f_n\|_{RBV^{p(\cdot)}([a,b])}\xrightarrow[m,n\to\infty]{}0.$$

We choose a representative from each equivalence class using the rule that $f_m(a)=0$ for all $m\in\mathbb{N}$, from which we get

$$|f_m(x)-f_n(x)|=|f_m(x)-f_n(x)-(f_m(a)-f_n(a))|.$$

Let $\Pi^*_{\{x\}}$ be any tagged partition that contains the point x. We have

$$\sum_{i=1}^N\frac{|(f_m-f_n)(t_i)-(f_m-f_n)(t_{i-1})|}{(t_i-t_{i-1})^{\frac{1}{p'(x_i)}}}\cdot(t_i-t_{i-1})^{\frac{1}{p'(x_i)}}\leqslant 2\|\mathbf{X}\|_{\ell^N_{\{p_i\}}}\|\mathbf{Y}\|_{\ell^N_{\{p'_i\}}}$$

from which we get

$$|f_m(x)-f_n(x)|\leqslant 2\|\mathbf{X}\|_{\ell^N_{\{p_i\}}}\|\mathbf{Y}\|_{\ell^N_{\{p'_i\}}}$$

where $\mathbf{X} = \{X_i\}_{i=1}^N$, $\mathbf{Y} = \{Y_i\}_{i=1}^N$, with

$$X_i = \frac{|(f_m - f_n)(t_i) - (f_m - f_n)(t_{i-1})|}{(t_i - t_{i-1})^{\frac{1}{p'(x_{i-1})}}}, \; Y_i = (t_i - t_{i-1})^{\frac{1}{p'(x_{i-1})}}.$$

Using

$$\begin{aligned}\sigma_{\{p_j\}}(\mathbf{X}) &= \sum_{i=1}^{N} \frac{|(f_m - f_n)(t_i) - (f_m - f_n)(t_{i-1})|^{p(x_{i-1})}}{(t_i - t_{i-1})^{p(x_{i-1})-1}} \\ &\leqslant V_{[a,b]}^{p(\cdot)}(f_m - f_n)\end{aligned}$$

we get that

$$\sigma_{\{p_j\}}(\mathbf{X}) \xrightarrow[m,n\to\infty]{} 0 \tag{8.12}$$

since $V_{[a,b]}^{p(\cdot)}(f_m - f_n) \xrightarrow[m,n\to\infty]{} 0$ because $\|f_m - f_n\|_{RBV^{p(\cdot)}([a,b])} \xrightarrow[m,n\to\infty]{} 0$ and Proposition 8.3. From the same Proposition 8.3, we get that $\|\mathbf{X}\|_{\ell^N_{\{p_i\}}} \xrightarrow[m,n\to\infty]{} 0$ due to (8.12) which implies that $\{f_n(x)\}_{n\in\mathbb{N}}$ is a Cauchy sequence in $\mathbb{R}$, and we take $f(x) := \lim_{n\to+\infty} f_n(x)$. Since, for a fixed tagged partition, we have

$$\begin{aligned}&\sum_{i=1}^{N} \frac{|(f - f_m)(t_i) - (f - f_m)(t_{i-1})|^{p(x_{i-1})}}{(t_i - t_{i-1})^{p(x_{i-1})-1}} \\ &\quad = \sum_{i=1}^{N} \frac{|\lim_{n\to\infty}[(f_n - f_m)(t_i) - (f_n - f_m)(t_{i-1})]|^{p(x_{i-1})}}{(t_i - t_{i-1})^{p(x_{i-1})-1}} \\ &\quad = \lim_{n\to\infty} \sum_{i=1}^{N} \frac{|(f_n - f_m)(t_i) - (f_n - f_m)(t_{i-1})|^{p(x_{i-1})}}{(t_i - t_{i-1})^{p(x_{i-1})-1}} \\ &\quad = \lim_{n\to\infty} V_{[a,b]}^{p(\cdot)}(f_n - f_m) \\ &\quad \leqslant 1\end{aligned}$$

whenever m is sufficiently large, due to the fact that $\{f_m\}$ is a Cauchy sequence. From the above estimate, we get that $f - f_m \in RBV^{p(\cdot)}([a,b])$ which implies that $f \in RBV^{p(\cdot)}([a,b])$ due to the vector space structure of the space. Moreover,

$$\lim_{n\to\infty} \|f_m\|_{RBV^{p(\cdot)}([a,b])} = \|f\|_{RBV^{p(\cdot)}([a,b])}$$

by a standard continuity argument. □

A classical result for functions of bounded variation in the Riesz sense is the celebrated Riesz representation lemma, which relates the Riesz seminorm to the L^p-norm of the derivative. In the variable exponent case, an analogous phenomenon occurs, but the relation becomes an equivalence of norms.

Theorem 8.16 *Let $p \in \mathbb{P}^{\log}([a,b])$. If f is an absolutely continuous function on $[a,b]$ and $f' \in L^{p(\cdot)}([a,b])$, then $f \in RBV_{p(\cdot)([a,b])}$. Furthermore, we have*

$$\|f\|_{RBV^{p(\cdot)}([a,b])} \lesssim \|f'\|_{L^{p(\cdot)}([a,b])}.$$

Proof Let $\|f'\|_{L^{p(\cdot)}([a,b])} = 1$ and $t_{i-1} \leqslant x_{i-1} \leqslant t_i$ be elements of $[a,b]$. Since f is an absolutely continuous function on $[a,b]$, we have

$$\begin{aligned}\left|\frac{f(t_i)-f(t_{i-1})}{t_i-t_{i-1}}\right|^{p(x_{i-1})}(t_i-t_{i-1}) &= \left|\frac{1}{t_i-t_{i-1}}\int_{t_{i-1}}^{t_i} f'(t)dt\right|^{p(x_{i-1})}(t_i-t_{i-1})\\ &\leqslant \left(\frac{1}{t_i-t_{i-1}}\int_{t_{i-1}}^{t_i} |f'(t)|dt\right)^{p(x_{i-1})}(t_i-t_{i-1})\\ &\lesssim \left(\frac{1}{t_i-t_{i-1}}\int_{t_{i-1}}^{t_i} |f'(t)|^{p(t)}dt+1\right)(t_i-t_{i-1})\\ &\lesssim \int_{t_{i-1}}^{t_i} |f'(t)|^{p(t)}dt+(t_i-t_{i-1}),\end{aligned}$$

where the second inequality is due to Lemma 8.8. Hence,

$$\begin{aligned}\sum_{i=1}^{n}\frac{|f(t_i)-f(t_{i-1})|^{p(x_{i-1})}}{(t_i-t_{i-1})^{p(x_{i-1})-1}} &\lesssim \sum_{i=1}^{n}\int_{t_{i-1}}^{t_i} |f'(t)|^{p(t)}dt+b-a\\ &\lesssim \varrho_{p(\cdot)}(f')+b-a\\ &\lesssim 1,\end{aligned}$$

since $\varrho_{p(\cdot)}(f') = 1$. Thus, we have proved

$$\varrho_{p(\cdot)}(f') \leqslant 1 \quad \text{implies} \quad V_{[a,b]}^{p(\cdot)}(f) \leqslant C,$$

therefore by (8.10) we get

$$\|f\|_{RBV^{p(\cdot)}([a,b])} \leqslant C^{\nu},$$

and now by the homogeneity of the norm we conclude

$$\|f\|_{RBV^{p(\cdot)}([a,b])} \leqslant C^{\nu}\|f'\|_{L^{p(\cdot)}([a,b])},$$

which ends the proof. □

Theorem 8.17 *Let* $p \in C([a,b])$ *and* $f \in RBV^{p(\cdot)}([a,b])$. *Then* $f' \in L^{p(\cdot)}([a,b])$ *and*

$$V^{p(\cdot)}_{[a,b]}(f) \geqslant \int_a^b |f'(t)|^{p(t)}\, dt. \tag{8.13}$$

Moreover,

$$\|f'\|_{L^{p(\cdot)}([a,b])} \lesssim \|f\|_{RBV^{p(\cdot)}([a,b])}. \tag{8.14}$$

Proof By Theorem 8.14 we have that $RBV^{p(\cdot)}([a,b]) \subseteq AC([a,b])$, from which exists f' almost everywhere on the interval $[a,b]$. Consider a tagged partition $\Pi_n^* = \{a = t_{0,n} < t_{1,n} < \cdots < t_{n,n} = b\}$ of $[a,b]$ where $t_{i,n} \leqslant x_{i,n} \leqslant t_{i+1,n}$ and

$$t_{0,n} = a, \quad t_{1,n} = a + \frac{1}{n}(b-1), \cdots, \quad t_{n-1,n} = a + \frac{n-1}{n}(b-a), \quad t_{n,n} = b.$$

We define the function F_n by

$$F_n(t) = \begin{cases} \left|\dfrac{f(t_{i+1},n) - f(t_{i,n})}{t_{i+1,n} - t_{i,n}}\right|, & t_{i,n} \leqslant t < t_{i+1,n}, \\ 0, & \text{for } t = b, \end{cases}$$

and observe that $\lim_{n\to\infty} F_n(t) = |f'(t)|$ almost everywhere. We set $p_n(t) = p\left(t_{i,n}\right)$, for $t_{i,n} \leqslant t < t_{i+1,n}$. Since p is continuous on $[a,b]$, we observe that $\lim_{n\to\infty} p_n(t) = p(t)$. Therefore, by Fatou's lemma, we have

$$\begin{aligned}\int_a^b |f'(t)|^{p(t)} dt &\leqslant \liminf_n \int_a^b |F_n(t)|^{p_n(t)} dt \\ &= \liminf_n \sum_{i=1}^n \frac{|f(t_{i+1,n}) - f(t_{i,n})|^{p(x_i)}}{(t_{i+1,n} - t_{i,n})^{p(x_i)-1}}\end{aligned}$$

$$\leqslant \sup_{n\geqslant 1}\sum_{i=1}^{n}\frac{|f(t_{i+1,n})-f(t_{i,n})|^{p(x_i)}}{(t_{i+1,n}-t_{i,n})^{p(x_i)-1}}$$

$$\leqslant V_{[a,b]}^{p(\cdot)}(f).$$

Inequality (8.14) is a consequence of (8.13) and the definition of norm. □

8.3 Linear Functionals on Variable Exponent Bochner–Lebesgue Spaces

In this section we introduce variable exponent Riesz bounded variation spaces on vector-valued measures and study the linear functionals of variable Bochner–Lebesgue spaces $L^{p(\cdot)}([a,b],B)$ without imposing the Radon–Nikodym property on the Banach space B.

8.3.1 *Preliminaries*

Vector Measures and Bochner Integral

We now recall some basic concepts of vector measures and Bochner integration. We also introduce the variable Bochner–Lebesgue spaces and state some known lemmas for further usage. By doing so, we will mainly follow the approach of Diestel and Faires [75], Dinculeanu [76]. We shall consider vector measures defined in some σ-algebra Σ.

Let (Ω,Σ,μ) be a measure space and B a Banach space. A set function $G:\Sigma\to B$ is said to be a *vector measure* if $G(E_1\sqcup E_2)=G(E_1)+G(E_2)$, where the symbol $\sqcup$ denotes pairwise disjoint union. If, in addition, $G(\bigsqcup_{n=1}^{\infty}E_n)=\sum_{n=1}^{\infty}G(E_n)$, in the norm topology of B, then G is called a countably additive vector measure. The *variation* of a vector measure G on a set $A\in\Sigma$ is the extended real number defined by

$$\|G\|(A):=\sup_{J}\sum_{j\in J}\|G(E_j)\|_B, \tag{8.15}$$

where the supremum is taken over all the finite families $\{E_j\}_{j\in J}$ of mutually disjoint sets of Σ contained in A. If $\|G\|(\Omega)<\infty$, then G will be called a vector measure of *bounded variation.*

From now on, we will assume that $G:\Sigma\to B^*$, where B^* denotes the topological dual space of B. Given $x_1,\cdots,x_N$ elements of B, assume that $\{E_j\}$ is a finite sequence of mutually disjoint sets in Σ. Then, for a simple function

$s = \sum_{j=1}^{N} x_j \chi_{E_j}$, we define

$$\int_{\Omega} s dG = \sum_{j=1}^{N} \langle G(E_j), x_j \rangle.$$

Based on the previous definition, we proceed to define the integral of an arbitrary function with respect to the vector measure G. For further details and an in-depth analysis of integrability concerning vector measures, consult Dinculeanu [76].

Definition 8.18 (Bartle–Dunford–Schwartz Integral) A function $f : \Omega \to B$ is said to be integrable with respect to G if there exists a Cauchy sequence $\{s_n\}_{n\in\mathbb{N}}$ of simple functions converging μ-almost everywhere in B-norm to f. Furthermore, the integral of f with respect to G is defined as

$$\int_{\Omega} f dG := \lim_{n\to\infty} \int_{\Omega} s_n dG.$$

Now, we define the Bochner integral of a simple function s, with respect to a measure μ by

$$\int_{\Omega} s d\mu = \sum_{j=1}^{N} x_j \mu(E_j).$$

We say that a function $f : \Omega \to B$ is μ-measurable if there exists a sequence $\{s_n\}_{n\in\mathbb{N}}$ of simple functions converging μ-almost everywhere in B-norm to f.

Definition 8.19 (Diestel and Faires [75]) A μ-measurable function $f : \Omega \to B$ is said to be *Bochner integrable* if there exists a sequence $\{s_n\}_{n\in\mathbb{N}}$ of simple functions such that

$$\lim_{n\to\infty} \int_{\Omega} \|s_n - f\|_B d\mu = 0.$$

In this case, the integral $\int_E f d\mu$ is defined for each $E \in \Sigma$ by

$$\int_E f d\mu := \lim_{n\to\infty} \int_E s_n d\mu,$$

where $\int_E s_n d\mu := \sum_{j=1}^{N} x_j \mu(E_j \cap E)$ when $s_n = \sum_{j=1}^{N} x_j \chi_{E_j}$.

Variable Exponent Bochner–Lebesgue Spaces

Let I be a closed interval of $\mathbb{R}$ and E be a Lebesgue measurable set. The Lebesgue measure of E will be denoted by $|E|$. By an exponent function we mean a measurable function $p : I \to [1, \infty)$ such that $p_I^+ < \infty$; where $p_I^+ := \operatorname{ess\,sup}_{x\in I} p(x)$ and $p_I^- := \operatorname{ess\,inf}_{x\in I} p(x)$. For the sake of simplicity we write $p^+ = p_I^+$ and $p^- = p_I^-$.

Now, we introduce the Bochner Lebesgue spaces with variable exponent (the theory of variable exponent spaces in the framework of the Euclidean space can be found in Cruz-Uribe and Fiorenza [58], Diening et al. [74]).

Definition 8.20 For a B-valued measurable function $f : I \to B$, we define the *modular* $\varrho_{p(\cdot)}$ by

$$\varrho_{p(\cdot)}(f) := \int_I \|f(x)\|_B^{p(x)} dx,$$

and the *Luxemburg–Nakano norm* by

$$\|f\|_{L^{p(\cdot)}(I,B)} = \|f\|_{L^{p(\cdot)}(B)} := \inf\left\{\lambda > 0 : \varrho_{p(\cdot)}\left(\frac{f}{\lambda}\right) \leqslant 1\right\}. \tag{8.16}$$

The *variable Bochner–Lebesgue space* $L^{p(\cdot)}(I, B)$ is the set of all B-valued measurable functions $f : I \to B$ for which the modular is finite, i.e., $\varrho_{p(\cdot)}(f) < \infty$. Equipped with the Luxemburg–Nakano norm (8.2) this is a Banach space.

By $p \in \mathcal{B}(I)$ we denote the set of exponent functions such that the Hardy–Littlewood maximal operator is bounded on $L^{p(\cdot)}(I, \mathbb{R})$. By p_E we denote the *harmonic mean* of p, given by

$$\frac{1}{p_E} = \frac{1}{|E|}\int_E \frac{1}{p(x)} dx, \quad 0 < |E| < \infty. \tag{8.17}$$

If p and q are conjugate exponent functions (i.e., if $1/p(x) + 1/q(x) = 1$), then $1/p_E + 1/q_E = 1$.

The Lemma 8.21 is well-known in the Euclidean case; see Cruz-Uribe and Fiorenza [58], Diening et al. [74]. The proof for the variable Bochner–Lebesgue space is essentially the same as in the Euclidean case, mutatis mutandis.

Lemma 8.21 *Let p, q be exponent functions such that $p(x) \leqslant q(x)$ almost everywhere. Then*

$$L^{q(\cdot)}(I, B) \hookrightarrow L^{p(\cdot)}(I, B). \tag{8.18}$$

A Riesz type result is also valid in this framework, namely:

Lemma 8.22 (Cheng and Xu [51]) *Let $\{f_n\}_{n\in\mathbb{N}}$ be a Cauchy sequence in $L^{p(\cdot)}(I, B)$. Then, $\{f_n\}_{n\in\mathbb{N}}$ has a subsequence that converges in B-norm almost everywhere.*

The following lemma is a reformulation of the well-known fact that the set of simple functions is dense in the variable exponent Lebesgue spaces $L^{p(\cdot)}(I, \mathbb{R})$ when $p_+ < \infty$ (see Cruz-Uribe and Fiorenza [58]). The proof in our framework does not involve any significant change.

Lemma 8.23 *Let p be an exponent function. Then the set of simple functions is dense in $L^{p(\cdot)}(I, B)$.*

We now give the definition of variable sequence Lebesgue spaces (see Edmunds and Nekvinda [90], Nekvinda [264]) with a notation adapted to our purposes.

Definition 8.24 Let $\mathcal{Q} = \{Q\}$ be a countable disjoint sequence of intervals and $\{x_Q\}_{Q\in\mathcal{Q}}$ be a sequence of real numbers. Given an exponent sequence $\{p_Q\}_{Q\in\mathcal{Q}}$, we define the variable exponent Lebesgue sequence space $\ell_{\mathcal{Q}}^{p(\cdot)}$ by

$$\ell_{\mathcal{Q}}^{p(\cdot)} := \left\{ \{x_Q\}_{Q\in\mathcal{Q}} \, : \sigma_{p(\cdot)}\left(\{x_Q\}_{Q\in\mathcal{Q}}\right) := \sum_{Q\in\mathcal{Q}} |x_Q|^{p_Q} < \infty \right\},$$

endowed with the Luxemburg–Nakano norm, viz.

$$\left\|\{x_Q\}_{Q\in\mathcal{Q}}\right\|_{\ell_{\mathcal{Q}}^{p(\cdot)}} := \inf\left\{\lambda > 0 : \ \sigma_{p(\cdot)}\left(\left\{\frac{x_Q}{\lambda}\right\}_{Q\in\mathcal{Q}}\right) \leqslant 1\right\}.$$

For a fixed disjoint sequence of intervals $\mathcal{Q} = \{Q\}$ the variable exponent Lebesgue sequence spaces satisfy the following Hölder inequality (cf. Cruz-Uribe and Fiorenza [58])

$$\sum_{Q\in\mathcal{Q}} |x_Q y_Q| \leqslant 2\left\|\{x_Q\}_{Q\in\mathcal{Q}}\right\|_{\ell_{\mathcal{Q}}^{p(\cdot)}} \left\|\{y_Q\}_{Q\in\mathcal{Q}}\right\|_{\ell_{\mathcal{Q}}^{q(\cdot)}}, \tag{8.19}$$

where p and q are conjugate exponents.

The next result can be found in Kopaliani [196].

Lemma 8.25 *Let $p \in \mathcal{B}(\mathbb{R})$. Then*

$$\left\|\sum_{Q\in\mathcal{Q}} x_Q \chi_Q\right\|_{L^{p(\cdot)}(\mathbb{R})} \approx \left\|\left\{x_Q \left\|\chi_Q\right\|_{p(\cdot)}\right\}_{Q\in\mathcal{Q}}\right\|_{\ell_{\mathcal{Q}}^{p(\cdot)}}, \tag{8.20}$$

for all sequences $\mathcal{Q} = \{Q\}$ of disjoint cubes and $x_Q \in \mathbb{R}$.

We will need the following lemma (see Diening et al. [74, Theorem 4.5.7] for a proof).

Lemma 8.26 *Let $p \in \mathcal{B}(\mathbb{R})$. Then $\left\|\chi_Q\right\|_{L^{p(\cdot)}(\mathbb{R})} \approx |Q|^{\frac{1}{p_Q}}$.*

We end this section with a result that relates the modular with the norm through some inequalities (see Cruz-Uribe and Fiorenza [58]).

Lemma 8.27 *Let p be an exponent function. Then, the following estimates hold:*

$$\left\|\{x_Q\}_{Q\in\mathcal{Q}}\right\|^{p_+}_{\ell^{p(\cdot)}_{\mathcal{Q}}} \leqslant \sigma_{p(\cdot)}\left(\{x_Q\}_{Q\in\mathcal{Q}}\right) \leqslant \left\|\{x_Q\}_{Q\in\mathcal{Q}}\right\|^{p_-}_{\ell^{p(\cdot)}_{\mathcal{Q}}}, \quad \left\|\{x_Q\}_{Q\in\mathcal{Q}}\right\|_{\ell^{p(\cdot)}_{\mathcal{Q}}} \leqslant 1,$$

$$\left\|\{x_Q\}_{Q\in\mathcal{Q}}\right\|^{p_-}_{\ell^{p(\cdot)}_{\mathcal{Q}}} \leqslant \sigma_{p(\cdot)}\left(\{x_Q\}_{Q\in\mathcal{Q}}\right) \leqslant \left\|\{x_Q\}_{Q\in\mathcal{Q}}\right\|^{p_+}_{\ell^{p(\cdot)}_{\mathcal{Q}}}, \quad \left\|\{x_Q\}_{Q\in\mathcal{Q}}\right\|_{\ell^{p(\cdot)}_{\mathcal{Q}}} \geqslant 1.$$

8.3.2 Variable Exponent Bounded Variation in the Riesz Sense

We define the $p(\cdot)$-Riesz bounded variation over the set of vector measures and also introduce an auxiliary sequence function space to define an auxiliary norm in the set of vector measures of $p(\cdot)$-Riesz variation. From now on, we shall deal with vector measures $G : \mathbb{B}_{[a,b]} \to B^*$, where $\mathbb{B}_{[a,b]}$ is the Borel σ-algebra restricted to the interval $[a, b]$.

Definition 8.28 Let $\mathcal{Q} = \{Q\}$ be a finite partition of $[a, b]$ by any kind of disjoint non-degenerate intervals and consider p an exponent function defined on $[a, b]$. We define the *$p(\cdot)$-Riesz variation of G on* $[a, b]$ by

$$V^{p(\cdot)}_{[a,b]}(G, B^*) = \sup_{\mathcal{Q}} \sum_{Q\in\mathcal{Q}} \frac{\|G(Q)\|^{p_Q}_{B^*}}{|Q|^{p_Q-1}},$$

and the space of vector measures of bounded $p(\cdot)$-variation in the Riesz sense is introduced as

$$\mathrm{RBV}_{p(\cdot)}\left([a, b], B^*\right) := \left\{G : V^{p(\cdot)}_{[a,b]}(G) < \infty\right\}.$$

The Lemma 8.29 can be proved in a similar fashion as Lemma 8.11.

Lemma 8.29 *The functional $V^{p(\cdot)}_{[a,b]}$ is a convex pseudomodular.*

Moreover, $V^{p(\cdot)}_{[a,b]}$ is a continuous pseudomodular. The left continuity follows from

$$\lambda^{p_+} V^{p(\cdot)}_{[a,b]}\left(G, B^*\right) \leqslant V^{p(\cdot)}_{[a,b]}\left(\lambda G, B^*\right) \leqslant \lambda^{p_-} V^{p(\cdot)}_{[a,b]}\left(G, B^*\right), \quad \lambda < 1,$$

whereas the right continuity can be seen from

$$\lambda^{p_-} V^{p(\cdot)}_{[a,b]}\left(G, B^*\right) \leqslant V^{p(\cdot)}_{[a,b]}\left(\lambda G, B^*\right) \leqslant \lambda^{p_+} V^{p(\cdot)}_{[a,b]}\left(G, B^*\right), \quad \lambda \geqslant 1.$$

By virtue of Lemma 8.29 we are able to introduce a norm on the quotient space

$$RBV^{p(\cdot)}([a,b], B^*) := \mathrm{RBV}_{p(\cdot)}([a,b], B^*)/\mathcal{C},$$

where $\mathcal{C}$ is the space of all constant vector measures, viz.

$$\|G\|_{RBV^{p(\cdot)}([a,b],B^*)} = \inf\left\{\lambda > 0 : V^{p(\cdot)}_{[a,b]}\left(\frac{G}{\lambda}\right) \leqslant 1\right\}. \tag{8.21}$$

For other properties of $RBV^{p(\cdot)}([a,b], B^*)$ in the case $B^* = \mathbb{R}$ see Castillo et al. [48], Kakochashvili and Zviadadze [160].

We now introduce the auxiliary sequence space $\lambda^{p(\cdot)}_{\mathcal{Q}}(B)$.

Definition 8.30 Let $\mathcal{Q} = \{Q\}$ be a sequence of disjoint intervals in $\mathbb{B}_{[a,b]}$ and $\{x_Q\}_{Q\in\mathcal{Q}}$ be a sequence in B. We define the space $\lambda^{p(\cdot)}_{\mathcal{Q}}(B)$ by

$$\lambda^{p(\cdot)}_{\mathcal{Q}}(B) = \left\{\{x_Q\}_{Q\in\mathcal{Q}} : \varsigma_{p(\cdot)}\left(\{x_Q\}_{Q\in\mathcal{Q}}\right) := \sum_{Q\in\mathcal{Q}} |Q| \|x_Q\|_B^{p_Q} < \infty\right\}.$$

Proposition 8.31 *The functional $\varsigma_{p(\cdot)}$ is a convex pseudomodular.*

Proof Clearly $\varsigma_{p(\cdot)}(\{0\}) = 0$. Given $\alpha \in \mathbb{R}$, such that $|\alpha| = 1$, it is also a straightforward calculation to show that $\varsigma_{p(\cdot)}\left(\{\alpha x_Q\}_{Q\in\mathcal{Q}}\right) = \varsigma_{p(\cdot)}\left(\{x_Q\}_{Q\in\mathcal{Q}}\right)$. Finally, given $0 < \beta \leqslant 1$, assume that $\{y_Q\}_{Q\in\mathcal{Q}}, \{x_Q\}_{Q\in\mathcal{Q}} \in \lambda^{p(\cdot)}_{\mathcal{Q}}(B)$. Therefore, the convexity of the function $t \mapsto t^p$ for $1 \leqslant p < \infty$, $t \geqslant 0$, and the homogeneity of the norm $\|\cdot\|_B$ yields

$$\begin{aligned}
\varsigma_{p(\cdot)}\left(\{\beta x_Q + (1-\beta) y_Q\}_{Q\in\mathcal{Q}}\right) &= \sum_{Q\in\mathcal{Q}} |Q| \|\beta x_Q + (1-\beta) y_Q\|_B^{p_Q} \\
&\leqslant \beta \sum_{Q\in\mathcal{Q}} |Q| \|x_Q\|_B^{p_Q} + (1-\beta) \sum_{Q\in\mathcal{Q}} |Q| \|y_Q\|_B^{p_Q} \\
&= \beta \varsigma_{p(\cdot)}\left(\{x_Q\}_{Q\in\mathcal{Q}}\right) + (1-\beta) \varsigma_{p(\cdot)}\left(\{y_Q\}_{Q\in\mathcal{Q}}\right),
\end{aligned}$$

which ends the proof. □

In virtue of the preceding lemma, we can introduce the norm

$$\left\|\{x_Q\}_{Q\in\mathcal{Q}}\right\|_{\lambda^{p(\cdot)}_{\mathcal{Q}}(B)} = \inf\left\{\lambda > 0 : \varsigma_{p(\cdot)}\left(\left\{\frac{x_Q}{\lambda}\right\}_{Q\in\mathcal{Q}}\right) \leqslant 1\right\}.$$

Theorem 8.32 *Let $\{x_Q\}_{Q\in\mathcal{Q}}$ be a sequence in $\lambda_{\mathcal{Q}}^{p(\cdot)}(B)$ and $\{y_Q\}_{Q\in\mathcal{Q}}$ be a sequence in $\lambda_{\mathcal{Q}}^{q(\cdot)}(B)$. Then*

$$\sum_{Q\in\mathcal{Q}} |Q| \left\| x_Q \right\|_B \left\| y_Q \right\|_B \lesssim \left\| \{x_Q\}_{Q\in\mathcal{Q}} \right\|_{\lambda_{\mathcal{Q}}^{p(\cdot)}(B)} \left\| \{y_Q\}_{Q\in\mathcal{Q}} \right\|_{\lambda_{\mathcal{Q}}^{q(\cdot)}(B)}. \tag{8.22}$$

Proof We assume, without loss of generality, that

$$\left\| \{x_Q\}_{Q\in\mathcal{Q}} \right\|_{\lambda_{\mathcal{Q}}^{p(\cdot)}(B)} = \left\| \{y_Q\}_{Q\in\mathcal{Q}} \right\|_{\lambda_{\mathcal{Q}}^{q(\cdot)}(B)} = 1.$$

By Young's inequality and Theorem 8.1, we have

$$\begin{aligned}
\sum_{Q\in\mathcal{Q}} |Q| \left\| x_Q \right\|_B \left\| y_Q \right\|_B &= \sum_{Q\in\mathcal{Q}} |Q|^{\frac{1}{p_Q}} \left\| x_Q \right\|_B |Q|^{\frac{1}{q_Q}} \left\| y_Q \right\|_B \\
&\lesssim \left(\frac{1}{p}\right)^+ \sum_{Q\in\mathcal{Q}} |Q| \left\| x_Q \right\|_B^{p_Q} + \left(\frac{1}{q}\right)^+ \sum_{Q\in\mathcal{Q}} |Q| \left\| y_Q \right\|_B^{q_Q} \\
&= \left(\frac{1}{p}\right)^+ \varsigma_{p(\cdot)}\left(\left\{x_Q\right\}_{Q\in\mathcal{Q}}\right) + \left(\frac{1}{q}\right)^+ \varsigma_{q(\cdot)}\left(\left\{y_Q\right\}_{Q\in\mathcal{Q}}\right) \lesssim 1,
\end{aligned}$$

and by the homogeneity of the norm $\|\cdot\|_B$, we conclude (8.22). □

We now introduce a functional in the space $RBV^{p(\cdot)}([a,b],B^*)$.

Definition 8.33 Let $G \in RBV^{p(\cdot)}([a,b],B^*)$. For a vector measure, we define the functional $\|\!|\cdot|\!\|_{RBV^{q(\cdot)}}$ by

$$\|\!| G |\!\|_{RBV^{q(\cdot)}([a,b],B^*)} = \sup_{\substack{\mathcal{Q}=\{Q\} \text{ finite partition of } [a,b] \\ \left\|\{x_Q\}_{Q\in\mathcal{Q}}\right\|_{\lambda_{\mathcal{Q}}^{p(\cdot)}(B)} \leqslant 1}} \left| \sum_{Q\in\mathcal{Q}} \langle G(Q), x_Q \rangle \right|. \tag{8.23}$$

It is straightforward to check that the functional $\|\!|\cdot|\!\|_{RBV^{q(\cdot)}}$ satisfies the following properties.

Lemma 8.34 *Given G, H vector measures, suppose $\alpha \in \mathbb{R}$. Then*

(a) $\|\!| G |\!\|_{RBV^{q(\cdot)}} \geqslant 0$ *and* $\|\!| G |\!\|_{RBV^{q(\cdot)}} = 0$ *if and only if* $G(Q) = 0$ *for all intervals* $Q \in \mathbb{B}_{[a,b]}$;
(b) $\|\!| G + H |\!\|_{RBV^{q(\cdot)}} \leqslant \|\!| G |\!\|_{RBV^{q(\cdot)}} + \|\!| H |\!\|_{RBV^{q(\cdot)}}$;
(c) $\|\!| \alpha G |\!\|_{RBV^{q(\cdot)}} = |\alpha| \, \|\!| G |\!\|_{RBV^{q(\cdot)}}$.

The next result follows immediately from Definition 8.33.

Lemma 8.35 *Let $\mathcal{Q} = \{Q\}$ be a finite disjoint sequence of intervals in $\mathbb{B}_{[a,b]}$ and $\{x_Q\}_{Q\in\mathcal{Q}}$ be a finite sequence in B. Then*

$$\left|\sum_{Q\in\mathcal{Q}} \langle G(Q), x_Q\rangle\right| \leqslant ||| G |||_{RBV^{q(\cdot)}} \left\|\{x_Q\}_{Q\in\mathcal{Q}}\right\|_{\lambda^{p(\cdot)}_{\mathcal{Q}}(B)}. \tag{8.24}$$

8.3.3 Main Results

We start by proving that the Luxemburg–Nakano norm (8.21) and the norm defined in (8.23) are equivalent.

Theorem 8.36 *Let G be a vector measure in $RBV^{q(\cdot)}([a,b], B^*)$. Then*

$$||| G |||_{RBV^{q(\cdot)}} \approx \|G\|_{RBV^{q(\cdot)}}. \tag{8.25}$$

Proof Let $\mathcal{Q} = \{Q\}$ be a finite partition of $[a,b]$ and $\{x_Q\}_{Q\in\mathcal{Q}}$ be a finite sequence of elements in B such that $\left\|\{x_Q\}_{Q\in\mathcal{Q}}\right\|_{\lambda^{p(\cdot)}_{\mathcal{Q}}(B)} \leqslant 1$. We may assume, without loss of generality, that $\|G\|_{RBV^{q(\cdot)}} = 1$. By Theorem 8.32 and Proposition 8.1, we have

$$\begin{aligned}
\sum_{Q\in\mathcal{Q}} |\langle G(Q), x_Q\rangle| &\leqslant \sum_{Q\in\mathcal{Q}} \|G(Q)\|_{B^*} \left\|x_Q\right\|_B \\
&= \sum_{Q\in\mathcal{Q}} \frac{\|G(Q)\|_{B^*}}{|Q|} \left\|x_Q\right\|_B |Q| \\
&\leqslant \left\|\{x_Q\}_{Q\in\mathcal{Q}}\right\|_{\lambda^{p(\cdot)}_{\mathcal{Q}}(\mathbb{R})} \left\| \left\{\frac{\|G(Q)\|_{B^*}}{|Q|}\right\}_{Q\in\mathcal{Q}} \right\|_{\lambda^{q(\cdot)}_{\mathcal{Q}}(\mathbb{R})} \\
&\lesssim \left(\sum_{Q\in\mathcal{Q}} \frac{\|G(Q)\|^{q_Q}_{B^*}}{|Q|^{q_Q-1}} + 1\right) \\
&\lesssim \left(V^{q(\cdot)}_{[a,b]}(G) + 1\right).
\end{aligned}$$

Since $||| \cdot |||_{RBV^{q(\cdot)}}$ is homogeneous, we deduce that

$$||| G |||_{RBV^{q(\cdot)}} \lesssim \|G\|_{RBV^{q(\cdot)}}. \tag{8.26}$$

On the other hand, let us assume, without loss of generality, that $V_{[a,b]}^{q(\cdot)}(G) = 1$. Given $\varepsilon > 0$, by the homogeneity of the linear functional $\langle G(Q), \cdot\rangle$, we may choose $x_Q \in B$ such that

$$\left\|x_Q\right\|_B = \left(\frac{\|G(Q)\|_{B^*}}{|Q|}\right)^{q_Q-1},$$

and

$$\|G(Q)\|_{B^*}\left\|x_Q\right\|_B - \frac{\varepsilon}{\#\mathcal{Q}} \leqslant |\langle G(Q), x_Q\rangle|,$$

where $\#\mathcal{Q}$ is the cardinality of $\mathcal{Q}$. Let us write $\iota_Q = \operatorname{sgn}\langle G(Q), x_Q\rangle$. Then, by Lemma 8.35 and Proposition 8.1, we estimate as follows:

$$\begin{aligned}
\sum_{Q\in\mathcal{Q}} \frac{\|G(Q)\|_{B^*}^{q_Q}}{|Q|^{q_Q-1}} - \varepsilon &= \sum_{Q\in\mathcal{Q}} \|G(Q)\|_{B^*}\left(\frac{\|G(Q)\|_{B^*}}{|Q|}\right)^{q_Q-1} - \varepsilon \\
&\leqslant \sum_{Q\in\mathcal{Q}} |\langle G(Q), x_Q\rangle| \\
&= \left|\sum_{Q\in\mathcal{Q}} \langle G(Q), \iota_Q x_Q\rangle\right| \\
&\leqslant \interleave G \interleave_{RBV^{q(\cdot)}} \left\|\{\iota_Q x_Q\}_{Q\in\mathcal{Q}}\right\|_{\lambda_{\mathcal{Q}}^{p(\cdot)}(B)} \\
&= \interleave G \interleave_{RBV^{q(\cdot)}} \left(\sum_{Q\in\mathcal{Q}} |Q|\left\|x_Q\right\|_B^{p_Q} + 1\right) \\
&= \interleave G \interleave_{RBV^{q(\cdot)}} \left(\sum_{Q\in\mathcal{Q}} \frac{\|G(Q)\|_{B^*}^{q_Q}}{|Q|^{q_Q-1}} + 1\right) \\
&\leqslant \interleave G \interleave_{RBV^{q(\cdot)}} \left(V_{[a,b]}^{q(\cdot)}(G) + 1\right).
\end{aligned}$$

Since $\varepsilon > 0$ is arbitrary, we have

$$V_{[a,b]}^{q(\cdot)}(G) \leqslant \interleave G \interleave_{RBV^{q(\cdot)}} \left(V_{[a,b]}^{q(\cdot)}(G) + 1\right),$$

thus

$$\frac{1}{2} \leqslant \interleave G \interleave_{RBV^{q(\cdot)}},$$

and, by the homogeneity of $\interleave \cdot \interleave_{RBV^{q(\cdot)}}$, we get $\|G\|_{RBV^{q(\cdot)}} \lesssim \interleave G \interleave_{RBV^{q(\cdot)}}$. □

We now characterize the linear functionals in variable exponent Bochner–Lebesgue spaces.

Theorem 8.37 *Let $p, q \in \mathcal{B}([a,b])$ be conjugate exponent functions. Then, the mapping $G \to \ell_G$, from the space $RBV^{q(\cdot)}([a,b], B^*)$ to $(L^{p(\cdot)}([a,b], B))^*$, defined by*

$$\ell_G(f) = \int_a^b f dG, \ f \in L^{p(\cdot)}([a,b], B) \tag{8.27}$$

is a linear isomorphism, and

$$\|\ell_G\|_{(L^{p(\cdot)}([a,b],B))^*} \approx \|G\|_{RBV^{q(\cdot)}([a,b],B^*)}. \tag{8.28}$$

Proof Let $\mathcal{Q}$ be a finite partition of $[a,b]$. First, we will assume without loss of generality that $\|G\|_{RBV^{q(\cdot)}} = 1$ to deduce one of the inequalities in (8.28) using the homogeneity of the norm $\|\cdot\|_{RBV^{q(\cdot)}}$. Consider the simple function

$$s = \sum_{Q \in \mathcal{Q}} x_Q \chi_Q, \quad x_Q \in B. \tag{8.29}$$

Then, by Lemma 8.26 and Hölder's inequality (8.19), we have

$$\begin{aligned}
|\ell_G(s)| &= \left| \int_a^b s dG \right| = \left| \sum_{Q \in \mathcal{Q}} \langle G(Q), x_Q \rangle \right| \leqslant \sum_{Q \in \mathcal{Q}} |\langle G(Q), x_Q \rangle| \\
&\leqslant \sum_{Q \in \mathcal{Q}} \|G(Q)\|_{B^*} \left\|x_Q\right\|_B = \sum_{Q \in \mathcal{Q}} \|G(Q)\|_{B^*} |Q|^{\frac{1}{q_Q}-1} \left\|x_Q\right\|_B |Q|^{\frac{1}{p_Q}} \\
&\lesssim \sum_{Q \in \mathcal{Q}} \|G(Q)\|_{B^*} |Q|^{\frac{1}{q_Q}-1} \left\|x_Q\right\|_B \left\|\chi_Q\right\|_{p(\cdot)} \\
&\lesssim \left\| \left\{ \|G(Q)\|_{B^*} |Q|^{\frac{1}{q_Q}-1} \right\}_{Q \in \mathcal{Q}} \right\|_{\ell_{\mathcal{Q}}^{q(\cdot)}} \cdot \left\| \left\{ \left\|x_Q\right\|_B \left\|\chi_Q\right\|_{p(\cdot)} \right\}_{Q \in \mathcal{Q}} \right\|_{\ell_{\mathcal{Q}}^{p(\cdot)}}.
\end{aligned} \tag{8.30}$$

Hence, Proposition 8.1.(iv) yields

$$\begin{aligned}
\left\| \left\{ \|G(Q)\|_{B^*} |Q|^{\frac{1}{q_Q}-1} \right\}_{Q \in \mathcal{Q}} \right\|_{\ell_{\mathcal{Q}}^{q(\cdot)}} &\leqslant \sigma_{q(\cdot)} \left(\left\{ \|G(Q)\|_{B^*} |Q|^{\frac{1}{q_Q}-1} \right\}_{Q \in \mathcal{Q}} \right) + 1 \\
&\leqslant V_{[a,b]}^{q(\cdot)}(G) + 1.
\end{aligned} \tag{8.31}$$

On the other hand, using Lemma 8.25, we get that

$$\begin{aligned}\left\|\left\{\|x_Q\|_B\|\chi_Q\|_{p(\cdot)}\right\}_{Q\in\mathcal{Q}}\right\|_{\ell^{p(\cdot)}_{\mathcal{Q}}} &\lesssim \left\|\sum_{Q\in\mathcal{Q}}\|x_Q\|_B\,\chi_Q\right\|_{L^{p(\cdot)}(\mathbb{R})}\\ &= \left\|\sum_{Q\in\mathcal{Q}} x_Q\chi_Q\right\|_{L^{p(\cdot)}(B)}\\ &= \|s\|_{L^{p(\cdot)}(B)}.\end{aligned} \tag{8.32}$$

Taking (8.30)–(8.32) into account, we get

$$|\ell_G(s)| \lesssim \|s\|_{L^{p(\cdot)}(B)}, \tag{8.33}$$

whenever s is a simple function and $\|G\|_{RBV^{q(\cdot)}} = 1$.

By Lemmas 8.22 and 8.23, for any function $f \in L^{p(\cdot)}([a,b],B)$, there exists a sequence $\{s_n\}$ of simple functions converging to f almost everywhere in B-norm. Thus, by Definition 8.18, we have

$$|\ell_G(f)| \lesssim \|f\|_{L^{p(\cdot)}(B)},$$

for all $f \in L^{p(\cdot)}(B)$ and $\|G\|_{RBV^{q(\cdot)}} = 1$, hence

$$\|\ell_G\|_{(L^{p(\cdot)}(B))^*} \lesssim 1.$$

Since $\ell_{\alpha G}(\cdot) = \alpha\ell_G(\cdot)$, for all $\alpha \in \mathbb{R}$, the homogeneity of $\|\cdot\|_{(L^{p(\cdot)}(B))^*}$ implies that

$$\|\ell_G\|_{(L^{p(\cdot)}(B))^*} \lesssim \|G\|_{RBV^{q(\cdot)}(B^*)},$$

for any $G \in RBV^{p(\cdot)}$. Thus ℓ_G is well defined on $L^{p(\cdot)}(B)$ and

$$RBV^{q(\cdot)}\left([a,b],B^*\right) \hookrightarrow \left(L^{p(\cdot)}([a,b],B)\right)^*.$$

Now, set $\ell \in (L^{p(\cdot)}([a,b],B))^*$ and define $G : \mathbb{B}_{[a,b]} \to B^*$ by

$$\langle G(P), x\rangle = \ell(x\chi_P), \qquad x \in B, \quad P \in \mathbb{B}_{[a,b]}.$$

Let $\mathcal{Q} = \{P\}$ be a finite disjoint collection of sets in $\mathbb{B}_{[a,b]}$. Thus

$$\left\langle G\left(\bigsqcup_{P\in\mathcal{Q}} P\right), x\right\rangle = \ell\left(x\chi_{(\sqcup_{P\in\mathcal{Q}}P)}\right) = \ell\left(x\sum_{P\in\mathcal{Q}}\chi_P\right)$$
$$= \sum_{P\in\mathcal{Q}} \ell(x\chi_P) = \sum_{P\in\mathcal{Q}} \langle G(P), x\rangle.$$

Therefore, G is finitely additive. Moreover, we get that

$$|\ell(x\chi_P)| \lesssim \|\ell\|\|x\|_B\|\chi_P\|_{L^{p(\cdot)}} < \infty.$$

It follows that G has its values on B^*. Without loss of generality, we assume that $\|\ell\| = 1$ and let $\mathcal{Q} = \{Q\}$ be a finite partition of $[a, b]$ by intervals. Due to Lemmas 8.25 and 8.27, it follows that

$$\begin{aligned}
\sum_{Q\in\mathcal{Q}} |\langle G(Q), x_Q\rangle| &= \sum_{Q\in\mathcal{Q}} |\ell\left(x_Q\chi_Q\right)| = \ell\left(\sum_{Q\in\mathcal{Q}} \iota_Q x_Q\chi_Q\right), \\
&\leqslant \|\ell\|\left\|\sum_{Q\in\mathcal{Q}} \iota_Q x_Q\chi_Q\right\|_{L^{p(\cdot)}(B)} = \|\ell\|\left\|\sum_{Q\in\mathcal{Q}} \|x_Q\|_B\chi_Q\right\|_{L^{p(\cdot)}(\mathbb{R})} \\
&\lesssim \|\ell\|\left\|\left\{\|x_Q\|_B\|\chi_Q\|_{p(\cdot)}\right\}_{Q\in\mathcal{Q}}\right\|_{\ell^{p(\cdot)}_{\mathcal{Q}}} \lesssim \|\ell\|\left(\sum_{Q\in\mathcal{Q}} \|x_Q\|_B^{p_Q}\|\chi_Q\|_{p(\cdot)}^{p_Q}\right)^{\alpha} \\
&\lesssim \|\ell\|\left(\sum_{Q\in\mathcal{Q}} \|x_Q\|_B^{p_Q}|Q|\right)^{\alpha}, \quad \alpha \in \{1/p_-, 1/p_+\},
\end{aligned} \tag{8.34}$$

where $\iota_Q = \operatorname{sgn}\ell(x_Q\chi_Q)$. Given $\varepsilon > 0$ and a finite sequence of non-negative numbers $\{b_Q\}_{Q\in\mathcal{Q}}$, we can choose $x_Q \in B$ such that $\|x_Q\|_B = b_Q$ and

$$|\langle G(Q), x_Q\rangle| > \|G(Q)\|_{B^*}\|x_Q\|_B - \frac{\varepsilon}{\#\mathcal{Q}},$$

where $\#\mathcal{Q}$ is the cardinality of $\mathcal{Q}$. Then, summing in both sides, we obtain

$$\sum_{Q\in\mathcal{Q}} |\langle G(Q), x_Q\rangle| > \sum_{Q\in\mathcal{Q}} \|G(Q)\|_{B^*}\|x_Q\|_B - \varepsilon. \tag{8.35}$$

By (8.34) and inequality (8.35), we have

$$\sum_{Q\in\mathcal{Q}} \|G(Q)\|_{B^*} b_Q < \|\ell\| \left(\sum_{Q\in\mathcal{Q}} b_Q^{p_Q} |Q| \right)^{\alpha} + \varepsilon,$$

which implies, since $\varepsilon > 0$ is arbitrary, that

$$\sum_{Q\in\mathcal{Q}} \|G(Q)\|_{B^*} b_Q < \|\ell\| \left(\sum_{Q\in\mathcal{Q}} b_Q^{p_Q} |Q| \right)^{\alpha}. \tag{8.36}$$

Using (8.36) and taking $b_Q = \left(\frac{\|G(Q)\|_{B^*}}{|Q|}\right)^{\frac{q_Q}{p_Q}}$, we obtain

$$\sum_{Q\in\mathcal{Q}} \frac{\|G(Q)\|_{B^*}^{q_Q}}{|Q|^{q_Q-1}} \leqslant \|\ell\| \left(\sum_{Q\in\mathcal{Q}} \frac{\|G(Q)\|_{B^*}^{q_Q}}{|Q|^{q_Q-1}} \right)^{\alpha}.$$

Thus

$$\left(\sum_{Q\in\mathcal{Q}} \frac{\|G(Q)\|_{B^*}^{q_Q}}{|Q|^{q_Q-1}} \right)^{1-\alpha} \leqslant \|\ell\| = 1.$$

Since $p \in \mathcal{B}([a,b])$ implies that $p_- > 1$, we get that $0 < \alpha < 1$. Therefore, by the arbitrariness of $\mathcal{Q}$ we conclude that $V_{[a,b]}^{q(\cdot)}(G) \leqslant 1$. So

$$\|G\|_{RBV^{q(\cdot)}(B^*)} \lesssim \|\ell\|,$$

by the homogeneity of the $RBV^{q(\cdot)}$ norm. Finally, observe that ℓ and ℓ_G coincide on $L^{p(\cdot)}(B)$. Take a simple function s as in (8.29), then

$$\ell(s) = \ell\left(\sum_{Q\in\mathcal{Q}} x_Q \chi_Q \right) = \sum_{Q\in\mathcal{Q}} \ell\left(x_Q \chi_Q\right) = \sum_{Q\in\mathcal{Q}} \langle G(Q), x\rangle = \int_{\Omega} s dG = \ell_G(s);$$

hence, $\ell = \ell_G$ on simple functions. Therefore, by the density of simple functions on $L^{p(\cdot)}(B)$, we conclude $\ell = \ell_G$ on $L^{p(\cdot)}(B)$ and $\|\ell\| = \|\ell_G\|_{(L^{p(\cdot)}(B))^*}$. □

8.4 Weighted Riesz Bounded Variation Spaces and the Nemytskii Operator

We define weighted Riesz bounded variation space $RBV_w^p(I)$ by the condition that

$$\sup \sum_j \left(\frac{|f(x_j) - f(x_{j-1})|}{|x_j - x_{j-1}|} \right)^p w([x_{j-1}, x_j]) < \infty,$$

where the supremum is taken over all partitions of I. We prove a weighted Riesz representation theorem for these spaces when the weight w lies in the Muckenhoupt class.

8.4.1 Preliminaries

By a weight we mean a non-negative, locally integrable real-valued function. We shall denote the weights by w and σ. Given a weight w and any measurable set $E \subset \mathbb{R}^n$, we define

$$w(E) := \int_E w\, dx.$$

If $0 < |E| < \infty$, where $|E|$ is the Lebesgue measure of E, we define the average of w on E by

$$⨍_E w dx := \frac{1}{|E|} \int_E w dx.$$

Let $I \subset \mathbb{R}$ be an interval and w a weight. The space of all measurable real-valued functions f that satisfy

$$\|f\|_{L_w^p} = \left(\int_I |f(x)|^p w dx \right)^{\frac{1}{p}} < \infty,$$

is called weighted Lebesgue space $L_w^p(I)$. We recall the definition of the Muckenhoupt class A_p, introduced by Muckenhoupt [254]; see Duoandikoetxea [81] for further information and references.

Definition 8.38 Given $1 < p, q < \infty$, assume that $1/p + 1/q = 1$. A non-negative locally integrable function w is say to be in the *Muckenhoupt class* A_p if $0 < w(x) < \infty$ a.e. and

$$[w]_{A_p} = \sup_Q \left(\fint_Q w dx \right) \left(\fint_Q w^{1-q} dx \right)^{p-1} < \infty, \tag{8.37}$$

where the supremum is taken over all cubes $Q \subset \mathbb{R}^n$.

Given $w \in A_p$, by $\sigma = w^{1-q}$, we denote the dual weight of w. Observe that $w \in A_p$ implies that $\sigma \in A_q$. Moreover, it follows from the Hölder inequality that

$$1 \leqslant \left(\fint_Q w dx \right) \left(\fint_Q w^{1-q} dx \right)^{p-1} \tag{8.38}$$

for every cube $Q \subset \mathbb{R}^n$.

Let $I = [a, b]$ be a closed interval, and let $\{x_j\}_{j=0}^n$ satisfy

$$a = x_0 < x_1 < x_2 < \cdots < x_n = b.$$

We define the finite partition $\Pi = \{\pi_j\}_{j=1}^n$ to be the collection of closed intervals $\pi_j = [x_{j-1}, x_j]$. For simplicity, from now on we write $\Delta f(\pi_j) = f(x_j) - f(x_{j-1})$.

A function f is in the Lipschitz class $\text{Lip}(I)$ if for any points $x, y \in I$, we have $|f(x) - f(y)| \leqslant C|x - y|$. A function f is absolutely continuous, denoted $f \in AC(I)$, if for every $\epsilon > 0$, there exists $\delta > 0$ such that given any finite collection $\{(a_j, b_j)\}_{j=1}^n$ in I,

$$\sum_{j=1}^n (b_j - a_j) < \delta \quad \text{implies} \quad \sum_{j=1}^n |f(b_j) - f(a_j)| < \epsilon.$$

For $1 < p < \infty$ we have the well-known inclusions

$$\text{Lip}(I) \subset RBV^p(I) \subset AC(I) \subset BV(I).$$

8.4.2 *Weighted Riesz Bounded Variation*

We define weighted Riesz variation and prove that the collection of functions of weighted Riesz bounded variation form a Banach space, and we prove some embedding results relating it to Lipschitz functions and functions of bounded variation.

Definition 8.39 Given $1 < p < \infty$, a weight w and $I = [a, b]$, a function f is said to be in the *weighted Riesz bounded variation space* $RBV_w^p(I)$ if

$$V_w^p(f, I) = \sup_{\Pi} \sum_{\pi_j \in \Pi} \left(\frac{|\Delta f(\pi_j)|}{|\pi_j|} \right)^p w(\pi_j) < \infty,$$

where the supremum is taken over all finite partitions Π of $[a, b]$.

Note that if $w \in A_p$, the above definition can be equivalently stated—up to a constant depending on p—as follows:

$$\begin{aligned} V_w^p(f, I) &= \sup_{\Pi} \sum_{\pi_j \in \Pi} \left(\frac{|\Delta f(\pi_j)|}{|\pi_j|} \right)^p w(\pi_j) \\ &\approx \sup_{\Pi} \sum_{\pi_j \in \Pi} \frac{|\Delta f(\pi_j)|^p}{\sigma(\pi_j)^{p-1}} \end{aligned}$$

where $\sigma = w^{1-q}$. As in the classical case, functions in $RBV_w^p(I)$ are (uniformly) continuous.

Lemma 8.40 *Given $w \in A_p$ and $1 < p < \infty$, then*

$$RBV_w^p(I) \subset C(I).$$

Proof Given a sequence $\{x_k\}_{k \in \mathbb{N}}$ in I converging to $x \in I$, let $[x, x_k]$ denote the closed interval with endpoints x and x_k and σ the dual weight of w. Then for each x_k we have

$$|f(x) - f(x_k)|^p = \frac{|f(x) - f(x_k)|^p}{\sigma([x, x_k])^{p-1}} \sigma([x, x_k])^{p-1} \lesssim V_w^p(f, I) \sigma([x, x_k])^{p-1} \to 0;$$

since this is true for every such sequence, f is a continuous function. □

Though the functional V_w^p is not a norm, if we define

$$\|f\|_{RBV_w^p} = |f(a)| + V_w^p(f, I)^{\frac{1}{p}}, \tag{8.39}$$

then this is a norm on $RBV_w^p(I)$, and with respect to this norm it is a Banach space.

Theorem 8.41 *Let $1 \leqslant p < \infty$ and $w \in A_p$. Then, the space $RBV_w^p(I)$ with norm $\|\cdot\|_{RBV_w^p}$ is a Banach space.*

Proof We will first prove that $RBV_w^p(I)$ is a normed linear space. Suppose $f = 0$ on I; then it is immediate that $\|f\|_{RBV_w^p} = 0$. On the other hand, if $\|f\|_{RBV_w^p} = 0$, then we have that $f(a) = 0$ and for any $x \in I, x > a$,

$$0 = V_w^p(f, I) \geqslant \left(\frac{|f(a) - f(x)|}{|[a, x]|}\right)^p w([x, a]).$$

Therefore, $f(x) = f(a) = 0$ for all $x \in I$.

Next, fix $f, g \in RBV_w^p(I)$. Let Π be a finite partition of I and $\tilde{f}_j = \frac{\Delta f(\pi_j)}{|\pi_j|}$, $\tilde{g}_j = \frac{\Delta g(\pi_j)}{|\pi_j|}$. By the Minkowski inequality in $L_w^p(I)$ and the Minkowski inequality in the sequence space ℓ_p, we have

$$\begin{aligned}
&\sum_{\pi_j \in \Pi} \left(\frac{|\Delta(f+g)(\pi_j)|}{|\pi_j|}\right)^p w(\pi_j) \\
&= \sum_{\pi_j \in \Pi} \left(\frac{|\Delta f(\pi_j) + \Delta g(\pi_j)|}{|\pi_j|}\right)^p w(\pi_j) = \sum_{\pi_j \in \Pi} |\tilde{f}_j + \tilde{g}_j|^p w(\pi_j) \\
&= \sum_{\pi_j \in \Pi} \int_{\pi_j} |\tilde{f}_j + \tilde{g}_j|^p w(t) dt = \sum_{\pi_j \in \Pi} \left\| \tilde{f}_j + \tilde{g}_j \right\|_{L_w^p(\pi_j)}^p \\
&\leqslant \sum_{\pi_j \in \Pi} \left(\left\| \tilde{f}_j \right\|_{L_w^p(\pi_j)} + \|\tilde{g}_j\|_{L_w^p(\pi_j)} \right)^p \\
&\leqslant \left[\left(\sum_{\pi_j \in \Pi} \left\| \tilde{f}_j \right\|_{L_w^p(\pi_j)}^p \right)^{\frac{1}{p}} + \left(\sum_{\pi_j \in \Pi} \|\tilde{g}_j\|_{L_w^p(\pi_j)}^p \right)^{\frac{1}{p}} \right]^p \\
&= \left[\left(\sum_{\pi_j \in \Pi} \int_{\pi_j} |\tilde{f}_j|^p w(t) dt \right)^{\frac{1}{p}} + \left(\sum_{\pi_j \in \Pi} \int_{\pi_j} |\tilde{g}_j|^p w(t) dt \right)^{\frac{1}{p}} \right]^p.
\end{aligned}$$

Thus,

$$\left[\sum_{\pi_j \in \Pi} \left(\frac{|\Delta(f+g)(\pi_j)|}{|\pi_j|}\right)^p w(\pi_j) \right]^{\frac{1}{p}} \leqslant \left(V_w^p(f, I)\right)^{\frac{1}{p}} + \left(V_w^p(g, I)\right)^{\frac{1}{p}},$$

from which it follows that

$$\|f + g\|_{RBV_w^p} \leqslant \|f\|_{RBV_w^p} + \|g\|_{RBV_w^p}.$$

Similarly, if we fix $f \in RBV_w^p(I)$ and $a \in \mathbb{R}$, then

$$\sum_{\pi_j \in \Pi} \left(\frac{|\Delta(af)(\pi_j)|}{|\pi_j|} \right)^p w(\pi_j) = |a| \sum_{\pi_j \in \Pi} \left(\frac{|\Delta(f)(\pi_j)|}{|\pi_j|} \right)^p w(\pi_j),$$

and so $\|af\|_{RBV_w^p} = |a| \|f\|_{RBV_w^p}$. Thus $RBV_w^p(I)$ is a normed linear space.

We now prove that it is a Banach space. Fix a Cauchy sequence $\{f_j\}_{j\in\mathbb{N}}$ in $RBV_w^p(I)$. Then, given a partition Π of $[a, b]$, for any $\epsilon > 0$ there exists $N \geqslant 0$ such that for $n, m \geqslant N$,

$$\sum_{\pi_j \in \Pi} \left(\frac{|\Delta(f_n - f_m)(\pi_j)|}{|\pi_j|} \right)^p w(\pi_j) \leqslant V_w^p(f_n - f_m, I) < \epsilon,$$

and $|f_n(a) - f_m(a)| < \epsilon$. Therefore, the sequence $\{f_n(a)\}$ converges. Similarly, for $x > a$,

$$\begin{aligned} |f_n(x) - f_m(x)|^p &\leqslant 2^p |f_n(a) - f_m(a)|^p + 2^p \frac{|\Delta(f_n - f_m)([a, x])|^p}{\sigma([a, x])^{p-1}} \sigma([a, x])^{p-1} \\ &\lesssim 2^p |f_n(a) - f_m(a)|^p + 2^p V_w^p(f_n - f_m, I) \sigma(I)^{p-1}. \end{aligned}$$

Thus, the sequence $\{f_n\}$ is uniformly Cauchy, and so by Lemma 8.40, it converges uniformly to a continuous function f. Consequently, we have that for $m, n \geqslant N$,

$$\sum_{\pi_j \in \Pi} \left(\frac{|\Delta(f - f_m)(\pi_j)|}{|\pi_j|} \right)^p w(\pi_j) < \epsilon;$$

since this is true for all such partitions Π, we have that

$$V_w^p(f - f_m, I) \leqslant \epsilon.$$

Therefore,

$$\|f - f_m\|_{RBV_w^p} \to 0,$$

as $m \to 0$. By the triangle inequality,

$$\|f\|_{RBV_w^p} \leqslant \|f - f_m\|_{RBV_w^p} + \|f_m\|_{RBV_w^p} < \infty,$$

so $f \in RBV_w^p(I)$. Thus, $RBV_w^p(I)$ is a Banach space. □

We now consider the relationship between $RBV_w^p(I)$ and the spaces $\mathrm{Lip}(I)$ and $BV(I)$.

Proposition 8.42 *Let $w \in A_p$ and $1 < p < \infty$. Then*

$$RBV_w^p(I) \subseteq BV(I). \tag{8.40}$$

Proof Let $\Pi = \{\pi_j\}$ be a finite partition of $I = [a, b]$ and σ the dual weight of w. Then, by Hölder's inequality, we have that

$$\begin{aligned}
\sum_{\pi_j} |\Delta f(\pi_j)| &= \sum_{\pi_j} |\Delta f(\pi_j)| \frac{\sigma(\pi_j)^{(p-1)/p}}{\sigma(\pi_j)^{(p-1)/p}} \\
&\lesssim \left[\sum_{\pi_j \in \Pi} \left(\frac{|\Delta f(\pi_j)|}{|\pi_j|} \right)^p w(\pi_j) \right]^{\frac{1}{p}} \left(\sum_{\pi_j \in \Pi} \sigma(\pi_j)^{(p-1)q/p} \right)^{1/q} \\
&\lesssim V_w^p(f, I)^{\frac{1}{p}} \left(\sum_{\pi_j \in \Pi} \sigma(\pi_j) \right)^{1/q} \leqslant V_w^p(f, I)^{\frac{1}{p}} \sigma(I)^{1/q} < \infty.
\end{aligned}$$

Taking the supremum over all the finite partition Π of I, we obtain the desired result. □

Proposition 8.43 *Let $1 < p < \infty$ and $w \in A_p$. Then*

$$\mathrm{Lip}(I) \subseteq RBV_w^p(I).$$

Proof Let $\Pi = \{\pi_j\}$ be a finite partition of $[a, b]$. Then, by the Lipschitz condition we have

$$\sum_{\pi_j \in \Pi} \left(\frac{|\Delta f(\pi_j)|}{|\pi_j|} \right)^p w(\pi_j) \lesssim \sum_{\pi_j \in \Pi} w(\pi_j) \lesssim w(I) < \infty.$$

If we take the supremum over all the finite partition Π of I, we get the desired result. □

Finally, we show that the spaces RBV_w^p are naturally embedded within one another.

Proposition 8.44 *Let $1 \leqslant p \leqslant q < \infty$ and $w \in A_p$. Then*

$$RBV_w^q(I) \subseteq RBV_w^p(I). \tag{8.41}$$

Proof The case $p = 1$ follows from Proposition 8.42, so we may assume that $1 < p < q < \infty$. Let

$$r = \frac{q}{p}, \quad s = \frac{q}{q-p},$$

and observe that $1/r + 1/s = 1$. Therefore, by the Hölder inequality we have that

$$\begin{aligned}
\sum_{\pi_j \in \Pi} \frac{|\Delta f(\pi_j)|^p}{\sigma(\pi_j)^{p-1}} &= \sum_{\pi_j \in \Pi} \frac{|\Delta f(\pi_j)|^p}{\sigma(\pi_j)^{p-1+1/s}} \sigma(\pi_j)^{1/s} \\
&\leqslant \left(\sum_{\pi_j \in \Pi} \frac{|\Delta f(\pi_j)|^{pr}}{\sigma(\pi_j)^{(p-1+1/s)r}} \right)^{1/r} \left(\sum_{\pi_j \in \Pi} \sigma(\pi_j) \right)^{1/s} \\
&\lesssim \left[\sum_{\pi_j \in \Pi} \left(\frac{|\Delta f(\pi_j)|}{|\pi_j|} \right)^p w(\pi_j) \right]^{1/r} \sigma(I)^{1/s} < \infty,
\end{aligned}$$

hence, $V_w^p(f, I) < \infty$. □

8.4.3 *Riesz Theorem: Weighted Version*

In this section we state and prove a weighted generalization of the Riesz theorem. As a first step, we establish that functions in $RBV_w^p(I)$ are absolutely continuous.

Proposition 8.45 *Let* $1 < p < \infty$ *and* $w \in A_p$. *Then*

$$RBV_w^p(I) \subset AC(I).$$

Proof Since $\sigma \in L^1(I)$, we have by the continuity of the integral that for any $\epsilon > 0$, there exists $\delta > 0$ such that if $|E| < \delta$, then $\sigma(E) < \epsilon$. Fix any collection $\{(a_j, b_j)\}_{j=1}^n$ of disjoint open intervals in I such that

$$\sum_{j=1}^n (b_j - a_j) < \delta.$$

Let

$$E = \bigcup_{j=1}^n [a_j, b_j];$$

then $\sigma(E) < \epsilon$. If we let $1/p + 1/q = 1$, then by Hölder's inequality in sequence spaces:

$$\begin{aligned}
&\sum_{j=1}^{n} |f(b_j) - f(a_j)| \\
&= \sum_{j=1}^{n} \frac{|f(b_j) - f(a_j)|}{\sigma([a_j, b_j])^{1/q}} \sigma([a_j, b_j])^{1/q} \\
&\lesssim \left[\sum_{j=1}^{n} \left(\frac{|f(b_j) - f(a_j)|}{|[a_j, b_j]|} \right)^p w([a_j, b_j]) \right]^{\frac{1}{p}} \left(\sum_{j=1}^{n} \sigma([a_j, b_j]) \right)^{1/q} \\
&\leqslant V_w^p(f, I)^{\frac{1}{p}} \epsilon^{1/q}.
\end{aligned}$$

Since $\epsilon > 0$ is arbitrary, it follows that $f \in AC(I)$. □

We can now state and prove our weighted version of the Riesz theorem.

Theorem 8.46 *Let $1 < p < \infty$. Given a weight $w \in A_p$, we have that $f \in RBV_w^p(I)$ if and only if $f \in AC(I)$ and $f' \in L^p(w)$. Moreover,*

$$V_w^p(f, I)^{\frac{1}{p}} \approx \|f'\|_{L^p(w)}. \tag{8.42}$$

Proof Suppose first that $f \in AC(I)$ and $f' \in L^p(w)$. Taking q as the conjugate exponent of p, fixing a finite partition $\Pi = \{\pi_j\}$ of $I = [a, b]$, and using the fact that $f \in AC(I)$, we can estimate as follows:

$$\begin{aligned}
\left(\frac{|\Delta f(\pi_j)|}{|\pi_j|} \right)^p w(\pi_j) &= \left| \int_{\pi_j} f'(t) dt \right|^p \sigma(\pi_j)^{1-p} \\
&\leqslant \left(\int_{\pi_j} |f'(t)| w^{\frac{1}{p}} w^{-1/p} dt \right)^p \sigma(\pi_j)^{1-p} \\
&\leqslant \left[\left(\int_{\pi_j} |f'(t)|^p w dt \right)^{\frac{1}{p}} \left(\int_{\pi_j} \sigma dt \right)^{1/q} \right]^p \sigma(\pi_j)^{1-p}
\end{aligned}$$

$$= \left(\int_{\pi_j} |f'(t)|^p w dt \right) \sigma\left(\pi_j\right)^{p/q} \sigma\left(\pi_j\right)^{1-p}$$

$$= \int_{\pi_j} |f'(t)|^p w dt.$$

If we sum over all the sub-intervals, we obtain

$$\sum_{\pi_j \in \Pi} \left(\frac{|\Delta f(\pi_j)|}{|\pi_j|} \right)^p w(\pi_j) \leqslant \int_I |f'(t)|^p w dt,$$

which implies that

$$V_w^p(f, I) \leqslant \|f'\|_{L^p(w)}^p.$$

To prove the reverse inequality, let $f \in RBV_w^p(I)$. Define the regular partition $\Pi_N = \{[t_{k,N}, t_{k+1,N}]\}_{k=0}^{N-1}$, where $t_{k,N} = a + \frac{k}{N}(b-a)$. For brevity, let $\pi_{j,N} = [t_{j-1,N}, t_{j,N}]$. By Proposition 8.45, $f \in AC(I)$, so f' exists in $L^1(I)$. Therefore, we may define

$$g_N(t) = \sum_{\pi_{j,N} \in \Pi_N} \fint_{\pi_{j,N}} f'(x) dx \, \chi_{\pi_{j,N}}(t).$$

Then, by the Lebesgue differentiation theorem, we obtain

$$\lim_{N \to \infty} g_N(t) = f'(t),$$

for almost every $t \in I$. Consequently, by Fatou's lemma applied with respect to the measure wdx and using the fact that $w \in A_p$, we get

$$\begin{aligned}
\int_I |f'(t)|^p w dt &\leqslant \liminf_{N \to \infty} \int_I |g_N(t)|^p w dt \\
&= \liminf_{N \to \infty} \sum_{\pi_{j,N} \in \Pi_N} \int_{\pi_{j,N}} |g_N(t)|^p w dt \\
&\leqslant \liminf_{N \to \infty} \sum_{\pi_{j,N} \in \Pi_N} \left(\frac{|\Delta f(\pi_{j,N})|}{|\pi_{j,N}|} \right)^p \int_{\pi_{j,N}} w(t) dt
\end{aligned}$$

$$= \liminf_{N\to\infty} \sum_{\pi_{j,N}\in\Pi_N} \left(\frac{|\Delta f(\pi_{j,N})|}{|\pi_{j,N}|}\right)^p w\left(\pi_{j,N}\right)$$

$$\lesssim V_w^p(f, I) < \infty.$$

This completes the proof of inequality (8.42). □

8.4.4 Nemytskii Operator in Weighted Riesz Bounded Variation Spaces

As an application of Theorem 8.46, we characterize the globally Lipschitz Nemytskii operators defined on $RBV_w^p(I)$. More precisely, we will consider these operators on

$$RBV_w^{p,0}(I) = \{f \in RBV_w^p : \ f(a) = 0\},$$

which is a Banach space with the norm

$$\|f\|_{RBV_w^{p,0}} = V_w^p(f, I)^{\frac{1}{p}}.$$

Moreover, given $w \in A_p$,

$$|f(x)|^p = |f(x) - f(a)|^p \leqslant \frac{|f(x) - f(a)|^p}{\sigma([a, x])^{p-1}}\sigma([a, x])^{p-1} \lesssim V_w^p(f, I)w(I)^{p-1},$$

yields

$$\|f\|_\infty \lesssim \|f\|_{RBV_w^{p,0}}. \tag{8.43}$$

Using this fact we will be able to prove that $RBV_w^{p,0}(I)$ is a Banach algebra. We will rely on the so-called Orlicz–Maligranda criterion, as presented in Maligranda and Orlicz [219], namely:

Proposition 8.47 (Orlicz–Maligranda Criterion) *Let* $(X, \|\cdot\|)$ *be a Banach space whose elements are bounded functions, which is closed under pointwise multiplication of functions. Let us assume that* $fg \in X$ *such that* $\|fg\| \leqslant \|f\|_\infty\|g\| + \|f\|\|g\|_\infty$. *Then* $(X, \|\cdot\|)$, *equipped with the norm* $\|\!| f |\!\|= \|f\|_\infty + \|f\|$, *is a Banach algebra.*

We need to show, keeping the Orlicz–Maligranda criterion in mind, that $RBV_w^{p,0}(I)$ is closed under pointwise multiplication.

Proposition 8.48 *Let* $1 \leqslant p < \infty$*,* $f, g \in RBV_w^{p,0}(I)$*, and* $w \in A_p$*. Then*

$$\|f \cdot g\|_{RBV_w^{p,0}} \leqslant \|f\|_\infty \|g\|_{RBV_w^{p,0}} + \|f\|_{RBV_w^{p,0}} \|g\|_\infty .$$

Proof Fix a finite partition Π of I. By the triangle inequality on ℓ_p, we get

$$\begin{aligned} &\left(\sum_{\pi_j \in \Pi} \frac{|\Delta(f \cdot g)(\pi_j)|^p}{\sigma(\pi_j)^{p-1}} \right)^{\frac{1}{p}} \\ &\quad \leqslant \left[\sum_{\pi_j \in \Pi} \frac{\left(\|g\|_\infty |\Delta f(\pi_j)| + \|f\|_\infty |\Delta g(\pi_j)| \right)^p}{\sigma(\pi_j)^{p-1}} \right]^{\frac{1}{p}} \\ &\quad \leqslant \|g\|_\infty \left(\sum_{\pi_j \in \Pi} \frac{|\Delta f(\pi_j)|^p}{\sigma(\pi_j)^{p-1}} \right)^{\frac{1}{p}} + \|f\|_\infty \left(\sum_{\pi_j \in \Pi} \frac{|\Delta g(\pi_j)|^p}{\sigma(\pi_j)^{p-1}} \right)^{\frac{1}{p}} \\ &\quad \leqslant \|g\|_\infty \|f\|_{RBV_w^{p,0}} + \|f\|_\infty \|g\|_{RBV_w^{p,0}} . \end{aligned}$$

Since Π is arbitrary, we get the desired inequality. □

By the Orlicz–Malingranda criterion from Proposition 8.47, we immediately get the following corollary.

Corollary 8.49 *The Banach space* $RBV_w^{p,0}(I)$ *endowed with the norm*

$$\|\!| f |\!\|_{RBV_w^{p,0}} = \|f\|_\infty + \|f\|_{RBV_w^{p,0}},$$

is a Banach algebra.

Note that by Corollary 8.49 and (8.43), the norms $\|\cdot\|_{RBV_w^{p,0}}$ and $\|\!| \cdot |\!\|_{RBV_w^{p,0}}$ are equivalent.

We now recall the definition of a *Nemytskii operator*, also known as a *superposition operator*.

Definition 8.50 Let $I \subset \mathbb{R}$ be a closed interval. We say that a function $f : I \times \mathbb{R} \to I$ is a *Carathéodory function* if:

(1) $f(\cdot, x)$ is a Lebesgue measurable function for all $x \in \mathbb{R}$;
(2) $f(t, \cdot)$ is a continuous function a.e. in $\mathbb{R}$ for all $t \in I$.

Definition 8.51 (Nemytskii Operator) Let $f : I \times \mathbb{R} \to I$ be a Carathéodory function. The operator F_f, defined by $(F_f\varphi)(t) = f(t, \varphi(t))$, for all $t \in I$ and $\varphi : I \to \mathbb{R}$, is called the Nemytskii operator based on f.

To state our main result, we give one more definition.

Definition 8.52 We say that an operator $T : X \to Y$, where X and Y are normed spaces, is *globally Lipschitz* if there exists a global constant C such that

$$\|Tx_1 - Tx_2\|_Y \leqslant C\|x_1 - x_2\|_X,$$

for all $x_1, x_2 \in X$.

We can now characterize globally Lipschitz Nemytskii operators acting in weighted Riesz bounded variation spaces, where the weight belongs to the Muckenhoupt class.

Theorem 8.53 *Let $1 < q \leqslant p < \infty$ and $w \in A_p$. Then the Nemytskii operator F_f maps $RBV_w^{p,0}(I)$ into $RBV_w^{q,0}(I)$ and is globally Lipschitz if and only if there are functions $g, h \in RBV_w^{q,0}(I)$ such that*

$$f(t, y) = g(t)y + h(t), \quad t \in I, y \in \mathbb{R}. \tag{8.44}$$

Proof Let $g, h \in RBV_w^{q,0}(I)$ and define $f(t, y)$ by (8.44). Hereafter we can use either $\||\cdot\||_{RBV_w^{q,0}}$ or $\|\cdot\|_{RBV_w^{q,0}}$ since they are equivalent norms. By Corollary 8.49, Proposition 8.48, and embedding (8.41), we have that

$$\begin{aligned}\||F_f(u)\||_{RBV_w^{q,0}} &= \||g \cdot u + h\||_{RBV_w^{q,0}}\\ &\leqslant \||g \cdot u\||_{RBV_w^{q,0}} + \||h\||_{RBV_w^{q,0}}\\ &\leqslant \||g\||_{RBV_w^{q,0}} \||u\||_{RBV_w^{q,0}} + \||h\||_{RBV_w^{q,0}}\\ &\lesssim \||g\||_{RBV_w^{q,0}} \||u\||_{RBV_w^{p,0}} + \||h\||_{RBV_w^{q,0}} < \infty,\end{aligned}$$

from which it follows that $F_f : RBV_w^{p,0}(I) \to RBV_w^{q,0}(I)$.

Now fix $u_1, u_2 \in RBV_w^{p,0}(I)$. Again, by Corollary 8.49, we have

$$\begin{aligned}\||F_f(u_1) - F_f(u_2)\||_{RBV_w^{q,0}} &= \||f(\cdot, u_1) - f(\cdot, u_2)\||_{RBV_w^{q,0}}\\ &= \||g\,(u_1 - u_2)\||_{RBV_w^{q,0}}\\ &\lesssim \||g\||_{RBV_w^{q,0}} \||u_1 - u_2\||_{RBV_w^{q,0}}\\ &\lesssim \||u_1 - u_2\||_{RBV_w^{p,0}} .\end{aligned}$$

Therefore, F_f is globally Lipschitz.

To prove the converse, suppose F_f maps $RBV_w^{p,0}(I)$ into $RBV_w^{q,0}(I)$ and is globally Lipschitz. Observe that given $y \in \mathbb{R}$, the function $x \mapsto f(x, y) \in RBV_w^{q,0}(I)$. Consequently, $f(\cdot, y) \in AC(I)$ by Corollary 8.45.

Let $t, t' \in [a, b]$, $t < t'$, $y_1, y_2, y_1', y_2' \in \mathbb{R}$. Define u_1, u_2 by

$$u_i(s) = \begin{cases} y_i & a \leqslant s \leqslant t, \\ \dfrac{y_i' - y_i}{t' - t}(s - t) + y_i & t < s \leqslant t', \\ y_i' & t' < s \leqslant b, \end{cases}$$

Since $u_1 - u_2 \in AC(I)$, $(u_1 - u_2)'$ exists a.e. and it is given by

$$(u_1 - u_2)'(s) = \begin{cases} 0 & a \leqslant s < t, \\ \dfrac{y_1' - y_1 - (y_2' - y_2)}{t' - t} & t \leqslant s \leqslant t', \\ 0 & t' < s \leqslant b, \end{cases}$$

and so

$$\begin{aligned} \left\|(u_1 - u_2)'\right\|_{L^p(w)} &= \left(\int_t^{t'} \left| \frac{y_1' - y_1 - y_2' + y_2}{t' - t} \right|^p w(s)ds \right)^{\frac{1}{p}} \\ &\lesssim \left| \frac{y_1' - y_1 - y_2' + y_2}{t' - t} \right| w([t, t'])^{\frac{1}{p}} \\ &\lesssim \left| \frac{y_1' - y_1 - y_2' + y_2}{t' - t} \right| w(I)^{\frac{1}{p}}. \end{aligned}$$

Hence, by Theorem 8.46, we have that

$$\|u_1 - u_2\|_{RBV_w^{p,0}} \lesssim \left| \frac{y_1' - y_1 - y_2' + y_2}{t' - t} \right|. \tag{8.45}$$

Clearly, $F_f(u_1) - F_f(u_2) \in RBV_w^{q,0}$, since $RBV_w^{q,0}(I)$ is a normed linear space. Then, by the global Lipschitz condition on F_f, we see that

$$\begin{aligned} \frac{\left|\left(F_f(u_1) - F_f(u_2)\right)(t') - \left(F_f(u_1) - F_f(u_2)\right)(t)\right|^q}{w([t, t'])^{q-1}} &\lesssim \left\|F_f(u_1) - F_f(u_2)\right\|^q_{RBV_w^{q,0}} \\ &\lesssim \|u_1 - u_2\|^q_{RBV_w^{p,0}}. \end{aligned}$$

By the definition of u_1 and u_2 and (8.45), we get

$$\frac{\left|f(t', y_1') - f(t', y_2') - f(t, y_1) + f(t, y_2)\right|}{(t' - t)^{\frac{1}{p}}} \lesssim \left| \frac{y_1' - y_1 - y_2' + y_2}{t' - t} \right|.$$

Set $y_1' = w + z$, $y_2' = w$, $y_1 = z$, $y_2 = 0$. Then

$$f(t', w + z) - f(t', w) + f(t, 0) - f(t, z) = 0.$$

Since $f(\cdot, x)$ is continuous, if we take the limit as $t' \to t$, we get

$$f(t, w + z) - f(t, 0) = (f(t, w) - f(t, 0)) + (f(t, z) - f(t, 0)).$$

Now, define $P_t(\cdot) := f(t, \cdot) - f(t, 0)$. With this definition, P_t is a linear map; that is, $P_t(w + z) = P_t(w) + P_t(z)$. Since $P_t(\cdot)$ is continuous, the Cauchy functional equation has a unique continuous solution given by $P_t(y) = g(t)y$ with $g : [a, b] \to \mathbb{R}$. Now define the function h on $[a, b]$ by $h(t) = f(t, 0)$. Then $h \in RBV_w^{q,0}(I)$ and $f(t, y) = g(t)y + h(t)$. Since for $t \in I$,

$$f(t, 1) - f(t, 0) = (P_t(1) + f(t, 0)) - f(t, 0) = g(t),$$

we have that $g \in RBV_w^{q,0}(I)$. This completes the proof. □

8.5 Characterization of Weighted Sobolev Spaces via Weighted Riesz Bounded Variation Spaces

The notion of Riesz bounded variation spaces has been extended to functions on general domains in $\mathbb{R}^n$ for $n > 1$; see, for instance, the works of Angeloni [20], Barza and Lind [24], Bojarski [33], Malý [220]. We are particularly interested in the latter paper, where $RBV^p(\Omega)$, $\Omega \subset \mathbb{R}^n$, is defined as the set of all functions f such that

$$V_p(f; \Omega) = \sup \left[\sum_{B_k \in \mathcal{D}} \left(\frac{\operatorname{osc}_{B_k}(f)}{r_k} \right)^p |B_k| \right]^{\frac{1}{p}} < \infty,$$

with the supremum taken over all countable collections $\mathcal{D} = \{B_k\}_{k=1}^{\infty}$ of disjoint balls of radius r_k contained in Ω. It is shown that, if $p > n$, then $f \in RBV^p(\Omega)$ if and only if $f \in W^{1,p}(\Omega)$.

We will extend this result and prove a version of the Riesz theorem characterizing weighted Sobolev spaces in $\mathbb{R}^n$, $W_w^{1,p}(\Omega)$, for weights w in the Muckenhoupt class A_p, using an appropriate weighted version of RBV^p. As an application, using Rubio de Francia's extrapolation theory, a similar characterization of the variable exponent Sobolev spaces via variable exponent Riesz bounded variation spaces is obtained.

8.5.1 Preliminaries

In the sequel, we regard Ω as an open subset of $\mathbb{R}^n$. For an open ball with center x and radius $r > 0$, we write $B(x, r)$. Throughout this section all cubes $Q \subset \mathbb{R}^n$ will have their sides parallel to the coordinate axes. We denote the integral average of f over the measurable subset E by

$$\langle f \rangle_E := \frac{1}{|E|} \int_E f(x)dx = \fint_E f(x)dx, \quad |E| > 0,$$

where $|E|$ denotes the Lebesgue measure of the set E.

Muckenhoupt A_p Weights

By a weight we mean a non-negative, locally integrable function. The measure of E induced by the weight w is $w(E) := \int_E w(x)dx$.

A weight w belongs to the Muckenhoupt A_p class, $1 < p < \infty$, denoted $w \in A_p$, if

$$[w]_{A_p} = \sup_Q \fint_Q w dx \left(\fint_Q w^{\frac{1}{1-p}} dx \right)^{p-1} < \infty,$$

where the supremun is taken over all cubes (or balls) in $\mathbb{R}^n$. When $p = 1$, we say that $w \in A_1$ if

$$[w]_{A_1} = \sup_Q \operatorname{ess\,sup}_{x \in Q} w(x)^{-1} \fint_Q w dx < \infty.$$

The class A_∞ is defined by

$$A_\infty = \bigcup_{p \geqslant 1} A_p.$$

For more information on A_p weights, and for proofs of the following results, see Duoandikoetxea [81], Grafakos [119]. It follows from Hölder's inequality and the definition that if $1 \leqslant p_1 < p_2 < \infty$, then $A_{p_1} \subset A_{p_2}$. Moreover, the A_p classes are left-open: if $w \in A_p$ for some $p > 1$, then there exists $\varepsilon > 0$ such that $w \in A_{p-\varepsilon}$ (see Duoandikoetxea [81, Corollary 7.6]).

Given a weight $w \in A_\infty$, we define r_w by

$$r_w := \inf\{q > 1 : w \in A_q\}. \tag{8.46}$$

Consequently, if $w \in A_p$, then $r_w < p$, and for every $q > r_w$, $w \in A_q$.

Lemma 8.54 *Let $1 \leqslant p < \infty$ and $w \in A_p$. Then, for every measurable subset E of Q, we have*

$$\left(\frac{|E|}{|Q|}\right)^p \leqslant [w]_{A_p} \frac{w(E)}{w(Q)}.$$

We say that $w \in RH_s$, for some $s > 1$, if

$$[w]_{RH_s} = \sup_Q \frac{\langle w^s \rangle_Q^{1/s}}{\langle w \rangle_Q} < \infty, \tag{8.47}$$

where the supremun is taken over all cubes in $\mathbb{R}^n$.

Lemma 8.55 *Assume that $w \in A_\infty$. Then there exists $s > 1$, depending on $[w]_{A_p}$, such that $w \in RH_s$.*

Some Necessary Function Spaces

For $1 \leqslant p < \infty$, let the *weighted Sobolev space* $W_w^{1,p}(\Omega)$ denote the collection of functions $f \in L_w^p(\Omega)$ whose weak derivatives $D_j f$ belong to $L_w^p(\Omega)$ and endowed with the norm

$$\|f\|_{W_w^{1,p}(\Omega)} = \left(\int_\Omega |f(x)|^p w(x)dx\right)^{\frac{1}{p}} + \left(\int_\Omega |\nabla f(x)|^p w(x)dx\right)^{\frac{1}{p}}. \tag{8.48}$$

The space $W_w^{1,p}(\Omega)$, $1 \leqslant p < \infty$, is a Banach space (see Turesson [326, Proposition 2.1.2]). Moreover, when $w \in A_p$, $W_w^{1,p}(\Omega)$ is the completion of $C^\infty(\Omega) \cap W_w^{1,p}(\Omega)$ with respect to the norm (8.48) (see Kilpeläinen [166, Theorem 2.5]).

Given a function $f \in L^{p(\cdot)}(\Omega)$, it belongs to the *variable exponent Sobolev space* $W^{1,p(\cdot)}(\Omega)$ if its weak derivatives $D_j f$, $1 \leqslant j \leqslant n$, exist and belong to $L^{p(\cdot)}(\Omega)$. $W^{1,p(\cdot)}(\Omega)$ is a Banach space when equipped with the norm

$$\|f\|_{W^{1,p(\cdot)}(\Omega)} = \|f\|_{L^{p(\cdot)}(\Omega)} + \|\nabla f\|_{L^{p(\cdot)}(\Omega)}$$

(see Cruz-Uribe and Fiorenza [58, Theorem 6.6]). Though $C^\infty(\Omega) \cap W^{1,p(\cdot)}(\Omega)$ need not be dense in $W^{1,p(\cdot)}(\Omega)$ for general functions $p \in \mathbb{P}(\Omega)$, it is when $p \in \mathbb{P}_\infty^{\log}(\Omega)$, cf. Cruz-Uribe and Fiorenza [58, Theorem 6.14].

For $f : \Omega \to \mathbb{R}$, its *oscillation* on a set $E \subset \Omega$ is defined as

$$\operatorname{osc}_E(f) := \sup_{x,y \in E} |f(x) - f(y)|.$$

Given a weight w and $1 \leqslant p < \infty$, define the *weighted Riesz p-variation* of f on the set Ω by

$$V_p(f; \Omega, w) := \sup \left[\sum_{B_k \in \mathcal{D}} \left(\frac{\operatorname{osc}_{B_k}(f)}{r_k} \right)^p w(B_k) \right]^{\frac{1}{p}}, \tag{8.49}$$

where the supremum is taken over all the countable collections $\mathcal{D} = \{B_k\}_{k=1}^{\infty}$ of disjoint balls of radius r_k contained in Ω. If $V_p(f; \Omega, w) < \infty$, we say that $f \in RBV_w^p(\Omega)$.

Remark 8.56 When $w = 1$, since $|B| \approx r^n$, we recover the definition of Riesz variation given in Barza and Lind [24], up to a dimensional constant.

For functions f and g, and $\alpha \in \mathbb{R}$, by Minkowski's inequality for sequence spaces, we have

$$V_p(f+g; \Omega, w) \leqslant V_p(f; \Omega, w) + V_p(g; \Omega, w),$$

and

$$V_p(\alpha f; \Omega, w) = |\alpha| V_p(f; \Omega, w);$$

hence, $V_p(f; \Omega, w)$ is a seminorm. It is not a norm since $V_p(f; \Omega, w) = 0$ for every constant function f.

Lemma 8.57 *Given w a weight, assume that Ω is a bounded set. If $1 < p_1 < p_2 < \infty$, then*

$$RBV_w^{p_2}(\Omega) \hookrightarrow RBV_w^{p_1}(\Omega).$$

Proof Fix $f \in RBV_w^{p_2}(\Omega)$ and let $\mathcal{D} = \{B_k\}_{k=1}^{\infty}$ be a disjoint collection of countable balls contained in Ω. Fix $s > 1$ such that $1/p_2 + 1/s = 1/p_1$. Then, by Hölder's inequality, we have

$$\begin{aligned}\sum_{B_k \in \mathcal{D}} \left(\frac{\operatorname{osc}_{B_k}(f)}{r_k} \right)^{p_1} w(B_k) &\leqslant \left(\sum_{B_k \in \mathcal{D}} \left(\frac{\operatorname{osc}_{B_k}(f)}{r_k} \right)^{p_2} w(B_k) \right)^{\frac{p_1}{p_2}} \left(\sum_{B_k \in \mathcal{D}} w(B_k) \right)^{\frac{p_1}{s}} \\ &\leqslant V_{p_2}(f; \Omega, w)^{p_1} w(\Omega)^{\frac{p_1}{s}} \\ &< \infty.\end{aligned}$$

Taking the supremum over all such collections $\mathcal{D}$, we get the desired embedding. □

8.5.2 *Main Results*

Our results cover Lebesgue and bounded variation spaces in both classical and variable exponent settings.

Classical Setting

Theorem 8.58 *Given $\Omega \subseteq \mathbb{R}^n$ an open set, assume that $p > n$. Let $w \in A_p$ be such that $p > nr_w$, with r_w defined in* (8.46). *Then $f \in W_w^{1,p}(\Omega)$ if and only if the function f is continuous (perhaps after being redefined on a null set) and $f \in RBV_w^p(\Omega) \cap L_w^p(\Omega)$. Furthermore,*

$$\|\nabla f\|_{L_w^p(\Omega)} \lesssim V_p(f; \Omega, w) \lesssim \|\nabla f\|_{L_w^p(\Omega)}, \tag{8.50}$$

where the implicit constants depend on p, n, and $[w]_{A_p}$.

Proof Fix $w \in A_p$, $p > nr_w$. We first prove that if $f \in W_w^{1,p}(\Omega)$, then $f \in RBV_w^p(\Omega)$. To do so, we prove a weighted version of the well-known *Morrey inequality* (see Brezis [35, Theorem 9.12]). Suppose that $f \in C^\infty(\Omega) \cap W_w^{1,p}(\Omega)$. Fix $x_0 \in \Omega$ and $R > 0$ such that the open ball $B(x_0, 2R)$ is contained in Ω. Fix $y, z \in B(x_0, R)$ and let $B_r = B(x, r) \subset B(x_0, 2R)$ be any ball containing y and z. Then

$$f(z) - f(y) = \int_0^1 \nabla f(tz - (1-t)y) \cdot (z - y)dt.$$

Observe that $B(tz + (1-t)y, tr) \subset B(x, r)$ and $|z_j - y_j| \leqslant r$. Since $p > nr_w$, we can fix $\delta > 0$ such that $n < 1 + \delta < \frac{p}{r_w}$; therefore, if $q = \frac{p}{1+\delta}$, then $nr_w < nq < p$. Hence, by Fubini's theorem, a change of variables, and Hölder's inequality applied twice, we have

$$|f(z) - \langle f\rangle_{B_r}| \leqslant \fint_{B_r} \int_0^1 \sum_{j=1}^n |D_j f(tz - (1-t)y)||z_j - y_j|dtdy$$

$$\lesssim \sum_{j=1}^n \frac{1}{r^{n-1}} \int_0^1 \left(\int_{B_r} |D_j f(y)|^{1+\delta} dy \right)^{1/(1+\delta)} |B(tx + (1-t)y, tr)|^{\delta/(1+\delta)} \frac{dt}{t^n}$$

$$\lesssim \sum_{j=1}^n r^{1-\frac{n}{1+\delta}} \int_0^1 \left(\int_{B_r} |D_j f(y)|^{1+\delta} dy \right)^{1/(1+\delta)} \frac{dt}{t^{n/(1+\delta)}}$$

$$\lesssim r^{1-\frac{nq}{p}} \sum_{j=1}^{n} \left(\int_{B_r} |D_j f(y)|^p w(y) dy \right)^{\frac{1}{p}} \left(\int_{B_r} w^{1/(1-q)}(y) dy \right)^{(q-1)/p}.$$

Note that in the final estimate we use that $n < 1 + \delta$ to evaluate the integral. Hence, if we set $\sigma := w^{1/(1-q)}$, the preceding estimate for $z,\, y \in B_r$ leads to

$$\begin{aligned} |f(z) - f(y)| &\leqslant |f(z) - f_{B_r}| + |f(y) - f_{B_r}| \\ &\lesssim r^{1-n\frac{q}{p}} \sigma(B_r)^{(q-1)/p} \|\nabla f\|_{L^p_w(B_r)} \\ &\leqslant r^{1-n\frac{q}{p}} \sigma(B_{2R})^{(q-1)/p} \|\nabla f\|_{L^p_w(B_R)}. \end{aligned}$$

Moreover, since $w \in A_q$, then $\sigma(B_R) < \infty$. This is true for all such balls B_r, so if we fix the ball $B(x, r)$ so that $r = 2|z - y|$, we get that

$$|f(z) - f(y)| \leqslant C|x - y|^{1-n\frac{q}{p}} \sigma(B_{2R})^{(q-1)/p} \|\nabla f\|_{L^p_w(B_R)}. \tag{8.51}$$

Since $C^\infty(\Omega) \cap W^{1,p}_w(\Omega)$ is dense in $W^{1,p}_w(\Omega)$ (see Gol'dshtein and Ukhlov [117, Theorem 1]), a standard argument shows that inequality (8.51) holds for every $f \in W^{1,p}_w(\Omega)$ and almost every $y,\, z \in B(x_0, R)$. Consequently, arguing as in the proof of Brezis [35, Theorem 9.12], we get that f has a continuous representative (that is to say, by redefining f on a null set, we have that f is continuous on $B(x_0, R)$). Since this is true for all $x_0 \in \Omega$, we have that f has a continous representative on Ω.

To estimate the $RBV^p_w(\Omega)$ norm of f, fix any collection $\mathcal{D} = \{B_k\}_{k=1}^\infty$ of disjoint balls contained in Ω. Since $w \in A_q$,

$$w(B_k)\sigma(B_k)^{q-1} \lesssim r_k^{nq}.$$

Therefore, by (8.51), we have that

$$\begin{aligned} \sum_{B_k} \left(\frac{\operatorname{osc}_{B_k}(f)}{r_k} \right)^p w(B_k) &= \sum_{B_k} \left(\frac{\operatorname{osc}_{B_k}(f)}{r_k} \right)^p w(B_k) \frac{\sigma(B_k)^{q-1}}{\sigma(B_k)^{q-1}} \\ &\lesssim \sum_{B_k} \frac{\operatorname{osc}_{B_k}(f)^p}{r_k^{p-qn}} \frac{1}{\sigma(B_k)^{q-1}} \\ &\lesssim \sum_{B_k} \|\nabla f\|^p_{L^p_w(B_k)} \\ &\lesssim \|\nabla f\|^p_{L^p_w(\Omega)}. \end{aligned}$$

If we take the supremum over all such collections $\mathcal{D}$, we get that $V_p(f; \Omega, w) \lesssim \|\nabla f\|_{L^p_w(\Omega)}$.

To prove the converse, suppose $f \in RBV_w^p(\Omega)$. We will adapt the construction in Barza and Lind [24, Theorem 1.1]. Fix an open set $\Omega_0 \Subset \Omega$ and set

$$0 < R < \frac{d\left(\Omega_0, \Omega^c\right)}{6\sqrt{n}}.$$

Let $\{\alpha_k\}_{k=1}^{\infty}$ be an enumeration of the lattice $2R\mathbb{Z}^n$. Form an index set $K \in \mathbb{N}$ so that the collection of cubes $\mathcal{D} = \{Q_k\}_{k \in K}$ consists of all cubes such that each cube Q_k has center α_k, side-length $2R$, and satisfies $Q_k \cap \Omega_0 \neq \emptyset$. We claim that

$$\Omega_0 \subset \bigcup_{k \in K} Q_k \subset \Omega.$$

The first inclusion is immediate. If the second did not hold, then for some $Q \in \mathcal{D}$ we would have that $Q \cap \Omega^c \neq \emptyset$. Hence, there exists $x \in Q \cap \Omega_0$ and $y \in Q \cap \Omega^c$ such that

$$d\left(\Omega_0, \Omega^c\right) \leqslant |x - y| \leqslant 2\sqrt{n}R,$$

which is a contradiction. Denote by $\mathcal{B} = \{B(\alpha_k, 3\sqrt{n}R)\}_{k \in K}$ the collection of balls with center α_k and radius $3\sqrt{n}R$ and observe that, arguing as before, we have that $Q_k \subset B(\alpha_k, 3\sqrt{n}R) \subset \Omega$. Since the centers of the balls in $\mathcal{B}$ lie on a lattice and they have uniform radii, the collection $\mathcal{B}$ can be partitioned into N collections $\{\mathcal{B}_i\}_{i=1}^N$ of disjoint balls, $\mathcal{B}_i \subseteq \mathcal{B}$, where N depends only on the dimension. For brevity, let $B_k = B(\alpha_k, 3\sqrt{n}R)$.

Let $\varphi \in C_c^\infty(B(0,1))$ be a non-negative, radially decreasing function such that $\|\varphi\|_1 = 1$. For $R > 0$ define the sequence of mollifiers $\varphi_R(x) = \frac{1}{R^n}\varphi(x/R)$. For all $x \in Q_k$, $y \in B(0,R)$ implies $x - y \in B_k$. Moreover, we have that for $1 \leqslant j \leqslant n$, $D_j\varphi$ is uniformly bounded and has $\int_{B(0,1)} D_j\varphi(x)dx = 0$ (the latter is well-known and follows, for instance, from the divergence theorem). Therefore, for each j we can estimate as follows:

$$\begin{aligned}
\left\|D_j(\varphi_R * f)\right\|_{L_w^p(\Omega_0)}^p &\leqslant \sum_{k \in K} \int_{Q_k} |D_j\varphi_R * f(x)|^p w(x)dx \\
&= \sum_{k \in K} \int_{Q_k} \left| \int_{B(0,R)} D_j\varphi_R(y)(f(x-y) - f(x))dy \right|^p w(x)dx \\
&\lesssim \sum_{k \in K} \int_{Q_k} R^{-(n+1)p} (\operatorname{osc}_{B_k}(f))^p |B(0,R)|^p w(x)dx \\
&\lesssim \sum_{k \in K} \left(\frac{\operatorname{osc}_{B_k}(f)}{3\sqrt{n}R}\right)^p w(Q_k)
\end{aligned}$$

$$\leqslant \sum_{k\in K} \left(\frac{\operatorname{osc}_{B_k}(f)}{3\sqrt{n}R} \right)^p w(B_k)$$

$$= \sum_{i=1}^{N} \sum_{B_k \in \mathcal{B}_i} \left(\frac{\operatorname{osc}_{B_k}(f)}{3\sqrt{n}R} \right)^p w(B_k)$$

$$\lesssim V_p(f; \Omega, w)^p,$$

where the last inequality follows since the balls in $\mathcal{B}_i$ are disjoint.

Because $w \in A_p$, the Hardy–Littlewood maximal operator is bounded on $L^p_w(\Omega)$, which yields $\|\varphi_R * f\|_{L^p_w(\Omega_0)} \lesssim \|f\|_{L^p_w(\Omega)}$, since $|\varphi_R * f(x)| \lesssim Mf(x)$. Thus, $\{\varphi_R * f\}$ is a bounded sequence in $W^{1,p}(\Omega_0, w)$ which converges pointwise almost everywhere to f. Thus, by Fatou's lemma, we have that

$$\|\nabla f\|_{L^p_w(\Omega_0)} \lesssim V_p(f; \Omega, w).$$

Since the implicit constants do not depend on Ω_0, we can again apply Fatou's lemma to get that the left-hand side inequality in (8.50) holds. Since we also have that $f \in L^p_w(\Omega)$, it follows that $f \in W^{1,p}_w(\Omega)$. □

The unweighted version of Theorem 8.58 (i.e., when $w = 1$ so $r_w = 1$) was proved in Barza and Lind [24]. We note, in passing, an omission in the statement of their main result: the hypothesis that $f \in L^p(\Omega)$ is missing. It is straightforward to give a counter-example without this assumption. Let $\Omega = \mathbb{R}^n$ and let $f \equiv 1$. Then f is continuous and $f \in RBV^p(\mathbb{R}^n)$ for any $p > n$, but $f \notin W^{1,p}(\mathbb{R}^n)$.

As a corollary to the proof of Theorem 8.58, we have that the left-hand inequality in (8.50) is always true.

Corollary 8.59 *Let* $1 \leqslant p < \infty$ *and* w *be a weight. If* $f \in RBV^p_w(\Omega)$*, then* $\|\nabla f\|_{L^p_w(\Omega)} \lesssim V_p(f; \Omega, w)$.

Proof It suffices to note that in the proof of the left-hand side of inequality (8.50), we did not use the A_p condition, and this estimate holds for any weight w and for all p. □

Although we are primarily concerned with spaces defined with A_p weights, Corollaries 8.59 and 8.61 below show that some of our results are true with weaker assumptions on the weights. Many classical results for Sobolev spaces can be extended to the weighted case by using a wider class of weights: e.g., see Heinonen et al. [137, Chapter 1] for a discussion of Sobolev spaces defined using the so-called p-admissible weights. It is an interesting question to determine the weakest hypotheses on the weights to define a *rich* theory of weighted Riesz p-variation.

We can also give a weak-type estimate for the local Lipschitz constant of a function. Recall that a function f is said to be *Lipschitz continuous* at x if

$$L_f(x) = \limsup_{y\to x} \frac{|f(x) - f(y)|}{|x - y|} < \infty.$$

Theorem 8.60 *Let $1 \leqslant p < \infty$, $w \in A_p$, and $f \in RBV_w^p(\mathbb{R}^n)$. Then, for all $t > 0$,*

$$w\left(\{x \in \mathbb{R}^n : L_f(x) > t\}\right) \lesssim \left(\frac{V_p(f; \mathbb{R}^n, w)}{t}\right)^p.$$

The implicit constants depend on p, n, and $[w]_{A_p}$.

Proof Fix $t > 0$. By the definition of L_f, for any x for which $L_f(x) > t$, there exists $\Delta_x \in \mathbb{R}^n$, $|\Delta_x| \leqslant 1$, such that

$$|f(x + \Delta_x) - f(x)| \geqslant t|\Delta_x|. \tag{8.52}$$

Since the balls $\{B(x, |\Delta_x|)\}$ cover the set $\{x : L_f(x) > t\}$ and have uniformly bounded radii, by the Vitali covering lemma, there exists a subcollection of disjoint balls $\{B_k\}_{k=1}^{\infty}$, $B_k := B(x_k, |\Delta_{x_k}|)$, such that

$$\{x \in \mathbb{R}^n : L_f(x) > t\} \subset \bigcup_k 5B_k.$$

By Lemma 8.54 (applied to balls instead of cubes), we have

$$\left(\frac{|B_k|}{|5B_k|}\right)^p \lesssim \frac{w(B_k)}{w(5B_k)}.$$

Hence,

$$w(\{x \in \mathbb{R}^n : L_f(x) > t\}) \leqslant w\left(\bigcup_k 5B_k\right) \leqslant \sum_k w(5B_k) \lesssim \sum_k w(B_k). \tag{8.53}$$

Set $r_k = |\Delta_{x_k}|$. If we combine (8.52) and (8.53) we get that

$$\begin{aligned} t^p w(\{x \in \mathbb{R}^n : L_f(x) > t\}) &\lesssim t^p \sum_k w(B_k) \\ &\lesssim \sum_k \left(\frac{\operatorname{osc}_{B_k}(f)}{r_k}\right)^p w(B_k) \leqslant V_p(f; \Omega, w)^p, \end{aligned}$$

which is the desired inequality. □

As a corollary to the proof of Theorem 8.60, we have that the A_p condition can be significantly weakened. Given any ball $B = B(x, r)$, let $2B := B(x, 2r)$. We say that a weight is *doubling* if $w(2B) \leqslant Cw(B)$ with a constant C independent of B.

Corollary 8.61 *Theorem 8.60 remains true when w is only assumed to be doubling.*

Proof To see that this result is true, it suffices to note that in the proof of Theorem 8.60, we only use the A_p condition to invoke Lemma 8.54, which in turn we only use to show that for each k, $w(5B_k)$ is uniformly bounded by $w(B_k)$. Thus, it is enough to assume doubling. □

A straightforward consequence of Theorems 8.58 and 8.60 is the following result.

Corollary 8.62 *Let* $1 \leqslant p < \infty$ *and* $w \in A_p$. *If* $f \in RBV_w^p(\mathbb{R}^n)$, *then* f *is differentiable almost everywhere. Moreover, if* $p > nr_w$ *and* $f \in W_w^{1,p}(\mathbb{R}^n)$, *then* f *is differentiable almost everywhere.*

Proof Fix $f \in RBV_w^p(\Omega)$. Then, by Theorem 8.60, $|\{x \in \mathbb{R}^n : L_f(x) = \infty\}| = 0$. Given any bounded set E,

$$|E| = \int_E w(x)^{\frac{1}{p}} w(x)^{-\frac{1}{p}} dx \leqslant \left(\int_E w(x)dx\right)^{\frac{1}{p}} \left(\int_E w(x)^{1-p'} dx\right)^{1/p'};$$

since $w \in A_p$ the second integral is finite. Therefore, given any set E, if $w(E) = 0$, it follows by a standard approximation argument that $|E| = 0$. In particular, we have that $|\{x \in \mathbb{R}^n : L_f(x) = \infty\}| = 0$, and so by Stepanov's theorem, see Federer [99, Theorem 3.1.9], f is differentiable almost everywhere.

If $p > nr_w$ and $f \in W_w^{1,p}(\Omega)$, then by Theorem 8.58, $f \in RBV_w^p(\Omega)$ and so again is differentiable almost everywhere. □

Remark 8.63 A few remarks are in order:

(a) In the unweighted case, the differentiability of functions in $W^{1,p}(\mathbb{R}^n)$, $p > n$, is due to Calderón [44] (see also Heinonen [136, Theorem 6.17]). The unweighted version of Corollary 8.62 was proved by Barza and Lind [24]; for the proof of the differentiability of functions in $RBV^p(\mathbb{R}^n)$, they assume that $p > n$, but this is not needed in the proof.
(b) By Corollary 8.61, if w is doubling and mutually absolutely continuous with respect to Lebesgue measure, then $f \in RBV_w^p(\Omega)$ is again differentiable almost everywhere.
(c) When $p = n$, Malý [220] defined a space of functions of bounded n-variation that is equivalent to $RBV^n(\Omega)$ and proved that it is continuously embedded in $W^{1,n}(\Omega)$. Furthermore, he showed that the elements of $RBV^n(\Omega)$ are differentiable almost everywhere. By modifying the proofs of Theorems 8.58 and 8.60, we can extend this result to the weighted case.

Theorem 8.64 *Let* $w \in A_n$ *and* $f \in RBV_w^n(\Omega) \cap L_w^n(\Omega)$. *Then* $f \in W_w^{1,n}(\Omega)$ *with* $\|\nabla f\|_{L_w^n(\Omega)} \lesssim V_n(f; \Omega, w)$. *Moreover,* f *is differentiable almost everywhere.*

Proof Given a weight w and $f \in RBV_w^n(\Omega) \cap L_w^n(\Omega)$, by Corollary 8.59, we have that $f \in W_w^{1,n}(\Omega)$ and the gradient estimate holds. If $w \in A_n$, then by Theorem 8.60, f is differentiable almost everywhere. □

Variable Exponent Setting

As an application of Rubio de Francia's extrapolation theory (as developed in Cruz-Uribe and Wang [61]), we can extend Theorem 8.58 to the variable exponent Sobolev spaces. Recall that variable exponent Lebesgue space $L^{p(\cdot)}(\Omega)$ is the set of all functions f such that

$$\|f\|_{L^{p(\cdot)}(\Omega)} = \inf\left\{\lambda > 0 : \varrho_{p(\cdot),\Omega}(f/\lambda) \leqslant 1\right\}, \tag{8.54}$$

where

$$\varrho_{p(\cdot),\Omega}(f) = \varrho_{p(\cdot)}(f) := \int_{\Omega} |f(x)|^{p(x)} dx. \tag{8.55}$$

The variable exponent Sobolev space $W^{1,p(\cdot)}(\Omega)$ consists of all functions $f \in L^{p(\cdot)}(\Omega)$ that are weakly differentiable and $\nabla f \in L^{p(\cdot)}(\Omega)$.

We say $p \in \mathbb{P}^{\log}_{\infty}(\Omega)$, $p : \Omega \to [1, \infty)$, if it is log-Hölder continuous locally and at infinity, i.e., it satisfies (8.3) and (8.4), respectively.

Given a countable collection of disjoint balls $\mathcal{D} = \{B_k\}_{k=1}^{\infty}$ of radius r_k contained in Ω, define the modular

$$V_{\mathcal{D}}^{p(\cdot)}(f; \Omega) = \sum_k \left(\frac{\operatorname{osc}_{B_k}(f)}{r_k}\right)^{p_{B_k}} \|\chi_{B_k}\|_{L^{p(\cdot)}(\Omega)}^{p_{B_k}}.$$

We define the functional

$$\|f\|_{RBV_{\mathcal{D}}^{p(\cdot)}(\Omega)} = \inf\left\{\lambda > 0 : V_{\mathcal{D}}^{p(\cdot)}(f/\lambda; \Omega) \leqslant 1\right\}. \tag{8.56}$$

A function f belongs to $RBV^{p(\cdot)}(\Omega)$ if

$$\|f\|_{RBV^{p(\cdot)}(\Omega)} = \sup_{\mathcal{D}} \|f\|_{RBV_{\mathcal{D}}^{p(\cdot)}(\Omega)} < \infty,$$

where the supremum is taken over all the countable collections $\mathcal{D}$.

To prove our main results, we first prove an alternative way to estimate the seminorm in $RBV^{p(\cdot)}(\Omega)$. Given an exponent function p, a disjoint collection $\mathcal{D} = \{B_k\}_{k=1}^{\infty}$ of balls contained in Ω, and a sequence of real numbers $\{t_{B_k}\}_{B_k \in \mathcal{D}}$ associated to $\mathcal{D}$, define the variable exponent sequence space

$$\ell^{\mathcal{D},p} = \left\{\hat{t} = \{t_{B_k}\}_{B_k \in \mathcal{D}} : \sum_{B_k \in \mathcal{D}} |t_{B_k}|^{p_{B_k}} < \infty\right\},$$

and endow it with the Luxemburg norm

$$\|\hat{t}\|_{\ell^{\mathcal{D},p}} = \inf\left\{\lambda > 0 : \sum_{B_k\in\mathcal{D}} (|t_{B_k}|/\lambda)^{p_{B_k}} \leqslant 1\right\},$$

for more on variable sequence spaces, see Diening et al. [74]. The following result shows the connection between these sequence spaces and the variable exponent Lebesgue space $L^{p(\cdot)}(\Omega)$ (for a proof see Kopaliani [196, Lemma 2.1]).

Lemma 8.65 *Let $p \in \mathbb{P}^{\log}_{\infty}(\Omega)$ and $\{e_{B_k}\}_{B_k\in\mathcal{D}}$ be the canonical basis for $\ell^{\mathcal{D},p}$ (i.e., e_{B_k} has entry 1 at the index k and 0 otherwise). Then for every disjoint countable collection $\mathcal{D}$ of balls and every real sequence $\{t_{B_k}\}$,*

$$\left\|\sum_{B_k\in\mathcal{D}} t_{B_k}\chi_{B_k}\right\|_{L^{p(\cdot)}(\Omega)} \approx \left\|\sum_{B_k\in\mathcal{D}} t_{B_k}\|\chi_{B_k}\|_{L^{p(\cdot)}(\Omega)}e_{B_k}\right\|_{\ell^{\mathcal{D},p}}. \tag{8.57}$$

The implicit constants depend only on p.

Given a disjoint countable collection of balls $\mathcal{D} = \{B_k\}_{k=1}^{\infty}$, define the operator $G_{\mathcal{D}}$ by

$$G_{\mathcal{D}}f(x) = \sum_{B_k\in\mathcal{D}} \frac{\operatorname{osc}_{B_k}(f)}{r_k}\chi_{B_k}(x).$$

A key fact is that the $RBV^{p(\cdot)}_{\mathcal{D}}$ norm of f and the $L^{p(\cdot)}$ norm of the operator $G_{\mathcal{D}}f$ are comparable.

Proposition 8.66 *Let $p \in \mathbb{P}^{\log}_{\infty}(\Omega)$. For every function f we have*

$$\|f\|_{RBV^{p(\cdot)}(\Omega)} \approx \sup_{\mathcal{D}} \|G_{\mathcal{D}}f\|_{L^{p(\cdot)}(\Omega)}, \tag{8.58}$$

where the supremum is taken over all countable collections $\mathcal{D}$ of pairwise disjoint balls in Ω, and the implicit constants depend only on p.

Remark 8.67 Note that in the case of weighted Riesz bounded variation, we immediately have that $V_p(f;\Omega,w) = \sup_{\mathcal{D}} \|G_{\mathcal{D}}f\|_{L^p_w(\Omega)}$.

Proof of Proposition 8.66 Fix a collection $\mathcal{D}$. By Lemma 8.65, the definition of $\ell^{\mathcal{D},p}$, and the definition of $RBV^p_{\mathcal{D}}$ (8.56), we have that

$$\begin{aligned}\|G_{\mathcal{D}}f\|_{L^{p(\cdot)}(\Omega)} &= \left\|\sum_{B_k\in\mathcal{D}} \frac{\operatorname{osc}_{B_k}(f)}{r_k}\chi_{B_k}\right\|_{L^{p(\cdot)}(\Omega)}\\ &\approx \left\|\sum_{B_k\in\mathcal{D}} \frac{\operatorname{osc}_{B_k}(f)}{r_k}\|\chi_{B_k}\|_p\right\|_{\ell^{\mathcal{D},p}}\end{aligned}$$

$$= \|f\|_{RBV^{p(\cdot)}_{\mathcal{D}}(\Omega)}.$$

If we take the supremum over every collection $\mathcal{D}$, we get (8.58). □

Finally, for our proof we need a version of Rubio de Francia extrapolation into the scale of variable exponent Lebesgue spaces. More precisely, we need a version of limited range extrapolation that was proved in Cruz-Uribe and Wang [61, Theorem 2.14]. This result is stated in terms of an abstract family of extrapolation pairs $\mathcal{F} = \{(f, g)\}$.

Theorem 8.68 *Given a family of extrapolation pairs $\mathcal{F}$, let $1 < q_- < q_+ < \infty$ and assume that there exists p, $q_- < p < q_+$, such that for every $w \in A_{p/q_-} \cap RH_{(q_+/p)'}$,*

$$\|f\|_{L^p(w)} \lesssim \|g\|_{L^p(w)} \tag{8.59}$$

for every pair $(f, g) \in \mathcal{F}$ such that $\|f\|_{L^p(w)} < \infty$. Given $p \in \mathbb{P}^{\log}_\infty(\Omega)$, suppose $q_- < p_- \leqslant p_+ < q_+$. Then for $(f, g) \in \mathcal{F}$ such that $\|f\|_{L^{p(\cdot)}(\Omega)} < \infty$,

$$\|f\|_{L^{p(\cdot)}(\Omega)} \lesssim \|g\|_{L^{p(\cdot)}(\Omega)}. \tag{8.60}$$

Remark 8.69 We want to emphasize that the family of extrapolation pairs $\mathcal{F}$ must be chosen so that the left-hand side of (8.59) and (8.60) are finite. Given this assumption, to prove a desired inequality in general often requires an approximation argument; this is the case in our proof below. On the other hand, we do not need to assume that the right-hand side of either of these inequalities is finite. For a complete discussion of this approach to extrapolation, see Cruz-Uribe and Wang [61], Cruz-Uribe et al. [64].

Theorem 8.70 *Given an open set $\Omega \subset \mathbb{R}^n$, let $p \in \mathbb{P}^{\log}_\infty(\Omega)$, $n < p_- \leqslant p_+ < \infty$. Then $f \in W^{1,p(\cdot)}(\Omega)$ if and only if f is a continuous function (possibly after being redefined on a null set) and $f \in RBV^{p(\cdot)}(\Omega) \cap L^{p(\cdot)}(\Omega)$. Furthermore,*

$$\|f\|_{RBV^{p(\cdot)}(\Omega)} \approx \|\nabla f\|_{L^{p(\cdot)}(\Omega)}. \tag{8.61}$$

Proof First assume that $f \in W^{1,p}(\Omega)$. Let $B \subset \Omega$ be any ball. Then by the embedding of variable Lebesgue spaces on compact domains (see Cruz-Uribe and Fiorenza [58, Corollary 2.48]), we get

$$W^{1,p}(\Omega) \subset W^{1,p}(B) \subset W^{1,p_-}(B).$$

Since $p_- > n$, by Barza and Lind Barza and Lind [24, Theorem 1.1] the function f has a continuous representative on B. Since this is true for all balls contained in Ω, f has a continuous representative on Ω.

We will now prove that the right-hand inequality in (8.61) holds: i.e., that $\|f\|_{RBV^{p(\cdot)}(\Omega)} \lesssim \|\nabla f\|_{L^{p(\cdot)}(\Omega)}$. Fix a collection $\mathcal{D} = \{B_k\}_{k=1}^{\infty}$ of disjoint balls contained in Ω. Since the balls are disjoint, given any weight w, we get that

$$\|G_{\mathcal{D}} f\|_{L^p(w,\Omega)}^p = \int_{\Omega} (G_{\mathcal{D}} f(x))^p w(x)dx$$

$$= \sum_{B_k \in \mathcal{D}} \left(\frac{\operatorname{osc}_{B_k}(f)}{r_k} \right)^p w(B_k) = V_p(f;\Omega,w)^p. \tag{8.62}$$

To apply extrapolation, we need to guarantee that the left-hand side of (8.59) and (8.60) are finite. To ensure this, we define a truncation of the operator $G_{\mathcal{D}}$. For each $N \geqslant 1$, define $\mathcal{D}_N = \{B_k : B_k \in \mathcal{D}, 1 \leqslant k \leqslant N\}$ and the truncated operator $G_{\mathcal{D}_N} f$. Define the family of extrapolation pairs

$$\mathcal{F} = \left\{ (G_{\mathcal{D}_N} f, \nabla f) : f \in C^{\infty}(\Omega) \cap W^{1,p(\cdot)}(\Omega),\ N \geqslant 1 \right\}.$$

Fix $p \in \mathbb{P}_{\infty}^{\log}(\Omega)$ such that $n < p_- \leqslant p_+ < \infty$. Let $q_- = n$ and fix $q_+ > p_+$. Let w be any weight in $A_{p_-/n} \cap RH_{(q_+/p_-)'}$ (note that this class is not empty: if we take $w \in A_{p_-/n}$, by Lemma 8.55 we have $w \in RH_{(q_+/p_-)'}$ for all q_+ sufficiently large). Since $w \in A_{p_-/n}$, we have that $p_- > nr_w$ and $w \in A_{p_-}$. Fix $f \in C^{\infty}(\Omega) \cap W^{1,p(\cdot)}(\Omega)$. For all N, $G_{\mathcal{D}_N} f$ is compactly supported and in L^{∞} with $\|G_{\mathcal{D}_N}\|_{L_w^{p_-}(\Omega)} < \infty$, since w is locally integrable. Thus, by Theorem 8.58 and (8.62),

$$\|G_{\mathcal{D}_N} f\|_{L_w^{p_-}(\Omega)} \leqslant C\|\nabla f\|_{L_w^{p_-}(\Omega)}.$$

Similarly, we have that

$$\|G_{\mathcal{D}_N} f\|_{L^{p(\cdot)}(\Omega)} \leqslant \|G_{\mathcal{D}_N} f\|_{\infty} \|\chi_{\operatorname{supp}(G_{\mathcal{D}_N} f)}\|_{L^{p(\cdot)}(\Omega)} < \infty.$$

Therefore, by Theorem 8.68 applied to the family $\mathcal{F}$, for every $N \geqslant 1$ and $f \in C^{\infty}(\Omega) \cap W^{1,p(\cdot)}(\Omega)$,

$$\|G_{\mathcal{D}_N} f\|_{L^{p(\cdot)}(\Omega)} \leqslant C\|\nabla f\|_{L^{p(\cdot)}(\Omega)}.$$

If we apply Fatou's lemma in the variable Lebesgue spaces (see Cruz-Uribe and Fiorenza [58, Theorem 2.61]), we have that for all $f \in C^{\infty}(\Omega) \cap W^{1,p(\cdot)}(\Omega)$,

$$\|G_{\mathcal{D}} f\|_{L^{p(\cdot)}(\Omega)} \leqslant C\|\nabla f\|_{L^{p(\cdot)}(\Omega)}. \tag{8.63}$$

To complete the proof, take $f \in W^{1,p(\cdot)}(\Omega)$. Recall that $C^{\infty}(\Omega) \cap W^{1,p(\cdot)}(\Omega)$ is dense in $W^{1,p(\cdot)}(\Omega)$ (see Cruz-Uribe and Fiorenza [58, Theorem 6.14]). Fix a sequence $\{f_k\}_{k=1}^{\infty}$ of functions in $C^{\infty}(\Omega) \cap W^{1,p(\cdot)}(\Omega)$ that converges to f in the $W^{1,p(\cdot)}$ norm. Then we can find a subsequence (still denoted by f_k) which converges pointwise to f (see Cruz-Uribe and Fiorenza [58, Theorem 2.67]). Therefore, by Fatou's lemma in the variable exponent Lebesgue spaces and by inequality (8.63),

$$\|G_{\mathcal{D}} f\|_{L^{p(\cdot)}(\Omega)} \leqslant \liminf_{k\to\infty} \|G_{\mathcal{D}} f_k\|_{L^{p(\cdot)}(\Omega)} \lesssim \lim_{k\to\infty} \|\nabla f_k\|_{L^{p(\cdot)}(\Omega)} = \|\nabla f\|_{L^{p(\cdot)}(\Omega)}.$$

The desired inequality now follows from Proposition 8.66.

We will now prove the converse. Fix a function $f \in RBV^{p(\cdot)}(\Omega) \cap L^{p(\cdot)}(\Omega)$. We aim to show that $f \in W^{1,p}(\Omega)$ and that the left-hand inequality in (8.61) holds; in other words,

$$\|\nabla f\|_{L^{p(\cdot)}(\Omega)} \lesssim \|f\|_{RBV^{p(\cdot)}(\Omega)}.$$

For $\Omega_0 \Subset \Omega$, we have the embedding

$$RBV^{p(\cdot)}(\Omega) \subset RBV^{p(\cdot)}(\Omega_0) \subset RBV^{p_-}(\Omega_0).$$

The first inclusion is immediate; the second one does not follow directly from the definition of RBV^p, but is a consequence of Corollary 8.66 and the embedding of variable exponent Lebesgue spaces on compact domains; see Cruz-Uribe and Fiorenza [58, Corollary 2.48]. Since $p_- > n$, again by Barza and Lind [24, Theorem 1.1], we have that $RBV^{p(\cdot)}(\Omega) \subset W^{1,p_-}(\Omega_0)$. Thus, weak derivatives are well defined in $RBV^{p(\cdot)}(\Omega)$.

We now define the family of extrapolation pairs

$$\mathcal{F} = \big\{(\langle\nabla f\rangle_N, G_{\mathcal{D}} f) : f \in RBV^{p(\cdot)}(\Omega),\ N \geqslant 1, \mathcal{D}\big\},$$

where $\langle\nabla f\rangle_N = \min\{N, |\nabla f|\}$ and $\mathcal{D}$ is a countable family of disjoint balls contained in Ω that depends on f and N; the exact choice of $\mathcal{D}$ will be made below. Because w is locally integrable, $\|\langle\nabla f\rangle_N\|_{L^{p_-}_w(\Omega)} < \infty$; similarly, since $p_+ < \infty$, $\|\langle\nabla f\rangle_N\|_{L^{p(\cdot)}(\Omega)} < \infty$. We now argue as we did in the proof above, fixing q_-, q_+, and $w \in A_{\frac{p_-}{n}} \cap RH_{(\frac{q_+}{p_-})'}$. If $f \in RBV^{p_-}(\Omega, w)$, then by Theorem 8.58 and inequality (8.62)

$$\|\langle\nabla f\rangle_N\|_{L^{p_-}_w(\Omega)} \leqslant \|\nabla f\|_{L^{p_-}_w(\Omega)} \lesssim V_{p_-}(f, \Omega, w) \leqslant 2\|G_{\mathcal{D}} f\|_{L^{p_-}_w(\Omega)}, \tag{8.64}$$

where we choose $\mathcal{D}$ to saturate the supremum used to define V_{p_-} (see Remark 8.67). On the other hand, if $f \notin RBV^{p_-}(\Omega, w)$, then $V_{p_-}(f, \Omega, w) = \infty$, so we can choose $\mathcal{D}$ such that $\|G_{\mathcal{D}} f\|_{L^{p_-}_w(\Omega)}$ is either infinite or arbitrarily large; in particular, we can fix $\mathcal{D}$ so that $\|\langle\nabla f\rangle_N\|_{L^{p_-}_w(\Omega)} \lesssim \|G_{\mathcal{D}} f\|_{L^{p_-}_w(\Omega)}$ with the same constant as in (8.64). Since the hypotheses of Theorem 8.68 are satisfied for any pair $(\langle\nabla f\rangle_N, G_{\mathcal{D}} f)$ in $\mathcal{F}$, we conclude that

$$\|\langle\nabla f\rangle_N\|_{L^{p(\cdot)}(\Omega)} \lesssim \|G_{\mathcal{D}} f\|_{L^{p(\cdot)}(\Omega)} \lesssim \|f\|_{RBV^{p(\cdot)}(\Omega)};$$

the last inequality holds by Corollary 8.66. If we again apply Fatou's lemma in the variable exponent Lebesgue spaces, we get

$$\|\nabla f\|_{L^{p(\cdot)}(\Omega)} \leqslant \liminf_{N\to\infty} \|\langle\nabla f\rangle_N\|_{L^{p(\cdot)}(\Omega)} \lesssim \|f\|_{RBV^{p(\cdot)}(\Omega)},$$

which completes the proof. □

Remark 8.71 Though our goal was to prove Theorem 8.70 via extrapolation, we want to note that we can also prove the second half of this result by adapting the proof of Theorem 8.58 to the variable exponent setting. Here we sketch the proof.

We will follow the same construction and notation from the proof of Theorem 8.58. First, by homogeneity of the norm and by Corollary 8.66, without loss of generality, we may assume that $\|f\|_{RBV^{p(\cdot)}(\Omega)} = c_0$, for some constant c_0 small enough that $\|G_{\mathcal{D}} f\|_{L^{p(\cdot)}(\Omega)} \leqslant 1$ for any collection of balls $\mathcal{D}$. Therefore, by Cruz-Uribe and Fiorenza [58, Corollary 2.22], $\varrho_{p(\cdot),\Omega}(G_{\mathcal{D}} f) \leqslant 1$. Since $p_+ < \infty$ and $\mathcal{B}_i$ is a disjoint collection of balls, we get the estimate

$$\begin{aligned}
\varrho_{p(\cdot),\Omega_0}\left(D_j(\varphi_R * f)\right) &\leqslant \sum_{k\in K}\int_{Q_k} |D_j\varphi_R * f(x)|^{p(x)}dx \\
&= \sum_{k\in K}\int_{Q_k}\left|\int_{B(0,R)} D_j\varphi_R(y)(f(x-y)-f(x))dy\right|^{p(x)} dx \\
&\lesssim \sum_{k\in K}\int_{Q_k} R^{-(n+1)p(x)}(\operatorname{osc}_{B_k}(f))^{p(x)}|B(0,R)|^{p(x)}dx \\
&\lesssim \sum_{k\in K}\int_{B_k}\left(\frac{\operatorname{osc}_{B_k}(f)}{3\sqrt{n}R}\right)^{p(x)} dx \\
&= \sum_{i=1}^{N}\sum_{B_k\in\mathcal{B}_i}\int_{B_k}\left(\frac{\operatorname{osc}_{B_k}(f)}{3\sqrt{n}R}\right)^{p(x)} dx \\
&= \sum_{i=1}^{N}\int_{\Omega} (G_{\mathcal{B}_i} f)^{p(x)}\, dx \\
&\leqslant \sum_{i=1}^{N} \varrho_{p(\cdot),\Omega}\left(G_{\mathcal{B}_i} f\right) \\
&\leqslant N.
\end{aligned}$$

So, $\|D_j(\varphi_R * f)\|_{L^{p(\cdot)}(\Omega_0)}$ is uniformly bounded. Since $p \in \mathbb{P}_\infty^{\log}(\Omega)$, the Hardy–Littlewood maximal operator is bounded on $L^{p(\cdot)}(\Omega)$, so $\|\varphi_R * f\|_{L^{p(\cdot)}(\Omega_0)}$ is bounded. Hence, the sequence $\{\varphi_R * f\}_{R>0}$ is uniformly bounded in $W^{1,p(\cdot)}(\Omega_0)$ and $\{\varphi_R * f\}_{R>0}$ converges pointwise to f. We can adapt the proof of Heinonen et al. [137, Theorem 1.31] to the variable exponent setting (again using the fact that $p \in \mathbb{P}_\infty^{\log}(\Omega)$ and $p_+ < \infty$) to conclude that $f \in W^{1,p(\cdot)}(\Omega_0)$ for every $\Omega_0 \Subset \Omega$. Then, by Fatou's lemma, we get that $f \in W^{1,p(\cdot)}(\Omega)$.

Remark 8.72 Given the previous remark, it would be interesting to have a direct proof of the first half of Theorem 8.70 that did not depend on extrapolation. We think that such a proof is possible, though we have not been able to find it. A starting point would be the variants of the Morrey inequality that hold for variable Sobolev spaces; see, for example, Cruz-Uribe and Fiorenza [58, Theorem 6.36].

8.6 Comments to Chap. 8

Bounded variation spaces have been an active area of research since 1881, when Jordan [158] introduced them in his foundational paper while studying the convergence of Fourier series. Later on, several generalizations of bounded variation spaces appeared. In 1924, Wiener defined the so-called Wiener p-variation (see Wiener [334]). It was introduced as the set of functions $f : [a, b] \to \mathbb{R}$ that satisfy

$$V_p^w(f; [a, b]) = \sup_{\Pi} \sum_{j=1}^{N} |f(t_j) - f(t_{j-1})|^p < \infty, \quad 1 \leqslant p < \infty, \tag{8.65}$$

where the sup is taken over all the partitions Π of $[a, b]$. When $p = 1$ we recover the classical definition of bounded variation. In Riesz [286], Riesz defined the space $RBV^p([a, b])$ of functions $f : [a, b] \to \mathbb{R}$, now called Riesz bounded variation, satisfying

$$V_p^R(f; [a, b]) = \sup_{\Pi} \sum_{j=1}^{N} \frac{|f(t_j) - f(t_{j-1})|^p}{(t_j - t_{j-1})^{p-1}} < \infty, \quad 1 \leqslant p < \infty, \tag{8.66}$$

where the supremum is taken as in (8.65).

For more historical developments and alternative bounded variation spaces, see the monograph Appell et al. [21].

In Matkowski [231], it is shown that the Nemytskii operator maps the space of Lipschitz functions (Lip$[a, b]$) into itself and is globally Lipschitz if and only if there exist functions $G, H \in \text{Lip}[a, b]$ such that

$$f(t, y) = G(t)y + H(t), \quad t \in [a, b], \ y \in \mathbb{R}.$$

Since the Lipschitz functions are embedded in the space of Riesz bounded variation, this result has been extended to the scale of bounded variation functions and other function spaces (see, e.g., Bugajewski et al. [41], Castillo et al. [46, 49], Maćkowiak [218], Merentes and Rivas [237]). Furthermore, it seems that the Nemytskii operator has recently received attention in problems regarding solvability, existence, and uniqueness of integral equations (see, e.g., Ambrosio et al. [16], Bugajewska [38, 39], Bugajewski et al. [40]).

In 1938, Bochner and Taylor [30] obtained an equivalence between $(L^p[0, 1], B)^*$ and the spaces of Riesz bounded variation over functions with values in B^*, where B is a Banach space and B^* its dual space; see Blasco et al. [29], Bogdanowicz [31, 32], Rao [283, 284], Sánchez Pérez [297], Sánchez Pérez and Tradacete [298] for related topics.

On the setting of generalized Lebesgue spaces, particularly variable exponent Bochner–Lebesgue spaces, in Cheng and Xu [51] it was proved that $(L^{p(\cdot)}(B))^*$ is isomorphic to $L^{p'(\cdot)}(B^*)$, whenever B has the Radon–Nikodym property.

Bounded variation spaces with variable exponent where introduced in the Wiener sense in Castillo et al. [47] and in the Riesz sense in Castillo et al. [48, 49]. For an alternative definition of the latter space, see Kakochashvili and Zviadadze [160].

Sections 8.2 and 8.3 are based on Castillo et al. [48, 50], respectively. On the other hand, Sects. 8.4 and 8.5 follow Cruz-Uribe et al. [65, 66], respectively.

Bibliography

1. M.D. Acosta, A. Kamińska, M. Mastyło, The Daugavet property and weak neighborhoods in Banach lattices. J. Convex Anal. **19**(3), 875–912 (2012)
2. D.R. Adams, Traces of potentials arising from translation invariant operators. Ann. Scuola Norm. Sup. Pisa (3) **25**, 203–217 (1971)
3. D.R. Adams, A trace inequality for generalized potentials. Studia Math. **48**, 99–105 (1973)
4. D.R. Adams, A note on Riesz potentials. Duke Math. J. **42**(4), 765–778 (1975)
5. D.R. Adams, On the existence of capacitary strong type estimates in $\mathbb{R}^n$. Ark. Mat. **14**, 125–140 (1976). https://doi.org/10.1007/BF02385830
6. D.R. Adams, Morrey spaces, in *Applied and Numerical Harmonic Analysis* (Birkhäuser/Springer, Cham, 2015). https://doi.org/10.1007/978-3-319-26681-7
7. D.R. Adams, L.I. Hedberg, Function spaces and potential theory, in *Grundlehren der Mathematischen Wissenschaften*, vol. 314 (Springer, Berlin, 1996)
8. D.R. Adams, J. Xiao, Morrey spaces in harmonic analysis. Ark. Mat. **50**(2), 201–230 (2012). https://doi.org/10.1007/s11512-010-0134-0
9. M. Aguilar Cañestro, R.P. Ortega Salvador, C. Torreblanca, Weighted bilinear Hardy inequalities. J. Math. Ana. Appl. **387**(1), 320–334 (2012)
10. A.Ç. Alabalik, A. Almeida, S. Samko, On the invariance of certain vanishing subspaces of Morrey spaces with respect to some classical operators. Banach J. Math. Anal. **14**(3), 987–1000 (2020a). https://doi.org/10.1007/s43037-019-00049-7
11. A.Ç. Alabalik, A. Almeida, S. Samko, Preservation of certain vanishing properties of generalized Morrey spaces by some classical operators. Math. Methods Appl. Sci. **43**(16), 9375–9386 (2020b). https://doi.org/10.1002/mma.6235
12. A. Almeida, D. Drihem, Maximal, potential and singular type operators on Herz spaces with variable exponents. J. Math. Anal. Appl. **394**(2), 781–795 (2012). https://doi.org/10.1016/j.jmaa.2012.04.043
13. A. Almeida, S. Samko, Embeddings of local generalized Morrey spaces between weighted Lebesgue spaces. Nonlinear Anal., Theory Methods Appl., Ser. A, Theory Methods **164**, 67–76 (2017a). https://doi.org/10.1016/j.na.2017.08.006. hdl.handle.net/10773/18638
14. A. Almeida, S. Samko, Approximation in Morrey spaces. J. Funct. Anal. **272**(6), 2392–2411 (2017b). https://doi.org/10.1016/j.jfa.2016.11.015
15. A. Almeida, S. Samko, Approximation in generalized Morrey spaces. Georgian Math. J. **25**(2), 155–168 (2018). https://doi.org/10.1515/gmj-2018-0023

A. Meskhi et al., *Integral Operators in Non-Standard Function Spaces*,
Operator Theory: Advances and Applications 310,
https://doi.org/10.1007/978-3-032-03691-9

16. L. Ambrosio, N. Fusco, D. Pallara, *Functions of Bounded Variation and Free Discontinuity Problems*, in Oxford Mathematical Monographs. The Clarendon Press (Oxford University Press, New York, 2000)
17. K. F. Andersen, Weighted norm inequalities for Hilbert transforms and conjugate functions of even and odd functions. Proc. Am. Math. Soc. **56**, 99–107 (1976). https://doi.org/10.2307/2041584
18. K. F. Andersen, E. Sawyer, Weighted norm inequalities for the Riemann-Liouville and Weyl fractional integral operators. Trans. Amer. Math. Soc. **308**(2), 547–558 (1988). https://doi.org/10.2307/2001091
19. T. Andô, On compactness of integral operators. Indag. Math. **24**, 235–239 (1962). Nederl. Akad. Wetensch. Proc. Ser. A **6**, 5
20. L. Angeloni, A new concept of multidimensional variation in the sense of Riesz and applications to integral operators. Mediterr. J. Math. **14**(4), Art. 149, 21 (2017)
21. J. Appell, J. Banaś, N. Merentes, Bounded variation and around, in *De Gruyter Series in Nonlinear Analysis and Applications*, vol. 17 (De Gruyter, Berlin, 2014)
22. U. Ashraf, V. Kokilashvili, A. Meskhi, Weight characterization of the trace inequality for the generalized Riemann-Liouville transform in $L^{p(x)}$ spaces. Math. Inequal. Appl. **13**(1), 63–81 (2010). https://doi.org/10.7153/mia-13-06
23. A.A. Balinsky, W.D. Evans, R.T. Lewis. *The Analysis and Geometry of Hardy's Inequality*. Universitext (Springer, Cham, 2015). https://doi.org/10.1007/978-3-319-22870-9
24. S. Barza, M. Lind, A new variational characterization of Sobolev spaces. J. Geom. Anal. **25**(4), 2185–2195 (2015)
25. C. Bennett, R. Sharpley, Interpolation of operators, in *Pure and Applied Mathematics*, vol. 129 (Academic Press, Inc., Boston, MA, 1988)
26. E. I. Berezhnoi, Two-weighted estimations for the Hardy-Littlewood maximal function in ideal Banach spaces. Proc. Am. Math. Soc. **127**(1), 79–87 (1999). https://doi.org/10.1090/S0002-9939-99-04998-9
27. M.A. Bhat, G.S.R. Kosuru, Trace principle for Riesz potentials on Herz-type spaces and applications. J. Inequal. Appl. **2024**, 13 (2024). https://doi.org/10.1186/s13660-024-03192-4. Id/No 113
28. A. Björn, J. Björn, Nonlinear potential theory on metric spaces, in *EMS Tracts Mathematics*, vol. 17 (European Mathematical Society (EMS), Zürich, 2011). https://doi.org/10.4171/099
29. O. Blasco, J.M. Calabuig, E.A. Sánchez-Pérez, p-variations of vector measures with respect to vector measures and integral representation of operators. Banach J. Math. Anal. **9**(1), 273–285 (2015)
30. S. Bochner, A.E. Taylor, Linear functionals on certain spaces of abstractly-valued functions. Ann. of Math. (2) **39**(4), 913–944 (1938)
31. W. Bogdanowicz, A generalization of the Lebesgue-Bochner-Stieltjes integral and a new approach to the theory of integration. Proc. Nat. Acad. Sci. U.S.A. **53**, 492–498 (1965a)
32. W. Bogdanowicz, Integral representation of linear continuous operators from the space of Lebesgue-Bochner summable functions into any Banach space. Proc. Nat. Acad. Sci. U.S.A. **54**, 351–354 (1965b)
33. B. Bojarski, Remarks on the Bourgain-Brezis-Mironescu approach to Sobolev spaces. Bull. Polish Acad. Sci. Math. **59**(1), 65–75 (2011)
34. A. Bonfiglioli, E. Lanconelli, F. Uguzzoni, *Stratified Lie Groups and Potential Theory for Their sub-Laplacians*. Springer Monograph of Mathematics (Springer, New York, 2007). https://doi.org/10.1007/978-3-540-71897-0
35. H. Brezis, *Functional Analysis, Sobolev Spaces and Partial Differential Equations*. Universitext (Springer, New York, 2011)
36. Y.A. Brudnyĭ, Spaces defined by means of local approximations. Trans. Mosc. Math. Soc. **24**, 73–139 (1974)
37. S.M. Buckley, Estimates for operator norms on weighted spaces and reverse Jensen inequalities. Trans. Am. Math. Soc. **340**(1), 253–272 (1993). https://doi.org/10.2307/2154555

38. D. Bugajewska, On the superposition operator in the space of functions of bounded variation, revisited. Math. Comput. Modelling **52**(5-6), 791–796 (2010)
39. D. Bugajewska, A note on differential and integral equations in the spaces of functions of Λ-bounded variation. Nonlinear Anal. **75**(11), 4213–4221 (2012)
40. D. Bugajewski, J. Gulgowski, P. Kasprzak, On integral operators and nonlinear integral equations in the spaces of functions of bounded variation. J. Math. Anal. Appl. **444**(1), 230–250 (2016)
41. D. Bugajewski, K. Czudek, J. Gulgowski, J. Sadowski, On some nonlinear operators in ΛBV-spaces. J. Fixed Point Theory Appl. **19**(4), 2785–2818 (2017)
42. V.I. Burenkov, T.V. Tararykova, An analog of Young's inequality for convolutions of functions for general Morrey-type spaces. Proc. Steklov Inst. Math. **293**, 107–126 (2016). https://doi.org/10.1134/S0081543816040088
43. E. Burtseva, N. Samko, Weighted Adams type theorem for the Riesz fractional integral in generalized Morrey space. Fract. Calc. Appl. Anal. **19**(4), 954–972 (2016). https://doi.org/10.1515/fca-2016-0052
44. A.P. Calderón, On the differentiability of absolutely continuous functions. Riv. Mat. Univ. Parma **2**, 203–213 (1951). ISSN 0035-6298,2284-2578
45. S. Campanato, Proprieta di una famiglia di spazi funzionali. Ann. Sc. Norm. Super. Pisa, Sci. Fis. Mat., III. Ser. **18**, 137–160 (1964)
46. R. Castillo, H. Rafeiro, E. Trousselot, Nemytskii operator on generalized bounded variation space. Rev. Integr. Temas Mat. **32**(1), 71–90 (2014a)
47. R.E. Castillo, N. Merentes, H. Rafeiro, Bounded variation spaces with p-variable. Mediterr. J. Math. **11**(4), 1069–1079 (2014b). https://doi.org/10.1007/s00009-013-0342-5
48. R.E. Castillo, O.M. Guzmán, H. Rafeiro, Variable exponent bounded variation spaces in the Riesz sense. Nonlinear Anal. **132**, 173–182 (2016). https://doi.org/10.1016/j.na.2015.10.019. Corrigendum in Nonlinear Analysis: Theory, Methods & Applications **196**, 111809 (2020)
49. R.E. Castillo, O.M. Guzmán, H. Rafeiro, Nemytskii operator in Riesz-bounded variation spaces with variable exponent. Mediterr. J. Math. **14**(1), Art. 2, 11 (2017)
50. R.E. Castillo, O.M. Guzmán, H. Rafeiro, Linear functionals on variable exponent Bochner-Lebesgue spaces. Atti Accad. Naz. Lincei Rend. Lincei Mat. Appl. **30**(3), 583–597 (2019)
51. C. Cheng, J. Xu, Geometric properties of Banach space valued Bochner-Lebesgue spaces with variable exponent. J. Math. Inequal. **7**(3), 461–475 (2013)
52. F. Chiarenza, M. Franciosi, A generalization of a theorem by C. Miranda. Ann. Mat. Pura Appl. (4) **161**, 285–297 (1992). https://doi.org/10.1007/BF01759642
53. F. Chiarenza, M. Frasca, Morrey spaces and Hardy-Littlewood maximal function. Rend. Mat. Appl. (7) **7**, 273–279 (1987)
54. R. Coifman, Y. Meyer, On commutators of singular integrals and bilinear singular integrals. Trans. Am. Math. Soc. **212**, 315–331 (1975)
55. R. Coifman, Y. Meyer, Commutateurs d'intégrales singuliéres et opérateurs multilinéaires. Ann. Inst. Fourier, Grenoble **22**, 177–202 (1978a)
56. R. Coifman, Y. Meyer, Au delá des opérateurs pseudo-différentiels. Astérisque **57** (1978b)
57. D. Cruz-Uribe, J. Cummings, Weighted norm inequalities for the maximal operator on $L^{p(\cdot)}$ over spaces of homogeneous type. Ann. Fenn. Math. **47**(1), 457–488 (2022). https://doi.org/10.54330/afm.115059
58. D. Cruz-Uribe, A. Fiorenza, *Variable Lebesgue Spaces. Foundations and Harmonic Analysis* (Birkhäuser, Basel, 2013). https://doi.org/10.1007/978-3-0348-0548-3
59. D. Cruz-Uribe, F.I. Mamedov, On a general weighted Hardy type inequality in the variable exponent Lebesgue spaces. Rev. Mat. Complut. **25**(2), 335–367 (2012). https://doi.org/10.1007/s13163-011-0076-5
60. D. Cruz-Uribe, K. Moen, A fractional Muckenhoupt-Wheeden theorem and its consequences. Integral Equations Oper. Theory **76**(3), 421–446 (2013). https://doi.org/10.1007/s00020-013-2059-z

61. D. Cruz-Uribe, L.-A.D. Wang, Extrapolation and weighted norm inequalities in the variable Lebesgue spaces. Trans. Am. Math. Soc. **369**(2), 1205–1235 (2017). https://doi.org/10.1090/tran/6730
62. D. Cruz-Uribe, A. Fiorenza, C. J. Neugebauer, The maximal function on variable L^p-spaces. Ann. Acad. Sci. Fenn. Math. **28**(1), 223–238 (2003)
63. D. Cruz-Uribe, A. Fiorenza, J.M. Martell, C. Pérez, The boundedness of classical operators on variable L^p spaces. Ann. Acad. Sci. Fenn. Math. **31**(1), 239–264 (2006)
64. D. Cruz-Uribe, J.M. Martell, C. Pérez, Weights, extrapolation and the theory of Rubio de Francia, in *Operator Theory: Advances and Applications*, vol. 215. (Birkhäuser/Springer, Basel, 2011)
65. D. Cruz-Uribe, O.M. Guzmán, H. Rafeiro, Weighted Riesz bounded variation spaces and the Nemytskii operator. Azerb. J. Math. **10**(2), 125–139 (2020)
66. D. Cruz-Uribe, O. Guzmán, H. Rafeiro, A characterization of weighted Sobolev spaces via weighted Riesz bounded variation spaces. Studia Math. **274**(3), 287–304 (2024). https://doi.org/10.4064/sm230729-14-12
67. D.V. Cruz-Uribe, P. Shukla, The boundedness of fractional maximal operators on variable Lebesgue spaces over spaces of homogeneous type. Stud. Math. **242**(2), 109–139 (2018). https://doi.org/10.4064/sm8556-6-2017
68. D. Cruz-Uribe OFS, D. Suragan, Hardy-Leray inequalities in variable Lebesgue spaces. J. Math. Anal. Appl. **530**, 127747 (2024)
69. E.B. Davies, A.M. Hinz, Explicit constants for Rellich inequalities in $L_p(\Omega)$. Math. Z. **227**(3), 511–523 (1998). https://doi.org/10.1007/PL00004389
70. M.F. de Almeida, L.S.M. Lima, Adams' trace principle in Morrey-Lorentz spaces on β-Hausdorff dimensional surfaces. Ann. Fenn. Math. **46**(2), 1161–1177 (2021). https://doi.org/10.5186/aasfm.2021.4670
71. L. Diening, Maximal function on generalized Lebesgue spaces $L^{p(\cdot)}$. Math. Inequal. Appl. **7**(2), 245–253 (2004). https://doi.org/10.7153/mia-07-27
72. L. Diening, S. Samko, Hardy inequality in variable exponent Lebesgue spaces. Fract. Calc. Appl. Anal. **10**(1), 1–18 (2007)
73. L. Diening, S. Schwarzacher, On the key estimate for variable exponent spaces. Azerb. J. Math. **3**(2), 62–69 (2013)
74. L. Diening, P. Harjulehto, P. Hästö, M. Růžička, Lebesgue and Sobolev spaces with variable exponents. *Lecture Notes in Mathematics*, vol. 2017 (Springer, Heidelberg, 2011). https://doi.org/10.1007/978-3-642-18363-8
75. J. Diestel, B. Faires, On vector measures. Trans. Am. Math. Soc. **198**, 253–271 (1974)
76. N. Dinculeanu, Vector measures, in *International Series of Monographs in Pure and Applied Mathematics*, vol. 95 (Pergamon Press, Oxford-New York-Toronto, Ont.; VEB Deutscher Verlag der Wissenschaften, Berlin, 1967)
77. Y. Ding, D. Fan, Y. Pan, Weighted boundedness for a class of rough Marcinkiewicz integrals. Indiana Univ. Math. J. **48**(3), 1037–1055 (1999). https://doi.org/10.1512/iumj.1999.48.1696
78. P.G. Dodds, D.H. Fremlin, Compact operators in Banach lattices. Isr. J. Math. **34**, 287–320 (1979). https://doi.org/10.1007/BF02760610
79. O. Dragičević, L. Grafakos, M.C. Pereyra, S. Petermichl, Extrapolation and sharp norm estimates for classical operators on weighted Lebesgue spaces. Publ. Mat., Barc. **49**(1), 73–91 (2005). https://doi.org/10.5565/PUBLMAT_49105_03. mat.uab.es/pubmat/pdf/v49(1)/49105_03.pdf
80. D. Drihem, F. Seghiri, Notes on the Herz-type Hardy spaces of variable smoothness and integrability. Math. Inequal. Appl. **19**(1), 145–165 (2016). https://doi.org/10.7153/mia-19-11
81. J. Duoandikoetxea, Fourier analysis. Transl. from the Spanish and revised by David Cruz-Uribe, in *Graduate Studies in Mathematics*, vol. 29 (American Mathematical Society, Providence, RI, 2001). ISBN 0-8218-2172-5
82. D. Edmunds, A. Meskhi, On the Rellich inequality in $L^{p(\cdot)}(a,b)$. Georgian Math. J. **25**(2), 207–216 (2018). https://doi.org/10.1515/gmj-2018-0024

83. D. Edmunds, A. Meskhi, On the weighted Rellich-Sobolev and Hardy-Sobolev inequalities in variable exponent Lebesgue spaces. Trans. A. Razmadze Math. Inst. **178**(1), 155–159 (2024)
84. D. Edmunds, A. Meskhi, Weighted Rellich and Hardy inequalities in $L^{p(\cdot)}$ spaces. Georgian Math. J. **32**(2), 251–269 (2025). https://doi.org/10.1515/gmj-2024-2053
85. D.E. Edmunds, W.D. Evans, *Spectral Theory and Differential Operators*. Oxford Mathematical Monographs (Oxford University Press, Oxford, 1987)
86. D.E. Edmunds, W.D. Evans, The Rellich inequality. Rev. Mat. Complut. **29**(3), 511–530 (2016). https://doi.org/10.1007/s13163-016-0200-7
87. D.E. Edmunds, A. Meskhi, Potential-type operators in $L^{p(x)}$ spaces. Z. Anal. Anwend. **21**(3), 681–690 (2002). https://doi.org/10.4171/ZAA/1102
88. D.E. Edmunds, A. Meskhi, Two-weighted Hardy operators in $L^{p(\cdot)}$ spaces and applications. Stud. Math. **249**(2), 143–162 (2019). https://doi.org/10.4064/sm180204-20-8
89. D.E. Edmunds, A. Meskhi, A multilinear Rellich inequality. Math. Inequal. Appl. **24**(1), 265–274 (2021). https://doi.org/10.7153/mia-2021-24-19
90. D.E. Edmunds, A. Nekvinda, Averaging operators on $l^{\{p_n\}}$ and $L^{p(x)}$. Math. Inequal. Appl. **5**(2), 235–246 (2002)
91. D.E. Edmunds, P. Gurka, L. Pick, Compactness of Hardy-type integral operators in weighted Banach function spaces. Stud. Math. **109**(1), 73–90 (1994)
92. D.E. Edmunds, V. Kokilashvili, A. Meskhi, Bounded and compact integral operators, in *Mathematics and its Applications*, vol. 543 (Kluwer Academic Publishers, Dordrecht, 2002a)
93. D.E. Edmunds, V. Kokilashvili, A. Meskhi, On Fourier multipliers in weighted Triebel-Lizorkin spaces. J. Inequal. Appl. **7**(4), 555–591 (2002b). https://doi.org/10.1155/S1025583402000279
94. D.E. Edmunds, V. Kokilashvili, A. Meskhi, On the boundedness and compactness of weighted Hardy operators in spaces $L^{p(x)}$. Georgian Math. J. **12**(1), 27–44 (2005)
95. D.E. Edmunds, A. Meskhi, L. Natelashvili, Boundedness criteria for the multilinear Riemann-Liouville operators. Math. Methods Appl. Sci. (2025). https://doi.org/10.1002/mma.10855
96. Eridani, Y. Sawano, Fractional integral operators in generalized Morrey spaces defined on metric measure spaces. Proc. A. Razmadze Math. Inst. **158**, 13–24 (2012)
97. Eridani, H. Gunawan, E. Nakai, Y. Sawano, Characterizations for the generalized fractional integral operators on Morrey spaces. Math. Inequal. Appl. **17**(2), 761–777 (2014). https://doi.org/10.7153/mia-17-56
98. A. Eridani, V. Kokilashvili, A. Meskhi, Morrey spaces and fractional integral operators. Expo. Math. **27**(3), 227–239 (2009)
99. H. Federer, Geometric measure theory, in *Die Grundlehren der mathematischen Wissenschaften, Band 153* (Springer-Verlag New York Inc., New York, 1969)
100. C.L. Fefferman, The uncertainty principle. Bull. Am. Math. Soc. New Ser. **9**, 129–206 (1983). https://doi.org/10.1090/S0273-0979-1983-15154-6
101. R. Fefferman, E.M. Stein, Singular integrals on product spaces. Adv. Math. **45**(2), 117–143 (1982). https://doi.org/10.1016/S0001-8708(82)80001-7
102. H.G. Feichtinger, F. Weisz, Herz spaces and summability of Fourier transforms. Math. Nachr. **281**(3), 309–324 (2008)
103. A. Fiorenza, B. Gupta, P. Jain, The maximal theorem for weighted grand Lebesgue spaces. Stud. Math. **188**(2), 123–133 (2008). https://doi.org/10.4064/sm188-2-2
104. V. Fischer, M. Ruzhansky, Quantization on nilpotent Lie groups. *Progress in Mathematics*, vol. 314 (Birkhäuser/Springer, New York, 2016). https://doi.org/10.1007/978-3-319-29558-9
105. G.B. Folland, E.M. Stein, Hardy spaces on homogeneous groups, in *Mathematics of Notes (Princeton)*, vol. 28 (Princeton University Press, Princeton, NJ, 1982). https://doi.org/10.1515/9780691222455
106. Z.W. Fu, S.L. Gong, S.Z. Lu, W. Yuan, Weighted multilinear Hardy operators and commutators. Forum Math. **27**(5), 2825–2851 (2015). https://doi.org/10.1515/forum-2013-0064
107. N. Fujii, Weighted bounded mean oscillation and singular integrals. Math. Japon. **22**, 529–534 (1978)

108. M. Gabidzashvili, V. Kokilashvili, *Two Weight Weak Type Inequalities for Fractional Type Integrals*. preprint No 45 (Mathematical Institute of Czechoslovak Academy of Sciences, Prague, 1989).
109. M.A. Gabidzashvili, Weighted inequalities of potentials in spaces of a homogeneous type. Soobshch. Akad. Nauk Gruz. SSR **133**(1), 17–20 (1989)
110. J. García-Cuerva, A.E. Gatto, Boundedness properties of fractional integral operators associated to non-doubling measures. Studia Math. **162**, 245–261 (2004). https://doi.org/10.4064/sm162-3-5
111. J. García-Cuerva, J.M. Martell, Two-weight norm inequalities for maximal operators and fractional integrals on non-homogeneous spaces. Indiana Univ. Math. J. **50**(3), 1241–1280 (2001). https://doi.org/10.1512/iumj.2001.50.2100
112. J. García-Cuerva, J.L. Rubio de Francia, Weighted norm inequalities and related topics, in *North-Holland Mathematics Studies*, vol. 116 (North-Holland Publishing Co., Amsterdam, 1985)
113. A.E. Gatto, On fractional calculus associated to doubling and non-doubling measures, in *Harmonic Analysis. Contemporary Mathematics*, vol. 411, pp. 15–37 (American Mathematical Society, Providence, 2006)
114. I. Genebashvili, A. Gogatishvili, V. Kokilashvili, M. Krbec, Weight theory for integral transforms on spaces of homogeneous type, in *Pitman Monographs and Surveys in Pure and Applied Mathematics*, vol. 92 (Longman, Harlow, 1998)
115. M. Giaquinta, *Multiple Integrals in the Calculus of Variations and Non-linear Elliptic Systems* (Princeton University Press, Princeton, 1983)
116. J. Gilles, A. Meskhi, Sharp weighted bounds for the Hilbert transform of odd and even functions. Trans. A. Razmadze Math. Inst. **171**(1), 24–31 (2017). https://doi.org/10.1016/j.trmi.2016.07.005
117. V. Gol'dshtein, A. Ukhlov, Weighted sobolev spaces and embedding theorems. Trans. Am. Math. Soc. **361**(7), 3829–3850 (2009)
118. L. Grafakos, On multilinear fractional integrals. Stud. Math. **102**(1), 49–56 (1992). https://doi.org/10.4064/sm-102-1-49-56
119. L. Grafakos, Modern Fourier analysis, in *Graduate Texts in Mathematics*, 2nd edn., vol. 250 (Springer, New York, 2009). https://doi.org/10.1007/978-0-387-09434-2
120. L. Grafakos, A. Meskhi, Sharp Olsen's inequality for multilinear Riesz potentials. Trans. A. Razmadze Math. Inst. **175**(3), 433–436 (2021)
121. L. Grafakos, A. Meskhi, On sharp Olsen's and trace inequalities for multilinear fractional integrals. Potential Anal. **59**, 1039–1050 (2023). https://doi.org/10.1007/s11118-022-09991-y
122. L. Grafakos, R.H. Torres, Multilinear Calderón–Zygmund theory. Adv. Math. **165**(1), 124–164 (2002). https://doi.org/10.1006/aima.2001.2028
123. L. Grafakos, X. Li, D. Yang, Bilinear operators on Herz-type Hardy spaces. Trans. Am. Math. Soc. **350**(3), 1249–1275 (1998). https://doi.org/10.1090/S0002-9947-98-01878-9
124. A. Grothendieck, Sur les applications linéaires faiblement compactes d'espaces du type $C(K)$. Can. J. Math. **5**, 129–173 (1953). https://doi.org/10.4153/CJM-1953-017-4
125. V.S. Guliyev, *Integral Operators on Function Spaces on Homogeneous Groups and on Domains in* $\mathbb{R}^n$ *(in Russian)*. PhD thesis, Doctor's degree (Steklov Mathematical Institute, Moscow, 1994)
126. V.S. Guliyev, *Function Spaces, Integral Operators and Two Weighted Inequalities on Homogeneous Groups. Some Applications*. Baku (1999). (Russian)
127. V.S. Guliyev, Boundedness of the maximal, potential and singular operators in the generalized Morrey spaces. J. Inequal. Appl., Art. ID 503948, 20 (2009). https://doi.org/10.1155/2009/503948
128. V.S. Guliyev, P.S. Shukurov, On the boundedness of the fractional maximal operator, Riesz potential and their commutators in generalized Morrey spaces, in *Advances in Harmonic Analysis and Operator Theory. The Stefan Samko Anniversary Volume on the Occasion of his 70th Birthday. Mainly Based on the Presentations at Two Conferences, Lisbon and Aveiro,*

Portugal, in June–July, 2011, pp. 175–199 (Birkhäuser, Basel, 2013). https://doi.org/10.1007/978-3-0348-0516-2_10

129. V.S. Guliyev, J. Hasanov, S. Samko, Boundedness of the maximal, potential and singular operators in the generalized variable exponent Morrey spaces. Math. Scand. **107**, 285–304 (2010)
130. J. Gustavsson, A function parameter in connection with interpolation of Banach spaces. Math. Scand. **42**, 289–305 (1978). https://doi.org/10.7146/math.scand.a-11754
131. Y. Han, D. Müller, D. Yang, A theory of Besov and Triebel-Lizorkin spaces on metric measure spaces modeled on Carnot-Carathéodory spaces. Abstr. Appl. Anal. **2008**, 250 (2008). https://doi.org/10.1155/2008/893409. Id/No 893409
132. P. Harjulehto, P. Hästö, Orlicz spaces and generalized Orlicz spaces, *Lecture Notes in Mathematics*, vol. 2236 (Springer, Cham, 2019). https://doi.org/10.1007/978-3-030-15100-3
133. P. Harjulehto, P. Hästö, M. Koskenoja, Hardy's inequality in a variable exponent Sobolev space. Georgian Math. J. **12**(3), 431–442 (2005)
134. Q. He, D. Yan, Bilinear fractional integral operators on Morrey spaces. Positivity **25**(2), 399–429 (2021). https://doi.org/10.1007/s11117-020-00763-9
135. L.I. Hedberg, On certain convolution inequalities. Proc. Am. Math. Soc. **36**(2), 505–510 (1972)
136. J. Heinonen, *Lectures on Analysis on Metric Spaces*. Universitext (Springer, New York, 2001)
137. J. Heinonen, T. Kilpeläinen, O. Martio, *Nonlinear Potential Theory of Degenerate Elliptic Equations*. Oxford Mathematical Monographs (The Clarendon Press, Oxford University Press, New York, 1993)
138. E. Hernández, D. Yang, Interpolation of Herz spaces and applications. Math. Nachr. **205**(1), 83–87 (1999)
139. C.S. Herz, Lipschitz spaces and Bernstein's theorem on absolutely convergent Fourier transforms. J. Math. Mech. **18**, 283–323 (1968/69)
140. R. Hunt, B. Muckenhoupt, R. Wheeden, Weighted norm inequalities for the conjugate function and Hilbert transform. Trans. Am. Math. Soc. **176**, 227–251 (1973). https://doi.org/10.2307/1996205
141. R.A. Hunt, On L(p,q) spaces. Enseign. Math. (2) **12**, 249–276 (1966)
142. T. Hytönen, A framework for non-homogeneous analysis on metric spaces, and the $RBMO$ space of Tolsa. Publ. Mat. **54**(2), 485–504 (2010). https://doi.org/10.5565/PUBLMAT_54210_10
143. T. Hytönen, C. Pérez, Sharp weighted bounds involving A_∞. Anal. PDE **6**(4), 777–818 (2013). https://doi.org/10.2140/apde.2013.6.777
144. T. Hytönen, C. Pérez, E. Rela, Sharp reverse Hölder property for A_∞ weights on spaces of homogeneous type. J. Funct. Anal. **263**(12), 3883–3899 (2012). https://doi.org/10.1016/j.jfa.2012.09.013
145. T.P. Hytönen, The sharp weighted bound for general Calderón–Zygmund operators. Ann. Math. (2) **175**(3), 1473–1506 (2012). https://doi.org/10.4007/annals.2012.175.3.9
146. T.P. Hytönen, The two-weight inequality for the Hilbert transform with general measures. Proc. Lond. Math. Soc. (3) **117**(3), 483–526 (2018). https://doi.org/10.1112/plms.12136
147. T. Iida, E. Sato, Y. Sawano, H. Tanaka, Weighted norm inequalities for multilinear fractional operators on Morrey spaces. Stud. Math. **205**(2), 139–170 (2011). https://doi.org/10.4064/sm205-2-2
148. T. Iida, E. Sato, Y. Sawano, Multilinear fractional integrals on Morrey spaces. Acta Math. Sin., Engl. Ser. **28**(7), 1375–1384 (2012a). https://doi.org/10.1007/s10114-012-0617-y
149. T. Iida, E. Sato, Y. Sawano, H. Tanaka, Sharp bounds for multilinear fractional integral operators on Morrey type spaces. Positivity **16**(2), 339–358 (2012b). https://doi.org/10.1007/s11117-011-0129-5
150. G. Imerlishvili, A. Meskhi, A note on the trace inequality for Riesz potentials. Georgian Math. J. **28**(5), 739–745 (2021). https://doi.org/10.1515/gmj-2020-2077

151. G. Imerlishvili, A. Meskhi, Weighted inequalities for one-sided multilinear fractional integrals. Positivity **27**(1) (2023). https://doi.org/10.1007/s11117-022-00954-6
152. T. Iwaniec, C. Sbordone, On the integrability of the Jacobian under minimal hypotheses. Arch. Ration. Mech. Anal. **119**(2), 129–143 (1992). https://doi.org/10.1007/BF00375119
153. M. Izuki, Boundedness of vector-valued sublinear operators on Herz-Morrey spaces with variable exponent. Math. Sci. Res. J. **13**(10), 243–253 (2009)
154. M. Izuki, Boundedness of sublinear operators on Herz spaces with variable exponent and application to wavelet characterization. Anal. Math. **13**(36), 33–50 (2010)
155. M. Izuki, T. Noi, Boundedness of some integral operators and commutators on generalized Herz spaces with variable exponents. Preprint (2011)
156. R. Johnson, Temperatures, Riesz potentials and the Lipschitz spaces of Herz. Proc. London Math. Soc. **27**(2), 290–316 (1973). https://doi.org/10.1112/plms/s3-27.2.290
157. R. Johnson, Lipschitz spaces, Littlewood-Paley spaces, and convoluteurs. Proc. London Math. Soc. **29**(1), 127–141 (1974)
158. C. Jordan, Sur la series de fourier. C. R. Acad. Sci., Paris **92**, 228–230 (1881)
159. A. Kairema, Sharp weighted bounds for fractional integral operators in a space of homogeneous type. Math. Scand. **114**(2), 226–253 (2014). https://doi.org/10.7146/math.scand.a-17109
160. G. Kakochashvili, S. Zviadadze, On the theorem of F. Riesz in variable Lebesgue space. Trans. A. Razmadze Math. Inst. **170**(1), 56–61 (2016)
161. L.V. Kantorovich, G.P. Akilov, *Functional Analysis*, 2nd edn. (Pergamon Press, Oxford-Elmsford, 1982)
162. T. Kato, Strong solutions of the Navier-Stokes equation in Morrey spaces. Bol. Soc. Bras. Mat., Nova Sér. **22**(2), 127–155 (1992). https://doi.org/10.1007/BF01232939
163. C.E. Kenig, E.M. Stein, Multilinear estimates and fractional integration. Math. Res. Lett. **6**(1), 1–15 (1999); erratum ibid. **6**(3–4), 467 (1999). https://doi.org/10.4310/MRL.1999.v6.n1.a1
164. R. Kerman, E. Sawyer, The trace inequality and eigenvalue estimates for Schrödinger operators. Ann. Inst. Fourier **36**(4), 207–228 (1986). https://doi.org/10.5802/aif.1074
165. S.V. Khrushchev, A description of weights satisfying the A_∞ condition of Muckenhoupt. Proc. Am. Math. Soc. **90**, 253–257 (1984). https://doi.org/10.2307/2045350
166. T. Kilpeläinen, Weighted Sobolev spaces and capacity. Ann. Acad. Sci. Fenn. Ser. A I Math. **19**(1), 95–113 (1994)
167. V. Kokilashvili, Bisingular integral operators in weighted spaces. Soobshch. Akad. Nauk Gruzin. SSR **101**(2), 289–292 (1981)
168. V. Kokilashvili, Weighted estimates for classical integral operators, in *Nonlinear Analysis, Function Spaces and Applications*, pp. 86–103 (1990)
169. V. Kokilashvili, Boundedness criteria for singular integrals in weighted grand Lebesgue spaces. J. Math. Sci.(N. Y.) **170**(1), 20–33 (2010)
170. V. Kokilashvili, M. Krbec, *Weighted inequalities in Lorentz and Orlicz spaces* (World Scientific Publishing Co., Inc., River Edge, 1991)
171. V. Kokilashvili, A. Meskhi, Fractional integrals on measure spaces. Fract. Calc. Appl. Anal. **4**(1), 1–24 (2001)
172. V. Kokilashvili, A. Meskhi, A note on the boundedness of the Hilbert transform in weighted grand Lebesgue spaces. Georgian Math. J. **16**(3), 547–551 (2009)
173. V. Kokilashvili, A. Meskhi, Potentials with product kernels in grand Lebesgue spaces: one-weight criteria. Lith. Math. J. **53**(1), 27–39 (2013). https://doi.org/10.1007/s10986-013-9191-y
174. V. Kokilashvili, A. Meskhi, Fractional integrals with measure in grand Lebesgue and Morrey spaces. Integral Transforms Spec. Funct. **32**(9), 695–709 (2021). https://doi.org/10.1080/10652469.2020.1833003
175. V. Kokilashvili, A. Meskhi, Compactness of fractional type integral operators on spaces of homogeneous type. J. Math. Sci. (N.Y.) **268**(3, Problems in mathematical analysis. No. 118), 368–375 (2022)

176. V. Kokilashvili, S. Samko, On Sobolev theorem for the Riesz type potentials in Lebesgue spaces with variable exponent. Z. Anal. Anwend. **22**(4), 899–910 (2003)
177. V. Kokilashvili, A. Meskhi, L.-E. Persson, Weighted norm inequalities for integral transforms with product kernels, in *Mathematics Research Developments Series* (Nova Science Publishers, Inc., New York, 2010)
178. V. Kokilashvili, A. Meskhi, M.A. Zaighum, Positive kernel operators in $L^{p(x)}$ spaces. Positivity **17**(4), 1123–1140 (2013). https://doi.org/10.1007/s11117-013-0225-9
179. V. Kokilashvili, M. Mastyło, A. Meskhi, On the boundedness of the multilinear fractional integral operators. Nonlinear Anal. **94**, 142–147 (2014). https://doi.org/10.1016/j.na.2013.08.016
180. V. Kokilashvili, M. Mastyło, A. Meskhi, Fractional integral operators between Banach function lattices. Nonlinear Anal. **117**, 148–158 (2015a). https://doi.org/10.1016/j.na.2015.01.014
181. V. Kokilashvili, M. Mastyło, A. Meskhi, Two-weight norm estimates for multilinear fractional integrals in classical Lebesgue spaces. Fract. Calc. Appl. Anal. **18**(5), 1146–1163 (2015b). https://doi.org/10.1515/fca-2015-0066
182. V. Kokilashvili, A. Meskhi, H. Rafeiro, S. Samko, Integral operators in non-standard function spaces, vol. 1. Variable exponent Lebesgue and amalgam spaces, *Operator Theory: Advances and Applications*, vol. 248 (Birkhäuser/Springer, Cham, 2016a). https://doi.org/10.1007/978-3-319-21015-5
183. V. Kokilashvili, A. Meskhi, H. Rafeiro, S. Samko, Integral operators in non-standard function spaces, vol. 2. Variable exponent Hölder, Morrey-Campanato and grand spaces, *Operator Theory: Advances and Applications*, vol. 249 (Birkhäuser/Springer, Cham, 2016b). https://doi.org/10.1007/978-3-319-21018-6_1
184. V. Kokilashvili, A. Meskhi, M.A. Zaighum, Sharp weighted bounds for multiple integral operators. Trans. A. Razmadze Math. Inst. **170**(1), 75–90 (2016c). https://doi.org/10.1016/j.trmi.2015.12.001
185. V. Kokilashvili, A. Meskhi, M.A. Zaighum, Sharp weighted bounds for one-sided operators. Georgian Math. J. **24**(2), 227–240 (2017). https://doi.org/10.1515/gmj-2017-0016
186. V. Kokilashvili, A. Meskhi, M.A. Zaighum, Sharp weighted bounds for fractional integrals via the two-weight theory. Banach J. Math. Anal. **12**(3), 673–692 (2018). https://doi.org/10.1215/17358787-2017-0063
187. V. Kokilashvili, M. Mastyło, A. Meskhi, The measure of noncompactness of multilinear operators. Nonlinear Anal. **188**, 70–79 (2019a). https://doi.org/10.1016/j.na.2019.05.011
188. V. Kokilashvili, M. Mastyło, A. Meskhi, Compactness criteria for fractional integral operators. Fract. Calc. Appl. Anal. **22**(5), 1269–1283 (2019b). https://doi.org/10.1515/fca-2019-0067
189. V. Kokilashvili, M. Mastyło, A. Meskhi, On the boundedness of multilinear fractional integral operators. J. Geom. Anal. **30**(1), 667–679 (2020). https://doi.org/10.1007/s12220-019-00159-6
190. V. Kokilashvili, A. Meskhi, H. Rafeiro, S. Samko, Integral operators in non-standard function spaces, Advances in Grand Function Spaces, vol. 3, in *Operator Theory: Advances and Applications*, vol. 298 (Birkhäuser/Springer, Cham, 2024). https://doi.org/10.1007/978-3-031-64983-7
191. V. Kokilashvili, A. Meskhi, H. Rafeiro, S. Samko, Integral operators in non-standard function spaces. Progress in Morrey-type spaces and related topics, vol. 4. *Operator Theory: Advances and Applications* (Birkhäuser/Springer, Cham, 2025). https://link.springer.com/book/9783031997464
192. Y. Komori, Notes on singular integrals on some inhomogeneous Herz spaces. Taiwanese J. Math. **8**(3), 547–556 (2004)
193. H. König, Eigenvalue distribution of compact operators, in *Operator Theory: Advances and Applications*, vol. 16 (Birkhäuser, Cham, 1986)
194. T. Kopaliani, A characterization of some weighted norm inequalities for maximal operators. Z. Anal. Anwend. **29**(4), 401–412 (2010). https://doi.org/10.4171/ZAA/1414

195. T.S. Kopaliani, On some structural properties of Banach function spaces and boundedness of certain integral operators. Czechoslovak Math. J. **54**(129)(3), 791–805 (2004). https://doi.org/10.1007/s10587-004-6427-3
196. T.S. Kopaliani, Greediness of the wavelet system in $L^{p(t)}(\mathbb{R})$ spaces. East J. Approx. **14**(1), 59–67 (2008)
197. M.V. Korobkov, J. Kristensen, The trace theorem, the Luzin N- and Morse-Sard properties for the sharp case of Sobolev-Lorentz mappings. J. Geom. Anal. **28**(3), 2834–2856 (2018). https://doi.org/10.1007/s12220-017-9936-7
198. O. Kováčik, J. Rákosník, On spaces $L^{p(x)}$ and $W^{k,p(x)}$. Czechoslovak Math. J. **41**(116), 592–618 (1991)
199. W.M. Kozlowski, Modular function spaces, in *Pure and Applied Mathematics, Marcel Dekker*, vol. 122. (Marcel Dekker, Inc., New York etc., 1988)
200. M.A. Krasnosel'skii, P.P. Zabreiko, E.I. Pustyl'nik, P.E. Sobolevskii, *Integral Operators in Spaces of Summable Functions* (Noordhoff International Publishing, Leiden, 1976)
201. S.G. Krein, Y.I. Petunin, E.M. Semenov, Interpolation of linear operators, in *Translations of Mathematical Monographs*, vol. 54. (American Mathematical Society, Providence, R. I., 1982)
202. A. Kufner, L.-E. Persson, *Weighted Inequalities of Hardy Type* (World Scientific Publishing Co., Inc., River Edge, N. J., 2003)
203. T. Kühn, M. Mastyło, Eigenvalues of Hille-Tamarkin operators and geometry of Banach function spaces. Stud. Math. **207**(3), 275–296 (2011). https://doi.org/10.4064/sm207-3-4
204. M.T. Lacey, K. Moen, C. Pérez, R.H. Torres, Sharp weighted bounds for fractional integral operators. J. Funct. Anal. **259**(5), 1073–1097 (2010). https://doi.org/10.1016/j.jfa.2010.02.004
205. M.T. Lacey, E.T. Sawyer, C.-Y. Shen, I. Uriarte-Tuero, Two-weight inequality for the Hilbert transform: a real variable characterization, I. Duke Math. J. **163**(15), 2795–2820 (2014). https://doi.org/10.1215/00127094-2826690
206. A. Lebow, M. Schechter, Semigroups of operators and measures of noncompactness. J. Funct. Anal. **7**, 1–26 (1971). https://doi.org/10.1016/0022-1236(71)90041-3
207. P.G. Lemarié-Rieusset, Multipliers and Morrey spaces. Potential Anal. **38**(3), 741–752 (2013). https://doi.org/10.1007/s11118-012-9295-8
208. A.K. Lerner, An elementary approach to several results on the Hardy-Littlewood maximal operator. Proc. Am. Math. Soc. **136**(8), 2829–2833 (2008). https://doi.org/10.1090/S0002-9939-08-09318-0
209. A.K. Lerner, K. Moen, Mixed A_p–A_∞ estimates with one supremum. Stud. Math. **219**(3), 247–267 (2013). https://doi.org/10.4064/sm219-3-5
210. X. Li, D. Yang, Boundedness of some sublinear operators on Herz spaces. Ill. J. Math. **40**(3), 484–501 (1996)
211. J. Lindenstrauss, L. Tzafriri, Classical Banach spaces. II: Function spaces, in *Ergeb. Math. Grenzgeb.*, vol. 97 (Springer, Berlin, 1979)
212. L. Liu, J. Xiao, A trace law for the Hardy-Morrey-Sobolev space. J. Funct. Anal. **274**(1), 80–120 (2018). https://doi.org/10.1016/j.jfa.2017.10.004
213. M. Lorente, A characterization of two weight norm inequalities for one-sided operators of fractional type. Canad. J. Math. **49**(5), 1010–1033 (1997). https://doi.org/10.4153/CJM-1997-051-6
214. M. Lorente, A. de la Torre, Weighted inequalities for some one-sided operators. Proc. Am. Math. Soc. **124**(3), 839–848 (1996). https://doi.org/10.1090/S0002-9939-96-03089-4
215. G. Lu, S. Lu, D. Yang, Singular integrals and commutators on homogeneous groups. Anal. Math. **28**(2), 103–134 (2002). https://doi.org/10.1023/A:1016568918973
216. S. Lu, D. Yang, The decomposition of weighted Herz space on $\mathbb{R}^n$ and its applications. Sci. China, Ser. A **38**(2), 147–158 (1995)
217. T. Luque, C. Pérez, E. Rela, Optimal exponents in weighted estimates without examples. Math. Res. Lett. **22**(1), 183–201 (2015). https://doi.org/10.4310/MRL.2015.v22.n1.a10

218. P. Maćkowiak, On the continuity of superposition operators in the space of functions of bounded variation. Aequationes Math. **91**(4), 759–777 (2017)
219. L. Maligranda, W. Orlicz, On some properties of functions of generalized variation. Monatsh. Math. **104**(1), 53–65 (1987)
220. J. Malý, Absolutely continuous functions of several variables. J. Math. Anal. Appl. **231**(2), 492–508 (1999)
221. F. Mamedov, S. Mammadli, Compactness for the weighted Hardy operator in variable exponent spaces. C. R., Math., Acad. Sci. Paris **355**(3), 325–335 (2017). https://doi.org/10.1016/j.crma.2016.12.010
222. F. Mamedov, Y. Zeren, A necessary and sufficient condition for Hardy's operator in the variable Lebesgue space. Abstr. Appl. Anal. **2014**, 7 (2014). https://doi.org/10.1155/2014/342910. Id/No 342910
223. F.I. Mamedov, A. Harman, On a Hardy type general weighted inequality in spaces $L^{p(\cdot)}$. Integral Equations Oper. Theory **66**(4), 565–592 (2010). https://doi.org/10.1007/s00020-010-1765-z
224. F.I. Mamedov, F.M. Mammadova, A necessary and sufficient condition for Hardy's operator in $L^{p(\cdot)}(0, 1)$. Math. Nachr. **287**(5-6), 666–676 (2014)
225. F.I. Mamedov, Y. Zeren, On equivalent conditions for the general weighted Hardy type inequality in space $L^{p(\cdot)}$. Z. Anal. Anwend. **31**(1), 55–74 (2012). https://doi.org/10.4171/ZAA/1448
226. F.I. Mamedov, Y. Zeren, L. Akin, Compactification of the weighted Hardy operator in variable exponent Lebesgue spaces. Asian J. of Math. Comp. Res. **17**, 38–47 (2017)
227. F.J. Martín-Reyes, New proofs of weighted inequalities for the one-sided Hardy-Littlewood maximal functions. Proc. Am. Math. Soc. **117**(3), 691–698 (1993). https://doi.org/10.2307/2159130
228. F.J. Martín-Reyes, A. de la Torre, Two weight norm inequalities for fractional one-sided maximal operators. Proc. Am. Math. Soc. **117**(2), 483–489 (1993). https://doi.org/10.2307/2159186
229. F.J. Martín-Reyes, L. Pick, A. de la Torre, A_∞^+ condition. Can. J. Math. **45**(6), 1231–1244 (1993). https://doi.org/10.4153/CJM-1993-069-9
230. M. Mastyło, Interpolation of compact operators. Funct. Approximatio, Comment. Math. **26**, 293–311 (1998)
231. J. Matkowski, Functional equations and Nemytskiǐ operators. Funkcial. Ekvac. **25**(2), 127–132 (1982)
232. P. Mattila, Geometry of sets and measures in Euclidean spaces. Fractals and rectifiability, in *Cambridge Studies in Advanced Mathematics*, vol. 44 (University Press, Cambridge, 1995)
233. V.G. Maz'ya, On the theory of the n-dimensional Schrödinger operator. Izv. Akad. Nauk SSSR, Ser. Mat. **28**, 1145–1172 (1964)
234. V.G. Maz'ya, On some integral inequalities for functions of several variables. Probl. Mat. Analiza **3**, 33–68 (1972).
235. V.G. Maz'ya, Sobolev spaces, in *Springer Series in Soviet Mathematics* (Springer, Berlin, 1985)
236. V.G. Maz'ya, I.E. Verbitsky, Capacitary inequalities for fractional integrals, with applications to partial differential equations and Sobolev multipliers. Ark. Mat. **33**(1), 81–115 (1995)
237. N. Merentes, S. Rivas, On characterization of the Lipschitzian composition operator between spaces of functions of bounded p-variation. Czechoslovak Math. J. **45**(120)(4), 627–637 (1995)
238. A. Meskhi, Two-weight inequalities for potentials defined on homogeneous groups. Proc. A. Razmadze Math. Inst. **112**, 91–111 (1997)
239. A. Meskhi, Solution of some weight problems for the Riemann-Liouville and Weyl operators. Georgian Math. J. **5**(6), 565–574 (1998). https://doi.org/10.1023/B:GEOR.0000008132.69276.cc
240. A. Meskhi, *Measure of Non-compactness for Integral Operators in Weighted Lebesgue Spaces* (Nova Science Publishers, Inc., New York, 2009)

241. A. Meskhi, Weighted criteria for the Hardy transform under the B_p condition in grand Lebesgue spaces and some applications. J. Math. Sci. (N. Y.) **178**(6), 622–636 (2011). https://doi.org/10.1007/s10958-011-0574-5
242. A. Meskhi, Criteria for the boundedness of potential operators in grand Lebesgue spaces. Proc. A. Razmadze Math. Inst. **169**, 119–132 (2015)
243. A. Meskhi, Essential norm estimates for multilinear singular and fractional integrals. Commentat. Math. **59**(1-2), 81–93 (2019). https://doi.org/10.14708/cm.v59i1-2.6523
244. A. Meskhi, Boundedness weighted criteria for multilinear Riemann-Liouville integral operators. Trans. A. Razmadze Math. Inst. **177**(1), 147–148 (2023)
245. A. Meskhi, L. Natelashvili, Boundedness criteria for linear and multilinear fractional integral operators in Lorentz spaces. Trans. A. Razmadze Math. Inst. **178**(2), 331–333 (2024)
246. A. Meskhi, L. Natelashvili, Criteria for multilinear sobolev inequality with non-doubling measure in Lorentz spaces. Anal. Math. Phys. **15**, 87 (2025). https://doi.org/10.1007/s13324-025-01090-6
247. A. Meskhi, M.A. Zaighum, Weighted kernel operators in $L^{p(x)}(\mathbb{R}_+)$ spaces. J. Math. Inequal. **10**(3), 623–639 (2016). https://doi.org/10.7153/jmi-10-50
248. A. Meskhi, H. Rafeiro, M.A. Zaighum, Central Calderón–Zygmund operators on Herz-type Hardy spaces of variable smoothness and integrability. Ann. Funct. Anal. **9**(3), 310–321 (2018). https://doi.org/10.1215/20088752-2017-0030
249. A. Meskhi, H. Rafeiro, M.A. Zaighum, On the boundedness of Marcinkiewicz integrals on continual variable exponent Herz spaces. Georgian Math. J. **26**(1), 105–116 (2019). https://doi.org/10.1515/gmj-2017-0050
250. T. Mizuhara, Boundedness of some classical operators on generalized Morrey spaces, in *Harmonic Analysis*, ed. by S. Igari, pp. 183–189 (Springer, Berlin, 1991). ICM 90 Satellite Proceedings
251. K. Moen, Weighted inequalities for multilinear fractional integral operators. Collect. Math. **60**(2), 213–238 (2009). https://doi.org/10.1007/BF03191210
252. C.B. Morrey, On the solutions of quasi-linear elliptic partial differential equations. Trans. Am. Math. Soc. **43**, 126–166 (1938)
253. B. Muckenhoupt, Hardy's inequality with weights. Studia Math. **44**, 31–38 (1972a)
254. B. Muckenhoupt, Weighted norm inequalities for the Hardy maximal function. Trans. Am. Math. Soc. **165**, 207–226 (1972b)
255. B. Muckenhoupt, R.L. Wheeden, Weighted norm inequalities for singular and fractional integrals. Trans. Am. Math. Soc. **161**, 249–258 (1971). https://doi.org/10.2307/1995940
256. B. Muckenhoupt, R.L. Wheeden, Weighted norm inequalities for fractional integrals. Trans. Am. Math. Soc. **192**, 261–274 (1974). https://doi.org/10.2307/1996833
257. J. Musielak, Orlicz spaces and modular spaces. *Lecture Notes in Mathematics*, vol. 1034 (Springer, Berlin, 1983)
258. H. Nafis, H. Rafeiro, M.A. Zaighum, A note on the boundedness of sublinear operators on grand variable Herz spaces. J. Inequal. Appl. **2020**, 13 (2020). https://doi.org/10.1186/s13660-019-2265-6. Id/No 1
259. H. Nafis, H. Rafeiro, M.A. Zaighum, Boundedness of the Marcinkiewicz integral on grand variable Herz spaces. J. Math. Inequal. **15**(2), 739–753 (2021). https://doi.org/10.7153/jmi-2021-15-52
260. H. Nafis, H. Rafeiro, M.A. Zaighum, Boundedness of multilinear Calderón-Zygmund operators on grand variable Herz spaces. J. Funct. Spaces **2022**, 11 (2022). https://doi.org/10.1155/2022/4845507. Id/No 4845507
261. E. Nakai, Hardy-Littlewood maximal operator, singular integral operators and the Riesz potentials on generalized Morrey spaces. Math. Nachr. **166**, 95–103 (1994)
262. E. Nakai, A characterization of pointwise multipliers on the Morrey spaces. Sci. Math. **3**(3), 445–454 (2000)
263. H. Nakano, Modulared sequence spaces. Proc. Japan Acad. **27**, 508–512 (1951)
264. A. Nekvinda, Equivalence of $l^{\{p_n\}}$ norms and shift operators. Math. Inequal. Appl. **5**(4), 711–723 (2002). ISSN 1331-4343. https://doi.org/10.7153/mia-05-72

265. A. Nekvinda, A note on maximal operator on $l^{\{p_n\}}$ and $L^{p(x)}(\mathbb{R})$. J. Funct. Spaces Appl. **5**(1), 49–88 (2007)
266. P.A. Olsen, Fractional integration, Morrey spaces and a Schrödinger equation. Commun. Partial Differ. Equations **20**(11-12), 2005–2055 (1995). https://doi.org/10.1080/03605309508821161
267. G. Oniani, On non-compact operators in weighted ideal and symmetric function spaces. Georgian Math. J. **13**(3), 501–514 (2006)
268. B. Opic, A. Kufner, Hardy-type inequalities, in *Pitman Research Notes in Mathematics Series*, vol. 219 (Longman Scientific and Technical, Harlow, 1990)
269. J. Peetre, On the theory of $\mathcal{L}_{p,\lambda}$ spaces. J. Funct. Anal. **4**, 71–87 (1969)
270. L.E. Persson, M.A. Ragusa, N. Samko, P. Wall, Commutators of hardy operators in vanishing morrey spaces. AIP Conf. Proc. **1493**, 859–866 (2012)
271. L.-E. Persson, N. Samko, P. Wall, Calderón-Zygmund type singular operators in weighted generalized Morrey spaces. J. Fourier Anal. Appl. **22**(2), 413–426 (2016). https://doi.org/10.1007/s00041-015-9418-x
272. S. Petermichl, The sharp bound for the Hilbert transform on weighted Lebesgue spaces in terms of the classical A_p characteristic. Am. J. Math. **129**(5), 1355–1375 (2007). https://doi.org/10.1353/ajm.2007.0036
273. S. Petermichl, The sharp weighted bound for the Riesz transforms. Proc. Am. Math. Soc. **136**(4), 1237–1249 (2008). https://doi.org/10.1090/S0002-9939-07-08934-4
274. L. Pick, A. Kufner, O. John, S. Fučík, Function spaces. Volume 1, *de Gruyter Series in Nonlinear Analysis and Applications*, vol. 14, 2nd revised and extended edn. (de Gruyter, Berlin, 2013). https://doi.org/10.1515/9783110250428
275. D.V. Prokhorov, On the boundedness and compactness of a class of integral operators. J. London Math. Soc. (2) **61**(2), 617–628 (2000). https://doi.org/10.1112/S002461079900856X
276. E.I. Pustyl'nik, Interpolation theorems in massives of Banach spaces. Sov. Math., Dokl. **16**, 1359–1363 (1975)
277. H. Rafeiro, S. Samko, Dominated compactness theorem in Banach function spaces and its applications. Complex Anal. Oper. Theory **2**(4), 669–681 (2008). https://doi.org/10.1007/s11785-008-0072-z
278. H. Rafeiro, S. Samko, Hardy type inequality in variable Lebesgue spaces. Ann. Acad. Sci. Fenn., Math. **34**(1), 279–289 (2009). Corrigendum in Ann. Acad. Sci. Fenn. Math. **35** (2010)
279. H. Rafeiro, S. Samko, Riesz potential operator in continual variable exponents Herz spaces. Math. Nachr. **288**(4), 465–475 (2015). https://doi.org/10.1002/mana.201300270
280. H. Rafeiro, N. Samko, S. Samko, Morrey-Campanato spaces: an overview, in *Operator Theory, Pseudo-Differential Equations, and Mathematical Physics. Oper. Theory Adv. Appl.*, vol. 228, pp. 293–323. (Birkhäuser/Springer Basel AG, Basel, 2013). https://doi.org/10.1007/978-3-0348-0537-7_15
281. H. Rafeiro, S. Samko, S. Umarkhadzhiev, Grand Lebesgue sequence spaces. Georgian Math. J. **25**(2), 291–302 (2018). https://doi.org/10.1515/gmj-2018-0017
282. M.A. Ragusa, Commutators of fractional integral operators on Vanishing-Morrey spaces. J. Global Optim. **40**(1–3), 361–368 (2008)
283. M.M. Rao, Linear functionals on Orlicz spaces. Nieuw Arch. Wisk. (3) **12**, 77–98 (1964)
284. M.M. Rao, Linear functionals on Orlicz spaces: General theory. Pacific J. Math. **25**, 553–585 (1968)
285. F. Rellich, Halbbeschränkte Differentialoperatoren höherer Ordnung, in *Proceedings of the International Congress of Mathematicians 1954. Amsterdam, September 2–9. Vol. III. Stated addresses in sections. Symposia* (Erven P. Noordhoff N. V.: Groningen; North-Holland Publishing Co., Amsterdam, 1956)
286. F. Riesz, Untersuchungen über Systeme integrierbarer Funktionen. Math. Ann. **69**, 449–497 (1910)
287. M. Rosenthal, H. Triebel, Morrey spaces, their duals and preduals. Rev. Mat. Complut. **28**(1), 1–30 (2015). https://doi.org/10.1007/s13163-013-0145-z

288. M. Ruzhansky, D. Suragan, Hardy and Rellich inequalities, identities, and sharp remainders on homogeneous groups. Adv. Math. **317**, 799–822 (2017). https://doi.org/10.1016/j.aim.2017.07.020
289. M. Ruzhansky, D. Suragan, Hardy inequalities on homogeneous groups. 100 years of Hardy inequalities, in *Progress in Mathematics*, vol. 327 (Birkhäuser, Cham, 2019). https://doi.org/10.1007/978-3-030-02895-4
290. N. Samko, Weighted Hardy operators in the local generalized vanishing Morrey spaces. Positivity **17**(3), 683–706 (2013a). https://doi.org/10.1007/s11117-012-0199-z
291. N. Samko, Maximal, potential and singular operators in vanishing generalized Morrey spaces. J. Global Optim. **57**(4), 1385–1399 (2013b). https://doi.org/10.1007/s10898-012-9997-x
292. S. Samko, Convolution type operators in $L^{p(x)}$. Integral Transforms Spec. Funct. **7**(1-2), 123–144 (1998)
293. S. Samko, Mellin convolution operators in variable exponent Lebesgue spaces, in *Analytic Methods of Analysis and Differential Equations: AMADE-2012*, ed. by S.V. Rogosin, M.V. Dubatovskaya, pp. 163–172. (Cambridge Scientific Publishers, Cottenham, 2013c)
294. S. Samko, Variable exponent Herz spaces. Mediterr. J. Math. **10**(4), 2007–2025 (2013d). https://doi.org/10.1007/s00009-013-0285-x
295. S. Samko, Erratum to "Variable exponent Herz spaces", Mediterr. J. Math. (2013). https://doi.org/10.1007/s00009-013-0285-x. Mediterr. J. Math. **10**(4), 2027–2030 (2013e). https://doi.org/10.1007/s00009-013-0335-4
296. S. Samko, A.A. Kilbas, O.I. Marichev, *Fractional Integrals and Derivatives* (Gordon and Breach Science Publishers, Yverdon, 1993)
297. E.A. Sánchez Pérez, Spaces of integrable functions with respect to vector measures of convex range and factorization of operators from L_p-spaces. Pacific J. Math. **207**(2), 489–495 (2002)
298. E.A. Sánchez Pérez, P. Tradacete, Bartle-Dunford-Schwartz integration for positive vector measures and representation of quasi-Banach lattices. J. Nonlinear Convex Anal. **17**(2), 387–402 (2016)
299. Y. Sawano, Generalized Morrey spaces for non-doubling measures. NoDEA, Nonlinear Differ. Equ. Appl. **15**(4-5), 413–425 (2008). https://doi.org/10.1007/s00030-008-6032-5
300. Y. Sawano, H. Tanaka, Morrey spaces for non-doubling measures. Acta Math. Sin. (Engl. Ser.) **21**(6), 1535–1544 (2005). https://doi.org/10.1007/s10114-005-0660-z
301. Y. Sawano, H. Wadade, On the Gagliardo-Nirenberg type inequality in the critical Sobolev-Morrey space. J. Fourier Anal. Appl. **19**(1), 20–47 (2013). https://doi.org/10.1007/s00041-012-9223-8
302. Y. Sawano, S. Sugano, H. Tanaka, Olsen's inequality and its applications to Schrödinger equations. RIMS Kôkyûroku Bessatsu **B26**, 51–80 (2011a)
303. Y. Sawano, S. Sugano, H. Tanaka, Generalized fractional integral operators and fractional maximal operators in the framework of Morrey spaces. Trans. Am. Math. Soc. **363**(12), 6481–6503 (2011b). https://doi.org/10.1090/S0002-9947-2011-05294-3
304. Y. Sawano, H. Gunawan, V. Guliyev, H. Tanaka, Morrey spaces and related function spaces [Editorial]. J. Funct. Spaces, Art. ID 867192, 2 (2014). https://doi.org/10.1155/2014/867192
305. Y. Sawano, G. Di Fazio, D.I. Hakim, *Morrey Spaces. Introduction and Applications to Integral Operators and PDE's. Volume II.* Monographs Research Notes in Mathematics (CRC Press, Boca Raton, FL, 2020a). https://doi.org/10.1201/9781003029076
306. Y. Sawano, G. Di Fazio, D.I. Hakim, Morrey spaces. Introduction and applications to integral operators and PDE's. Volume I. Monographs Research Notes in Mathematics (CRC Press, Boca Raton, FL, 2020b). https://doi.org/10.1201/9780429085925
307. E. Sawyer, A characterization of a two-weight norm inequality for maximal operators. Stud. Math. **75**, 1–11 (1982). https://doi.org/10.4064/sm-75-1-1-11
308. E. Sawyer, Weighted inequalities for the one-sided Hardy-Littlewood maximal functions. Trans. Am. Math. Soc. **297**(1), 53–61 (1986). https://doi.org/10.2307/2000455
309. E. Sawyer, A characterization of two weight norm inequalities for fractional and Poisson integrals. Trans. Am. Math. Soc. **308**(2), 533–545 (1988). https://doi.org/10.2307/2001090

310. E. Sawyer, R.L. Wheeden, Weighted inequalities for fractional integrals on Euclidean and homogeneous spaces. Am. J. Math., 114(4), 813–874 (1992). https://doi.org/10.2307/2374799
311. I.I. Sharapudinov, The topology of the space $\mathcal{L}^{p(t)}([0, 1])$. Mat. Zametki **26**(4), 613–632 (1979). English transl. in Math. Notes. **26**(3-4), 796–806 (1979)
312. F. Soria, G. Weiss, A remark on singular integrals and power weights. Indiana Univ. Math. J. **43**(1), 187–204 (1994). https://doi.org/10.1512/iumj.1994.43.43009
313. E.M. Stein, On the functions of Littlewood-Paley, Lusin and Marcinkiewicz. Trans. Am. Math. Soc. **88**, 430–466 (1961). https://doi.org/10.2307/1993226
314. E.M. Stein, G. Weiss, *Introduction to Fourier analysis on Euclidean spaces*. Princeton Mathematical Series, No. 32. (Princeton University Press, Princeton, NJ, 1971)
315. V.D. Stepanov, Two-weight estimates for Riemann-Liouville integrals. Izv. Akad. Nauk SSSR Ser. Mat. **54**(3), 645–656 (1990). ISSN 0373-2436. https://doi.org/10.1070/IM1991v036n03ABEH002039
316. J.-O. Strömberg, A. Torchinsky. Weighted Hardy spaces, in *Lecture Notes in Mathematics*, vol. 1381 (Springer, Berlin, 1989)
317. L. Tang, Endpoint estimates for multilinear fractional integrals. J. Aust. Math. Soc. **84**(3), 419–429 (2008). https://doi.org/10.1017/S1446788708000724
318. X. Tao, S. He, The boundedness of multilinear operators on generalized Morrey spaces over the quasi-metric space of non-homogeneous type. J. Inequal. Appl. **2013**, 15 (2013). Id/No 330. https://doi.org/10.1186/1029-242X-2013-330
319. X. Tao, X. Yu, H. Zhang, Multilinear Calderón Zygmund operators on variable exponent Morrey spaces over domains. Appl. Math., Ser. B (Engl. Ed.) **26**(2), 187–197 (2011). https://doi.org/10.1007/s11766-011-2704-8
320. M.E. Taylor, Tools for PDE. Pseudodifferential operators, paradifferential operators, and layer potentials, in *Mathematical Surveys and Monographs*, vol. 81 (American Mathematical Society (AMS), Providence, RI, 2000)
321. C.M. Thiele, Multilinear singular integrals. Publ. Mat., Barc. **2002**, 229–274 (2002). https://doi.org/10.5565/PUBLMAT_Esco02_11
322. X. Tolsa, Analytic capacity, the Cauchy transform, and non-homogeneous Calderón-Zygmund theory, in *Progress in Mathematics*, vol. 307 (Birkhäuser/Springer, Cham, 2014). https://doi.org/10.1007/978-3-319-00596-6
323. H. Triebel, Theory of function spaces, in *Monographs in Mathematics, Basel*, vol. 78 (Birkhäuser, Cham, 1983)
324. H. Triebel, Local function spaces, heat and Navier-Stokes equations, EMS Tracts Mathematics, vol. 20 (European Mathematical Society (EMS), Zürich, 2013). https://doi.org/10.4171/123
325. H. Triebel, Hybrid function spaces, heat and Navier-Stokes equations, in EMS Tracts Mathematics, vol. 24 (European Mathematical Society (EMS), Zürich, 2015). https://doi.org/10.4171/150
326. B.O. Turesson, Nonlinear potential theory and weighted Sobolev spaces, in *Lecture Notes in Mathematics*, vol. 1736 (Springer, Berlin, 2000)
327. C. Vitanza, Functions with vanishing Morrey norm and elliptic partial differential equations, in *Proceedings of Methods of Real Analysis and Partial Differential Equations, Capri*, pp. 147–150 (Springer, Berlin, 1990)
328. C. Vitanza, Regularity results for a class of elliptic equations with coefficients in Morrey spaces. Ric. Mat. **42**(2), 265–281 (1993)
329. M. Křepela, Iterating bilinear hardy inequalities. Proc. Edinburgh Math. Soc. **60**(4), 955–971 (2017). https://doi.org/10.1017/S0013091516000602
330. A. Volberg, Calderón-Zygmund capacities and operators on nonhomogeneous spaces, in *CBMS Regional Conference Series in Mathematics*, vol. 100. Published for the Conference Board of the Mathematical Sciences, Washington, DC; by the (American Mathematical Society, Providence, R. I., 2003)

331. H. Wadade, Optimal embeddings of critical Sobolev-Lorentz-Zygmund spaces. Stud. Math. **223**(1), 77–95 (2014). https://doi.org/10.4064/sm223-1-5
332. H. Wang, Z. Liu, The Herz-type Hardy spaces with variable exponent and their applications. Taiwanese J. Math. **16**(4), 1363–1389 (2012). https://doi.org/10.11650/twjm/1500406739
333. R.L. Wheeden, A characterization of some weighted norm inequalities for the fractional maximal function. Stud. Math. **107**(3), 257–272 (1993). https://doi.org/10.4064/sm-107-3-257-272
334. N. Wiener, The quadratic variation of a function and its fourier coefficients. J. Math. Phys. **3**(2), 72–94 (1924)
335. J.M. Wilson, Weighted inequalities for the dyadic square function without dyadic A_∞. Duke Math. J. **55**, 19–49 (1987). https://doi.org/10.1215/S0012-7094-87-05502-5
336. J.M. Wilson, Weighted norm inequalities for the continuous square functions. Trans. Am. Math. Soc. **314**(2), 661–692 (1989). https://doi.org/10.2307/2001403
337. J. Xiao, A trace problem for associate Morrey potentials. Adv. Nonlinear Anal. **7**(3), 407–424 (2018). https://doi.org/10.1515/anona-2016-0069
338. J. Xu, X. Yang, Herz-Morrey-Hardy spaces with variable exponents and their applications. J. Funct. Spaces **2015**, 19 (2015). Id/No 160635. https://doi.org/10.1155/2015/160635
339. D. Yang, G. Hu, Herz Type Spaces and Their Applications. *Mathematics Monograph Series*, vol. 10 (Science Press, New York, 2008)
340. D. Yang, D. Yang, G. Hu, *The Hardy Space H^1 with Non-doubling Measures and Their Applications*. *Lecture Notes in Mathematics*, vol. 2084 (Springer, Berlin, 2013). https://doi.org/10.1007/978-3-319-00825-7
341. W. Yuan, W. Sickel, D. Yang, Interpolation of Morrey-Campanato and related smoothness spaces. Sci. China, Math. **58**(9), 1835–1908 (2015). https://doi.org/10.1007/s11425-015-5047-8
342. C.T. Zorko, Morrey space. Proc. Am. Math. Soc. **98**, 586–592 (1986). https://doi.org/10.2307/2045731

Index

A. Meskhi et al., *Integral Operators in Non-Standard Function Spaces*,
Operator Theory: Advances and Applications 310,
https://doi.org/10.1007/978-3-032-03691-9

GPSR Compliance

The European Union's (EU) General Product Safety Regulation (GPSR) is a set of rules that requires consumer products to be safe and our obligations to ensure this.

If you have any concerns about our products, you can contact us on ProductSafety@springernature.com

In case Publisher is established outside the EU, the EU authorized representative is:

Springer Nature Customer Service Center GmbH
Europaplatz 3
69115 Heidelberg, Germany

Batch number: 10755312

Printed by Printforce, the Netherlands